METAMORPHISM AND METAMORPHIC BELTS

Metamorphism and Metamorphic Belts

By

AKIHO MIYASHIRO

Professor of Geology, State University of New York at Albany

London
GEORGE ALLEN & UNWIN LTD
Ruskin House · Museum Street

First published in Great Britain in 1973

ISBN: 0 04 550017 7

Original Japanese language edition published by Iwanami Shoten, Publishers, Tokyo

Henseigan To Henseita © Akiho Miyashiro, 1965

Printed in Great Britain by William Clowes & Sons Ltd
London, Colchester and Beccles

Contents

Preface

My book *Metamorphic Rocks and Metamorphic Belts* (in Japanese) was published by Iwanami Shoten, Publishers, in Tokyo in 1965. A few years later, Mr D. Lynch-Blosse of George Allen & Unwin Ltd contacted me to explore the possibility of translating it into English. Thus, translation accompanied by rewriting of substantial parts of the book was made in subsequent years, resulting in the present book *Metamorphism and Metamorphic Belts*. This title was chosen to emphasize the tectonic significance of metamorphic belts.

Metamorphic geology has a long history. The microscopic description and classification of metamorphic rocks began in the late nineteenth century. The theory of equilibrium mineral assemblages began in the first half of the twentieth century. Detailed mineralogical studies and the experimental determination of the pressure-temperature conditions of metamorphism began in the 1950s. The importance of metamorphic petrology in our understanding of the tectonic processes has been realized only in the past decade. This book is intended to synthesize the mineralogic, petrologic and tectonic aspects of metamorphism. Advanced treatment of the thermodynamic and structural aspects is not intended.

Part I of this book treats the basic concepts of metamorphic petrology and geology. Part II deals mainly with the progressive mineral changes and their diversity in regional metamorphism. Part III deals with the tectonic aspects of regional, ocean-floor and transform-fault metamorphism in relation to the evolution of the crust and lithosphere. A short history of metamorphic geology is given in the Appendix to help readers understand the historical background of concepts and terms and to emphasize the great contribution of the Finno-Scandinavian school of metamorphic petrology in the first half of the twentieth century.

From the tectonic viewpoint, regional and ocean-floor metamorphism are the most important of all categories of metamorphism. Abundant data are available for regional metamorphism. Hence, the mineralogy, metamorphic facies, facies series, and pressure-temperature and tectonic conditions of regional metamorphism are discussed in detail. Available data, though they are not abundant, of ocean-floor, transform-fault and contact metamorphism are reviewed.

In 1960, the hardworking geologist could read virtually all the papers in metamorphic geology published in the world. At present, probably no one would be able to do so, because the amount of publications and the diversification of their contents have so drastically increased in the past decade. About 800 papers and books are listed toward the end of this book. Approximately 1 per cent of them were published in the period 1795–1899, 51 per cent in the period 1900–65, and 48 per cent in 1966–72 (i.e. in the 7 years since the publication of the Japanese edition). These figures will give an idea about the recent increase of publications and the extent of rewriting made for this edition. No doubt, many important publications escaped the attention they deserve.

The researches made by my former co-workers in Japan have played an essential role in this as well as in the Japanese edition. Among them Drs Yotaro Seki, Fumiko Shido, Shohei Banno, Masao Iwasaki, Mitsuo Hashimoto, Tatsujiro Uno, Shunro Ueta and Akira Ono made particularly important contributions, which helped me formulate some of the basic ideas presented in this book.

Some differences between the Japanese and this edition resulted because recent ocean-floor studies have given a much wider perspective to geology. My acquaintance with ocean-floor studies was acquired during my stay at Lamont-Doherty Geological Observatory of Columbia University, during which time half of this manuscript was written. I am most grateful to Dr Maurice Ewing, former Director of the Observatory, for his interest and help in my ocean-floor research and for his encouragement to write this book. The rest was written after I had moved to the State University of New York at Albany.

I am greatly indebted to Drs Fumiko Shido and Roger Mason, who read through the manuscript and gave a lot of helpful advice for its improvement. Drs Hugh Greenwood, Mineo Kumazawa and Peter Benedict read parts of the manuscript with friendly criticism.

I am most grateful to Mr Yujiro Iwanami, President of Iwanami Shoten, for his permission and encouragement to make this rewritten English edition from the Japanese version published by him, and also for his lasting friendship to me since our high school days. Messrs Yutaka Ogawa and Nobuyoshi Urabe of Iwanami Shoten were helpful in the publication of the Japanese edition. The late Mr David Lynch-Blosse and Mr David Grimshaw of George Allen & Unwin Ltd have been very helpful in the preparation of the present edition.

AKIHO MIYASHIRO

Albany, New York
October 1972

Acknowledgments for Figures

I am indebted to the following friends and organizations for permission for the use of figures they published:

Iwanami-Shoten for Figs. 1-3, 5-1 and 5-2.

Dr P. C. Hess for Fig. 3-9.

Dr R. L. Oliver and the Geological Society of Australia for Fig. 4-3.

Dr J. B. Thompson and the Mineralogical Society of America for Figs. 5-7 and -8.

Dr T. Matsuda for Fig. 6A-1.

Dr Y. Seki for Fig. 6A-3.

Universitetsforlaget (Norway) for Figs. 6B-1, -2, -3 and -4 and for Fig. 7B-11.

Dr A. Ono for Fig. 7A-1.

Dr J. P. Bard for Fig. 7A-3.

Drs M. R. W. Johnson and G. A. Chinner for Fig. 7A-10.

Dr E. Niggli and Fortschritte der Mineralogie for Fig. 7A-11.

Dr S. Banno for Fig. 7A-12.

Dr W. G. Ernst and the Geological Society of America for Fig. 7A-13.

Dr B. McKee and American Journal of Science for Fig. 7A-14.

Drs B. W. Evans and C. V. Guidotti and Springer-Verlag for Fig. 7B-3.

Dr E-an Zen and the Mineralogical Society of America for Fig. 7B-4.

Dr G. A. Chinner for Fig. 7B-8.

Dr L. C. Hsu and Clarendon Press for Figs. 7B-9 and -10.

Dr B. E. Leake for Fig. 8B-2.

Dr H. G. F. Winkler for Fig. 9-2.

Dr D. H. Green for Figs. 12-1 and -6.

Drs J. F. Lovering and A. J. R. White and Springer-Verlag for Figs. 12-4 and -5.

Dr J. G. Moore for Fig. 13-2.

Dr E. H. Bailey for Fig. 13-3.

Dr H. J. Zwart and Geological Society of Denmark for Figs. 14-2 and -3.

Dr A. Sugimura for Fig. 15-2.

Dr S. Uyeda and American Geophysical Union for Fig. 15-4.

Dr N. Kawai for Fig. 15-7.

Dr M. Hashimoto for Fig. 15-9.

Dr Y. Ueda for Fig. 15-11.

Dr D. S. Coombs and Elsevier Publishing Company for Fig. 16-2.

Note on the Transliteration of Japanese Names

A large number of Japanese geographical names appear in this book. There are two commonly used systems for the transliteration of Japanese names in western languages. The main differences between them are as follows:

Hepburn system:	shi	shu	sho	tsu	chi	fu	ji	mb
Governmental system:	si	syu	syo	tu	ti	hu	zi	nb

For example, the same names may be written in the two folowing ways:

Hepburn system:	Honshu	Shikoku	Kyushu	Sambagawa
Governmental system:	Honsyu	Sikoku	Kyusyu	Sanbagawa

Practically all the atlases published in English-speaking countries adopt the Hepburn system. In this book, however, both systems are used intentionally mixed in order to follow the commoner usages in the relevant papers.

I Basis of Metamorphic Petrology

Chapter 1

Introduction

1-1 CONCEPT OF METAMORPHISM

Threefold Classification of the Rocks

Since the last quarter of the nineteenth century, it has been a widespread custom among geologists to classify rocks into three categories, igneous, sedimentary and metamorphic. There are some difficulties in this classification, however. For example, pyroclastic rocks may be either igneous or sedimentary, while granitic rocks may be either igneous or metamorphic. Some authors proposed other systems of classification, which did not however come into common use. Therefore, I will accept the threefold classification as the starting point of our discussions.

Igneous rocks are aggregates of minerals that crystallized out of magmas, which were high-temperature melts mainly composed of silicates. The usual temperature of crystallization appears to be in the range of 600–1300 °C. The crystallization takes place either on the surface of the earth or at some depth, but magmas, if rapidly cooled, may produce glassy rocks.

Rocks may be weathered, broken down and chemically decomposed on or near the surface of the earth. The resultant products may be then transported by water or air, and their deposition on the surface of the earth will lead to the formation of *sedimentary rocks.* After deposition, the rocks usually become gradually more coherent and harder owing to the process of *diagenesis,* which includes compaction, cementation and recrystallization. Sedimentary rocks commonly contain clay minerals which are stable at low temperatures on or near the surface of the earth.

Some igneous and sedimentary rocks are subjected to conditions that differ from those under which the rocks were originally formed. As a result, changes in mineral composition and structure may take place in such rocks. Van Hise (1904) used the term 'metamorphism' to represent all such changes. In this meaning, metamorphism naturally includes weathering and diagenesis.

However, the use of the term metamorphism in such a broad sense is both unconventional and undesirable, and in normal usage, *metamorphism* does not include weathering and diagenesis. Rather, it refers to mineralogical and structural changes of rocks which take place in deeper parts of the earth, usually at higher temperatures than those encountered on its surface. The rocks that were subjected to such changes are called *metamorphic rocks* (Daly, 1917).

In metamorphism, the mineralogical reconstitution takes place in essentially solid rocks. This process is called *metamorphic recrystallization,* or simply *recrystallization.* (The term recrystallization is used in different meanings in metallurgy and chemistry.)

Aspects of the Concept of Metamorphism

Insofar as metamorphism is defined as a certain group of mineralogical and structural changes of rocks, it may be naturally regarded as a process comparable to, for example, weathering and cementation. From this viewpoint, the characteristics of metamorphism are displayed in such small-scale features as are observable in thin sections, hand-specimens or a few outcrops. In the study of small-scale metamorphism (such as pyrometamorphism to be discussed below), this aspect usually plays the only or the most important role.

However, certain kinds of metamorphism (such as regional metamorphism to be discussed below) are a direct result of large-scale tectonic processes. Though in this case also, the mineralogical changes observable in thin sections are of interest for themselves, many geologists are interested in such large-scale metamorphism for the reason that the metamorphosed geologic bodies preserve records of the characteristics, distribution and variations of P-T (pressure and temperature) and other conditions during the tectonic processes, and hence may serve as a clue to these processes themselves. From this viewpoint, the mineralogy of metamorphic rocks is of interest because it can give information of the conditions during the tectonic processes. When we emphasize this aspect of large-scale metamorphism, such metamorphism may be defined as the mineralogical and structural recording of the physico-chemical conditions in geologic bodies involved in tectonic processes.

Of course, small-scale metamorphism is also a recording of the physico-chemical conditions involved. In this case, however, the pertinent process has little or no tectonic significance, and petrologic studies of such metamorphism usually have no direct connection with tectonic problems.

The simple definition of metamorphism as mineralogical and structural changes of rocks, therefore, does not reveal the full meaning of the term. A metamorphic terrane has a definite distribution of P-T conditions and chemical composition as a result of its tectonic development. The study of such a terrane (or metamorphic complex) has its own *raison d'être* apart from that of individual minerals and mineral assemblages in it.

The problems related to minerals and mineral assemblages will be treated mainly in Part II of this book, while the problems related to the tectonic aspects of large-scale metamorphic complexes will be treated mainly in Part III.

Low-Temperature Limit of Metamorphism

In the 1950s, Coombs (1954) and others began to study the intermediate field between diagenesis and metamorphism, and the study of metamorphism came to merge with that of diagenesis.

We may define diagenesis to include those changes taking place at essentially the same temperature as that of the original deposition, and metamorphism to include those changes taking place at essentially higher temperatures (Coombs, 1961). The boundary should be defined by definite mineralogical changes. Since the stability of a mineral or a mineral assemblage is partly related to the bulk chemical composition of the rocks, the temperature for such a boundary should vary with the composition. Our present knowledge on the mineralogy of low-temperature metamorphic rocks is not ample enough for effective application of this definition.

As will be discussed in §§ 3-5 and 4-4, the low-temperature limit of metamorphism is probably around 150 °C. Ordinary crystalline schists would form at temperatures above 300 °C.

High-Temperature Limit of metamorphism

If a rock is heated until it is completely or largely melted within the earth, the melt should be regarded as a magma, and rocks formed from consolidation of this melt should be classed as igneous. It follows that metamorphism should be defined as processes taking place in essentially solid rocks which are roughly below the temperature of the beginning of melting. The possible existence of a small proportion of silicate melt or aqueous fluid in the interstices between mineral grains, however, is not prohibited.

The experimental determination of the melting temperature of rocks yields an estimate for the high-temperature limit of metamorphism. Within the pressure range realized in the typical continental crust, i.e. 0–10 kbar, granitic rocks begin to melt at a lower temperature than other common rocks. In the presence of an aqueous fluid, the temperature for the beginning of melting of granite is around 950 °C at 1 bar, and decreases with increasing pressure to about 620 °C at 10 kbars (Yoder and Tilley, 1962; Boettcher and Wyllie, 1968). Pelitic rocks begin to melt usually at slightly higher temperatures than granite as shown in fig. 1-1 (Wyllie and Tuttle, 1961). The temperature of the beginning of melting of basalts is about 1000–1100 °C at 1 bar, but decreases at a greater rate with increasing pressure than that of granite, approaching that of granite at 10 kbars (Yoder and Tilley, 1962). Therefore, the high-temperature limit of metamorphism may be estimated at about 700–900 °C for most rock compositions in the presence of an aqueous fluid.

It is an unsettled problem, however, whether there is an aqueous fluid in a large part of the crust. The high-temperature limit of metamorphism should be considerably raised under dry conditions.

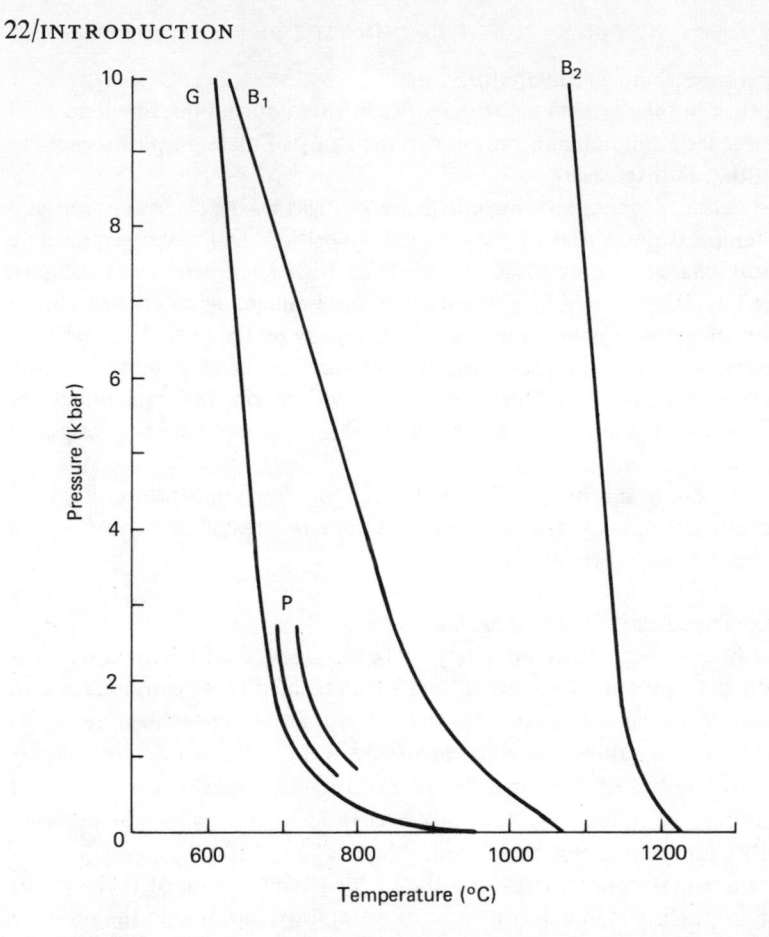

Fig. 1-1   Decrease of melting temperature of rocks with pressure in the presence of an aqueous fluid. Curves G, P. and B_1 represent the beginning of melting of granite, pelites and basalt (olivine tholeiite) respectively. Curve B_2 represents the complete melting of basalt (olivine tholeiite). (Wyllie and Tuttle, 1961; Yoder and Tilley, 1962.)

1-2   GEOLOGIC CLASSIFICATION OF METAMORPHISM

Metamorphism takes place under a great variety of geologic conditions. It is desirable to divide it into classes on this basis.

Since the classification of metamorphism began in the middle of the nineteenth century, metamorphic geology has been burdened with confused terminology related to it. Petrologists might well have some knowledge on the historical background of classification and terminology (as will be given in the appendix of this book). Here, however, we will confine ourselves to a present-day view of this problem.

Metamorphism and Basic Geologic Processes

The importance of the study of metamorphism depends largely on the information it can give toward the clarification of basic geologic processes. It appears that ocean-floor spreading and movements of the lithospheric plates are the most basic processes in the reconstruction of the crust and upper mantle of the earth (Vine, 1966; Morgan, 1968; Le Pichon, 1968; Isacks, Oliver and Sykes, 1968; Wilson, 1968). These processes are accompanied by metamorphism on a grand scale in the belts where the plates are created and disappear, that is, in mid-oceanic ridges and in orogenic belts. Grand-scale metamorphism in orogenic belts has been called *regional metamorphism,* whereas that in mid-oceanic ridges will be called *ocean-floor metamorphism* (Miyashiro, Shido and Ewing, 1971, p. 602; Miyashiro, 1972*a*). These two classes of metamorphism are of the greatest geologic significance. Metamorphism takes place along transform faults also and hence along almost all the plate boundaries (cf. Miyashiro, 1972*b*).

We have detailed data on the geology, petrology and mineralogy of regional metamorphic rocks, which form the sound basis of metamorphic petrology. On the other hand, the data now available on ocean-floor metamorphism and metamorphism along transform faults are scarce. In view of their geologic importance, however, special attention will be directed to them in chapters 19 and 20 respectively.

The stable continental crust (especially of Precambrian shields) is usually more than 30 km thick and the temperature in its lower part is probably a few or several hundred degrees Centigrade. It is natural to presume that recrystallization takes place widely in the lower crust even outside the regions which are undergoing orogeny. Such metamorphism may be an extremely slow process. No adequate name has been proposed for it, and available data are scarce. The metamorphic nature of the lower crust in general will be discussed in chapter 18. Probably some of the metamorphic changes which will be discussed in that chapter took place within the stabilized crust, and others had taken place during orogenies prior to the stabilization.

Metamorphism should not be considered as being confined to the crust. Some rocks in the upper mantle should have been subjected to metamorphism, though our knowledge in this matter is very meagre. Since the oceanic crust is thinner than the continental one, mantle materials appear on the surface of the earth more easily in oceanic regions (especially in mid-oceanic ridges) than in continents. The consideration of possible metamorphism in the upper mantle is, therefore, more pressing in the study of oceanic rocks.

Brief preliminary comments will be given below on regional and ocean-floor metamorphism together with a few other classes of minor geologic importance.

Regional or Orogenic Metamorphism

Since regional metamorphism is a part of orogeny, the name *orogenic metamorphism* could be more appropriate for it. However, we will use the former name simply because its use is well established in geology. The term *orogeny* has been used in a number of different meanings. In this book, it denotes an association in time and space of large-scale tectonic, metamorphic and magmatic events along convergent junctures of lithospheric plates, and does not necessarily include the formation of topographical mountains. We will not mind the etymological inadequacy of this usage.

In common cases, a cycle of orogeny includes the formation and development of a geosyncline. However, the formation of a geosyncline could not be an inevitable part of orogeny if the ultimate cause of orogeny lies in processes in the deeper part of a lithosphere and asthenosphere. Regional metamorphism may take place at a distinctly later time than the formation of the pertinent sediment piles, as in the case of the Ryoke metamorphism in Japan during which a contemporaneous geosyncline was present but was situated to the south of the belt of the regional metamorphism.

Rocks subjected to regional metamorphism usually occur in a great belt, hundreds or thousands of kilometres long and tens or hundreds of kilometres wide, within an ancient orogenic belt on a continent or an island arc. They include phyllites, schists and gneisses, being strongly deformed and more or less schistose. For example, a *regional metamorphic belt* extends from the Scottish Highlands to Norway along the Caledonian orogenic belt. Another belt runs almost along the full length of the Appalachian Mountains in North America.

In a geosynclinal pile subjected to orogeny, the temperature would generally rise with depth, though it should be influenced greatly by plutonic intrusions and the migration of any aqueous fluid. With increasing denudation, deeper parts of the orogenic belt which recrystallized at higher temperatures will be exposed on the surface. Such a belt of regional metamorphic rocks has one or more axes, representing the maximum temperature of recrystallization, generally running parallel with the trend of the orogenic belt. The temperature of metamorphism increases toward such a *thermal axis* from both sides, if later tectonic movements did not destroy the primary structure. The first successful demonstration of such a relation was made by Goldschmidt (1915) in the Trondheim (Trondhjem) area of the Caledonian belt of Norway. A simplified form of his geologic map is shown in fig. 1-2.

In most examples of regional metamorphism, granitic and/or ultramafic masses are intruded into metamorphic and adjacent terranes. The relationship between regional metamorphism and granitic masses is in dispute. The general distribution of the temperature of metamorphic recrystallization in such a terrane is on a grand scale, and is more or less independent of the shapes and distribution of individual intrusive masses (Harker, 1932, p. 177). Isothermal

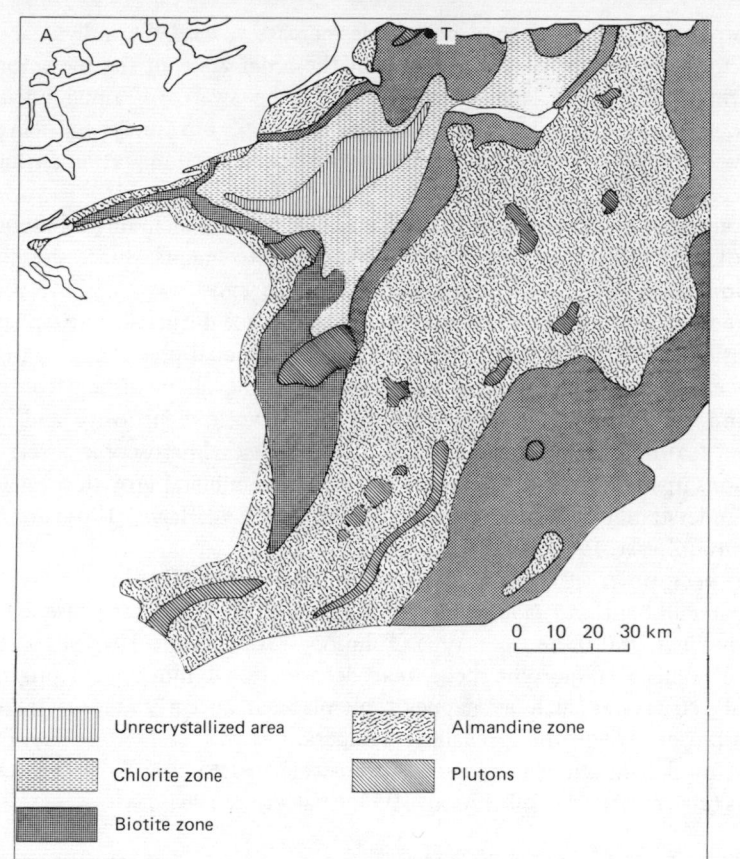

0 10 20 30 km

Unrecrystallized area

Chlorite zone

Biotite zone

Almandine zone

Plutons

Fig. 1-2   Classical metamorphic terrane of the Trondheim area, Norway. (Modified from Goldschmidt, 1915.) The terrane is mainly composed of Cambro–Silurian metapelites, and the zone boundaries are approximate isothermal curves. The thermal axis lies within the almandine zone. The blank area seen in the lower-right and left parts of the figure is composed of Precambrian gneisses. The Atlantic Ocean (A) is present in the upper-left part, with Trondheim (indicated by T) on the coast of a fjord.

curves plotted on a map of regional metamorphic terranes commonly tend to be parallel to the thermal axis of the metamorphic belt. The formation boundaries also tend to be parallel to the axis and hence to the isothermal curves.

In some cases of regional metamorphism, syn- or late-tectonic granite is abundant in higher temperature parts of terranes. In the Caledonides of the Scottish Highlands, for example, small granitic masses that formed simultaneously with the regional metamorphism are characteristically abundant in high-temperature parts of the terrane. Even if the temperature of metamorphism does not show a noticeable increase toward the individual small masses, the general rise of temperature in the axial zone of the metamorphic terrane may have been largely controlled by the swarm of small intrusives (Barrow, 1893). Conversely, some of the granitic rocks now under consideration may have been generated by incipient melting or chemical migration during the relevant regional metamorphism.

Regional metamorphism appears to be a long process including a number of phases of recrystallization and of deformational movements. Successive phases of deformation were usually separated from one another by a static period. Active recrystallization does not necessarily take place during an active phase of deformation. The deformational movements in individual phases appear to have definite characteristics and directions, possibly being distinguished from each other and put into a time sequence by field work. Schistosity and other structural features may be assigned to individual phases. Microscopic observation of relationships between the structural features and mineral growth contributes to the understanding of such time relations (Sturt and Harris, 1961; Johnson, 1962, 1963; Zwart, 1962, 1963).

It has been observed in some metamorphic belts that folds in an early phase are of a recumbent and isoclinal type, whereas those in a later phase are of a steeply inclined and more open type (Johnson, 1963; Fyson, 1971). It was also found in some metamorphic belts that deformational movements producing large-scale structures such as nappes took place at an early stage of regional metamorphism, when the prevailing temperature was still low. They were followed by deformations producing smaller-scale structures and by temperature increases (e.g. Zwart, 1962; Johnson, 1963; Chatterjee, 1961).

Ocean-Floor Metamorphism

Until quite recently our knowledge of metamorphic rocks was confined to the realm of continents and islands. In the last few years, however, it became clear that metamorphic rocks are probably widespread in the deep ocean floors. Dredging on mid-ocean ridges has yielded a great variety of metamorphic rocks.

The surface layer of the hard crust in the deep ocean floor is largely composed of basaltic rocks. Judging from the observed patterns of strong magnetic anomalies, however, the basaltic layer is probably only 0·5–2·0 km thick. The underlying layer is virtually demagnetized and is probably mainly composed of metamorphosed mafic rocks (Miyashiro, Shido and Ewing, 1970a, 1971). The ocean-floor metamorphic rocks are mostly of basic and ultrabasic compositions, being formed by the recrystallization of deeper parts of volcanic piles and associated basic and ultrabasic intrusions. The oceanic crust beneath normal ocean basins is too low in temperature to cause intensive metamorphic recrystallization. Ocean-floor metamorphism would take place mainly beneath the crest of mid-ocean ridges as the geothermal gradient would be relatively high there. The metamorphic rocks thus produced are probably moved laterally by ocean-floor spreading, leading to the formation of the oceanic crust.

Coombs (1961) coined the term *burial metamorphism* to represent large-scale recrystallization of deeply buried rocks without marked penetrative movements. The resultant rocks are lacking in schistosity. This name was originally intended to apply to zeolite-facies rocks of New Zealand, which represent the lowest temperature part of the regional metamorphic belt in the New Zealand geosyncline. Large-scale recrystallization without marked penetrative movements, however, is common in ocean-floor metamorphism also. Therefore the term burial metamorphism is considered in this book to represent a category to include parts of both regional and ocean-floor metamorphism.

Other Classes of Metamorphism

Contact metamorphism is recrystallization of rocks in an aureole around an intrusive igneous body due to rise in temperature. The width of contact-metamorphic aureoles varies, but is in most cases in the range of several metres to a few kilometres. The most typical example of contact-metamorphic rock is the non-schistose rock called hornfels. However, schistose rocks are occasionally present. The temperature rise in the aureole is due partly to heat conduction and partly to permeation of an aqueous fluid derived from the igneous body.

Comparison of contact metamorphism with regional metamorphism contributes to a better understanding of the latter. Contact metamorphism will be discussed mainly in § 4-5 and chapter 10.

Pyrometamorphism is recrystallization that shows effects of particularly high temperatures and takes place commonly in xenolithic fragments included in volcanic rocks or in the wall rocks of some minor intrusions. Partial melting is common, and in this respect pyrometamorphism may be regarded as being intermediate between metamorphism and igneous processes. Such a partially melted rock derived from shale or sandstone is called buchite. Pyrometamorphism may be regarded as an extraordinary kind of contact metamorphism (chapter 10).

Hydrothermal metamorphism is the process of recrystallization which takes place under the influence of a hydrothermal solution introduced from the outside. When the recrystallization is confined to small areas such as walls of veins, hydrothermal alteration appears to be a more familiar designation. However, in some geothermal fields, recrystallization takes place over such wide areas as to justify the name of hydrothermal metamorphism (§ 1-5).

Cataclastic metamorphism (or cataclasis) means crushing and grinding of rocks which usually take place as a result of fault movement. Though this process has widely been called dislocation or dynamic metamorphism, these terms were initially coined to represent what is now called regional metamorphism, and are still used in this meaning by some authors (see appendix). In order to avoid misunderstanding, the name 'cataclastic metamorphism' will be used in this book. This category of metamorphism takes place at low temperatures where recrystallization is not active. If the process takes place near the surface, the resultant rocks are incoherent, whereas if it takes place at a considerable depth, the resultant rocks remain coherent in spite of crushing.

Major faults are commonly accompanied by a belt of cataclastic rocks, up to several kilometres wide. Cataclastic metamorphism along major transform faults will be treated in chapter 20, because it is of some importance from the tectonic viewpoint. Our knowledge of cataclastic metamorphism is, however, very limited.

1-3   COMPOSITIONAL GROUPS OF METAMORPHIC ROCKS

When a metamorphic rock has been derived from a particular type of original rock, we may call the metamorphic rock by the name of the original rock with the prefix *meta,* as is exemplified by the terms metasediment, metaclastics, metapelite, metagraywacke and metavolcanics.

If the original rock had an igneous porphyritic texture, which was modified by later metamorphic recrystallization, but was not entirely obliterated, the resultant metamorphic texture is called blastoporphyritic. In general, the prefix *blasto* denotes relict textures, while in contrast the suffix *blast* refers to textural features which are newly formed by metamorphic recrystallization. For example, the porphyroblastic texture means a texture that resembles an igneous porphyritic one but is produced by metamorphic recrystallization. The pseudo-porphyritic crystals in such rocks are called *porphyroblasts.*

In the course of metamorphism, some degree of material migration takes place, resulting in metasomatism, which may be very intense on a small scale. On a large scale, however, metasomatism is usually not so marked as to efface the chemical features of the original rocks beyond recognition. Hence, we may classify most metamorphic rocks into the following five broad groups based on the composition of the original rocks.

Pelitic Metamorphic Rocks
Metamorphic rocks derived from pelitic sediments are characterized by relatively high contents of Al_2O_3 and K_2O, which during metamorphism usually give rise to the generation of abundant micas (tables 1-1 and 3-1). Muscovite is common in low-temperature metapelites, but tends to decompose at high temperatures. Biotite begins to form usually at a higher temperature than muscovite. Owing to the abundance of micas and other flaky minerals, metapelites commonly show strong schistosity, resulting in *mica schists.* Highly aluminous minerals such as

andalusite, kyanite, sillimanite and cordierite are common. Almandine, garnet and staurolite may also occur.

Rocks of this group are widespread in geosynclinal piles and are susceptible to changes in temperature and pressure during metamorphism. In other words, small changes in temperature and pressure commonly cause mineral reactions.

Quartzo-Feldspathic Metamorphic Rocks

Rocks of this group are usually derived either from quartzose sedimentary rocks or from felsic igneous rocks. They are characterized by a high content of SiO_2 and low contents of FeO and MgO. Schistosity, if present, is usually weak. They are abundant in geosynclinal piles, but are not susceptible to changes in temperature and pressure.

Metamorphic Rocks Derived from Calcareous Sediments

If the original rock is composed of nearly pure $CaCO_3$, metamorphic recrystallization changes it to calcite marble. If the original rock contains in addition some MgO and SiO_2, metamorphism produces tremolite, diopside, wollastonite and other calc silicates. Were the original rock to contain abundant Al_2O_3 also, plagioclase, epidote and hornblende would be produced, leading to generation of a rock type mineralogically similar to metamorphic derivatives of mafic igneous ones. They are susceptible to changes in temperature and pressure.

Basic Metamorphic Rocks

It is a well-established custom in petrology to call igneous rocks with about 40–45 per cent of SiO_2 ultrabasic, those with about 50 per cent SiO_2 basic, those with about 60 per cent SiO_2 intermediate, and those with about 70 per cent SiO_2 acidic. Basaltic rocks, for example, are basic. This custom may be extended to metamorphic derivatives of igneous rocks, as is exemplified by such terms as basic metamorphic rocks. The term *metabasites* was originally proposed by V. Hackman in Finland in 1907 for basic metamorphic rocks, and will be adopted in this book.

Many authors like the term mafic better than basic, but we will not be as scrupulous about this trifle.

Basaltic volcanics are commonly included in sedimentary piles of orogenic belts. We can correlate the mineral assemblages of metabasites with those of associated metasediments which were recrystallized under practically the same temperature and pressure.

Since the basic rocks are rich in MgO, FeO, CaO and Al_2O_3, they commonly produce chlorite, actinolite and epidote in metamorphism at low temperatures (greenschists), and hornblende at higher temperatures (amphibolites). At still higher temperatures, clino- and orthopyroxene-plagioclase rocks are formed.

Metabasites are very susceptible to changes in temperature and pressure, and hence are of great petrologic interest.

Ultrabasic Metamorphic Rocks

Ultrabasic rocks occur in many metamorphic terranes. Serpentinites, for example, are common in glaucophane-bearing metamorphic terranes, such as the Franciscan Formation of California. In many cases, they have been metamorphosed together with the surrounding rocks.

In contrast to other classes of metamorphic rocks, however, no detailed petrographic data are available on the progressive changes to take place in this class.

1-4 NOMENCLATURE OF METAMORPHIC ROCK TYPES

The textural feature is emphasized in the definitions of such names as schist and gneiss, whereas the mineral composition is delineated in the definitions of such names as amphibolite. Texture is, however, more or less related to mineral composition. For example, abundant flaky minerals could produce strong schistosity. If this relationship is emphasized, most rock names are related to both texture and mineral composition.

The same names have been used in different meanings by different authors. The meanings of most names in common usage have changed to some extent with time. In recent years, the concepts of metamorphic facies have become so popular as to modify the meaning of some rock names. For example, there is a strong tendency to limit the use of the names granulite and eclogite to rocks of strictly defined granulite and eclogite facies. In this respect, the existing textbooks of descriptive petrography do not give information necessary for the proper application of terms. In view of this situation, brief comments on main rock names and related textural terms will be given below.

Metamorphic rocks which contain abundant flaky minerals showing parallel disposition, tend to split into thin irregular plates or lenticles. This property is called *schistosity*.

Well-recrystallized metamorphic rocks showing strong schistosity are called *schists*. Many old petrographers liked to limit the use of the name schists to rocks poor or practically lacking in feldspars. They called more feldspathic metamorphic rocks gneisses. This usage is, however, obsolete. *Mica schists* are a variety of schists mainly composed of quartz, feldspar and micas, with or without garnet, kyanite and so on, usually being derived from pelitic sedimentary rocks. *Greenschists* (to be spelled as a single word) are mainly composed of feldspar, chlorite and epidote with or without quartz and actinolite, being formed by the low-temperature metamorphism of basic rocks.

Foliation is an unfortunate term which is widely used in America to denote planar stuctures of metamorphic and plutonic rocks, usually due to parallel arrangement of minerals. In this meaning, schistosity and gneissosity are a typical foliation. Harker (1932) and many British authors used the term to denote compositional layering (banding) in metamorphic rocks. Many but not all schists show compositional layering. Schistosity, rock cleavage, joints and bedding produce planar structures in rocks. Any kind of such planar structures may be called *s-surfaces* as proposed by Bruno Sander (1930) regardless of its origin. An elaborate compilation of such structural terms with critical comments is given by Dennis (1967).

Table 1-1 *Average chemical compositions of metamorphic rocks (anhydrous basis)*

	1 Phyllites	2 Mica schists	3 Two-mica gneisses	4 Quartzo-feldspathic gneisses	5 Amphibolites
SiO_2	60·0	64·3	67·7	70·7	50·3
TiO_2	1·1	1·0	—	0·5	1·6
Al_2O_3	20·7	17·5	16·6	14·5	15·7
Fe_2O_3	3·0	2·1	1·9	1·6	3·6
FeO	4·8	4·6	3·4	2·0	7·8
MnO	0·1	0·1	—	0·1	0·2
MgO	2·9	2·7	1·8	1·2	7·0
CaO	1·2	1·9	2·0	2·2	9·5
Na_2O	2·0	1·9	3·1	3·2	2·9
K_2O	4·0	3·7	3·5	3·8	1·1
P_2O_5	0·2	0·2	—	0·2	0·3
	100·0	100·0	100·0	100·0	100·0

Note: 1, 2, 4, 5 after Poldervaart (1955); 3 after Lapadu–Hargues (1945).

If the parallel disposition of flaky or prismatic minerals is more or less disturbed, or if the proportion of flaky or prismatic minerals decreases with a complementary increase of quartz and feldspar, the tendency of the rocks to split up into plates decreases. Such rocks commonly show compositional layering. These textural properties are termed *gneissosity.* Medium- or coarse-grained metamorphic rocks with gneissosity are called *gneisses.* The term gneiss has been most commonly applied to metapelites and quartzo-feldspathic rocks, but may be used for other compositional groups as well.

Schists and gneisses are widespread in regional metamorphic terranes. Typical pelitic rocks tend to become schists owing to the abundance of micas, whereas quartzo-feldspathic rocks tend to become gneisses. With increasing temperature, the parallel disposition of flaky minerals tends to be disturbed, and some flaky minerals such as chlorite and muscovite may be broken down. Hence, even pelitic rocks tend to change from schists to gneisses at high temperatures.

Pelitic rocks that were recrystallized during regional metamorphism at lower temperatures than schists are usually very fine-grained, showing a lustre on cleavage surfaces due to parallel arrangement of muscovite and chlorite. Such rocks are called *phyllites*.

As the degree of recrystallization decreases, phyllites would grade into *slates* (clay slates). The constituent minerals of slates are too fine-grained to be distinguished by the unaided eye. Slates are indurated rocks with a tendency to split into thin plates, the direction of which is independent of the original bedding. Usually slates are a link between unmetamorphosed and meta-morphosed terranes.

Thus, we have the following series of common pelitic metamorphic rocks with rising temperature: slates → phyllites → mica schists → gneisses. Average chemical compositions of these rocks are shown in table 1-1. Since quartzo-feldspathic rocks tend to become gneisses rather than schists, the average chemical compositions of gneisses show higher SiO_2 and Na_2O contents than that of schists.

Some quartzo-feldspathic rocks recrystallized at low temperatures show a gneissose texture. According to the traditional petrographic nomenclature, there can be no objection to calling such rocks gneisses. In recent years, however, most authors prefer to confine the use of the term gneiss to medium- and high-temperature metamorphic rocks.

Some gneissic rocks are megascopically heterogeneous and look like mixed rocks composed of seemingly metasedimentary and granitic materials. Such rocks are called *migmatites*. Though this name was coined by Sederholm (1907) to denote truly mixed rocks, it is now widely used in the above-mentioned non-genetic meaning. The apparently granitic material in migmatite may have been introduced from the outside, formed by partial melting in situ, or segregated by a metamorphic process. Migmatites usually occur in association with granitic rocks in the higher temperature part of regional metamorphic terranes.

Amphibolite is mainly composed of hornblende and plagioclase, but in some cases plagioclase may be scarce. A subordinate amount of quartz, almandine, biotite, clinopyroxene or cummingtonite may also be present. Amphibolite may or may not be schistose, and may or may not show compositional banding. When the presence of schistosity or gneissosity is emphasized, the term hornblende-plagioclase schist or gneiss is more appropriate. Most amphibolites are derived from basic igneous rocks, but some apparently form from impure calcareous sediments.

Amphibolite is stable in medium-temperature metamorphism. If the tempera-ture is lower, *epidote amphibolite* which is mainly composed of hornblende, sodic plagioclase and epidote forms instead. Formerly epidote amphibolite was treated as a subgroup of the amphibolites. In and prior to the 1930s, it was common to call all the basic metamorphic rocks rich in actinolite or hornblende by the name of amphibolite and to include actinolite in hornblendes. Since the concept of metamorphic facies became popular, however, it has become common to use the name amphibolite for amphibolite-facies metabasites and the name epidote amphibolite for epidote-amphibolite facies metabasites, as defined above. Greenschist-facies metabasites, even if they may contain abundant actinolite, are not called amphibolite in the current usage.

Granulites originally meant metamorphic rocks with granulitic texture. The *granulitic* texture (as distinct from the granular) is characterized by alternation of megascopic or microscopic streaks and bands of different mineral compositions, where the individual streaks and bands are made up of more or less equigranular and equidimensional grains and are nearly or entirely lacking in flaky and prismatic minerals. Naturally this texture occurs typically in rocks rich in quartz, feldspars, garnets and/or pyroxenes. This old usage of the term granulites is still accepted by many geologists.

However, Eskola (1939) proposed the name of granulite facies to include a group of metamorphic rocks recrystallized under a range of high-temperature conditions. Since then, it has become gradually common to call some or all of the rocks of this facies by the name of granulite. Many but not all of the rocks of this facies have a granulitic texture (White, 1971).

There is some confusion in the usage of the three textural terms granulitic, granulose and granular. The term granulitic is recommended here for the metamorphic texture characteristic of granulites as described above. The terms granulose and granular (or granoblastic) should be avoided, because their meanings are obscure. Granular, for example, was used sometimes as a synonym of equigranular (i.e. non-porphyritic), sometimes in the meaning of equidimensional, and sometimes for rocks composed of rounded grains.

There are wide exposures of Precambrian crystalline rocks characterized by the presence of orthopyroxene in Peninsular India. This group of rocks with diverse compositions was named the Charnockite Series. Acidic members of the Series were called *charnockite* by Holland (1900). Some later authors, however, used this rock name to designate all the rocks, acidic and basic, of the Series. Similar rocks throughout the world came to be called by this name. Charnockites were considered to be igneous by many authors, but are now widely believed to be metamorphic rocks of the granulite facies. They are usually equigranular and xenomorphic with or without parallel structure. The most characteristic feature is a dark bluish or brownish colour of quartz and feldspars, owing to which not only the basic but also the acidic charnockites are very dark coloured.

Eclogite, in a recent common usage, is a basic rock composed mainly of omphacite (clinopyroxene) and rather magnesian garnet. Eskola (1939) proposed the name of eclogite facies for eclogites and related rocks. However, some modification of his definition is necessary as will be discussed in chapter 12. Certain recent workers prefer to confine the use of the name eclogite to clinopyroxene-garnet rocks belonging to a more strictly defined eclogite facies (§ 12-2).

Hornfels is typically a fine-grained, non-schistose, contact-metamorphic rock of pelitic composition, sometimes containing porphyroblasts of andalusite or cordierite. It may be entirely massive, but may show parallel structure inherited from the original sedimentary rock. Such a parallel structure, however, does not give the rock a tendency to split into thin plates. This is the diagnostic property of hornfels as distinct from schist. It is believed that this property is a result of the absence of penetrative movements. The term hornfelses may be used in a wider sense to include non-schistose contact metamorphic rocks of any other compositions.

Some geologists consider that the names of rock types should be defined so as to be independent of genetic hypotheses. Since descriptions are a mere preliminary to genetic discussions in many cases, however, it is natural that the definitions of rock names change with genetic views so that genetic discussions are facilitated and confusions are avoided.

The nomenclature of cataclastic metamorphic rocks will be treated in chapter 20, since it has no relation to other chapters.

1-5   DISTRIBUTION OF METAMORPHIC ROCKS

Precambrian Shields

Shield areas were subjected to phases of orogeny and metamorphism in Precambrian, and became stabilized in Palaeozoic and later time. They usually occur as cores of continents. The proportions of areal extents of various rock types have been estimated for the Baltic Shield in Finland (Sederholm, 1925) and in Norway (Barth, 1961) and for the Canadian Shield (Engel, 1963; Reilly and Shaw, 1967). A rough estimate of the average relative areal extents of the rock types exposed on shields is as follows:

Granitic rocks and quartzo-feldspathic gneisses	70-80%
Other metasediments	10% or a little more
Metabasites	10% or less

For average chemical compositions, see table 18-2.

Large parts of the metamorphic rocks exposed on the shields are in the amphibolite and the epidote-amphibolite facies. It is to be noted that the intensity or temperature of metamorphism does not tend to increase with

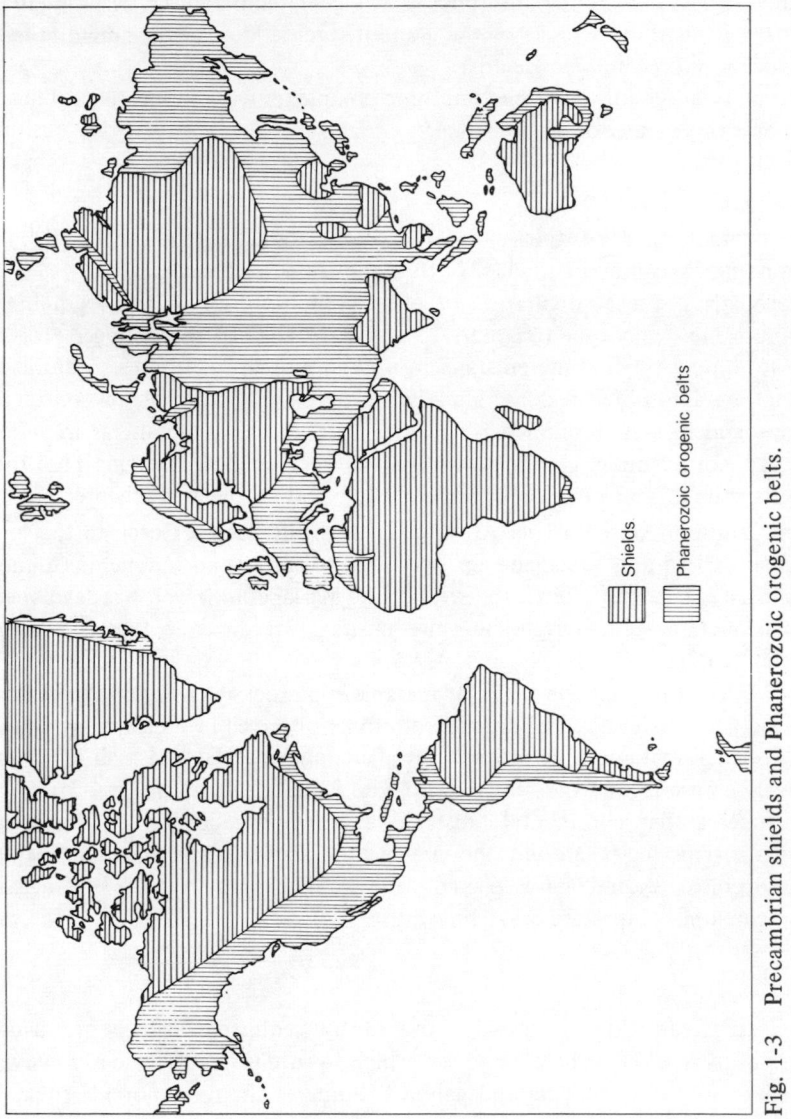

Fig. 1-3 Precambrian shields and Phanerozoic orogenic belts.

geologic age within shields. The Superior Province is the oldest part of the Canadian Shield, but generally speaking the intensity of metamorphism observed there is not high. The Grenville Province is the youngest part of the same shield, but in general has been metamorphosed at higher temperatures, resulting in the abundant formation of rocks of the granulite facies. Most of the granulite facies rocks occur in Precambrian shields.

Precambrian granitic and metamorphic complexes were found to exist also as basements of younger orogenic belts.

Phanerozoic Orogenic Belts

Phanerozoic (i.e. Palaeozoic and later) orogenic belts tend to be disposed surrounding Precambrian shields. Such disposition becomes especially clear, if we reconstruct the configuration of old continents prior to the beginning of drifting in Mesozoic time (chapters 13 and 14). Some of the younger orogenic belts are on continental margins and in well-developed island arcs. Granitic and regional metamorphic rocks are abundant in most orogenic belts on continental margins and in well-developed island arcs, but not so abundant as in shields. In Japan, for example, granite rocks and well-recrystallized metamorphic rocks underlie about 13 and 4 per cent respectively of the whole area (chapter 15).

The Izu-Bonin and Mariana Arcs jut out into the Pacific Ocean to the south of Japan. These arcs are made up of young volcanics and sediments (unmetamorphosed). The Yap Arc to the west of the Mariana, however, has developed a regional metamorphic terrane in spite of its great distance from continents (chapter 16).

It is interesting that the kinds of metamorphic rocks exposed in Phanerozoic orogenic belts differ to some extent from those observed in Precambrian shields. In the former, metamorphic rocks of the glaucophane-schist and zeolite facies are relatively common, and those of the granulite facies are very rare (§ 4-3).

It appears that continental crusts have grown by addition of successively newer orogenic belts around the pre-existing sialic crustal masses. Regional metamorphism should have played an essential part in transforming soft geosynclinal piles into hard crystalline crust.

Deep Ocean Floors

The deep ocean floor is mostly covered by sediments, which are usually underlain by a layer of basaltic rocks, which in turn probably grades downward into metamorphosed basalts and gabbros. Some of the metamorphic rocks are brought up to the surface of the ocean floor by tectonic processes and as inclusions in peridotites or serpentinites. Virtually all the oceanic metamorphic rocks appear to have been derived from basic and ultrabasic igneous rocks. As stated before, the metamorphism appears to take place in mid-oceanic ridges and the resultant metamorphic rocks would be transported laterally by ocean-floor spreading (chapter 19).

Present-Day Metamorphism, Particularly of Geothermal Fields

Metamorphism is probably taking place now on a grand scale beneath some active orogenic regions in island arcs and beneath mid-oceanic ridges (chapters 15–17 and 19). On a smaller scale, present-day metamorphism is taking place in many active geothermal fields.

In the Salton Sea area to the north of the Gulf of California, a Pliocene and Quaternary sedimentary deposit, which reaches a thickness of 6 km in some places, was formed as a part of the delta of the Colorado River. Under the influence of an ascending hydrothermal fluid rich in carbon dioxide as well as in sodium, calcium and potassium chlorides, metamorphic recrystallization is taking place within the deposit. The sediments on and near the surface are unmetamorphosed clays, silts and sands, containing montmorillonite, illite, kaolinite, calcite, dolomite and quartz. With an increase of depth, montmorillonite and kaolinite disappear, whereas ankerite, chlorite, and furthermore epidote appear. Rocks recrystallized at temperatures around 300°C have the quartz+epidote+chlorite+potassium feldspar+albite assemblage characteristic of the greenschist facies, though schistosity is lacking (Muffler and White, 1969; Keith, Muffler and Cromer, 1968).

In some active geothermal areas of New Zealand and Japan, metamorphic recrystallization is taking place, producing zeolite-bearing assemblages at shallow depths, and albite-bearing assemblages at greater depths (Steiner, 1953; Coombs *et al.* 1959; Seki *et al.* 1969*b*). A detailed review of such areas is given in § 6A-3. The presence of zeolite-bearing assemblages in this case and their absence in the Salton Sea area are considered to be due in part to a lower concentration of CO_2 here than in the Salton Sea (§ 4-4).

Chapter 2

Basic Characteristics of Metamorphic Reactions

2-1 AIM OF THIS CHAPTER AND UNITS OF MEASUREMENT

The mineral changes in metamorphism are chemical reactions from the viewpoint of thermodynamics. Thermodynamics and crystal chemistry are important for a better understanding of such reactions. There is an extensive literature in this field. The present book, however, is not intended to cover them in any way, but is directed toward the mineralogic, petrographic and tectonic aspects of metamorphism. However some elementary knowledge of the thermodynamic characteristics of metamorphic reactions is needed even for understanding mineralogic and petrographic problems. In view of this need, some of the basic characteristics are summarized in a simple way in this chapter, largely as a preliminary to the discussion of progressive metamorphism to be treated in the next chapter. In this respect, this chapter is essentially a part of the introduction of this book. Readers particularly interested in the thermodynamic problems in metamorphism are referred to other books and papers (e.g. Kern and Weisbrod, 1967; Thompson, 1955; Fyfe, Turner and Verhoogen, 1958; Korzhinskii, 1959).

Since our direct concern here is with semi-quantitative relations, fluids will always be regarded as ideal gases. Four main classes of metamorphic reactions will be treated, in this order:

1. Solid–solid reactions, i.e. reactions among solid phases, involving no liberation of volatiles.
2. Dehydration reactions, i.e. reactions involving liberation of H_2O with rise in temperature. The majority of metamorphic reactions belong to this class.
3. Decarbonation reactions, i.e. reactions involving liberation of CO_2 with rise in temperature.
4. Oxidation-reduction reactions. Since many rock-forming minerals contain iron, this class of reactions is also important.

The units of pressure and heat most commonly used in the past were the bar, kilobar, calorie and kilocalorie. In future, however, SI units, pascal (Pa) and joule (J), will become common for pressure and heat respectively (Pa = N m^{-2}; J

= N m). The conversion factors between them are given in table 2-1. The bar and J will be mainly used in this chapter.

The internationally agreed thermochemical standard state is 25 °C (298·15 K) at 1 atm (1·013 bars). Most thermochemical data have been given for these conditions (e.g. Robie and Waldbaum, 1968). The equilibrium pressure directly calculated from such data is in atmospheres. However, the difference between the atmosphere and the bar is small (table 2-1), and may be neglected for the purposes of this chapter. Therefore, such calculated pressures will be regarded below as being in bars.

Table 2-1   *Some important symbols and units*

Symbol	Physical quantity	Unit
T	Temperature	K, °C
P	Pressure	
P_s	Rock-pressure	
P_{fluid}	Pressure of a fluid phase	bar, kilobar
$P_{H_2O}, P_{CO_2}, P_{O_2}$	Pressure of H_2O, CO_2 and O_2 within an imaginary semi-permeable cell in a rock	
$\Delta G^{\ominus}$	Standard free energy of a reaction	J (joules)

1 bar $= 10^5$ Pa $= 0.987$ atm $= 14.5$ p.s.i.
1 kbar $= 0.1$ GPa (gigapascal)
1 cal $= 4.184$ J $= 41.8$ cm^3 bar
1 J $= 10$ cm^3 bar $= 10^7$ ergs

Note: The pressure at a depth of h km in the crust with density of d g cm^{-3} is about $98dh$ bar. Approximate pressures within a continental crust are as follows:

Depth (km)	10	35	50
Pressure (kbar)	2·6	10	15

2-2   SOLID–SOLID REACTIONS

As a simple and important case, we consider phase transformations between the three polymorphs of Al_2SiO_5 (andalusite, kyanite and sillimanite):

$$kyanite \rightleftharpoons sillimanite \qquad (2\text{-}1)$$
$$andalusite \rightleftharpoons sillimanite. \qquad (2\text{-}2)$$

The phase relations corresponding to these transformations are observed in many metamorphic terranes, and are of prime importance in determining the temperature and pressure of metamorphism. The right-hand sides of these equations represent a higher temperature. Such reactions may take place with or without the catalytic action of a fluid. Since these equations involve solid phases only, however, the equilibrium $P-T$ relations are independent of the presence and absence of a fluid.

A number of experimental studies have been carried out, revealing that there exists a triple point (fig. 2-1). The $P-T$ values of the point, however, have not

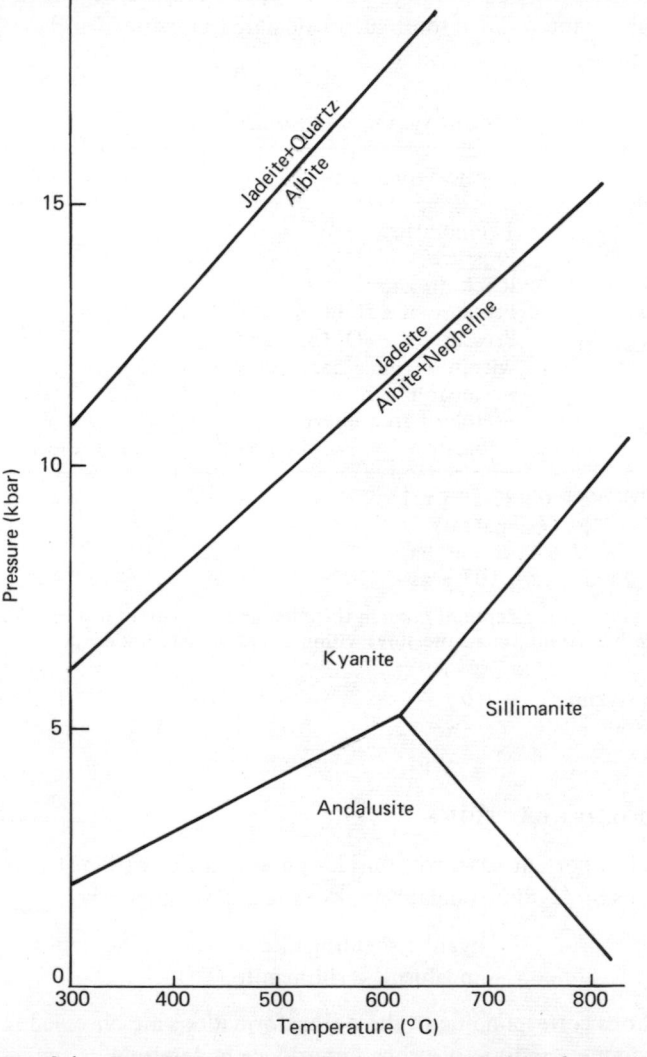

Fig. 2-1   Equilibrium curves for polymorphic transformations of Al_2SiO_5 (Richardson *et al.* 1969) and for solid–solid reactions involving jadeite. (Birch and Le Comte, 1960; Newton and Kennedy, 1968.)

been well established yet. A synopsis of recently-obtained values is given in table 2-2. Althaus (1967) and Richardson, Gilbert and Bell (1969) gave values near 600 °C and 6 kbar, whereas Newton (1966*b*) and Holdaway (1971) gave values near 510 °C and 4 kbar. The natural minerals are not of pure Al_2SiO_5, and may show strain and structural disorder. The stability relations may be influenced by these factors as well as the surface free energy when the relevant minerals are fine-grained.

Table 2-2   *Temperature and pressure of the triple point of Al_2SiO_5*

Author and year	Temperature (°C)	Pressure (kbar)
Newton, 1966*b*	520	4·0
Weill, 1966	410	2·4
Althaus, 1967	595	6·5
Pugin and Khitarov, 1968	540	7·6
Richardson *et al*. 1969	622	5·5
Holdaway, 1971	501	3·76

As examples of a more common kind of solid–solid reactions, we consider reactions to form jadeite from albite.

$$NaAlSi_2O_6 + SiO_2 \rightleftharpoons NaAlSi_3O_8 \qquad (2\text{-}3)$$
$$\text{jadeite} \qquad \text{quartz} \qquad \text{albite}$$

$$2\,NaAlSi_2O_6 \rightleftharpoons NaAlSi_3O_8 + NaAlSiO_4 \qquad (2\text{-}4)$$
$$\text{jadeite} \qquad \text{albite} \qquad \text{nepheline}$$

The right-hand side of these equations represents higher temperatures and lower pressures (Birch and Le Comte, 1960; Newton and Kennedy, 1968), as shown in fig. 2-1.

Natural jadeite and albite do not have the idealized compositions given above, but are solid solutions containing some other components, which should have effects on their stability relations (e.g. Ramberg, 1952, pp. 18–37). For example, natural jadeite may contain diopside and acmite components, which should decrease the pressure necessary for the stabilization of jadeite.

In all reactions, the high-temperature side has a greater entropy than the low-temperature side, and the high-pressure side has a smaller volume than the low-pressure side.

The slope of such equilibrium curves could be calculated by the Clausius–Clapeyron equation:

$$\frac{dP}{dT} = \frac{10\cdot0\,\Delta H}{T\Delta V} = \frac{10\cdot0\,\Delta S}{\Delta V}. \qquad (2\text{-}5)$$

Here, P and T represent the pressure (bars) and the absolute temperature, and

ΔH, ΔV, and ΔS represent the heat of reaction (J) and the changes in volume (cm^3) and in entropy (J K^{-1} mol^{-1}) respectively. (When ΔH is expressed in cal and ΔS in cal K^{-1} mol^{-1}, a coefficient of 41·8 should be used instead of 10.0.)

In solid–solid reactions, the change of heat capacity is usually very small. By neglecting it, we have the following approximate relation:

$$\Delta G = \Delta H^{\circ}_{298} - T\Delta S^{\circ}_{298} + (P - 1)\Delta V/10\cdot0, \qquad (2\text{-}6)$$

where ΔG = free energy of the reaction (J) as a function of absolute temperature T and pressure P(bars), ΔH°_{298} and ΔS°_{298} = heat (J) and entropy (J K^{-1} mol^{-1}) of the reaction in the standard state, and ΔV = volume change (cm^3) of the reaction. The values of ΔH°_{298} and ΔS°_{298} can be calculated from data given in tables of thermodynamic constants (e.g. Robie and Waldbaum, 1968).

The equation of the equilibrium curve for the reaction is obtained by the condition $\Delta G = 0$ in equation (2-6).

Solid–solid reactions are important for the estimation of temperature and pressure, since their equilibrium curves are independent of the chemical potentials of H_2O and CO_2.

2-3 OPEN SYSTEM AND MODELS FOR THE BEHAVIOUR OF H_2O

P_s (Rock-Pressure) and P_{H_2O}

Pelitic sediments contain a large amount of H_2O, part as interstitial water between mineral grains, and part as a constituent of minerals. Metamorphism causes marked dehydration of them. On the other hand, basaltic rocks are originally poor in H_2O. Their metamorphic transformation to greenschists involves marked hydration. Since dehydration and hydration are widespread occurrences, H_2O must be mobile to some degree within a rock complex undergoing metamorphism. Similarly, since the metamorphism of calcareous sediments involves decarbonation, CO_2 also should be mobile to some degree. In other words, a rock complex undergoing metamorphism is probably *open* to H_2O and CO_2.

The degree of mobility, however, is not clear. It could differ in different cases. As an ideal case, we may imagine that a component can move perfectly freely through a rock complex. In such a case, the chemical potential of the perfectly mobile component is uniform throughout, and is controlled by external conditions just like the temperature and pressure acting on the solid phases. In reality, the degree of mobility is probably much less than in such a case (e.g. Zen, 1961a; Carmichael, 1970; Guidotti, 1970). The behaviour of H_2O and other mobile components is of vital importance for understanding metamorphic reactions. This problem has been discussed in general forms by Ramberg (1952), Thompson (1955), Korzhinskii (1959) and Greenwood (1961).

The pressure acting on the solid phases will be denoted as P_s and called *rock-pressure* in this book.

Chemical potential may be regarded as the free energy that one mole of a component has in a solution. Chemical potential is an intensive property, being of the same value for the same component in all phases in a system in equilibrium. The chemical potential of a pure substance is equal to its molar free energy. The molar free energy or chemical potential of an ideal gas is related to pressure P by the following equation:

$$\mu = \mu^\ominus + RT \ln P. \qquad (2\text{-}7)$$

Here, $\mu^\ominus$, R and T represent the chemical potential at unit pressure, the gas constant, and absolute temperature respectively. The same relationship holds for a component gas of an ideal gas mixture when P is regarded as its partial pressure. In other words, the chemical potential of a component gas increases with its partial pressure in a gas mixture.

Let us imagine a small cell made up of rigid walls permeable only to H_2O. This cell is contained in a rock. H_2O molecules should diffuse through the walls until osmotic equilibrium is reached between the inside and outside of the cell. The chemical potential of H_2O should be equal between them. The pressure within the cell will be denoted as P_{H_2O}. It is related to the chemical potential of H_2O by equation (2-7).

Four Simple Models

Four models for the behaviour of H_2O in a rock complex undergoing metamorphic recrystallization will be discussed below. Each of these models could be approximately applied to certain cases of metamorphism.

Model (A). A rise in temperature should cause progressive dehydration of minerals. The liberated H_2O will move gradually out of the rock. If the mobility is not great in comparison to the rate of the dehydration reactions, an aqueous fluid phase would continue to exist between mineral grains. It may be assumed that the temperature and pressure of the fluid are equal to those of the adjacent solid phases. The fluid is not of pure H_2O composition, but contains a small amount of dissolved mineral substance. If we neglect this for a first approximation, we have the relation: $P_s = P_{fluid} = P_{H_2O}$.

Before the middle of the 1930s, the existence of such an intergranular fluid was advocated by practically all the authorities in metamorphic petrology, including Becke, Niggli, Harker and Eskola. The fluid was assumed to behave as a solvent to help metamorphic reactions and as a carrier of dissolved substances. This view is supported by many contemporary authors also (e.g. Korzhinskii, 1959; Winkler, 1967). The metamorphic reactions in such a case are identical in character to those observed in simple hydrothermal experiments.

Model (B). If pelitic sediments are mixed with calcareous ones, metamorphic reactions should produce an intergranular fluid containing both H_2O and CO_2. The temperature and pressure of the fluid are assumed to be equal to those of the adjacent solids.

The chemical potential of H_2O and P_{H_2O} should decrease with a decreasing proportion of the component in the mixture. In this model, $P_s = P_{fluid} > P_{H_2O}$.

Model (C) The intergranular spaces containing a fluid may be mutually connected for a vertical distance. In an extreme case, they may be continuous up to the surface of the earth. If equilibrium is approached in such a channel, the rate of downward increase of pressure therein should depend on the density of the fluid column. It should be much less than the rate of pressure increase in the surrounding solid phases. In this model, if the dissolved mineral substances are neglected and the fluid is composed of pure H_2O, $P_s > P_{fluid} = P_{H_2O}$.

This model appears to be applicable at least to incipient recrystallization of sediments and hydrothermal metamorphism in active geothermal fields. The situation represented by this model is mechanically and thermodynamically unstable. Metamorphic recrystallization would tend to disconnect the intergranular spaces. However, penetrative deformations in regional metamorphism may continually produce new channels.

Model (D). Since the 1940s, a number of petrologists have concluded that an intergranular fluid phase is not necessary for metamorphic reactions to take place, and that probably it does not exist in many or most cases of metamorphism. This view was particularly emphasized by radical advocates of granitization around 1950 (Holmes and Reynolds, 1947; Ramberg, 1952).

It must be noted that the absence of a fluid phase in a rock does not necessarily mean the absence of H_2O and CO_2 molecules. Such molecules can be present in the grain boundaries between minerals, and move by diffusion down the gradients of their chemical potentials, as was clearly discussed by Ramberg (1952).

The total mass of H_2O and CO_2 molecules in grain boundaries should be negligible as compared with that of the rest of the rock. If a rock is closed to such molecules, their effects on mineral changes should be negligible. If the rock is open, however, such components may be successively liberated within the rock and diffuse away, or may be successively introduced from the outside and react with the existing minerals within the rock. Continued reactions with continuous migration of such mobile components would result in a great extent of change. Therefore, this model has significance only when the pertinent rock is open for H_2O and/or CO_2. This situation is quite different from that of ordinary experimental studies.

In this model, $P_s > P_{H_2O}$, and the chemical potential of H_2O is smaller than the molar free energy of pure H_2O under the pertinent temperature and pressure. If the chemical potential is increased to approach the molar free energy of pure H_2O, a fluid phase of virtually pure H_2O composition may form.

In the early phase of regional metamorphism, the temperature should increase progressively. H_2O and CO_2 should be liberated from hydrous and carbonate minerals through a sequence of reactions which take place with rising temperature. The models for this case are particularly important, and will be treated in § 3-1.

2-4   DEHYDRATION REACTIONS

Most metamorphic reactions occurring with rising temperature cause liberation of H_2O from solid phases as exemplified below.

$$Al_2Si_4O_{10}(OH)_2 = Al_2SiO_5 + 3\ SiO_2 + H_2O \qquad (2\text{-}8)$$
$$\quad\ \text{pyrophyllite}\qquad\ \text{kyanite}\quad\ \text{quartz}$$

$$KAl_3Si_3O_{10}(OH)_2 + SiO_2 = Al_2SiO_5 + KAlSi_3O_8 + H_2O \qquad (2\text{-}9)$$
$$\quad\ \text{muscovite}\qquad\ \text{quartz}\ \ \text{sillimanite}\ \ \text{K-feldspar}$$

$$KAl_3Si_3O_{10}(OH)_2 = Al_2O_3 + KAlSi_3O_8 + H_2O \qquad (2\text{-}10)$$
$$\quad\ \text{muscovite}\qquad\ \text{corundum}\ \ \text{K-feldspar}$$

Equation (2-8) represents a possible reaction to produce kyanite in metapelites. Equations (2-9) and (2-10) represent the breakdown of muscovite in quartz-bearing and quartz-free rocks. The side having liberated H_2O in such equations represents higher temperatures in almost all cases, because the entropy of the liberated H_2O is usually very great and the side with a greater entropy represents higher temperatures.

Equilibrium Curves in Model (A)
We assume that the liberated H_2O forms an aqueous fluid phase at the same temperature and pressure as the solid phase in contact. This corresponds to model (A) discussed in the preceding section. The total volume and entropy of the products in such reactions are usually greater than those of the reactants in the pressure range up to several kilobars, because the volume and entropy of the liberated H_2O are great. The Clausius–Clapeyron equation (2-5) holds for equilibrium curves in this case. Here, ΔV means an increase in the total volume.

Since both ΔV and ΔS are usually positive, dP/dT is positive. In other words, the equilibrium curves usually have a positive slope (fig. 2-2). At pressures below 1 kbar, ΔV is very great and hence dP/dT is very small. An increase of pressure to about 3 kbar markedly reduces the volume of the liberated H_2O, and hence dP/dT increases. Thus, the equilibrium curves for dehydration reactions usually show a sharp bend in the pressure range of $0\cdot1$–$3\cdot0$ kbar. At pressures above 3 kbar, they are nearly straight and are steeper than the equilibrium curves of most

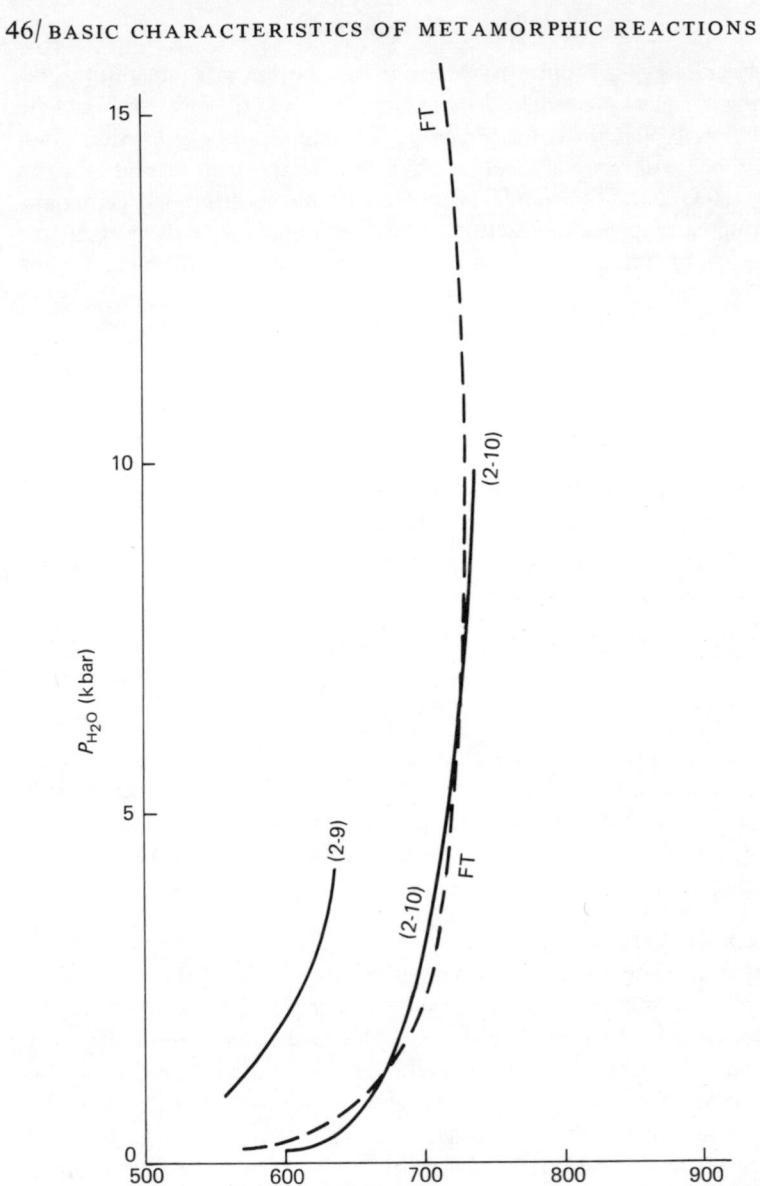

Fig. 2-2   Equilibrium curves for dehydration reactions in the presence of an aqueous fluid. Full lines represent equilibria for the reaction of muscovite and quartz (2-9), and for the breakdown of muscovite (2-10) (Evans, 1965; Velde, 1966). The dashed line FT represents equilibria for the reaction: forsterite+talc = enstatite +H_2O. This curve shows a negative slope at pressure above 10 kbar. (Kitahara and Kennedy, 1967.)

solid–solid reactions. In other words, the effect of pressure on equilibrium temperature is small under such pressures.

It has been found empirically that the total volume of the solids commonly decreases in such reactions (table 5 in Thompson, 1955). At pressures higher than 10 kbar, an aqueous fluid is highly compressed and the volume of the liberated H_2O is small. If the volume decrease of the solids is relatively large, the total volume of the system including both solid and fluid may decrease. In such a case, an equilibrium curve should take on a negative slope. An example of the change of a curve from positive to negative slopes with increasing pressure is shown in fig. 2-2 (curve FT).

In the pressure range of more than 30 kbar (corresponding to the upper mantle), most dehydration curves may have a negative slope (e.g. Fry and Fyfe, 1969). In such a case, increasing fluid pressure should cause dehydration.

Some reactions representing the breakdown of zeolites have a large decrease in volume of the solid phases. Hence, their equilibrium curves come to have a negative slope at relatively low pressures (Coombs et al. 1959), as is illustrated by the breakdown of analcime in the presence of quartz (fig. 6B-5.

Some breakdown reactions of zeolite involve not only a decrease of volume but also a decrease of entropy. The following are examples: analcime = jadeite + H_2O, and laumontite = lawsonite + 2 quartz + 2 H_2O (Fyfe and Valpy, 1959; Thompson, 1970b). The equilibrium curves of these reactions have a positive slope, as shown in fig. 6B-5.

Equilibrium Curves in More General Cases

The equilibrium curves for any dehydration reaction can be calculated by the following approximate relation:

$$\Delta G^{\ominus} = -19.14\, T \log P_{H_2O} - P_s \Delta V_s/10{\cdot}0. \qquad (2\text{-}11)$$

Here, $\Delta G^{\ominus}$ and ΔV_s represent the free energy of reaction (J) under 1 bar and the volume change of solid phases (cm^3) respectively, both per mole of liberated H_2O, and hence $\Delta G^{\ominus}$ is a function of absolute temperature T. P_{H_2O} and P_s are in bars and log indicates the common logarithm. If we accept model (A), $P_s = P_{H_2O}$ in this equation.

In models (B), (C) and (D), $P_s > P_{H_2O}$. The dehydration temperature in this case is lower than the corresponding one in model (A) under the same rock-pressure, and decreases with decreasing P_{H_2O}. Equation (2-11) is applicable to this case also.

The general shapes of equilibrium curves are shown in fig. 2-3. The curves for P_{H_2O} = constant shown there meet the curve for $P_s = P_{H_2O}$ at their lower end where P_s is equal to the pertinent constant value of P_{H_2O}. The difference between P_{H_2O} and P_s becomes greater upwards along the curves of constant

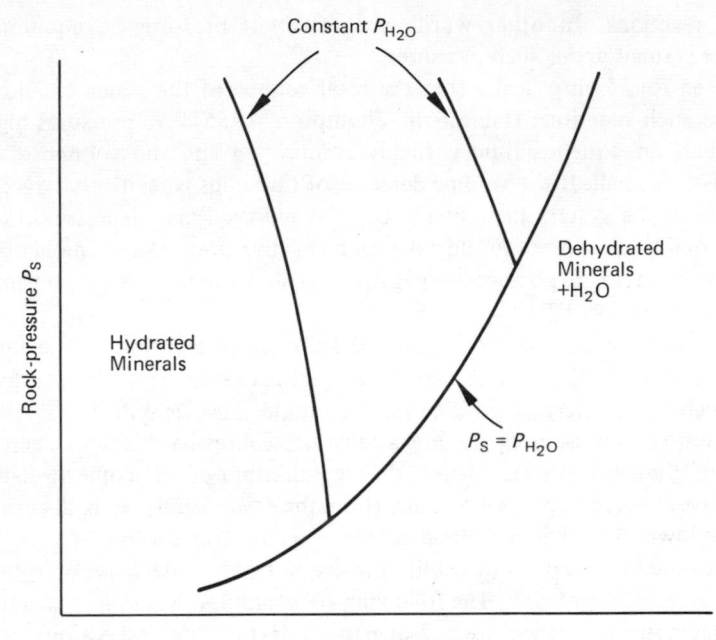

Fig. 2-3   Common forms of equilibrium curves for dehydration reactions under the conditions of $P_s = P_{H_2O}$ and of constant P_{H_2O}. (Thompson, 1955; Greenwood, 1961.)

P_{H_2O}, which usually have a negative slope. In other words, increasing P_s when P_{H_2O} = constant tends to dehydrate minerals.

The relation between the slope of a curve under a constant P_{H_2O} and that of the corresponding curve under $P_s = P_{H_2O}$ is given by the following relation (Greenwood, 1961):

$$\left(\frac{\partial P_s}{\partial T}\right)_{P_{H_2O}} = \frac{10.0\,\Delta S}{\Delta V_s} = \left(\frac{dP}{dT}\right)_{P_s = P_{H_2O}} \times \frac{\Delta V}{\Delta V_s} \tag{2-12}$$

When $(dP/dT)_{P_s = P_{H_2O}}$ is known from a synthetic study, the slope of the curve for a constant P_{H_2O} can be directly calculated by this equation. In this equation, P and ΔS are in bars and J K^{-1} mol^{-1} respectively.

2-5   MINERAL ASSEMBLAGES AND ELEMENT DISTRIBUTION

It is instructive here to call attention to the effect of the presence of quartz on the P-T relations of metamorphic reactions. Equilibria for equations (2-4) and

(2-10) represent the high temperature stability limit of jadeite and muscovite respectively. If jadeite or muscovite is associated with quartz, reactions would take place as shown by equations (2-3) and (2-9) respectively. Thus, the stability fields of these minerals are reduced, as shown in figs. 2-1 and 2-2.

Generally speaking, if quartz can participate in a reaction, its presence will change the equilibrium temperature. This effect is very important in metamorphism, since quartz is widespread and is able to participate in a large number of reactions (Fyfe, Turner and Verhoogen, 1958; Miyashiro, 1960).

The stability field of a mineral is widest when the mineral is present alone. A lot of synthetic studies have been made to clarify the breakdown temperature and pressure of single minerals. Such studies yield only the maximum stability fields of the pertinent minerals. The stability fields of mineral assemblages containing these minerals are the same as or smaller than the maximum stability fields.

The recent progress of metamorphic petrology has enhanced the importance of mineral assemblages. Description of metamorphic minerals without due regard to their paragenetic relations is now becoming less important. If two coexisting minerals in equilibrium have a component in common, Nernst's *partition law* holds for the distribution of the component as follows:

$$\frac{x}{y} = K \tag{2-13}$$

where x and y represent the mole fractions of the component in the two minerals, which are regarded as ideal solutions. The partition coefficient K is a function of temperature and rock-pressure.

The chemical compositions of solid solution minerals should vary with the bulk chemical composition of their host rock. If two ferromagnesian minerals coexist, an increase in the $Mn/(Mg+Fe^{2+}+Mn)$ ratio of the rock, for example, should induce an increase of the same ratio in both minerals. The Mg–Fe distribution between associated ortho- and clinopyroxenes, for example, can be treated from the viewpoint of equilibrium for the reaction:

$$MgSiO_3 + CaFeSi_2O_6 = FeSiO_3 + CaMgSi_2O_6. \tag{2-14}$$

The values of atomic ratio Fe/(Mg + Fe) in ortho- and clinopyroxenes are designated as x and y respectively. If the two minerals are regarded as ideal solutions, we have the following relation:

$$\ln\left(\frac{x}{1-x}\right)\bigg/\left(\frac{y}{1-y}\right) = -\frac{\Delta G}{RT}. \tag{2-15}$$

Here, ΔG represents the free energy change for equation (2-14) under the condition that each term indicates an end-member mineral. This relation may be rewritten as follows:

$$\frac{\text{(Fe/Mg) in orthopyroxene}}{\text{(Fe/Mg) in clinopyroxene}} = \exp(-\Delta G/RT) = K(T, P_s) \qquad (2\text{-}16)$$

where (Fe/Mg) represents an atomic ratio, and the equilibrium constant K is a function of temperature and rock-pressure (Mueller, 1960; Kretz, 1961).

The graphical analysis of the relationship between mineral paragenesis and bulk chemical composition is important for a better understanding of thermodynamic relations. This problem will be treated in chapter 5.

2-6 DECARBONATION REACTIONS

The metamorphism of calcareous rocks liberates CO_2, as is exemplified by the following equation:

$$CaCO_3 + SiO_2 = CaSiO_3 + CO_2 \qquad (2\text{-}17)$$
$$\text{calcite} \quad \text{quartz} \quad \text{wollastonite}$$

The equilibrium curve for this reaction under the condition of $P_{CO_2} = P_s$ is shown in fig. 2-4.

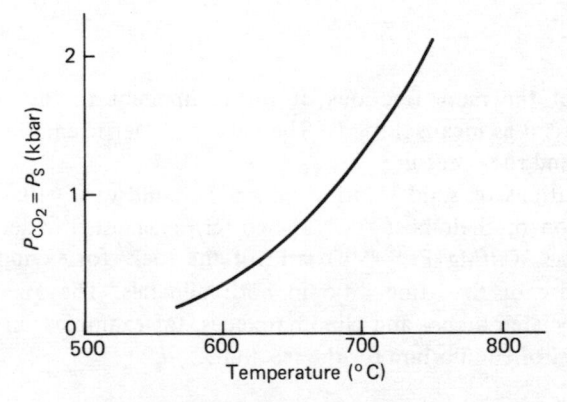

Fig. 2-4 Equilibrium curve for the reaction to form wollastonite (2-17) in the presence of a fluid phase with the composition of pure CO_2. (Harker and Tuttle, 1956.)

The equilibrium curves for decarbonation reactions show a marked bend in the pressure range of 0·1–2·0 kbar just as those for dehydration reactions do. Calcite has a low density, and hence, the ΔV_s of decarbonation reactions usually has a large negative value. When $P_{CO_2} = P_s$, the equilibrium curves may turn to have a negative slope at pressures above several kilobars.

The thermodynamics of decarbonation reactions has been discussed by Danielsson (1950) and Weeks (1956). The entropy increase of decarbonation is roughly similar to that of dehydration. At 298 K and 1 bar, the former is about 170 J K^{-1} mol^{-1} of liberated CO_2, whereas the latter is about 150 J K^{-1} mol^{-1} of liberated H_2O. The equilibrium curves can be calculated by equation (2-11) in which P_{CO_2} is substituted for P_{H_2O}.

In calcareous rocks undergoing metamorphism, a fluid phase if any would usually be a mixture of CO_2 and H_2O. Equilibria for mixtures of CO_2 and H_2O will be treated in detail in chapter 9.

2-7   OXIDATION AND REDUCTION IN METAMORPHISM

Iron Oxides and O_2

In the atmosphere, oxygen is mostly in the free molecular state with a partial pressure of about 0·2 bar near the surface of the earth, whereas within the crust and uppermost mantle it is mostly in combined forms as silicates and oxides. Just as in the case of H_2O, we can imagine here a small cell with rigid walls, permeable only to O_2, which is contained in a rock. The oxygen in the cell is at a pressure P_{O_2}, being in osmotic equilibrium with that outside.

The most useful indicators for P_{O_2} in the crust are iron oxides, i.e. haematite, magnetite and wüstite. These minerals are related to one another by the following equations:

$$6 \, Fe_2O_3 = 4 \, Fe_3O_4 + O_2 \qquad (2\text{-}18)$$

haematite   magnetite

$$2 \, Fe_3O_4 = 6 \, FeO + O_2 \qquad (2\text{-}19)$$

magnetite   wüstite

$$2 \, FeO \; = 2 \, Fe \; + O_2 \qquad (2\text{-}20)$$

wüstite   native iron

$$\tfrac{1}{2} \, Fe_3O_4 \; = \tfrac{3}{2} Fe \; + O_2. \qquad (2\text{-}21)$$

magnetite   native iron

P_{O_2} for these equilibria can be calculated by the following equation:

$$\Delta G^{\ominus} = -19{\cdot}14 \, T \log P_{O_2} - P_s \Delta V_s / 10{\cdot}0 \qquad (2\text{-}22)$$

Here, $\Delta G^{\ominus}$, T, and ΔV_s represent the free energy of reaction (J) under 1 bar, absolute temperature, and the volume change of solid phases (cm^3) respectively. P_{O_2} and P_s are in bars. The last term is small compared with the others, and may usually be neglected (Ernst, 1960; Miyashiro, 1964). Note that equation (2-22) has the same form as equation (2-11). The calculated equilibrium curves are shown in fig. 2-5.

As is clear from the figure, haematite is stable at high P_{O_2}. As P_{O_2} decreases, magnetite, wüstite, and finally native iron become stable in turn. Haematite occurs in some and magnetite in many metamorphic rocks, but wüstite and native iron do not. This suggests that P_{O_2} in rocks undergoing metamorphism is usually of the order of 10^{-10} to 10^{-40} bar, that is, much lower than the partial pressure of O_2 in the atmosphere.

Fig. 2-5 indicates that the magnetite + haematite assemblage show a definite

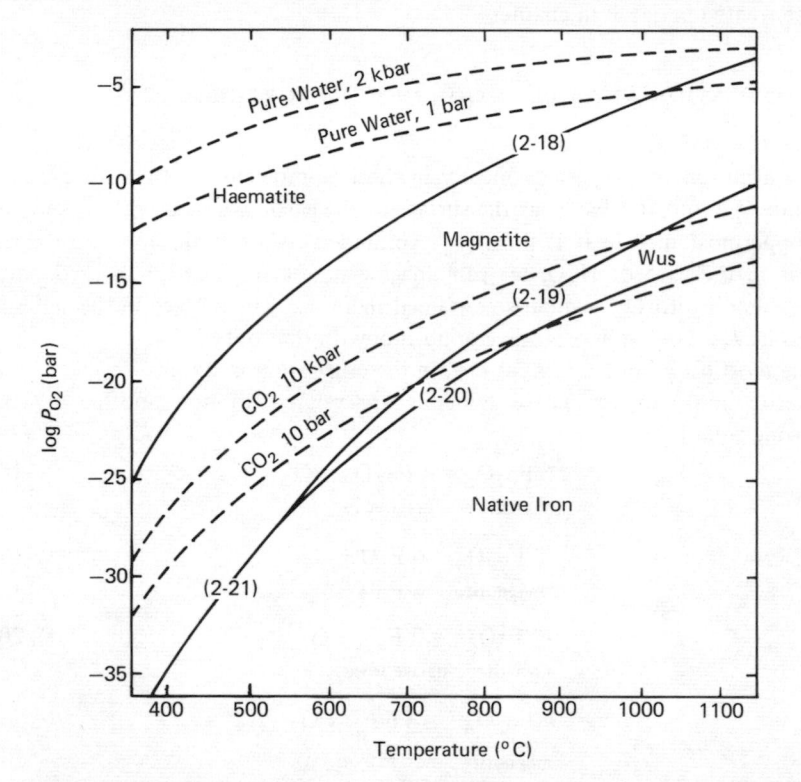

Fig. 2-5 Stability relations of oxides of iron and native iron. Curves (2-18), (2-19), (2-20) and (2-21) represent equilibria for the equations which are denoted in the text with these numbers. The effect of pressure on solid phases is neglected. (Miyashiro, 1964.) 'Wus' means the wüstite field.

P_{O_2} at a definite temperature (and rock-pressure), so far as the two minerals belong to the Fe–O system. Natural magnetite and haematite contain some TiO_2 (see e.g. fig. 8B-6). Under such a situation, the three-mineral assemblage magnetite + haematite + ilmenite has a definite P_{O_2} at a specific temperature and rock-pressure. The effect of TiO_2, however, will be neglected here for simplicity of discussion.

Dissociation of H_2O

The H_2O present in rocks undergoing metamorphism would be approximately in dissociation equilibrium with H_2 and O_2 as follows:

$$H_2O \rightleftharpoons H_2 + \tfrac{1}{2}O_2 \qquad (2\text{-}23)$$

$$\Delta G^{\ominus} = -19 \cdot 14\, T \log[(P_{H_2} \cdot \sqrt{P_{O_2}})/P_{H_2O}]. \qquad (2\text{-}24)$$

Here, P_{H_2}, P_{O_2} and P_{H_2O} are in bars. If we assume that an aqueous fluid has the pure water composition, i.e. an atomic ratio of H : O equal to 2, P_{O_2} becomes a function of P_{H_2O} and T. The calculated curves of P_{O_2} for $P_{H_2O} = 1$ bar and 2 kbar are shown in fig. 2-5. These curves indicate that if an aqueous fluid has the pure water composition, haematite is to be stable for almost the entire temperature range of metamorphism so long as P_{H_2O} is higher than 1 bar. This is not the case, since magnetite is very widespread (§ § 7B-21 and 8B-12).

Therefore, P_{O_2} in such rocks should be lower than that produced by dissociation of pure water. In other words, H_2 must be present in excess of the pure water composition. Such a reducing condition may have been fostered partly by the conceivable ascent of H_2 and CO from the mantle of the earth. However, it is more likely that the reducing condition is produced within the crust by organic materials and graphite. Such substances are almost omnipresent in metapelites.

Role of Graphite

Black carbonaceous matter in metamorphic rocks has usually been called graphite in the geologic literature. Recent investigations have shown, however, that the carbonaceous matter in low-temperature metamorphic rocks are organic materials comparable to semi-anthracite and anthracite. Their composition and atomic structure change gradually with increasing metamorphism, until graphite is formed in the amphibolite facies (Izawa, 1968). Since the individual compounds in them have not been identified, assignment of numerical values to their reducing power is not possible. As an approximation, we will take up graphite.

The following equilibrium would hold between graphite and O_2:

$$C + O_2 = CO_2. \qquad (2\text{-}25)$$

By neglecting the effect of P_s, we have

$$\Delta G^{\ominus} = -19 \cdot 14\, T \log \frac{P_{CO_2}}{P_{O_2}}. \qquad (2\text{-}26)$$

Here, P_{CO_2} and P_{O_2} are in bars, and $\Delta G^{\ominus}$ is in J. Thus, P_{O_2} is a function of P_{CO_2} and T. The value of P_{CO_2} within the crust could not be greater than 10 kbar. The calculated curves of P_{O_2} for the conditions $P_{CO_2} = 10$ bar and 10 kbar are shown in fig. 2-5. These curves indicate that the values of P_{O_2} in the presence of graphite is within the magnetite field over almost the entire temperature range of metamorphism (Miyashiro, 1964).

In rocks containing graphite (or organic matter) and H_2O, not only CO_2, H_2 and O_2 but also CO and CH_4 (methane) are formed by such reactions as $CO_2 = CO + \frac{1}{2}O_2$ and $C + 2H_2 = CH_4$. Under a relatively reducing condition, the amount of CH_4 can be considerable (French and Eugster, 1965). The presence of CH_4 has been detected in bore-holes made through metamorphic rocks in the Bessi and other mines in the Sanbagawa belt (Izawa, 1968).

Effect and Mobility of O_2

An FeO-bearing compound having a definite composition is stable over a range of P_{O_2} at a definite temperature and rock-pressure. For example, fayalite (Fe_2SiO_4) is stable in the range of P_{O_2} = about $10^{-16} - 10^{-26}$ bar at 700 °C (with a slight variation with rock-pressure). If P_{O_2} is higher or lower than this range, fayalite decomposes to produce magnetite or native iron, as follows:

$$2\,Fe_3O_4 + 3\,SiO_2 = 3\,Fe_2SiO_4 + O_2 \qquad (2\text{-}27)$$
$$\text{magnetite} \quad \text{quartz} \qquad \text{fayalite}$$

$$Fe_2SiO_4 = 2\,Fe + SiO_2 + O_2. \qquad (2\text{-}28)$$
$$\text{fayalite} \quad \text{iron} \quad \text{quartz}$$

Similarly, a silicate solid solution containing FeO is stable in a range of P_{O_2} at a definite temperature and rock-pressure. As P_{O_2} increases, the FeO-bearing component gradually decomposes to produce magnetite and other products. Therefore, the FeO/MgO ratio of a silicate solid solution associated with magnetite and other decomposition products is an indicator of P_{O_2}. The higher the P_{O_2}, the poorer in FeO becomes the silicate solid solution (Mueller, 1960).

In contrast to H_2O and CO_2, which are relatively highly mobile during metamorphism, O_2 and H_2 are not as mobile (Thompson, 1957; Eugster, 1959; Mueller, 1960; Chinner, 1960). In metamorphosed iron formations in the Canadian Shield, a haematite-bearing layer is found to be in contact with a magnetite-bearing layer. The actinolite in the former layer is lower in FeO/MgO ratio than that in the latter. This suggests a higher P_{O_2} in the former. The boundaries between such layers are sharp in some examples and gradational in others. At any rate, such relations indicate that the mobility of O_2 in rocks undergoing metamorphism is extremely small.

Equation (2-24) indicates the dependence of P_{H_2} on P_{O_2} and P_{H_2O}. If P_{O_2} varies greatly across a short distance, P_{H_2} also must do so. Accordingly, the mobility of H_2 was also very small in the above case. Such small mobilities of O_2 and H_2 are probably a result of a very low value of P_{O_2} and of P_{H_2}. The gradients of concentration for O_2 and H_2 were probably too small to cause effective diffusion.

It has been shown in some terranes that the Fe^{3+}/Fe^{2+} ratio of metamorphic rocks tends to decrease with increase in temperature of metamorphism (e.g. Shaw, 1956; Miyashiro, 1958). Probably this is not due to long-distance flow or diffusion of O_2. In pelitic rocks undergoing metamorphism, Fe_2O_3-bearing minerals lose some oxygen by reaction with organic matter or graphite to form CO_2. The CO_2 would flow or diffuse over a long distance, leading to a general decrease of the Fe^{3+}/Fe^{2+} ratio of the rocks. Flow of an intergranular fluid or diffusion would have spread the reducing condition of metapelites to the surrounding rocks. Such a flow or diffusion should not be uniform, and the degree of reduction should be variable in metamorphic rocks. It would cause diversity in mineral composition. Some metapelites were found to contain no graphite and hence to show highly variable degrees of reduction with resultant generation of variable mineral compositions (Chinner, 1960).

In experimental work, P_{O_2} can be controlled by using a set of oxygen buffers (Eugster and Wones, 1962; Ernst, 1960). For example, a buffer composed of a mixture of magnetite and iron keeps P_{O_2} at a value corresponding to the equilibria of reaction (2-21). The curves of P_{O_2} for common buffers are shown in fig. 7B-10. The fayalite-magnetite-quartz buffer, which corresponds to the equilibria of (2-27), gives P_{O_2} values within the magnetite field, and hence relatively close to those in ordinary pelites undergoing metamorphism.

Chapter 3

Basic Concepts of Metamorphic Petrology

3-1 PROGRESSIVE REGIONAL METAMORPHISM

Concept of Progressive Metamorphism

Metamorphism usually takes place in response to a rise of temperature. Progressive changes brought about in rocks by rising temperature are called *progressive metamorphism*. Successive rises of temperature bring about a series of reactions, which might be understood as changes of mineral assemblages approximately at equilibrium. In many cases, however, some minerals which formed at a lower temperature remain as unstable relics in the core of minerals that formed later.

The progressive nature of metamorphic changes was strongly emphasized by Alfred Harker (1932) in order to characterize his line of approach or attitude to the study of metamorphism. He regarded the old German school of petrography, particularly Grubenmann (1904-6), as being too static, classificatory and descriptive. Grubenmann classified metamorphic rocks according to their chemical composition and the assumed depths of formation. Genetic relations were mostly ignored. In contrast, Harker characterized his line of approach as genetic, and attempted to treat metamorphism as a sequence of progressive changes in response to changing conditions. The Harker line of approach is generally adopted in this book.

The effectiveness of this approach results mainly from the empirical fact that metamorphic rocks, recrystallized at progressively increasing temperature, are exposed side by side so that the sequence of changes can be confidently followed in the field. It has been found in many regional metamorphic terranes that the temperature of metamorphism is distributed in a relatively simple, regular pattern, as exemplified by fig. 1-2. In a metamorphic belt as a whole, the temperature increases toward the thermal axis, and a series of zones can usually be delineated, each one representing a fairly definite range of temperature.

Since metamorphism is a long process involving a number of phases, it is important to recognize which phase is recorded in the mineral composition of metamorphic rocks. It is widely believed that reactions take place easily with rising temperature, but do not occur as easily with declining temperature (e.g. Harker, 1932; Fyfe, Turner and Verhoogen, 1958). Insofar as this holds true, the mineral compositions of metamorphic rocks are usually representative of the state of the highest temperature reached by the individual rocks.

However, it is also common for some mineral changes to take place in the declining phases of metamorphism. We can recognize such changes usually without great difficulty through observation of the textural relations of the minerals.

Barrovian Zones as a Classical Example

George Barrow, a British geologist, was the first to succeed in showing the existence of progressive mineral zoning on a regional scale in a metamorphic terrane. His work was carried out toward the end of the nineteenth century in a part of the Dalradian Series in the Scottish Highlands, about 50 km southwest of Aberdeen. As shown in fig. 3-1, granitic rocks are exposed in the north of the area, and pelitic schists and gneisses in the south. He found that sillimanite-bearing metapelites were exposed in a particular zone in the northernmost part

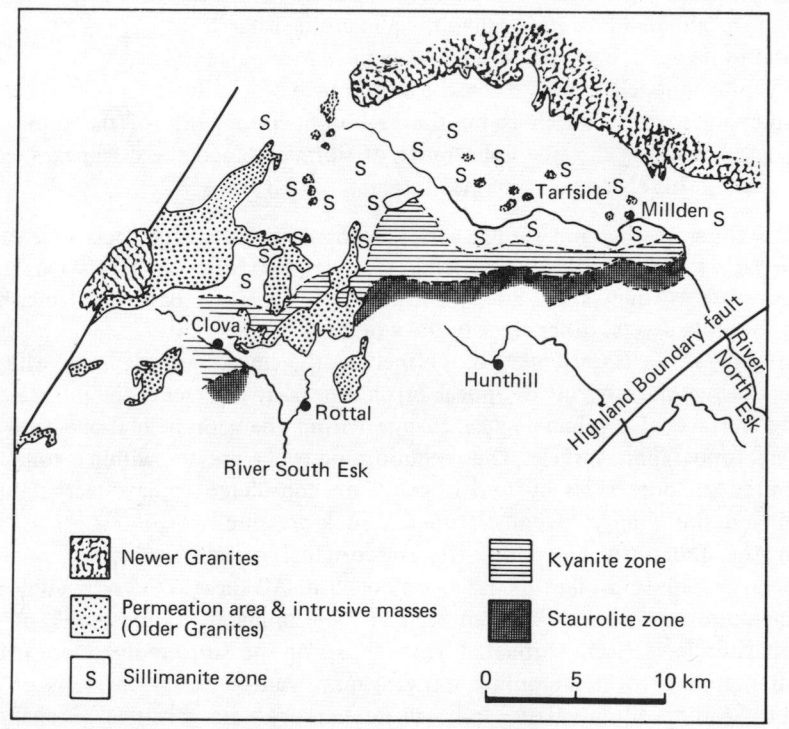

Fig. 3-1 Classical metamorphic terrane mapped by Barrow (1893) in the Scottish Highlands. (Simplified from the original map.)

of the metamorphic terrane, kyanite-bearing metapelites occurred in an adjacent zone to the south, and staurolite-bearing ones still further south (Barrow, 1893). He considered that the prevailing temperature had been successively higher toward the north.

About twenty years later, Barrow extended his survey of progressive metamorphic zones to the southern limit of the terrane, the Highland Boundary fault (Barrow, 1912). The general validity of his work was later confirmed by Tilley (1925) and Harker (1932). The following series of metamorphic zones (fig. 3-2) have been delineated on the basis of mineral changes in metapelites in response to increasing temperature:

1. Chlorite zone, characterized by chlorite-muscovite phyllite or schist.
2. Biotite zone, characterized by the appearance of biotite, which persists to all the higher zones. The boundary between the chlorite and the biotite zone, i.e. the line that marks the first appearance of biotite is called the *biotite isograd*.
3. Almandine zone (garnet zone), characterized by the appearance of almandine garnet, which persists to the higher zones. The boundary between the biotite and the almandine zone is called the almandine isograd.
4. Staurolite zone, characterized by the appearance of staurolite.
5. Kyanite zone, characterized by the appearance of kyanite.
6. Sillimanite zone, which is on the high-temperature side of the sillimanite isograd marking the first appearance of sillimanite and the disappearance of kyanite. This change corresponds to reaction (2-1) of § 2-2.

The zones in this and similar series, recognized for pelitic rocks, are called *Barrovian zones*. Pelitic rocks tend to show textural changes with rising temperature as follows: phyllites → mica schists → gneisses. Barrovian zones have been found in several other areas in the world.

In general, an *isograd* may be defined as the line of outcrops on which a mineral assemblage begins to appear or disappear. It represents the intersection of the surface of a mineralogical change within the geosynclinal pile with the present topographic surface. The metamorphic rocks present within a zone that is limited on both sides by two isograds are considered to have recrystallized within a definite range of temperature and rock-pressure.

In the Dalradian metamorphics, the original geosynclinal pile contained pyroclastic beds and dikes of basic composition. All these rocks were subjected to metamorphism (e.g. Wiseman, 1934). The mineral compositions of the metabasites have been correlated with those of the surrounding metapelites. Zonal mapping of metamorphic terranes may well be based on rocks of any chemical composition. Basic and calcareous rocks are sensitive to changing temperature, and no less appropriate for this purpose than pelitic ones, if they are widespread in the terrane under consideration.

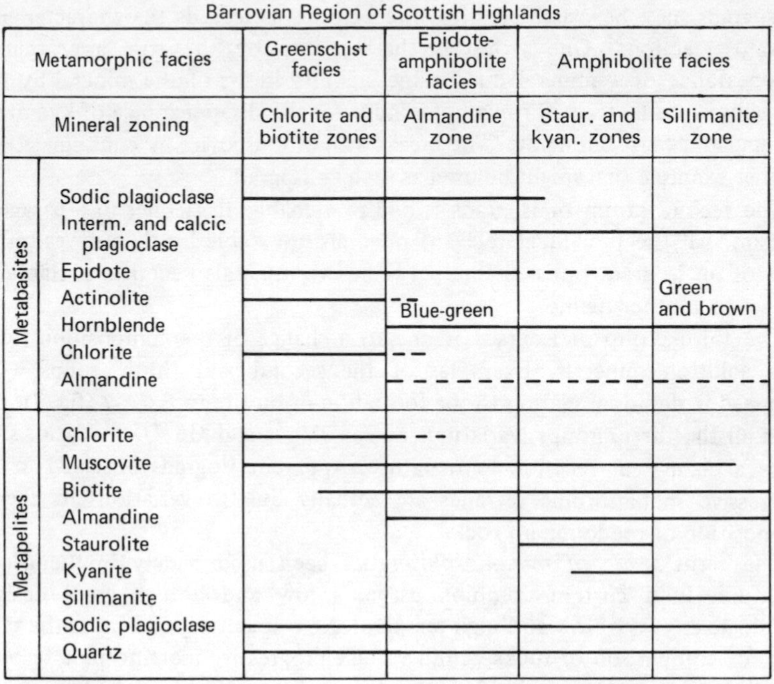

Barrovian Region of Scottish Highlands

Metamorphic facies		Greenschist facies	Epidote-amphibolite facies	Amphibolite facies	
Mineral zoning		Chlorite and biotite zones	Almandine zone	Staur. and kyan. zones	Sillimanite zone
Metabasites	Sodic plagioclase				
	Interm. and calcic plagioclase				
	Epidote				
	Actinolite		Blue-green		Green and brown
	Hornblende				
	Chlorite				
	Almandine				
Metapelites	Chlorite				
	Muscovite				
	Biotite				
	Almandine				
	Staurolite				
	Kyanite				
	Sillimanite				
	Sodic plagioclase				
	Quartz				

Fig. 3-2   Progressive mineral changes in the region of the Barrovian zones in the Scottish Highlands. (Harker, 1932; Wiseman, 1934.)

Significance of Isograds

In the progressive metamorphism of pelitic sediments, a series of reactions proceeds with successive rises of temperature. The medium- and high-temperature parts of regional metamorphic terranes usually approach a state of chemical equilibrium. Isograds are defined in terms of mineral variations in the near-equilibrium state.

Toward the lower temperature, a metamorphic terrane grades imperceptibly into an unrecrystallized area. The low-temperature limit of a metamorphic terrane depends on the rate of, and the time available for, recrystallization. Such a boundary is not an isograd in the proper sense.

Isograds give a general picture of the temperature distribution in a metamorphic terrane. Strictly speaking, however, a definite isograd (e.g. the sillimanite isograd) does not necessarily represent a definite temperature. The sillimanite isograd in the Barrovian zones is due to a solid–solid reaction and represents a definite temperature only under a definite rock-pressure (fig. 2-1). The isograds defined in terms of dehydration reactions depend on P_s and P_{H_2O}, both of which may vary within a terrane.

Isograds may be classified into three groups as regards the character of the relevant reactions. One group is characterized by the true appearance or disappearance of a mineral due to the stability limits of the mineral by itself. The sillimanite isograd, for example, indicates the disappearance of kyanite and the appearance of sillimanite. The breakdown of muscovite by equation (2-10) is another example that might be used as such an isograd.

The second group of isograds is due to a change in tielines in a paragenesis diagram. All the individual relevant minerals are stable by themselves on both sides of an isograd, but a tieline between two minerals becomes unstable with respect to another tieline.

The third group of isograds is due to a change of the composition field of solid solution minerals. Examples of the second and third groups will be discussed in detail in relation to the formation of biotite in § 7B-7 (fig. 7B-6).

In all the three groups, variations in Fe^{2+}/Mg^{2+} and Mn^{2+}/Fe^{2+} ratios should cause a change in reaction temperature. Apparent isograds observed in some progressive metamorphic terranes are actually due to variations in chemical composition of predominant rocks.

The term *grade of metamorphism* has been used widely in the geologic literature. In a current common usage, a low and a high grade designate petrologic effects of low and high temperatures respectively. So far as the term is used for comparison of rocks within a single progressive metamorphic terrane, it has a clear and empirically determinable meaning. When applied to rocks in different regions, its meaning is not always clear, since entirely reliable values of recrystallization temperature are not known for most metamorphic rocks. Many geologists called metamorphic rocks recrystallized at higher rock-pressures, high-grade rocks. If this usage were to be accepted, we could not tell which of the following is higher in grade: rocks recrystallized at high pressure and low temperature, or rocks recrystallized at low pressure and high temperature.

Compositional Changes in Solid–Solution Minerals

In the chlorite and biotite zones of the Dalradian and other similar metamorphic areas, the stable plagioclase is albite regardless of the composition of the host-rocks. Some of the original rocks (e.g. basaltic ones) must have contained intermediate or calcic plagioclase. When they were subjected to low-temperature metamorphism, the anorthite molecule of the plagioclase was converted into epidote with the resultant formation of the albite+epidote assemblage. With a rise in temperature of metamorphism, the albite reacts with epidote to form intermediate or calcic plagioclase, as is the case with metabasites of the kyanite and sillimanite zones (§ 8B-3). If epidote is very scarce or absent, albite will persist into the high-temperature zones.

As epidote contains H_2O and Fe_2O_3, the reaction of albite and epidote to form more calcic plagioclase is affected by P_{H_2O} and P_{O_2}. As epidote is not strictly isochemical with anorthite, the compositional difference must be adjusted by a change in composition of associated minerals, and thus the reaction is affected by the associated minerals. Therefore, the anorthite content of plagioclase accompanied or unaccompanied by epidote does not increase regularly with rise in temperature in a progressive metamorphic terrane. Moreover, a discontinuous change in composition appears to take place in sodic plagioclase (§ 7B-8).

The monoclinic amphibole, stable in ordinary metabasites of the chlorite and biotite zones, is actinolite, whereas that of higher temperature zones is common hornblende with higher Al_2O_3 and Na_2O contents. The colour (Z) of common hornblende usually changes from blue–green to green or green–brown with rise in temperature (§ 8B-4).

In an isochemical series of metamorphic rocks, the amounts of minerals and the distribution relation of elements in a particular pair of minerals should change with temperature. If such rocks are in equilibrium, the chemical composition of a mineral is a function of the bulk chemical composition of the host rock, the amounts of the minerals present, and the distribution relations of elements. A progressive compositional change of a solid solution mineral takes place owing to a change in the amounts of minerals and in the distribution relations of elements. The progressive compositional change of some minerals is mainly controlled by such equilibrium relations (Miyashiro and Shido, 1973).

On the other hand, the composition of garnet, particularly formed at lower temperatures, appears to be mainly controlled by disequilibrium growth. Garnets in low-temperature metapelites are usually rich in MnO. The MnO content tends to decrease with rising temperature (§ 7B-10). Such garnets usually show strong compositional zoning with the MnO content decreasing to the rim. Hollister (1966) and Atherton (1968) have shown that the MnO decrease in some garnets can be semi-quantitatively explained by making the assumption that chemical diffusion in garnet crystals is negligible and only the outermost layer of garnet is in equilibrium with the surroundings. In such a case, garnet, once formed, is immediately removed from the reacting system resulting in a progressive change in the effective bulk chemical composition of the reacting system, and hence a progressive change in the composition of newly formed garnet. This will be the case even if there is no noticeable change in the distribution relation of MnO between garnet and associated minerals.

Migration of Materials

The kind, extent and mechanism of the migration of materials in rock complexes undergoing regional metamorphism have been subjects in dispute for many years. Some geologists envisaged large-scale migration of various materials, while others denied it. The migration, if it occurs, may cause a change in bulk chemical composition of rocks. The compositional change of rocks caused by the introduction of materials is called *metasomatism* (Goldschmidt, 1922).

Decreases in H_2O and CO_2 contents and in Fe_2O_3/FeO ratio are the most commonly observed compositional changes in progressive regional metamorphism (Shaw, 1956; Miyashiro, 1958; Engel and Engel, 1958, 1962a; Vallance, 1960). H_2O molecules liberated by dehydration of minerals should either flow or diffuse out of the rock complex. Generally, most of the H_2O would migrate upwards to the surface of the earth.

Not only liberated H_2O but also H_2O of magmatic or mantle origin may move through a rock complex undergoing regional metamorphism. In the progressive metamorphic area of Dutchess County in southeastern New York it has been demonstrated that the $^{18}O/^{16}O$ ratios of whole-rock samples of metapelites tend to become lower with an increase in temperature of metamorphism, ultimately approaching that of pegmatitic granite and plutonic rocks in the highest temperature zone. The $^{18}O/^{16}O$ ratios of the original pelites were probably high, and simple dehydration does not change the ratio so markedly. Probably H_2O of magmatic, or some other deep-seated origin with low $^{18}O/^{16}O$, has been mixed and moved through the metamorphic terrane, resulting in a significant change of the $^{18}O/^{16}O$ ratio of metapelites on a regional scale. It is interesting that a quartz-rich pod enclosed in limestone, and an amphibolite, showed a much higher and lower $^{18}O/^{16}O$ ratio respectively than the surrounding metapelites. Probably the moving H_2O did not permeate into the limestone and amphibolite sufficiently to homogenize their $^{18}O/^{16}O$ ratios.

Similar relationships have been noticed in some other metamorphic terranes also. In other areas, however, metapelites show a more or less irregular distribution of $^{18}O/^{16}O$ ratios probably owing to an incomplete exchange reaction of oxygen isotopes (Garlick and Epstein, 1967; Epstein and Taylor, 1967).

CO_2 which is mainly produced by decarbonation would also move through rock complexes during metamorphism. A decrease in Fe_2O_3/FeO would be due to reduction of ferric iron by organic matter or graphite to result in the production of CO_2 and H_2O. There is evidence suggesting that the mobility of O_2 is very low (§ 2-7).

It has been claimed in a number of metamorphic terranes that introduction of Na_2O and/or K_2O on a regional scale has taken place in the process of transformation of slates and phyllites to gneisses (e.g. Goldschmidt, 1921; MacGregor and Wilson, 1939; Read, 1957; Miyashiro, 1958). Such introduction may be valid in some cases, and a possible mechanism was proposed by Orville (1962). The hitherto available evidence for the introduction is, however, mostly equivocal and the observed variation in composition can be well or better accounted for by the original diversity of the relevant rocks (e.g. Müller, 1970; Müller and Schneider, 1971; Miyashiro, 1967a). A marked constancy in the chemical composition of metamorphic rocks has been demonstrated in some terranes ranging from a slate to a gneiss zone (e.g. Shaw, 1956; Uno, 1961; Vallance, 1960, 1967; R. W. White, 1966).

Table 3-1 shows average chemical compositions of unmetamorphosed and metamorphosed pelitic rocks. Between two groups of pelitic rocks which have undergone different intensities of metamorphic recrystallization, significant differences are noticed only in H_2O, CO_2 and CaO contents, and in Fe_2O_3/FeO ratio. The difference in CaO is probably due to collecting bias of samples (Shaw, 1956).

Table 3-1 *Average compositions of pelitic rocks of the world. After Shaw (1956)*

	Clays, shales, and slates (85 analyses)		Phyllites, schists, and gneisses (70 analyses)	
	Average	Standard deviation	Average	Standard deviation
SiO_2	59·93	6·33	63·51	8·94
TiO_2	0·85	0·57	0·79	0·67
Al_2O_3	16·62	3·33	17·35	5·08
Fe_2O_3	3·03	2·08	2·00	1·66
FeO	3·18	1·84	4·71	2·44
MgO	2·63	1·98	2·31	1·82
CaO	2·18	2·54	1·24	0·92
Na_2O	1·73	1·27	1·96	1·06
K_2O	3·54	1·33	3·35	1·31
H_2O	4·34	2·38	2·42	1·53
CO_2	2·31[a]	2·60	0·22[b]	0·22

[a] Determined in only 43 analyses.
[b] Determined in only 19 analyses.

Na_2O, K_2O, CaO, Al_2O_3 and SiO_2 can move in solution in an aqueous fluid at relatively low temperatures. If a fluid is not available, migration may take place by diffusion in solids. However, the effect of such diffusion would be very limited. In the Abukuma plateau, Japan, grains of corundum and quartz, only 0·3 mm apart, were found to remain unchanged in an amphibolite facies zone, other incompatible assemblages commonly occur at a distance of a few millimetres in a low amphibolite facies zone and of a few centimetres in a higher amphibolite facies zone in the same region (Miyashiro, 1958, p. 262). The common occurrence of zoned plagioclase also suggests a narrow range of diffusion in solids.

At temperatures higher than the middle of the amphibolite facies, however, partial melting would take place. The resultant melts are generally of more or less granitic composition and may move to permeate or inject at some places, leaving solid residues behind. This may cause drastic changes in the bulk chemical composition of the relevant rocks.

In regional metamorphic terranes, quartz veins occur commonly in low temperature zones (especially in those of the greenschist facies), whereas pegmatite usually occurs in high-temperature zones (more particularly in those of the amphibolite facies). Such relationships suggest that their formation is due to migration associated with the metamorphism of the surrounding rocks (Ramberg, 1948).

In Dutchess County, New York, quartz veins and pegmatites tend to occur in lower and higher temperature zones respectively of regional metamorphism. The $^{18}O/^{16}O$ ratio of quartz in the veins and pegmatites decreases with temperature rise in accordance with that of the surrounding metamorphic rocks. This suggests that quartz veins and pegmatites are genetically related to their surrounding rocks (Garlick and Epstein, 1967). Quartz veins could form through the activity of an aqueous fluid, whereas many pegmatites could form by the consolidation of magma produced by the partial melting of pelitic and psammitic metamorphic rocks.

Possible Behaviour of H_2O in Progressive Metamorphism

Newly deposited pelitic sediments have a high proportion of open spaces filled with an aqueous fluid. With increasing depth of burial, the porosity decreases, reaching a value of several per cent usually at a depth of few kilometres. The decrease is mainly due to mechanical squeezing-out of fluid combined with cementation. Rise of temperature in burial and incipient metamorphism induces recrystallization to cause generally upward expulsion not only of intergranular fluid but also of potential water that is contained in minerals as OH^-, H_2O and H_3O^+. The intergranular fluid would be continuous through long narrow channels for a great vertical distance, possibly even to the surface of the earth. This situation corresponds to model (C) of § 2-3.

Some rocks in a geosynclinal pile such as unaltered basalt do not have such a high initial content of H_2O. However, pelitic and other hydrous rocks are the main constituents of ordinary geosynclinal piles. Hence, a geosynclinal pile as a whole should usually have a high original H_2O content. In the process of recrystallization at low temperatures, H_2O should move from hydrous rocks to some less hydrous ones in the vicinity to produce new hydrous mineral assemblages in the latter.

Progressive rise in temperature during metamorphism should cause gradual dehydration of minerals and reduce intergranular spaces, thus tending to cut off channels of fluid. Individual intergranular spaces would be isolated from one another. The fluid filling them would be nearly at the same P and T as the surrounding solids. This situation corresponds to models (A) and (B) of § 2-3.

Deformational movement and volume increase by dehydration should tend to produce new paths of flow for fluid during metamorphism. Mechanical flow combined with chemical diffusion should lead to a considerable extent of regional migration of H_2O, as suggested by oxygen isotope studies as well as to a progressive decrease of H_2O content. The migrating H_2O should tend to move generally upwards, partly reaching the surface of the earth.

Prolonged metamorphism at relatively high temperatures may cause a virtually complete disappearance of the intergranular fluid phase, though discrete H_2O molecules should still be present in an adsorbed state along the intergranular surfaces. Under this condition, P_{H_2O} within the rock may become lower than the value that would be obtained in the presence of an aqueous fluid phase. This situation corresponds to model (D) of § 2-3.

In some cases the temperature of metamorphism may become high enough to cause partial melting of rocks. Such a melt phase should absorb a large part of H_2O present in the rock if the system is nearly closed with respect to H_2O. This should result in an abrupt decrease of P_{H_2O}, which may lead to the breakdown of hydrous minerals. The breakdown should to some extent give a buffering effect to P_{H_2O}. If the system is ideally open for H_2O, such a change in P_{H_2O} does not take place. It is conceivable that the formation of strongly dehydrated granulite facies mineral assemblages observed in high-temperature regional metamorphic terranes is usually promoted by a decrease of P_{H_2O} due to the beginning of partial melting.

If a rock mass formerly metamorphosed at medium or high temperatures is subjected repeatedly to metamorphic recrystallization at a later geologic time, the initial H_2O content of the mass should be small and this would cause a condition different from that in the case discussed above.

Scrutiny of Complex History

The high-temperature zones of progressive metamorphism must have passed through low temperatures in the course of temperature rise. However, recrystallization may not have taken place at the low temperatures if the rise was rapid. Even if recrystallization took place, the equilibrium state in such cases may not have been identical to the state of the low-temperature zones now observed of the same region. Different location in a geosynclinal pile could have resulted in different P_s and P_{H_2O}.

In certain metamorphic terranes, original sedimentary structures such as false bedding are known to be completely destroyed in low-temperature zones, but to be preserved in some high-temperature zones. A greater degree of deformational movements may have taken place in the lower-temperature zones (Read, 1957, pp. 24-5, 31-2).

It appears common that metasomatism is more intense in low-temperature zones than in the higher temperature ones of the same area. The existence of a flowing fluid phase in the low-temperature part may cause intense chemical migration. The higher temperature zones could have passed low-temperature conditions too rapidly to undergo any appreciable metasomatism.

The history of a progressive metamorphic terrane may be much more complicated than might be considered on a cursory investigation. Detailed investigations of apparently progressive metamorphic terranes in the Akaishi and Tanzawa Mountains in Japan have revealed their complicated and auto-cannibalistic histories (Matsuda and Kuriyagawa, 1965; Seki, Oki, Matsuda, Mikami and Okumura, 1969a). Higher temperature metamorphic rocks had formed at an early stage of an orogeny. They were later exposed and eroded, and their fragments were transported, to be deposited in an adjacent area. Then the new deposits were metamorphosed at a later stage of the same orogeny at lower temperatures, with a resultant generation of a zeolite facies zone which is apparently continuous with the older higher temperature ones (§ 6A-2).

Recent analysis of deformation phases and accumulation of radiometric age data have also revealed that some apparently simple metamorphic terranes have complex histories. Different parts of a terrane may have different histories and an apparently simple progressive metamorphic terrane may be the product of overprinting of two or more episodes of metamorphism. Even the Dalradian Series may be polymetamorphic (e.g. Brown, Miller, Soper and York, 1965). If the temperature of recrystallization is highest in the last phase, however, minerals formed in earlier phases may be entirely reconstructed. In such a case, it may be justified to ignore the earlier phases as far as mineral reactions are concerned.

3-2   OUTLINE OF METAMORPHIC FACIES

Metamorphic Facies

The concept of metamorphic facies was first proposed by Eskola (1920, 1939) and was later modified by other writers, especially by Korzhinskii (1959). A brief outline of the concept will be given here. Detailed discussions on it and related problems will be made in chapter 11.

It is assumed that the temperature, rock-pressure and the chemical potential of H_2O (or P_{H_2O}) in a metamorphic complex are controlled by external conditions. Under a given set of values for these variables, a rock with a chemical composition definite except for H_2O content shows a definite mineral assemblage. With a slight change of any externally controlled variable, there would be a slight variation in the compositional relations of coexisting minerals. At some value of the variable, a new phase would appear. Thus, the paragenetic relations of minerals serve as indicators of the externally controlled variables.

The rocks recrystallized within definite ranges of temperature, rock-pressure and chemical potential of H_2O may be designated as belonging to a *metamorphic facies*. It is conventional to define a metamorphic facies using boundaries on which major mineral changes occur.

Each of the zones, based on progressive mineral changes, corresponds to definite ranges of temperature, rock-pressure and chemical potential of H_2O. In practice, therefore, we may treat the rocks of such a zone as belonging to a metamorphic facies. Individual zones are usually limited by major mineral changes. Thus, a progressive series of metamorphic facies can be defined on the basis of mineral zoning in a terrane.

Metamorphic rocks in other areas which show similar mineral assemblages for similar chemical compositions as rocks in a zone of a standard terrane, are regarded as belonging to the same metamorphic facies as the rocks of the zone.

Eskola (1939) adopted eight metamorphic facies, which are accepted in this book with some modifications. Many later authors proposed a number of new facies and facies names, which however are not used in this book (for the reason given in § 11-1), except for the two new important facies established by Coombs (1960, 1961; Coombs *et al.* 1959).

Four Common Metamorphic Facies
Three of Eskola's metamorphic facies will be introduced here in terms of the Barrovian zones of the Scottish Highlands, and one more facies of Eskola will be added at the high-temperature end. The occurrence of these four facies is very common in the world, suggesting that each of them represents a wide range of rock-pressure.

The progressive mineral changes in the Scottish Highlands are summarized in fig. 3-2.

Greenschist Facies. The metabasites occurring in the chlorite and biotite zones commonly show the mineral assemblage actinolite + chlorite + epidote + albite, which is characteristic of greenschists. Actinolite may be absent and quartz may be present. This assemblage is diagnostic of the greenschist facies. Not only such metabasites but also all the other rocks including pelitic and quartzo-feldspathic ones within the two zones belong to the greenschist facies. Pelitic rocks contain no biotite in the chlorite zone, but do contain it in the biotite zone. When desirable, the greenschist facies may be divided into the chlorite-zone and biotite-zone subfacies.

If the metabasites in another area show the greenschist assemblage, those rocks and the associated ones recrystallized within the same ranges of externally controlled variables belong to the greenschist facies.

Epidote-Amphibolite Facies. The metabasites in the almandine zone commonly show the mineral assemblage albite + epidote + hornblende, which is diagnostic of the epidote–amphibolite facies. All the rocks in this zone belong to the epidote–amphibolite facies.

Oligoclase may substitute for albite, and quartz may join the assemblage. Almandine may occur in metapelites and metabasites, but is not regarded as essential to the definition of this facies.

Epidote amphibolites are distinguished from greenschists by the presence of hornblende instead of actinolite, and from amphibolites by the presence of the albite + epidote assemblage instead of intermediate or calcic plagioclase.

Amphibolite Facies. The metabasites in the kyanite and sillimanite zones commonly show the mineral assemblage intermediate or calcic plagioclase + hornblende, which is diagnostic of the amphibolite facies. All the rocks in these zones as well as amphibolites and associated metamorphic rocks in other areas belong to this facies.

Granulite Facies. In the Dalradian metamorphics, amphibolite is stable to the highest observed temperature of metamorphism. In certain other metamorphic regions, however, a zone of the amphibolite facies grades with increasing temperature into another zone belonging to Eskola's granulite facies. In this facies, ortho- and clinopyroxenes take the place of hornblende in metabasites, and garnets of the almandine–pyrope series may be common. Since granulite-facies rocks are highly dehydrated, their formation should be promoted by lower P_{H_2O}.

Thus, generally in order of increasing temperature, we have the following series of metamorphic facies: greenschist → epidote–amphibolite → amphibolite → granulite.

Such a series corresponding to progressive changes is called *metamorphic facies series* (Miyashiro, 1961a). Each progressive metamorphic terrane has its own facies series. The facies series observed throughout the world are very diverse in character. Nonetheless they can be classified according to aspects of similarity and difference, as will be discussed in § 3-3.

All these metamorphic facies have been defined primarily on the basis of mineralogical variations in metabasites, which are sensitive to changes in temperature, rock-pressure and chemical potential of H_2O. If, for simplicity, we neglect the effect of the chemical potential of H_2O, each metamorphic facies represents a definite field in a P_s-T diagram.

Metamorphic Facies Characterized by High Rock-Pressure
Eskola (1939) proposed two facies as representing relatively high pressures.

Glaucophane–Schist Facies. Some metamorphic rocks of ordinary basaltic compositions contain glaucophane. The P-T conditions for their formation should differ from those of any of the facies so far discussed. Hence, such rocks are said to belong to the glaucophane-schist facies. Some authors prefer the name of *blueschist facies* instead.

It has been demonstrated that glaucophane by itself is stable under a $P-T$ range much wider than that here considered for this facies (Ernst, 1961, 1963*a*). Moreover, the stability of glaucophane should vary with its composition. Therefore, a proper definition of the facies must be based on a rigorous paragenetic analysis, which, however, is not available at present.

In this book, we proceed under the provisional definition that glaucophane-bearing metabasites and associated isophysical metamorphic rocks belong to the glaucophane-schist facies, so long as there are no positive indications against it. Under this definition, a rock complex only rarely containing glaucophane is also classed in this facies. It may really be intermediate between the more typical glaucophane-schist facies and some other facies (e.g. greenschist facies).

The occurrence of lawsonite and of the jadeite + quartz assemblage is confined to the typical part of the glaucophane-schist facies. Thus, we may subdivide this facies into a typical and atypical subfacies. If necessary, we may well use such names as lawsonite-glaucophane-schist subfacies for the typical part.

Jadeite not associated with quartz is stable over a wider $P-T$ range than the jadeite + quartz assemblage. So far as I am aware, however, such jadeite occurs only in glaucophane-bearing metamorphic areas and associated serpentinites.

With rising temperature, a glaucophane-schist facies zone usually grades into a greenschist-facies or epidote–amphibolite facies zone. Glaucophane schist-facies rocks are widespread in Mesozoic and Cenozoic metamorphic belts of the world.

Eclogite Facies. This facies is characterized by eclogites which show the assemblage clinopyroxene (omphacite) + magnesian garnet + quartz (or kyanite), devoid of feldspar. (The original definition of Eskola has been modified here. Detailed discussions will be given in chapter 12.) The rocks of this facies do not show a regular zonal distribution in progressive metamorphic terranes, but occur in irregular distribution in some gneiss and glaucophane-schist areas, or as inclusions in peridotite and kimberlite.

Metamorphic Facies Characteristic of Contact and Pyrometamorphism
These facies were proposed by Eskola for contact and pyrometamorphism under relatively low rock-pressures.

Pyroxene-Hornfels Facies. This facies is characterized by the presence of ortho- and clinopyroxenes and the absence of almandine and pyrope in rocks of common chemical compositions. This facies represents high temperature combined with relatively low rock-pressure, as is common in high-temperature contact metamorphism around shallow intrusions. The type areas are the inner contact aureoles of alkali plutonic masses near Oslo, Norway (Goldschmidt, 1911). Contact metamorphic rocks recrystallized at relatively low temperatures usually belong to the amphibolite facies and not to the pyroxene-hornfels facies.

Sanidinite Facies. This facies represents the highest temperature combined with low rock-pressure as is common in pyrometamorphism of inclusions within volcanic rocks and minor basic intrusions.

It is important to realize that the definitions of metamorphic facies are based only on T, P_s, and P_{H_2O}, and on the paragenetic relations of minerals, and not on the geologic settings. If the same paragenetic relations are observed in groups of regional, ocean-floor and contact metamorphic rocks, all these rocks are regarded as belonging to the same facies. The amphibolite facies, for example, is common to all the three kinds of metamorphism. In the past, some authors used different facies names between regional-metamorphic and contact-metamorphic rocks, even when all the rocks were considered to show practically the same paragenetic relations. This is not consistent with the concept of metamorphic facies, as will be discussed again in § 4-5 and chapter 11.

Two Metamorphic Facies Common in Burial Metamorphism

In many progressive metamorphic terranes (e.g. the Ryoke belt and Dalradian Series), the lowest temperature zone is in the greenschist facies, whereas in others the lowest temperature zone is in the glaucophane-schist facies. In either case, the greenschist or glaucophane-schist facies zone grades into an unre-crystallized area with decreasing temperature.

On the other hand, it has been found in some other regional as well as more localized metamorphic areas that the zone of the greenschist facies grades with decreasing temperature into a zone with prehnite and/or pumpellyite, which in turn grades into another zone containing various zeolites, which then grades into an unrecrystallized area. Examples have been documented from the regional metamorphic terrane on the South Island of New Zealand (Coombs *et al.* 1959) and in the central Kii Peninsula of Japan (Seki, Onuki, Oba and Mori, 1971) and from a smaller scale metamorphic area of the Tanzawa Mountains in Japan (Seki *et al.* 1969*a*).

The metamorphic facies corresponding to these prehnite- and/or pumpellyite-bearing zones is called the *prehnite–pumpellyite facies,* whereas that corresponding to the zeolite-bearing zone is called the *zeolite facies* after Coombs (1960, 1961).

The rocks in these two facies are usually non-schistose and preserve textural features of their original sedimentary or igneous character owing to lack of penetrative movements. In other words, the metamorphism may be said to be of the burial type (§ 1-2). Recrystallization of the rocks is usually incomplete.

Within the zeolite facies, the stable zeolite assemblages vary with the chemical composition of the rocks as well as T, P_s, and P_{H_2O}. With rising temperature, progressive dehydration takes place. Thus, in the presence of quartz and an aqueous fluid, the stable calcium zeolites appear to vary with rising temperature in the order: stilbite $\rightarrow$ heulandite $\rightarrow$ laumontite $\rightarrow$ wairakite (Miyashiro and Shido, 1970). The stages of stilbite and of wairakite were observed only in a few zeolite areas, probably of high geothermal gradients. The analcime–quartz assemblage is stable in the heulandite stage, whereas albite becomes stable in the laumontite and wairakite stage.

3-3   BARIC TYPES AND PAIRED METAMORPHIC BELTS

Concept of Baric Types

Before World War II, British petrologists took the lead in the study of progressive metamorphism through elaborate descriptions of the Barrovian zones in the Scottish Highlands (e.g. Tilley, 1925; Harker, 1932; Wiseman, 1934). Under their influence, petrologists throughout the world at that time considered that regional metamorphism as observed in the Barrovian zones was normal and widespread in the world, and that other kinds of regional metamorphism, if any, might well be regarded as anomalous and unimportant. Moreover, it was also widely believed that glaucophane schists were formed by local metasomatism due to materials coming from associated ultramafic and mafic intrusions (e.g. Taliaferro, 1943).

After the war, however, progress in metamorphic petrology in the circum-Pacific region and especially in Japan has indicated that the above view is not valid. Glaucophane-producing metamorphism was shown to have taken place on a regional scale. This type as well as the andalusite-producing type of regional metamorphism was found to be very widespread in the circum-Pacific region. These findings have led to a marked expansion of the concept of regional metamorphism, resulting in a threefold classification of regional metamorphism in terms of rock-pressure (Miyashiro, 1961a). Each of the three categories contains a number of characteristic series of metamorphic facies covering a wide range of temperature.

The geothermal gradient varies from place to place on the earth. The depth at which a given temperature is reached differs from region to region. Hence, the rock-pressure therein also differs. In a P_s–T diagram, a series of Barrovian zones would represent a certain definite curve (or band) showing increasing rock-pressure with increasing temperature. Since kyanite is stable in the lower temperature part and sillimanite in the higher, the curve should be located as shown in fig. 3-3 in relation to the stability fields of Al_2SiO_5 polymorphs. The slope of the curve represents the pertinent geothermal gradient.

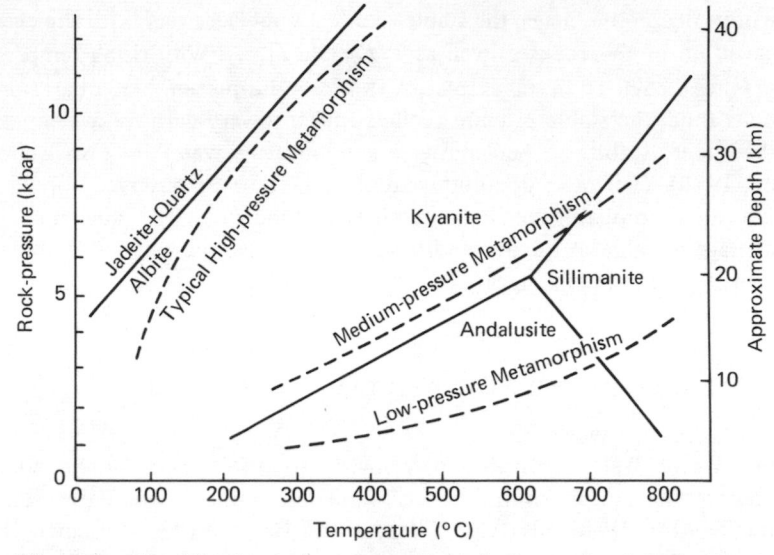

Fig. 3-3   Classification of metamorphic facies series in relation to the stability fields of Al_2SiO_5 minerals and jadeite.

Now we know that there are many metamorphic terranes that are recrystallized at lower rock-pressures and others that are recrystallized at higher rock-pressures than the Barrovian zones.

Therefore, we term the kind of metamorphism as observed in the Barrovian zones as the *medium-pressure type,* and the other two categories as the *low-pressure type* and the *high-pressure type* respectively (table 3-2). These three types should correspond to curves with different slopes in a P_s-T diagram. These three will be called the *baric types.* They represent a classification of metamorphic facies series and hence a classification of metamorphism and of metamorphic terranes.

It is to be noted that such names as 'high-pressure type' and 'high-pressure metamorphism' are somewhat misleading, since the baric types as defined here refer to the slopes of the geothermal curves corresponding to facies series, and not the values of P_s themselves. As is clear from fig. 3-3, for example, low-temperature zones in high-pressure metamorphism may represent lower pressures than high-temperature zones in low-pressure metamorphism. It is logically more appropriate to use such names as high P/T type (Miyashiro, 1967*b*) and *high-pressure series* metamorphism, instead. In this book, however, high-pressure type, high-pressure metamorphism and analogous names will be used only for simplicity of expression.

In this section, the threefold classification will be discussed mainly in terms of regional metamorphism. Since the classification is based only on the P_s-T

Table 3-2 *Threefold classification of facies series of regional metamorphism*

Baric type	Characterized by	Common minerals	Common metamorphic facies series	Associated magmatism
Low-pressure	Andalusite	Biotite, cordierite, staurolite, sillimanite	Greenschist → amphibolite	Geosynclinal volcanics, ranging from basic to acidic, are present but usually scarce. Granites are very abundant, being accompanied in some cases by andesite and rhyolite
Medium-pressure	Presence of kyanite and absence of glaucophane	Biotite, almandine, staurolite, sillimanite	Greenschist → epidote-amphibolite → amphibolite → granulite	Both ophiolites and granites are present
High-pressure	Glaucophane, jadeite, lawsonite	Almandine, barroisite, stilpnomelane	Glaucophane schist → epidote amphibolite; Glaucophane schist → greenschist; Prehnite-pumpellyite → glaucophane schist	Ophiolites, ranging from ultrabasic to basic, are abundant. Granites are usually absent

Note: For a greater detail, refer to tables 4-1, 7A-1 and 8A-1.

relations, it also may be applied to other geologic classes of metamorphism such as ocean-floor and contact ones. Some cases of contact metamorphism appear to have a metamorphic facies series identical to that of the low-pressure regional one, but others appear to have a different series. If the rock-pressure in such contact metamorphism is the same as, or lower than that of low-pressure regional metamorphism, we will include it in the category of low-pressure type.

Low-Pressure Type

This type is characterized by andalusite, which is stable at lower rock-pressures than kyanite at the same temperature. Andalusite occurs in the lower and sillimanite in the higher temperature part of metamorphic terranes of this series, if the chemical compositions of the rocks permit it. Hence, Miyashiro (1961a) designated this series as the andalusite–sillimanite type. A P_s-T curve for it is shown in fig. 3-3. The occurrence of cordierite and the rarity of pyralspite garnet are among the common features of this type.

The progressive mineral changes in low-pressure metamorphism in the Shiojiri area of the Ryoke belt, Japan, are shown in fig. 3-4. Since metabasites are relatively rare in this area, it is not clear to what extent a zone of the epidote–amphibolite facies is developed between those of the greenschist and amphibolite facies.

Miyashiro (1961a) originally used the central Abukuma Plateau in Japan as the type terrane for the low-pressure type. It also shows the facies series: greenschist → amphibolite (fig. 3-5). Recently, however, kyanite has been found to occur occasionally besides andalusite and sillimanite in this area (Hara et al. 1969). The P_s-T curve corresponding to this area would pass on or near the triple point of Al_2SiO_5 (Tagiri, 1971). In other words, the metamorphism of this area appears to be transitional between the low- and medium-pressure types.

The orthopyroxene-bearing metamorphic areas in Lapland (Eskola, 1952) and Uusimaa (Parras, 1958), both in Finland, appear to represent the highest temperature part of low-pressure regional metamorphism. They fall into an intermediate state between the granulite facies of medium-pressure meta-morphism and the pyroxene–hornfels facies characteristic of low-pressure contact metamorphism. In this book, such a P_s-T condition will be included within the range of the granulite facies. Thus, we have a common low-pressure facies series of regional metamorphism: greenschist → amphibolite → granulite.

There is a considerable diversity in the metamorphic facies series belonging to this type, conceivably owing to the wide ranges of geothermal gradients and P_{H_2O}. Other observed facies series belonging to this type will be described in §§ 4-4 and 4-5 and chapters 10 and 11.

Medium-Pressure Type

This type corresponds to the Barrovian zones as stated above. Since kyanite

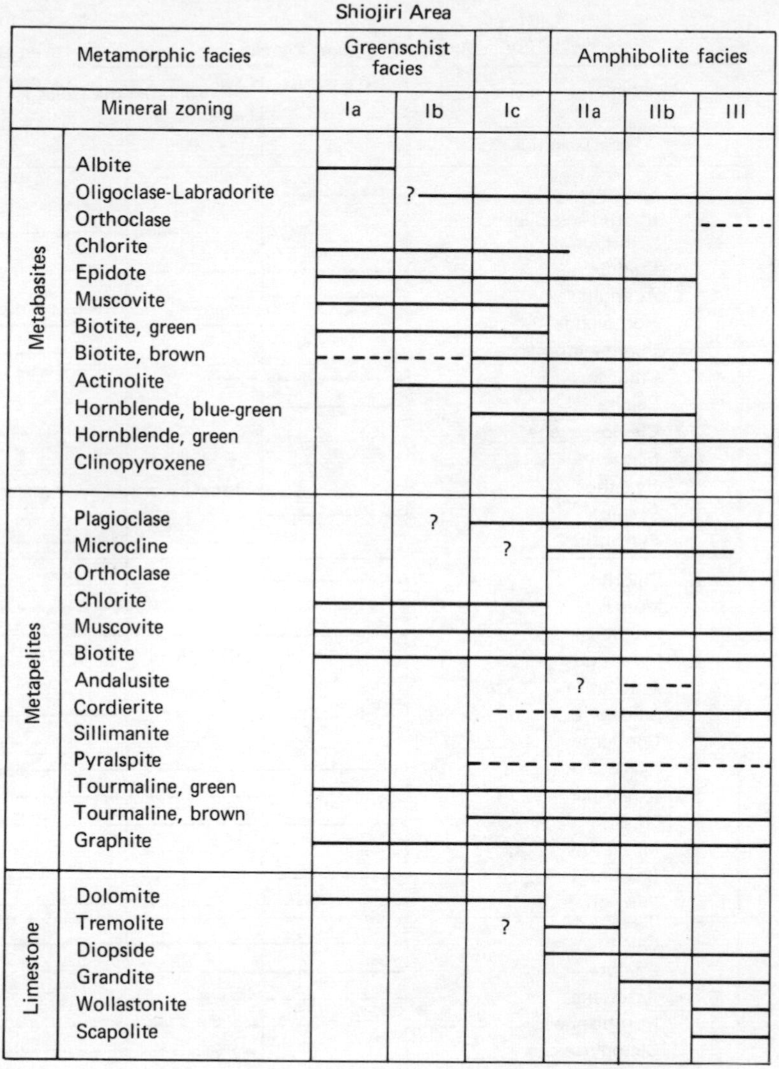

Shiojiri Area

Metamorphic facies		Greenschist facies			Amphibolite facies		
Mineral zoning		Ia	Ib	Ic	IIa	IIb	III
Metabasites	Albite						
	Oligoclase-Labradorite		?				
	Orthoclase						
	Chlorite						
	Epidote						
	Muscovite						
	Biotite, green						
	Biotite, brown						
	Actinolite						
	Hornblende, blue-green						
	Hornblende, green						
	Clinopyroxene						
Metapelites	Plagioclase		?				
	Microcline			?			
	Orthoclase						
	Chlorite						
	Muscovite						
	Biotite						
	Andalusite				?		
	Cordierite						
	Sillimanite						
	Pyralspite						
	Tourmaline, green						
	Tourmaline, brown						
	Graphite						
Limestone	Dolomite						
	Tremolite			?			
	Diopside						
	Grandite						
	Wollastonite						
	Scapolite						

Fig. 3-4   Progressive mineral changes in the Ryoke terrane of the Shiojiri area (northern Kiso), Japan. (Katada, 1965.) See fig. 7A-1.

occurs in the lower, and sillimanite in the higher temperature part, Miyashiro (1961a) designated the series as the kyanite-sillimanite type. Cordierite is usually absent, and almandine garnet is common.

It usually shows the metamorphic facies series: greenschist → epidote-amphibolite → amphibolite → granulite. The greenschist, amphibolite and granulite facies are shared in common between the low- and medium-pressure

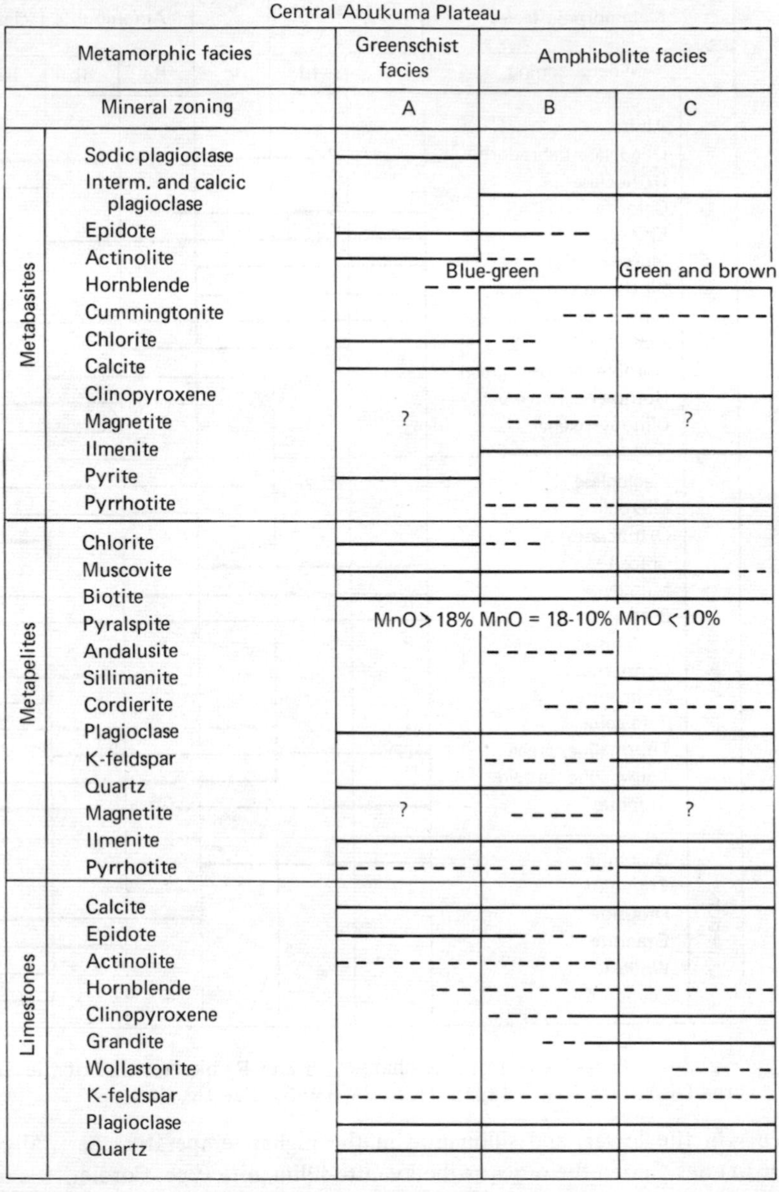

Central Abukuma Plateau

Metamorphic facies		Greenschist facies	Amphibolite facies	
Mineral zoning		A	B	C
Metabasites	Sodic plagioclase			
	Interm. and calcic plagioclase			
	Epidote			
	Actinolite			
	Hornblende		Blue-green	Green and brown
	Cummingtonite			
	Chlorite			
	Calcite			
	Clinopyroxene			
	Magnetite	?		?
	Ilmenite			
	Pyrite			
	Pyrrhotite			
Metapelites	Chlorite			
	Muscovite			
	Biotite			
	Pyralspite	MnO > 18%	MnO = 18-10%	MnO < 10%
	Andalusite			
	Sillimanite			
	Cordierite			
	Plagioclase			
	K-feldspar			
	Quartz			
	Magnetite	?		?
	Ilmenite			
	Pyrrhotite			
Limestones	Calcite			
	Epidote			
	Actinolite			
	Hornblende			
	Clinopyroxene			
	Grandite			
	Wollastonite			
	K-feldspar			
	Plagioclase			
	Quartz			

Fig. 3-5 Progressive mineral changes in the central Abukuma Plateau, Japan. (Miyashiro, 1958; Banno and Kanehira, 1961.)

types. When necessary, we can designate the metamorphic facies of a particular case by the combination of a baric type and a metamorphic facies; e.g. low-pressure amphibolite facies.

It is to be noted that the occurrence of kyanite does not warrant assignment to the medium-pressure type because the mineral occurs in some high-pressure type terranes also (Niggli and Niggli, 1965; Banno, 1964). Hence, the lack of glaucophane, jadeite and lawsonite is a necessary condition for assignment to the medium-pressure type.

High-Pressure Type

This type is characterized by jadeite, lawsonite and glaucophane. Jadeite, especially in association with quartz, is a well-demonstrated high-pressure mineral as shown in fig. 3-3. However, the equilibrium curve shown refers to pure jadeite $NaAlSi_2O_6$. The jadeites that occur in association with quartz in metamorphic rocks usually contain significant amounts of diopside and acmite components. It follows that the pressure necessary for their formation should be slightly lower than that shown by the equilibrium curve for pure jadeite. Such impure jadeites may usually form on the low-pressure side of, but in the vicinity of the curve (Ernst and Seki, 1967; Ernst, Seki, Onuki and Gilbert 1970; Coleman and Clarke, 1968). The P_s-T curve for the high-pressure type in fig. 3-3 is based on this view.

In some high-pressure regional metamorphic terranes, jadeite, lawsonite and glaucophane all occur. However, there are many other areas where glaucophane occurs but jadeite and lawsonite do not. As was discussed in § 3-2, glaucophane is not necessarily a reliable indicator of high pressures. When it occurs in rocks of ordinary basic compositions, however, it may be regarded as representing relatively high pressures (i.e. the glaucophane-schist facies). Since the precise definition of 'ordinary basic compositions' is very difficult, this criterion may not give an unequivocal assignment.

All the facies series that contain the glaucophane-schist facies in a broad sense as defined in § 3-2, may be regarded as belonging to the high-pressure type. Since glaucophane is a familiar mineral, high-pressure metamorphism was sometimes called *glaucophanitic metamorphism*.

The facies series of the high-pressure type are diversified. When the jadeite + quartz assemblage is stable in a part or the whole of a facies series, this will be designated as a *typical high-pressure series*. When it is not, the name of an *atypical high-pressure series* will be used.

Figs. 3-6, 3-7, and 3-8 indicate three examples of high-pressure terranes. They appear to show three successive steps in the passage from the medium-pressure to the typical high-pressure type. In the first two examples, an epidote–amphibolite or a greenschist facies zone occurs on the high-temperature side of a

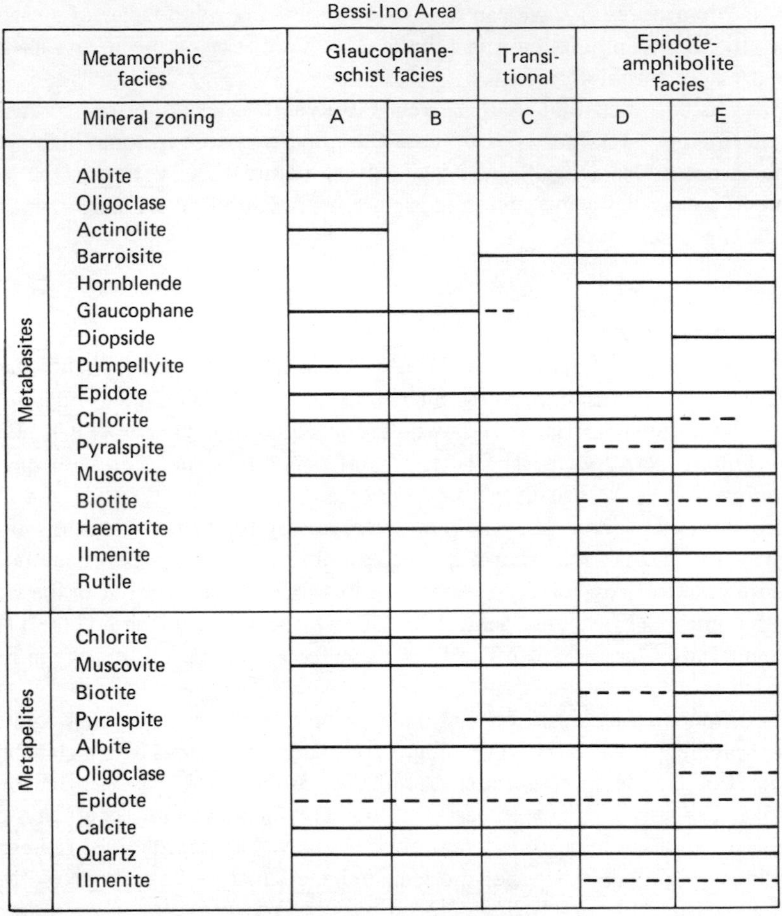

Bessi-Ino Area

Metamorphic facies		Glaucophane-schist facies		Transi-tional	Epidote-amphibolite facies	
Mineral zoning		A	B	C	D	E

Metabasites:
Albite, Oligoclase, Actinolite, Barroisite, Hornblende, Glaucophane, Diopside, Pumpellyite, Epidote, Chlorite, Pyralspite, Muscovite, Biotite, Haematite, Ilmenite, Rutile

Metapelites:
Chlorite, Muscovite, Biotite, Pyralspite, Albite, Oligoclase, Epidote, Calcite, Quartz, Ilmenite

Fig. 3-6 Progressive mineral changes in the Sanbagawa terrane of the Bessi-Ino area, Japan. (Banno, 1964.) See fig. 7A-12.

glaucophane–schist facies zone. Other facies series of this type will be discussed in § 4-4 and especially in chapter 11.

Petrogenetic Grids

The three baric types have been defined here on the basis of Al_2SiO_5 polymorphs and glaucophane-schist facies minerals. Table 3-2 indicates, however, that cordierite is common in low-pressure metamorphic terranes, whereas barroisite and stilpnomelane are common in high-pressure ones. Chloritoid, staurolite and garnet also show some extent of preferential occurrence in certain baric types. Such problems will be discussed in detail in chapters 7B and 8B.

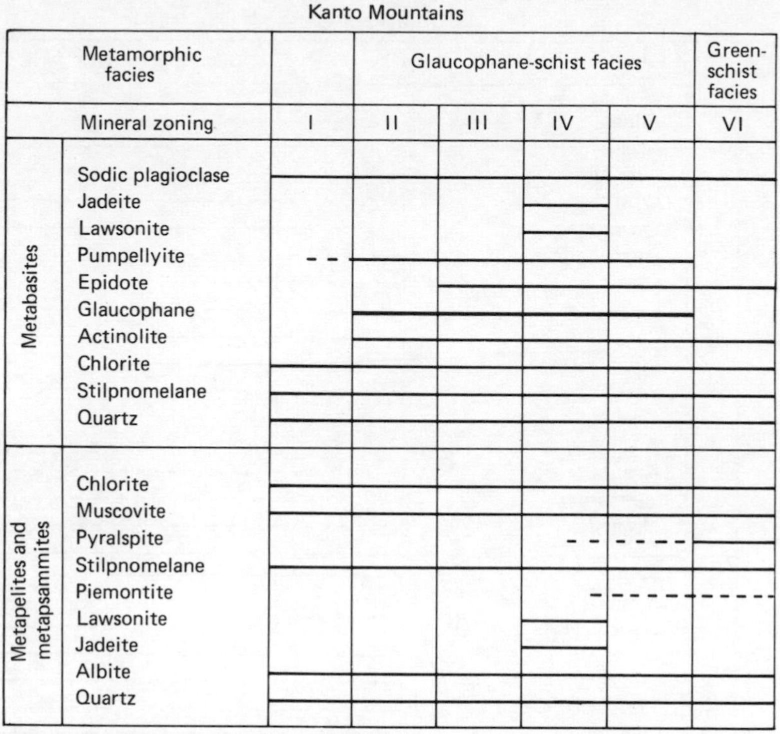

Kanto Mountains

Metamorphic facies			Glaucophane-schist facies				Green-schist facies
Mineral zoning		I	II	III	IV	V	VI
Metabasites	Sodic plagioclase						
	Jadeite						
	Lawsonite						
	Pumpellyite						
	Epidote						
	Glaucophane						
	Actinolite						
	Chlorite						
	Stilpnomelane						
	Quartz						
Metapelites and metapsammites	Chlorite						
	Muscovite						
	Pyralspite						
	Stilpnomelane						
	Piemontite						
	Lawsonite						
	Jadeite						
	Albite						
	Quartz						

Fig. 3-7 Progressive mineral changes in the Sanbagawa terrane of the Kanto mountains, Japan (Seki, 1958, 1960, 1961a). Recrystallization is very poor in Zone I. Some of the 'actinolite' in Zone VI may actually be barroisite.

It is important to emphasize that the occurrence of these minerals depends not only on T, P_s and P_{H_2O} but also on the bulk chemical compositions of the metamorphic rocks. The formation and composition of cordierite and garnet, for example, are strongly controlled by the $Al_2O_3/(MgO + FeO)$ and FeO/MgO ratios and the oxidation condition of the rocks. The presence or absence of such minerals is not a reliable indicator of baric types and P_s-T conditions, until a rigorous analysis of their paragenetic relations becomes available. Theoretical efforts in this direction have been made by Albee (1965a), Hoschek (1969), Hess (1969), and Hensen (1971).

Hess (1969) has given a semi-quantitative analysis of the P-T relations of cordierite-bearing parageneses observed in metapelites with quartz and muscovite (or quartz and K-feldspar). The topology of the univariant curves was derived from Schreinemakers rule (§ 5-4) and observed paragenetic relations. Approximate slopes of the curves were calculated by the Clausius–Clapeyron equation from the estimated volume and entropy changes. This system of curves was

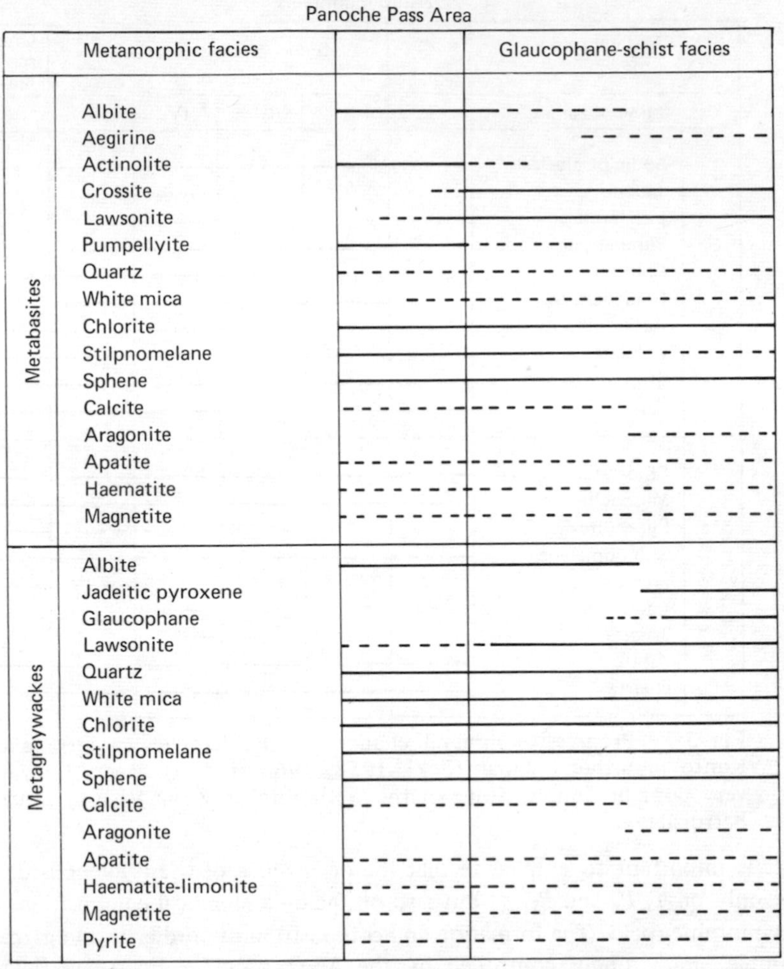

Panoche Pass Area

Metamorphic facies		Glaucophane-schist facies

Metabasites

- Albite
- Aegirine
- Actinolite
- Crossite
- Lawsonite
- Pumpellyite
- Quartz
- White mica
- Chlorite
- Stilpnomelane
- Sphene
- Calcite
- Aragonite
- Apatite
- Haematite
- Magnetite

Metagraywackes

- Albite
- Jadeitic pyroxene
- Glaucophane
- Lawsonite
- Quartz
- White mica
- Chlorite
- Stilpnomelane
- Sphene
- Calcite
- Aragonite
- Apatite
- Haematite-limonite
- Magnetite
- Pyrite

Fig. 3-8   Progressive mineral changes in the Franciscan terrane of the Panoche Pass area, California. (Ernst, 1965.) See Fig. 7A-13.

placed on the *P–T* diagram with reference to the experimentally determined curves for the stability relations of Al_2SiO_5 polymorphs and for the reaction of muscovite with quartz. The Fe^{2+}/Mg ratios of ferromagnesian minerals were somewhat arbitrarily fixed in order to simplify the calculation. The results (for metapelites containing an aqueous fluid phase) is shown in fig. 3-9. It is noticeable in this diagram that increasing pressure causes the transformation of assemblages containing cordierite into those containing andalusite, sillimanite, kyanite, biotite, staurolite and chlorite.

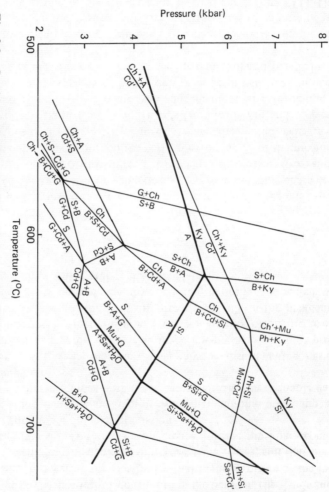

Fig. 3-9 Semi-quantitative stability relations of cordierite and related minerals in association with muscovite (or K-feldspar) and quartz in metapelites. (After Hess, 1969; with modification based on his personal communication, 1970.)

Abbreviations:

A = Andalusite
B = Biotite
Cd = Cordierite
Cd' = Mg-cordierite
Ch = Chlorite
Ch' = Mg-chlorite
G = Garnet
Ky = Kyanite
Mu = Muscovite
Ph = Phlogopite
Q = Quartz
Sa = Sanidine
S = Staurolite
Si = Sillimanite
H = Hypersthene

Facies Series of Metamorphic Belts

The observed facies series varies to some extent from area to area within a single regional metamorphic belt. For example, the Caledonian metamorphic terrane in the Scottish Highlands, including the Barrovian zones of the Dalradian, shows predominantly the medium-pressure facies series, whereas its northeastern end in Aberdeenshire and Banffshire belongs to the low-pressure series (fig. 7A-10). A part of the northern Appalachians belongs to the medium-pressure series, and the rest to the low-pressure ones (fig. 7A-6).

A low-pressure metamorphic area appears to grade into a medium-pressure one without any structural discontinuity in the Scottish Highlands and northern Appalachians. It suggests the possibility that metamorphism of the two types in such areas occurs by the same or similar geologic causes, and that the difference between the two simply reflects a difference in some subsidiary conditions (see chapter 17 for a more detailed discussion).

On the other hand, some high-pressure metamorphic areas appear to grade into medium-pressure areas within the same belt, though available data for this are not complete.

It is not inconceivable that there exist only two distinct categories of regional metamorphism as regards geologic settings: low-pressure and high-pressure, and that medium-pressure metamorphic areas are variants belonging to one or the other of them. This possibility is important especially in the problem of paired metamorphic belts and of the magmatism genetically connected with metamorphism, both of which will be discussed later.

Paired Metamorphic Belts and their Significance

In Japan and many other parts of the circum-Pacific regions, two regional metamorphic belts of similar geologic ages but of contrasting characters run side by side, forming a pair (Miyashiro, 1961a). One belt is of the low-, and the other is of the high-pressure type, though parts of each may be of the medium-pressure type, as shown in fig. 3-10. Outside Japan, the paired belts in California (Hamilton, 1969a), in Chile (González-Bonorino, 1971) and in New Zealand (Landis and Coombs, 1967) are well documented. The formation of paired metamorphic belts appears to be a result of underthrusting of an oceanic plate beneath an island arc or a continental margin (Miyashiro, 1961a, b, 1967b; Takeuchi and Uyeda, 1965).

The high-pressure metamorphic belt of a pair usually lies on the oceanic side of the associated low-pressure belt. The underthrusting of an oceanic plate would bring a sediment pile in the trench zone to great depth (fig. 3-11). The tectonic descent should cause an unusually low geothermal gradient, leading to high-pressure metamorphism.

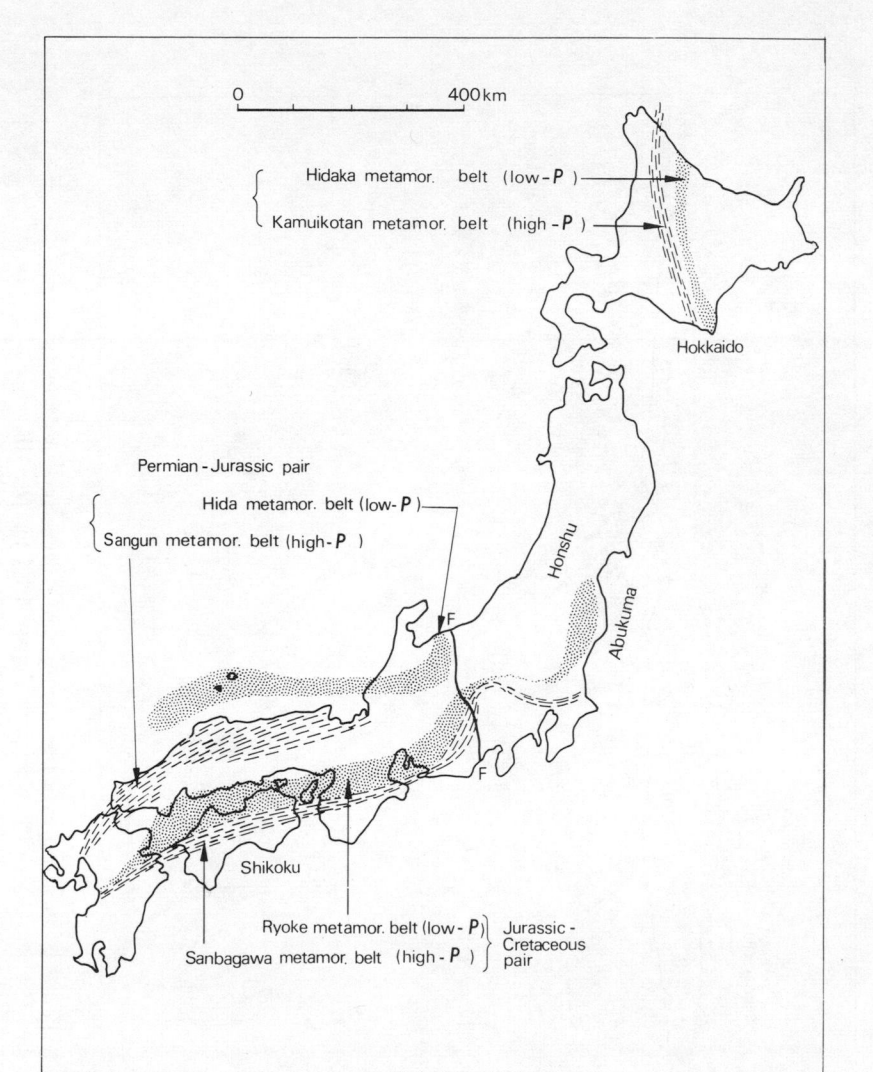

Fig. 3-10   Three pairs of regional metamorphic belts in Japan (Miyashiro, 1961*a*, 1972*b*). A pair is made up of the Hida and Sangun belts, another pair of the Ryoke and Sanbagawa belts, and a third pair of the Hidaka and Kamuikotan belts. The eastward extension of the Ryoke belt is regarded as being exposed in the Abukuma Plateau. F–F = Itogawa-Shizuoka Line (a Tertiary fault). For a detailed description see chapter 15.

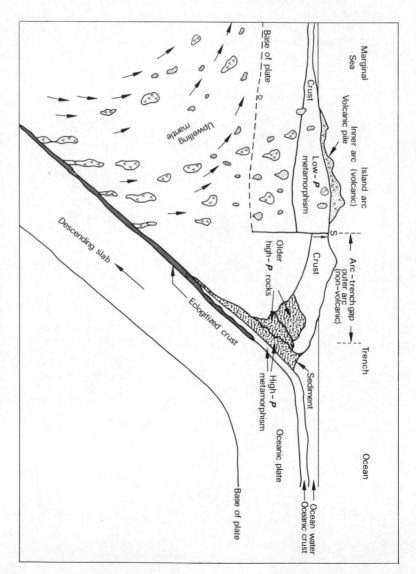

Fig. 3-11   Schematic figure representing the origin of paired metamorphic belts with special reference to the Northeast Japan Arc from the Miocene to the present (Miyashiro, 1972b). For a detailed account see chapter 17.

Paired metamorphic belts are not confined to the circum-Pacific regions. The Scottish Caledonides, for example, appear to have paired belts in which one belt is only poorly developed (§ 14-3).

When the velocity of underthrusting is not rapid enough to cause high-pressure metamorphism, medium-pressure metamorphism would take place instead, and the contrasting characters of two parallel belts, if present, may become obscure, or one belt of the two may not develop well (Miyashiro, 1972b). At any rate, most metamorphic belts in the Atlantic region are apparently unpaired. For example, the Hercynian belts in Europe are mostly of the low-pressure type, whereas the Alpine belt is of the high-pressure type. It is likely, however, that some of them are actually paired and one belt of the pair is very poorly developed or has not been exposed yet (§ 17-1).

Metamorphic belts of the low-pressure type are characteristically accompanied by abundant granitic plutons (§ 4-1). If erosion has been less advanced, well-recrystallized regional metamorphic terranes may not be exposed, but high-level granitic plutons and associated contact aureoles may appear on the surface. It is likely that chains of andesitic volcanoes in island arcs and continental margins are a surface manifestation of low-pressure metamorphic complexes and associated granitic rocks as will be discussed in § 4-1. In this meaning, the low-pressure metamorphic belt may be regarded as a *belt of granitic plutonism and andesitic volcanism*.

Most of the typical paired metamorphic belts in the circum-Pacific regions are Mesozoic and Cenozoic in age. Most of typical glaucophane schists in the world were formed in Mesozoic and Cenozoic time. Conceivably, lithospheric plates may have become thicker and/or have come to move more rapidly since the beginning of Mesozoic time, particularly in the Pacific regions (Miyashiro, 1972b). More rapid underthrusting of plates should create lower geothermal gradients and should tend to form more typical high-pressure metamorphic belts and more typical paired belts (§ § 3-4 and 17-1).

3-4   BARIC TYPES AND GEOTHERMAL GRADIENTS

Average Geothermal Gradients for Regional Metamorphism
If the triple point of Al_2SiO_5 lies at about 600 °C and 6·0 kbar (Althaus, 1967; Richardson, Gilbert and Bell, 1969), and if we assume that the temperature rises at a constant rate with depth, the gradient of the geothermal curve which passes through the triple point is about 25 °C km^{-1}.

If the existing data on the phase relations of jadeite (Newton and Kennedy, 1968) is accepted and a zone having the jadeite + quartz assemblage in typical high-pressure metamorphism is assumed to have been recrystallized at 400 °C or a lower temperature (§ 3-5), the average geothermal gradient should be about 10 °C km^{-1} or less.

Thus, we obtain a very rough estimation of average geothermal gradients as follows:

Low-pressure regional metamorphism: greater than 25 °C km^{-1},
Medium-pressure regional metamorphism: about 20 °C km^{-1},
High-pressure regional metamorphism: about 10 °C km^{-1} or less.

The P-T values adopted above for phase relations and the temperature stated above for the jadeite + quartz zone may be in error. If they are, the estimates of average geothermal gradients must be modified. The triple point of Al_2SiO_5 minerals, for example, was located at 510 °C and 4·0 kbar by Newton (1966b) and Holdaway (1971). In case we accept these values, the necessary modification in the estimate of the average geothermal gradient that divides the low- and medium-pressure types is very slight, though the estimates of the temperature of the sillimanite isograd must be decreased by about 100 °C.

Possible General Shapes of Geothermal Curves

In the high-pressure area of the Kanto Mountains (fig. 3-7), albite occurs throughout the metamorphic terrane, whereas the jadeite + quartz assemblage occurs only in a zone of medium temperature. The configuration of the P-T diagram of fig. 3-3 suggests that this relation can be explained by assuming a geothermal curve convex upward. If this is the case, the geothermal gradient would have been very low (say, 5 °C km^{-1}) to a depth of about 20 km, and then increased at greater depths.

This curvature is contrary to that of generally accepted ordinary geothermal curves, as is given, for example, by Ringwood (1969, fig. 3). The ordinary geothermal curve is controlled mainly by the production and conduction of heat within the crust, whereas the curve of the Kanto Mountains must be a result of some other factors. Possible factors include the rapid accumulation of overlying sediments, rapid tectonic sinking of the metamorphic complex to greater depths and endothermic reactions in metamorphism. Other high-pressure metamorphic terranes would have geothermal curves similar to that of the Kanto Mountains, if high-pressure metamorphism generally occurs by a similar mechanism.

No evidence is available for the general shapes of geothermal curves for medium- and low-pressure regional metamorphism. However, abundant granitic rocks are usually associated with such metamorphic regions and especially with low-pressure terranes. The temperature rise in such regional metamorphism would not simply be due to the heat conduction, but would be largely related to the upward movement of magma, H_2O and other materials through the crust (§§ 4-1 and 4-2). High-temperature zones are commonly accompanied by syn- or late-tectonic granitic masses. As is clear from oxygen isotope studies, some oxygen-containing material (probably mainly composed of H_2O) migrates through such regional metamorphic terranes, but the migration is not uniform.

Thus, the geothermal gradient would vary more or less irregularly in the lateral and vertical directions.

If the temperature distribution at depth in the medium- and low-pressure metamorphic belts is largely controlled by ascending magma and H_2O, the geothermal curves in fig. 3-3 should be concave upward in contrast to those of the high-pressure belts.

In contact metamorphic aureoles, the temperature rise would take place toward the pluton, and hence downwards, horizontally, or even upwards. If so, the average geothermal gradient calculated down to a point lying within a rock complex undergoing contact metamorphism should have no relation to the actual geothermal gradient at that point.

Estimation of Depths and Geothermal Gradients

There are many metamorphic terranes where none of the diagnostic minerals for baric types occur. In some such cases, we may be able to get indirect evidence suggesting a presumptive type.

Such evidence could be feasible in particular in some zeolite-facies terranes, since the deformation there is relatively weak so that the depths of recrystallization can be estimated with some degree of reliability. The temperature reached at a depth may be inferred from the occurrence of metamorphic reactions such as analcime + quartz = albite + H_2O. In this way, we may obtain an estimate of the average geothermal gradient. (Examples will be given in chapter 6A.) However, the reliability of such an estimate is very low.

The reliable estimation of the depths of metamorphism is very difficult in most regional metamorphic terranes where deformation is intense. The depth at which a rock lies is quite different from the thickness of an overlying stratigraphic column in tectonically disturbed terranes. Zwart (1963) has estimated the depths of two mineral zones in a low-pressure terrane in the Pyrenees on a structural basis (§ 7A-4), giving a geothermal gradient of about $150\,^\circ C\,km^{-1}$.

Width of Metamorphic Belts

In areas of higher geothermal gradients, the widths of individual mineral zones tend to be narrower, though they depend not only on the gradient but also on the angle between the surface of the earth and the isothermal surfaces. However, the reliable determination of this angle is usually difficult, since isograds are not sharp lines but broad zones depending upon the variation in bulk chemical composition of the rocks.

The width of a metamorphic belt, as measured from the line of beginning of marked recrystallization to the thermal axis (line of maximum temperature) is commonly of the order of 10–20 km in the low-pressure facies series, and 50–150 km in the medium- and high-pressure series. Wide metamorphic terranes of the low-pressure type such as the Hercynides in western Europe and the northern Appalachians contain two or more metamorphic belts or thermal axes running parallel. Fig. 7A-6 shows that the wide metamorphic belt of the northern Appalachians has three thermal axes running approximately in a NE-SW direction, that is, along the elongation of the pertinent orogenic belt. Moreover, along one of the thermal axes, there is an unusually wide zone representing nearly uniformly high temperatures (about 80 km in width). In this zone, the present erosion surface is probably roughly parallel to an original isothermal surface.

On the other hand, the high-pressure Alpine metamorphic belt in Switzerland contains only one thermal axis.

3-5 TEMPERATURE AND PRESSURE CORRESPONDING TO INDIVIDUAL METAMORPHIC FACIES

Difficulties in P-T Estimation

The low- and high-temperature limits of metamorphism were defined in § 1-1 as the lowest temperature forming minerals unstable on the earth's surface and as the incipient melting temperature of rocks respectively. Since a lot of experimental data on the phase relations of minerals have accumulated in the last twenty years, we can now estimate the temperature and pressure of individual metamorphic facies. This estimation, however, involves various difficulties, as will be outlined below.

Many published synthetic studies give metastable relations, and hence are not directly applicable to metamorphic reactions. This difficulty is especially serious at low temperatures. For example, most studies on zeolites have given metastable phase relations (e.g. Koizumi and Roy, 1960).

Even if true stability relations may be revealed, almost all experiments have been carried out on samples having greatly simplified compositions. For example, breakdown $P-T$ curves have been determined for many minerals under the condition that they are not accompanied by any other minerals. In reality, mineralogical changes in a rock take place commonly by reaction between coexisting minerals. The lower and upper breakdown temperatures of a mineral that is present by itself represent only its maximum possible stability limits (§ 2-5).

Experiments are usually carried out first on end members of solid solution series. The mixing of other components may sometimes exert a profound influence not only on the reaction temperature but also on the type of stable assemblages. Many such difficulties will be removed in the near future with increasing experimental data.

Most hydrothermal syntheses have been carried out in the presence of an aqueous fluid of nearly pure H_2O composition. This condition corresponds to model (A) of § 2-3. Decarbonation reactions have been studied with a fluid of nearly pure CO_2 composition, or with a mixture of CO_2 and H_2O. The latter corresponds to model (B) of the same section. It is not clear to what extent these models represent real metamorphic processes. It is likely that models (C) and (D) of the same section represent some or many cases of metamorphism. If so, the experimental data cannot be applied quantitatively. Solid–solid reactions, which do not suffer from this difficulty, accordingly, play an important role in the estimation.

Another difficulty results from the lack of relevant field data. For example, we have some experimental data on the reaction of kaolinite and quartz to form pyrophyllite and of the breakdown of pyrophyllite to form an Al_2SiO_5 mineral and quartz (fig. 7B-1). However, we have little petrographic data on the distribution of kaolinite and pyrophyllite in metamorphic rocks. Hence, such experimentally determined equilibrium curves, reliable or not, cannot be effectively used in coordinating metamorphic facies in the P-T diagram.

P-T Estimation by Solid–Solid Reactions

Relatively reliable, pertinent equilibrium curves have been compiled in fig. 3-12. Let us consider solid–solid reactions first.

The phase relations of andalusite, kyanite and sillimanite that have been determined by Althaus (1967) and Richardson, Gilbert and Bell (1969), give an important framework for the P_s-T relations of metamorphism (figs. 2-1 and 7B-1). The triple point for the polymorphs has been located approximately at 5·5 kbar and 620 °C by the latter authors. Petrographic experience shows that this triple point lies in the low amphibolite facies as shown in fig. 3-12. By definition, the triple point separates the geothermal curves of low-pressure metamorphism from those of medium-pressure.

The equilibrium curve for the calcite–aragonite transformation is shown as curve (3) in fig. 3-12 (Crawford and Fyfe, 1964; Johannes and Puhan, 1971). Aragonite appears to be stable in a high-pressure part of the glaucophane-schist facies and an adjacent range of conditions.

The equilibrium curves for the reactions: jadeite + quartz = albite, and jadeite = albite + nepheline, have been determined by Birch and Le Comte (1960) and Newton and Kennedy (1968). Nearly pure jadeite occurs in serpentinite in some high-pressure metamorphic terranes. Such jadeite is not associated with quartz (e.g. Coleman and Clark, 1968), and hence is considered to have formed under conditions which lie on the high-pressure side of the curve for the reaction: jadeite = albite + nepheline, which in turn lies on the low-pressure side of the curve for jadeite + quartz = albite.

On the other hand, jadeites in ordinary metamorphic rocks are commonly associated with quartz, but usually contain a considerable amount of other pyroxene components. These components should decrease the pressure necessary for the formation of jadeites. Hence, it is widely belived that such jadeites formed under the condition on the low-pressure side of the equilibrium curve for the reaction: pure jadeite + quartz = albite (Ernst and Seki, 1967; Coleman and Clark, 1968). Thus, even the conditions for the most typical high-pressure metamorphism as in the Franciscan Formation may not have reached the high-pressure side of the equilibrium curve of the reaction: pure jadeite + quartz = albite, shown as curve (4) in fig 3-12.

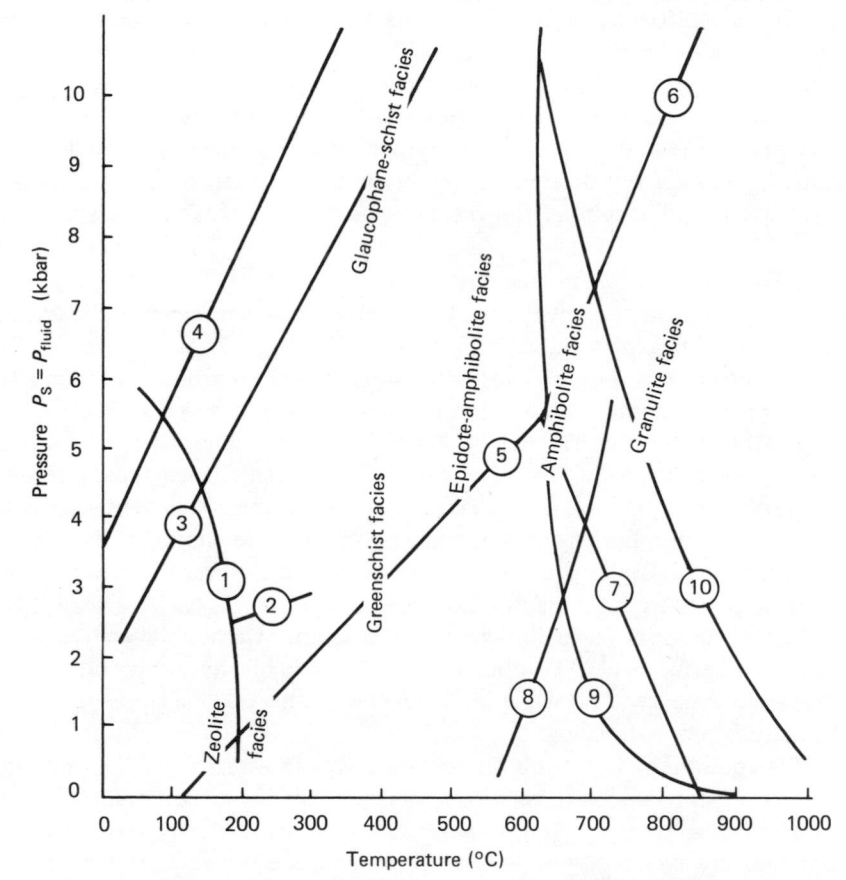

Fig. 3-12 Temperature and pressure of metamorphic facies. The equilibrium curves are for the following reactions. (The left-hand side of each equation is stable on the low-temperature side of the corresponding curve.)

(1) analcime + quartz = albite + H_2O; (2) lawsonite + 2 quartz + $2H_2O$) = laumontite; (3) aragonite = calcite; (4) jadeite + quartz = albite; (5) kyanite = andalusite; (6) kyanite = sillimanite; (7) andalusite = sillimanite; (8) muscovite + quartz = K-feldspar + Al_2SiO_5 + H_2O; (9) beginning of melting of granite; (10) beginning of melting of olivine tholeiite.

P-T Estimation by Dehydration Reactions

Here we will provisionally accept model (A) of § 2-3. In other words $P_s = P_{fluid}$ $= P_{H_2O}$.

The curves for the formation and decomposition of pyrophyllite are of little use, since relevant petrographic data are too meagre.

The reaction: analcime + quartz = albite + H_2O takes place in the middle of the zeolite facies, as shown by curve (1) in fig. 3-12. The low-pressure limit of the lawsonite + quartz assemblage is shown as curve (2) on the basis of recent work by A. B. Thompson (1970*b*).

The equilibrium curve of the reaction: muscovite + quartz = Al_2SiO_5 + K-feldspar + H_2O (fig. 7B-1) is relatively reliable (Evans, 1965). This reaction takes place within the amphibolite facies in some areas of regional metamorphism (§ 7B-15), but is said to take place approximately at the boundary between the amphibolite and granulite facies in the northern Appalachians (Thompson and Norton, 1968).

The experimental data outlined above, combined with our knowledge of the facies series of metamorphic rocks enable us to locate metamorphic facies roughly as shown in fig. 3-12. The curves for the beginning of melting of granite and olivine tholeiite determined by Yoder and Tilley (1962) are also shown.

It is impressive that all these dehydration curves, which were determined in the presence of an aqueous fluid, when plotted together with curves for solid-solid reactions on a *P-T* diagram, give relations generally consistent with our knowledge of the sequence of reactions in progressive metamorphism. This means that we may provisionally accept model (A) for dehydration reactions, though it does not prove the universal validity of the model.

Decarbonation reactions will not be used in our *P-T* estimation, because even if we assume the presence of a fluid phase during metamorphism, we cannot specify the concentration of CO_2 in the fluid.

Oxygen Isotope Geothermometry

The oxygen isotopes ^{18}O and ^{16}O show a measurable extent of systematic fractionation between coexisting minerals in metamorphic rocks. In most cases, there is a close approach to isotopic equilibrium in a rock specimen. Since the equilibrium distribution of isotopes depends mainly on the temperature of formation of mineral assemblages, the $^{18}O/^{16}O$ ratios measured on coexisting minerals combined with experimentally determined fractionation curves can be used as a geothermometer.

Experimentally determined fractionation curves are available for the pairs quartz–muscovite, quartz–calcite, quartz–magnetite and so on. The fractionation becomes less marked with rising temperature, and hence the precision of temperature determination usually becomes very poor at temperatures above 500 °C.

A considerable number of temperature determinations have been made by this method on the regional metamorphic rocks of the medium-pressure type in the northern Appalachians and the high-pressure rocks of the Franciscan group in California (Epstein and Taylor, 1967; Garlick and Epstein, 1967; Taylor and Coleman, 1968). The results are summarized as follows:

Glaucophane-schist facies {	Franciscan, type III	270-315 °C
	Franciscan, type IV	410-535 °C
	Biotite zone	350-400 °C
Appalachians {	Almandine zone	410-490 °C
	Staurolite-kyanite zone	490-600 °C
	Sillimanite zone	600-750 °C

(For the meaning of types III and IV, refer to § 7A-10.)

These figures are mutually consistent and are fairly close to the temperature values inferred from phase equilibrium data. This fact suggests that the isotope distribution obtained at the maximum temperature has been approximately preserved to the present.

The wide range of temperature of the glaucophane-schist facies is noted. This facies covers at least the same temperature range as the greenschist and epidote–amphibolite facies combined, and probably extends toward a lower temperature.

Chapter 4

Some Important Problems in Metamorphic Geology

4-1 IGNEOUS ROCK ASSOCIATIONS RELATED TO REGIONAL META-MORPHISM

Contrasting Igneous Rocks between Low- and High-Pressure Terranes

In a classical view, one cycle of orogeny includes a long series of events beginning with geosynclinal sedimentation, followed by phases of deformation, regional metamorphism, and plutonic intrusion, and ending with post-metamorphic disturbances. The duration of a cycle is usually of the order of tens or hundreds of million years. In some regions, the geosynclinal sedimentaton is followed immediately by tectonic disturbances and metamorphism. In other regions, there is so long a lapse of time between the end of sedimentation and metamorphism that the existence of genetic connection may well be doubted.

Many authors believed that three distinct phases of magmatism exist in a cycle of orogeny, namely: basaltic magmatism in the geosynclinal stage, granitic intrusion in the orogenic stage, and post-orogenic magmatism (Tyrrell, 1955; de Sitter, 1956; Aubouin, 1965). However, they paid no attention to the relationship of igneous rocks to the associated metamorphism. The characteristics of igneous rocks vary with the baric type of associated regional metamorphism as emphasized by Miyashiro (1961*a*, 1967*b, c*) and by Zwart (1967*a,b*, 1969). The associated igneous rocks are mainly of acidic and intermediate compositions in low-pressure metamorphic terranes, but of basic and ultrabasic compositions in high-pressure terranes. Medium-pressure terranes tend to be intermediate in character in this respect also (table 3-2).

Table 4-1 summarizes the contrasting petrologic and tectonic characteristics of orogenic belts with low-pressure and high-pressure metamorphic terranes, particularly on the basis of the available data from the Hercynian and Alpine belts.

We may add that a relationship also exists between the baric types of regional metamorphism and the sedimentation in relevant metamorphic terranes. Typical aluminous pelites and limestones are usually widespread in low-pressure metamorphic terranes, but are scarce in high-pressure ones. In the latter, the most

Table 4-1   *Contrasting characters of orogenies. After Zwart (1967a)*

	Hercynotype orogeny	Alpinotype orogeny
Metamorphism	1. Shallow, low-pressure metamorphism	1. Deep, high pressure metamorphism
	2. Narrow progressive mineral zones	2. Wide progressive mineral zones
	3. Mineral changes due to increasing temperature commonly observed	3. Mineral changes due to decrease of rock-pressure tend to take place possibly owing to rapid erosion resulting from rapid isostatic uplift
Igneous rocks	1. Abundant granites and migmatites	1. Few granites and migmatites.
	2. Few ophiolites. Ultrabasic rocks practically lacking	2. Abundant ophiolites with a considerable amount of ultrabasic rocks
Width of orogen	Very wide (thousands of kilometres)	Relatively narrow (hundreds of kilometres)
Isostatic uplift after disappearance of compressive force	Small and slow uplift. (Average rate of uplift is about 1 mm per 10–25 years)	Large and rapid uplift. (Average rate of uplift is about 1 mm per year)
Structure	Nappe structures missing or rare	Nappe structures predominant

typical sedimentary rocks are graywackes. The K_2O/Na_2O ratio and the maturity tend to be higher in clastic sediments of the former than in those of the latter (e.g. Miyashiro and Haramura, 1966).

Ophiolites. Abundant metabasalts and metagabbros occur in high-pressure regional metamorphic terranes such as the Franciscan and Sanbagawa. The structural features suggest that originally some of the metabasalts were pillow lavas and others were pyroclastics. They are usually accompanied by serpentinite. The basic and ultrabasic rocks of such an association have traditionally been called ophiolites (e.g. Bailey, Irwin and Jones, 1964).

Some high-pressure metamorphic terranes are so chaotically disturbed that reliable interpretation of the field relations of ophiolites are not possible. In the Sanbagawa terrane of central Shikoku, however, disturbance was not so strong and the modes of occurrence of the ophiolite have been clarified to a considerable extent. There are many layers of basic schists which are apparently conformably interbedded with pelitic schists. Some layers can be traced for a distance of many kilometres within the deformed geosynclinal pile. This is clearly shown in the geologic maps, for example, by Kawachi (1968, fig. 3) and by Ernst *et al.* (1970, plates 4 and 5). Thus these layers appear to have been derived from pyroclastics (and possibly also from associated lavas) which were erupted in the geosyncline concurrently with sedimentation. Basic rock layers in only slightly recrystallized parts of the terrane commonly preserve the original pyroclastic structure. Serpentinite usually occurs as independent masses in the neighbourhood. There is no definite stratigraphic sequence in the rocks. Such basic and ultrabasic rocks may well be referred to as *non-sequence type ophiolites*.

There is another group of ophiolites, here called the *sequence-type ophiolites* for convenience, which usually form layered masses showing a more or less regular stratigraphic sequence of rocks: ultrabasic rocks → gabbros → basalts from the bottom upwards. The Troodos massif, Cyprus (Gass, 1968; Moores and Vine, 1971) and the Vourinos complex, Greece (Moores, 1969) are well-documented examples of this type. These two examples occur in unmetamorphosed parts of the Alpine terranes and probably represent basaltic volcanoes in ancient island arcs (A. Miyashiro, in preparation).

Most of the well-developed high-pressure metamorphic belts are accompanied by abundant ophiolites. This suggests the existence of some genetic connection between ophiolites and high-pressure metamorphism. Such ophiolites may have been derived from rocks in oceanic regions.

In the last several years, a large number of papers have been published on the origin of ophiolites. Most of them have supported the hypothesis that ophiolites represent fragments of oceanic crust and upper mantle pushed into a pre-existing geosynclinal sediment pile during strong disturbances caused by plate under-thrusting (e.g. Dietz, 1963; Coleman, 1971; Moores and Vine, 1971). This view may be valid for some sequence-type ophiolites which represent conceivable structures of an oceanic crust and the underlying mantle, but may not be valid for many of the non-sequence-type ophiolites, as discussed above.

Some ophiolites, such as those in Puerto Rico, show a marked similarity in chemical composition to common ocean-floor rocks (Donnelly, Rogers, Pushkar and Armstrong, 1971). However, most metabasaltic rocks in the ophiolitic associations in high-pressure terranes differ in chemical composition from common ocean-floor igneous rocks. No one has examined whether or not the difference can be ascribed to the metamorphism in mid-oceanic ridges and orogenic belts as well as to ocean-floor weathering. Thus, we are not prepared to make any reliable comprehensive evaluation of the relevant hypotheses.

Spilite-Keratophyre group and Granitic Rocks. Metamorphosed basaltic rocks are present usually only in a small quantity in low-pressure metamorphic terranes (e.g. the Ryoke and Hercynian belts). There, intermediate and acidic volcanics may be relatively common. An association of slightly metamorphosed basic, intermediate and acidic volcanic rocks forms a *spilite–keratophyre group* (Rocci and Juteau, 1968). In rare cases, ultrabasic rocks also occur in the association.

A large amount of granitic rocks, ranging from quartz diorite to granite in the strictly-defined petrographic sense, occurs in all low-pressure regional metamorphic terranes. It appears that granitic magmas and associated fluids are important carriers of heat in low- (and medium) pressure regional metamorphism. Low-temperature zones of regional metamorphism would be formed by a general rise of temperature in thick geosynclinal piles, but high-temperature zones might be mostly formed through heat transfer by upward movement of granitic masses and associated fluids. In deeper levels, such regional metamorphism would cause partial melting of rocks, resulting in the formation of granitic magma, whose rise would carry up a large amount of heat to higher levels.

The cooperative heat effect of granitic masses and regional metamorphism can take place only where they are synchronous. In most low-pressure regional metamorphic terranes, granitic masses appear to have been emplaced at a late phase of and after the end of regional metamorphism.

Examples of Granitic Rocks

Finnish Precambrian Terrane. Granitic rocks are widely exposed in the Svecofennian low-pressure metamorphic terrane of the Baltic Shield (§ 14-2). Some granites are gneissose, but others are not. Such differences were ascribed to the different relationships of the granites to the tectonic movements in a cycle of orogeny. Thus, the idea of the tectonic classification of granitic rocks was formulated by Eskola (1932), Simonen (1960a, b), Marmo (1962) and others in Finland, as follows:

(a) *Synkinematic granites.* These are gneissose and concordant with the structure of the surrounding metamorphic rocks, and contain abundant basic inclusions. They are presumed to have been emplaced at an early stage of an orogenic movement, and the gneissosity has been ascribed to orogenic deformation. The so-called Older Granite of Finland belongs to this group. Rocks of this group are mostly granodioritic and quartz–dioritic in composition, though some ultrabasic and gabbroic rocks also occur.

(b) *Late-kinematic granites.* These are seldom gneissose, and clearly cross-cut the country rocks or show a diapir-like mode of intrusion within the axial culminations of metamorphosed terranes. They are presumed to have been emplaced at a later stage of orogenic movement. Rocks of this group in Finland are commonly called Younger Granite. They are microcline-rich granites or aplites, and are not accompanied by ultrabasic and basic rocks. Sederholm (1907, 1923, 1926, 1967) investigated the process of migmatization related to the Younger Granite in southwestern Finland.

(c) *Post-kinematic granites.* These occur as discordant bodies entirely devoid of gneissosity, and are presumed to have been emplaced after all the deformational movements had ceased. The Rapakivi Granite of southern Finland belongs to this group.

Recent radiometric dating has shown that both Older and Younger Granites were formed about 1800 m.y. ago, and the Rapakivi Granite about 1650 m.y. ago (Eskola, 1963). No detailed analysis has been made of the tectonic movements. The meaning of syn- and late-kinematic, therefore, is not definitive in this case.

Scottish Caledonides. The granitic rocks of the Scottish Highlands are usually classified into two groups: the Older and the Newer Granites (§ 14-3). The Older Granites are gneissose and occur in many small concordant bodies in high-temperature parts of regional metamorphic areas. No contact metamorphism is noticed. Barrow (1893) ascribed the regional metamorphism to the temperature rise resultant from intrusion of the Older Granites. Recent investigations, however, suggest that the Older Granites are syntectonic masses resulting from regional metamorphism, metasomatism and migmatization (Harry, 1958; Dalziel, 1966).

In the Dalradian metamorphics, four (or five) phases of deformation have been distinguished. The temperature of metamorphism increased progressively, reaching the maximum value between the third and fourth phases. The regional migmatization and the formation of the Older Granites and sillimanite zone probably took place in this period (Chinner, 1966a, b; Johnson, 1963; Harte and Johnson, 1969), probably about 500 m.y. ago or earlier (Brown, Miller, Soper and York, 1965).

The Newer Granites occur commonly in large discordant masses, including not only granitic but also dioritic and gabbroic rocks. They show clear contact metamorphism producing a hornfels zone. The time of intrusion is about 415–390 m.y. ago, that is, Late Silurian to Early Devonian (Brown, Miller and Grasty, 1968). At the same time, many masses of granitic rocks were intruded into unmetamorphosed terranes in the Midland Valley and the Southern Uplands to the south of the Highlands (Read, 1961). This period of intrusion roughly agrees with that of the last phase of deformation.

Northern Appalachians. Three plutonic series, named the Oliverian, the New Hampshire and the Late Devonian, have been distinguished among the Devonian Plutons of the northern Appalachians. These series show systematic differences in relation to the wall rocks, metamorphic effects, modes and place of emplacement, internal structure, mineralogy and so on. Page (1968) regards the three series as representing pretectonic, syn- and early post-tectonic, and late post-tectonic intrusions respectively in relation to the Acadian (Devonian) orogeny. The Oliverian and a part of the New Hampshire plutonic series are gneissose. The masses of the New Hampshire series occur characteristically in high-temperature zones of the regional metamorphic terrane, suggesting the existence of a close genetic connection with the metamorphism. The masses of the Late Devonian series produced contact-metamorphic aureoles.

However, many other authors consider that the rocks of the Oliverian series are actually metamorphic derivatives of volcanic and associated plutonic rocks, and form the cores of mantled gneiss domes. The cores are encircled concordantly by a mantle of well-stratified metamorphic rocks (especially of the Ammonoosuc Volcanics). The volcanic rocks within the cores are regarded as stratigraphically underlying the envelope rocks (Ordovician). During the metamorphism, the core rocks would have risen to form domes owing to their lower densities (Thompson, Robinson, Clifford and Trask, 1968; Naylor, 1968).

Cooma Granite, New South Wales. The Cooma granite is exposed over an area of about 31 km^2 roughly along the thermal axis of a Palaeozoic low-pressure regional metamorphic belt in southeastern Australia (§§ 7A-3 and 16-2). The granite is gneissose and contains quartz, feldspars, biotite and muscovite, together with subordinate sillimanite, andalusite and cordierite. The outer limit of the metamorphic belt is at a distance of about 10 km or less from the granite body (Joplin, 1942).

Pidgeon and Compston (1965) have determined the Rb-Sr ages and the initial $^{87}Sr/^{86}Sr$ ratios (for total-rock samples) of the granite and associated metasediments with the following results.

	Rb-Sr age (m.y.)	Initial $^{87}Sr/^{86}Sr$
Cooma granite	415 ± 12	0·7179 ± 0·0005
Migmatites (surrounding the granite)	406	0·720
Metasediments of high-temperature zone	399 ± 29	0·719 ± 0·002
Metasediments of low-temperature zones	460 ± 11	0·710 ± 0·002

The Rb-Sr age of the Cooma granite is the same as that of the associated high-temperature metasediments within experimental error, whereas it is significantly smaller than that of the associated low-temperature metasediments. It is conceivable that the low-temperature metamorphism occurred about 460 m.y. ago, whereas the high-temperature event and granite emplacement occurred about 410 m.y. ago. Since the original sediments seem to be Ordovician, the 460 m.y. metamorphism would have immediately followed the sedimentation.

The $^{87}Sr/^{86}Sr$ ratio of the upper mantle is generally considered to be about 0·702–0·704. The initial $^{87}Sr/^{86}Sr$ ratios of most Phanerozoic granitic rocks are about 0·705–0·710, whereas the same ratio of the Cooma granite is much higher (about 0·718), and is virtually the same as that of the associated high-temperature metasediments. The $^{87}Sr/^{86}Sr$ ratio of the metasediments in the low-temperature zones should have gradually increased so as to become similar to the initial ratio in high-temperature zones and in the granite at the time of granite emplacement (410 m.y. ago). Therefore, a likely explanation for the high initial $^{87}Sr/^{86}Sr$ ratio of the granite is that the granite magma was formed by the melting of associated sediments, as far as the Sr isotope evidence is concerned.

Holum Granite, Norway. It is conceivable that some granites in orogenic belts are of magmatic origin, but others are of metamorphic origin, though the available evidence is mostly equivocal. As an example of granitic plutons that have been interpreted as being of metamorphic origin, the Holum granite in a Precambrian migmatite region at the southern tip of Norway will be described below (Smithson and Barth, 1967). Here, the term migmatite is used in the non-genetic meaning (§ 1-4).

Migmatites occur widely in the area. There is a zone of augen gneisses, 0–2 km wide, between the migmatite area and the Holum granite. The augen gneisses contain well-developed porphyroblasts of K-feldspar perthite together with smaller grains of K-feldspar, plagioclase, quartz, hornblende and biotite.

The Holum granite is a pluton elongated in the N–S direction with a maximum width of about 7 km. The rocks usually show a weak parallel texture, and are mainly composed of K-feldspar perthite, plagioclase, quartz, biotite and hornblende, just like the adjacent augen gneisses; petrographically, they are adamellite. The K-feldspars of the granite and the adjacent augen gneisses are usually microcline, but are nearly monoclinic in some specimens.

Mineralogically and probably chemically, the granite closely resembles the augen gneisses. The main difference between them lies in texture. The augen gneisses show a stronger parallel texture and a stronger tendency to xenomorphic crystal outlines. The trend of the parallelism in the granite, augen gneisses and migmatites generally follows the direction of elongation of the pluton. The augen gneisses grade into the granite with a transitional zone up to 500 m wide.

The main chemical variation within the granite is produced by different proportions of K-feldspar perthite ($Or_{80}Ab_{20}$) and plagioclase (An_{26-30}), where the composition of each feldspar remains nearly constant. The compositions of various parts of the granite fall mostly outside the low-temperature valley of the pertinent experimental systems. These data suggest that the Holum granite was formed not by magmatic crystallization, but by metamorphic recrystallization of the associated augen gneisses. The recrystallization appears to have taken place in a synform, possibly being accompanied by incipient partial melting.

4-2 VOLCANIC ARCS IN RELATION TO LOW-PRESSURE REGIONAL METAMORPHISM

Low-Pressure Metamorphic Belts and Volcanic Arcs

The Quaternary volcanism of continental margins and island arcs is characterized by the great abundance of intermediate and acidic rocks, though basaltic rocks also occur. (It will be shown later that this is due to the abundance of rocks of the calc-alkali series in a narrow sense.) The volcanism of the low-pressure terranes and granitic belts such as the Ryoke belt (§ 15-7) and the Sierra Nevada–Klamath zone (Dickinson, 1962) is also characterized by the abundance of intermediate and acidic rocks, though the ratios of intermediate rocks to acidic ones differ in different cases. It may be presumed that andesitic–dacitic volcanic chains on island arcs and active continental margins are surface manifestations of low-pressure metamorphic terranes and granitic belts.

The main differences in the ratio of volcanic to granitic rocks and in the grade of metamorphism in such areas could be ascribed to differences in the degree of erosion. Table 4-2 shows an attempt to explain the different characteristics of

Table 4-2   *Hypothetical series of increasing depth*

General depths	Examples	Andesitic-rhyolitic volcanics	Granites	Regional metamorphism
Surface	Quaternary volcanic arcs	Abundant	Absent	Absent
	Late Tertiary terrane on the Japan Sea side of northeast Japan	Abundant	Scarce	Zeolite and prehnite-pumpellyite facies
	Sierra Nevada-Klamath zone of N. America	Abundant	Abundant	Mainly green-schist facies(?)
Relatively deep	Ryoke belt of Japan	Present	Abundant	Mainly amphibo-lite facies

Note: Contact metamorphism by granitic plutons is ignored. For the Late Tertiary terrane of northeast Japan, refer to § 15-9.

four regions in this way (Miyashiro, 1972*b*). With increasing depth, the grade of metamorphism should increase and the ratio of volcanic rocks should decrease. It is not clear whether the ratio of granitic rocks increases with depth, or whether it first increases and then becomes constant or even decreases.

Thus, a low-pressure regional metamorphic complex would be present beneath a belt of volcanic piles in island arcs and continental margins. Therefore, volcanic rocks in island arcs will be discussed in some detail below.

Igneous Petrology of Volcanic Arcs

Igneous Rock Series. In the first half of the twentieth century it was realized that the chains of volcanoes in island arcs and active continental margins contain large amounts of basalt, andesite, dacite and rhyolite, in which the SiO_2 and alkali contents tend to increase and the MgO and Fe_2O_3 + FeO contents tend to decrease from basalt to rhyolite in the above-named order. Bowen (1928) regarded this order as representing progressive fractional crystallization which is controlled by two main reaction series, one for colourless and the other for coloured minerals. The coloured minerals were considered to crystallize in the order: olivine → orthopyroxene → clinopyroxene → amphibole → biotite. Such volcanics were usually classed as subalkali or *calc-alkali rocks.*

Fenner (1929), and more spectacularly Wager and Deer (1939) demonstrated the existence of the tholeiitic series of igneous rocks which is not alkaline but is distinct from the above calc-alkali rocks. The tholeiitic series shows little or no increase in SiO_2 content and instead a considerable increase in Fe_2O_3 + FeO content in the early (or main) part of its course of crystallization. This series has olivine and pyroxenes as the main coloured minerals but has little or no amphibole and biotite. The tholeiitic series is usually composed of abundant mafic, less abundant intermediate and few acidic rocks, whereas the calc-alkali series usually includes abundant intermediate and acidic rocks.

Thus, Nockolds and Allen (1953, 1954, 1956) have discussed the compositional variation of igneous rocks in terms of a threefold classification into the tholeiitic, calc-alkali and alkali series. Kuno (1959), Taylor and White (1965), Jakeš and White (1972) have accepted this classification with certain modifications in their studies of island arc volcanism in Japan and southwest Pacific regions. Alkali basalts in some areas have K_2O/Na_2O ratios much lower than unity, whereas those in others have K_2O/Na_2O ratios near unity. Joplin (1964) and Jakeš and White (1972) have accepted the name *shoshonite* for the latter group of alkali basalts.

Textbooks of petrography may define the name *andesite* mainly in terms of SiO_2 content, colour index, or the ratios of plagioclase-alkali feldspar-silica minerals. In such definitions, all the three series of volcanic rocks include andesites as members. However, the 'andesites' so defined show different chemical and mineralogical characteristcs in the three different series. This situation is undesirable as the rock series classification has great genetic implications. Thus, Macdonald (1960) has proposed to use the name hawaiite and mugearite for the so-called 'andesites' of the alkali series. Carmichael (1964, p. 442) coined the name icelandite for 'andesites' of Iceland, and this name may well be accepted for the 'andesites' of the tholeiitic series. In this way, the use of the name andesite can be limited to the 'andesites' of the calc-alkali series.

Thus, we have the following three series of volcanic rocks in island arcs and active continental margins.

1. *Tholeiitic series,* including tholeiitic basalt, icelandite and some dacite. The SiO_2 contents are mostly in the range 48–63 per cent by weight.

2. *Calc-alkali series,* including abundant andesite and dacite, and some rhyolite. The SiO_2 contents are mostly in the range 52–70.

Series (1) and (2) are 'calc-alkalic' in the broader meaning that they both have Peacock's (1931) alkali-lime indices as high as 56–57.

3. *Alkali series,* which may well be subdivided into: (a) the sodic alkali group, including alkali olivine basalt, hawaiite, mugearite, trachyte and alkali rhyolite, and (b) the shoshonite group, including shoshonite, latite and leucite-bearing rocks.

Compositional Variation Across Volcanic Arcs. In island arcs which have a Benioff zone inclined towards the continent as in northeast Japan, the Kurile Islands and Indonesia, the volcanic rocks tend to be increasingly alkaline toward the continent. In other words, the K_2O and $Na_2O + K_2O$ contents and the K/Na ratio of volcanic rocks tend to increase towards the continent, if we compare rocks with the same SiO_2 content (Sugimura, 1960, 1968; Kuno, 1959, 1960, 1966; Katsui, 1961; Dickinson, 1968; Dickinson and Hatherton, 1967; Kawano, Yagi and Aoki, 1961; Yagi, Kawano and Aoki, 1963).

In active island arcs in mature stages of development such as northeast Japan and Kamchatka, petrographic provinces of the tholeiitic series, calc-alkali series and alkali series are present successively in this order from the oceanic to the continental side of the volcanic belt, though there may be marked overlapping of provinces (fig. 15-3).

Nature of the Calc-Alkali Series. The status and character of basalt in the calc-alkali series are not clear. Many volcanoes made up of rocks of the calc-alkali series have no basalt. The series may begin with andesite. Nockolds and Allen (1953) found in variation diagrams that acidic and intermediate rocks of the calc-alkali series fall on a smooth curve for each area, whereas more basic rocks show more or less random scattering. They interpreted this as suggesting that the parental magmas were intermediate (i.e. andesite or dioritic) in composition (with SiO_2 = 52–56 per cent), and that more basic rocks of the series were formed by the accumulation of early crystals. Green and Ringwood (1966, 1968) demonstrated experimentally the possibility of the formation of primary andesitic magma.

Some basalt occurs in many andesitic volcanoes, however. Such andesitic rocks may have been derived from basaltic magmas (e.g. Jakeš and White, 1972; Aoki and Oji, 1966; Kuno, 1968). Osborn (1962) has emphasized the possible importance of oxygen fugacity in the derivation of the tholeiitic and the calc-alkali series from more or less similar basaltic magmas.

Volcanic rocks of the calc-alkali series are confined to orogenic regions (table 4-3). The occurrence of calc-alkali rocks, or of a great amount of andesitic and

Table 4-3 *Younger volcanic rocks and tectonic environment*

	Orogenic belts				
Stable continents	Immature, very active island arcs	Mature, very active island arcs and active continental margins	Less active island arcs	Oceanic islands	Mid-oceanic ridges
Tholeiite series + +	+ +	+ +	+	+	+ +
Calc-alkali series		+ +	+ +		
Alkali series + +		+	+	+ +	+

+ +, abundant; +, subordinate

dacitic rocks may be regarded as suggesting the former existence of an island arc or an active continental margin.

The volcanic rocks of the calc-alkali series resemble granitic rocks in chemical composition and they both occur in orogenic belts. They could have been derived from the same andesitic (dioritic) magma (e.g. Nockolds and Allen, 1953; Dickinson, 1970). It is of interest to discover whether granitic rocks show a regular compositional variation across an island arc and a continental margin like the volcanic rocks. The existence of such a variation in granitic rocks has been demonstrated by Moore (1959) and Moore, Grantz and Blake (1961) on the west coast of North America, and by Taneda (1965) in Japan.

Factors Controlling the Diversity of Magmas. Magmas which cause volcanism in island arcs and active continental margins should be created in some genetic relationship to the descending lithospheric slab. Since volcanic rocks, in particular those of the calc-alkali series, are confined to volcanic arcs (table 4-3), they could be directly related to the descending slab. For example, the primary andesitic magma which gives rise to this series could be created by partial melting of the oceanic crust which forms the uppermost layer of the descending slab, as shown in fig. 3-11 (cf. Ringwood, 1969).

Basaltic magmas of the tholeiitic and alkali series could be produced by partial melting either in the descending slab or in the upper mantel overlying it. The oceanic crust with abyssal tholeiitic composition and the underlying presumably peridotitic layer of the descending lithospheric slab should undergo a series of phase changes with increasing depth, that is, with increasing pressure. The equilibrium relations between melt and solid residue should differ at different pressures. Recent experimental results suggest that the melt formed in peridotite under greater pressure tends to be more undersaturated with silica (e.g. Green and Ringwood, 1967; Kushiro, 1968). Even under the same $P-T$ conditions, the composition of the magmas tends to be less alkalic as the proportion of the melt increases. The pressure and the proportion of melt could vary regularly with depth in the descending slab, resulting in a regular compositional variation in volcanic rocks across the volcanic arc.

Even if we may assume with Ringwood (1969) that basaltic magmas are generated by partial melting not in the descending slab but in the overlying upper mantle where convective upwelling or diapiric rise occurs, the upwelling and the generation and separation of basaltic magmas should be controlled largely by the descending slab, and hence the resulting magmas could vary regularly across the arc.

Rate of Plate Motion and the Evolution of Island Arcs

There exists a clear correlation between the rate of plate motion and the activeness and petrography of island arcs (Miyashiro, 1972b), as summarized in table 4-4. Here, the activity of arcs may be represented by the depths of mantle earthquakes and the associated trenches. The depth of a trench should be influenced by the rate of burial by sediments, but nevertheless it shows a good correlation with the rate of plate motion (cf. Sugimura, 1968).

In table 4-4, group I of island arcs is characterized by a rate of relative motion (convergence) of plates as high as 8-9 cm/year. Here, the maximum depth of earthquakes is 600-700 km, that of the trench is 10-11 km, and volcanic rocks of the typical tholeiitic series are well developed. Rocks of the calc-alkali and alkali series are abundant only in arcs in the mature stage of development in this group. Fig. 15-3 illustrates petrographic provinces in such a mature stage.

In group II of island arcs, the rate of convergence of plates is slower and the maximum depths of earthquakes and of the trench are smaller, than those in group I. Here, typical tholeiite is absent, though some rocks present show an affinity with the tholeiitic series. Calc-alkali volcanic rocks are abundant. Alkali rocks, though usually present, are not abundant.

Group III is characterized by a still lower rate of plate convergence (about 0-2 cm/year). Alkalic rocks are well developed in such arcs, though a few atypical tholeiites and calc-alkali rocks may also occur. Thus, in the order of decreasing rate of plate convergence, the main volcanic rocks of island arcs vary from tholeiites to calc-alkali rocks, and then to alkali rocks.

Since the K_2O and usually also the Na_2O content decrease across an arc towards the ocean, the volcanic rocks in the oceanic-side zone within the volcanic belt usually have smaller K_2O and Na_2O contents than the volcanic rocks in the continental-side zone. Fig. 4-1 shows a comparison of the K_2O contents of volcanic rocks in the oceanic-side zone within the volcanic belts of

Table 4-4 *Correlation between the activeness and volcanic rock series in island arcs. After Miyashiro (1972b)*

Group	Arc and trench	Rate of plate convergence (cm/year)	Maximum depth of earthquakes (km)	Maximum depth of the trench (km)	Volcanic rock series
I	Tonga	9	700	11	Th
	Izu-Bonin	9	600	11	Th + (C) + (A)
	N.E. Japan	9	600	11	Th + C + (A)
	Kurile-Kamchatka	8	600	10	Th + C + A
II	Aleutian	6	300	8	(Th) + C + A
	Indonesia	5-6	600	7	(Th) + C + A
	Ryukyu	?	300	7	(Th) + C
	North Island (N.Z.)	Slow (3?)	300	4	C + A
	Hellenic (Aegean)	Slow (3?)	200	4	(Th) + C + A
III	Calabrian (Sicily)	Very slow (2?)	300	Buried	A
	Macquarie	Very slow	100	Shallow	(Th) + C + A

Note: The activeness of arcs decreases in the order of group I → II → III. Th = tholeiitic series, C = calc-alkali series, A = alkali series. The rocks of the series shown in parentheses are not typical of the series, and very small in quantity. The rates of plate convergence are mainly after Le Pichon (1968).

island arcs. The very active arcs (group I in table 4-4) always show lower K_2O contents than less active arcs (groups II and III).

In group I, the alkali basaltic rocks are sodic, whereas arcs of groups II and III appear to have both sodic and potassic (shoshonitic) types of alkali basaltic rocks. The Calabrian Arc (group III) has especially well-developed potassic basaltic rocks.

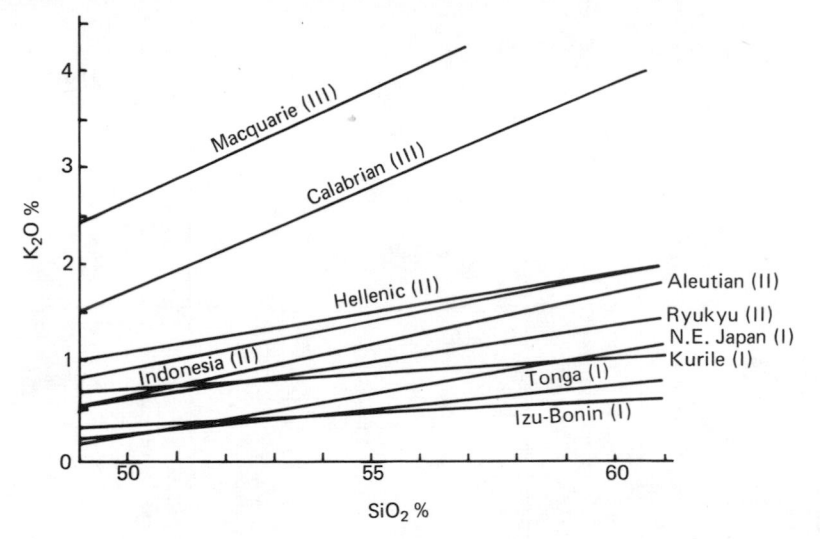

Fig. 4-1 K_2O versus SiO_2 relation in volcanic rocks of the oceanic-side (i.e. trench-side) zone within the volcanic belts of island arcs. The Roman numerals in parentheses represent the group numbers shown in table 4-4. (Miyashiro, 1972*b*.)

The activity of the northeast Japan Arc began in the early Miocene, and magmatism of the tholeiitic, calc-alkali and sodic alkali series took place from the beginning (Ozawa, 1968; Miyagi, 1964). Jakeš and White (1972) claimed that the occurrence of shoshonite begins at a later stage in the development of island arcs.

4-3   METAMORPHIC FACIES AND GEOLOGIC AGE

Distribution Rules

Some metamorphic facies show clear relationships to geologic age. Most of the glaucophane-schist facies rocks now observed in the world were formed by metamorphism in Cenozoic and Mesozoic time. Fig. 4-2 shows the distribution of glaucophane in the world. Its relationship with younger orogenic belts is clear (Eskola, 1939, p. 368; de Roever, 1956, 1964; van der Plas, 1959; Miyashiro, 1961*a*; Zwart, 1967*a*; Dobretsov, 1968; Ernst 1972).

There are some glaucophane schists in Palaeozoic metamorphic areas. However, such rocks occur extremely rarely in Precambrian metamorphic terranes (Soboler *et al.* 1967).

Zeolite-facies rocks appear to be much more common in Cenozoic and Mesozoic formations than in the older ones (e.g. Hay, 1966). On the other hand, most of the granulite facies rocks appear to have formed in Precambrian time. They usually occur in the shields, but are sometimes present in Phanerozoic orogenic belts as uplifted blocks of Precambrian basements (Oliver, 1969). A small proportion of granulite facies rocks, however, appears to be truly young. Fig. 4-3 shows the distribution of granulite facies terranes in the world.

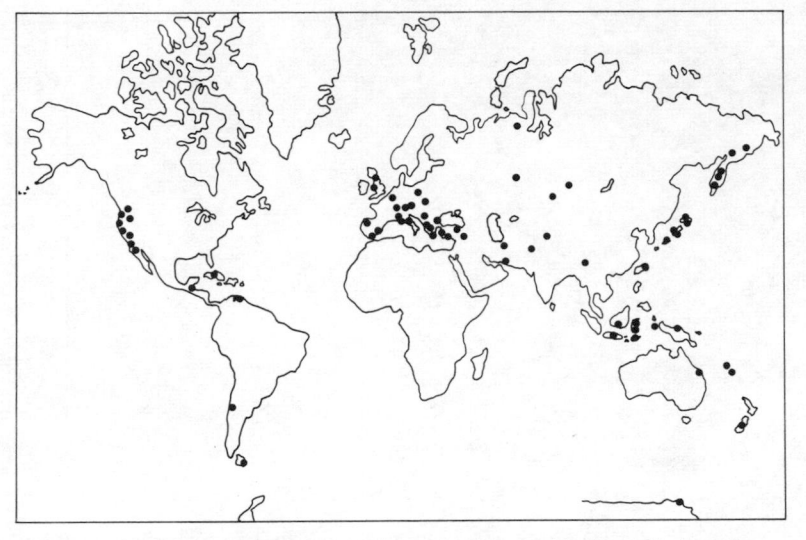

Fig. 4-2 Distribution of glaucophane in the world. (Largely based on van der Plas, 1959 and Sobolev *et al.* 1967.)

Many authors have pointed out the existence of a close relationship between anorthosites and geologic age. It has been claimed that there are at least three types of anorthosites. One type occurs as a member of differentiated gabbroic complexes such as the Bushveld of South Africa, and can be of any geologic age. Another type occurs in great dome-like or batholithic masses mainly composed of andesine or labradorite, and of middle Precambrian age (1100-1700 m.y. old). The third type occurs as a member of layered masses, being mainly composed of bytownite or anorthite, and of early Precambrian age (more than 3000 m.y. ago). The latter two types show close association with granulite or amphibolite facies metamorphic rocks (Isachsen, 1969; Windley, 1970).

On the other hand, low-pressure regional metamorphic rocks of the greenschist and amphibolite facies show no relationships to geologic age. Such rocks

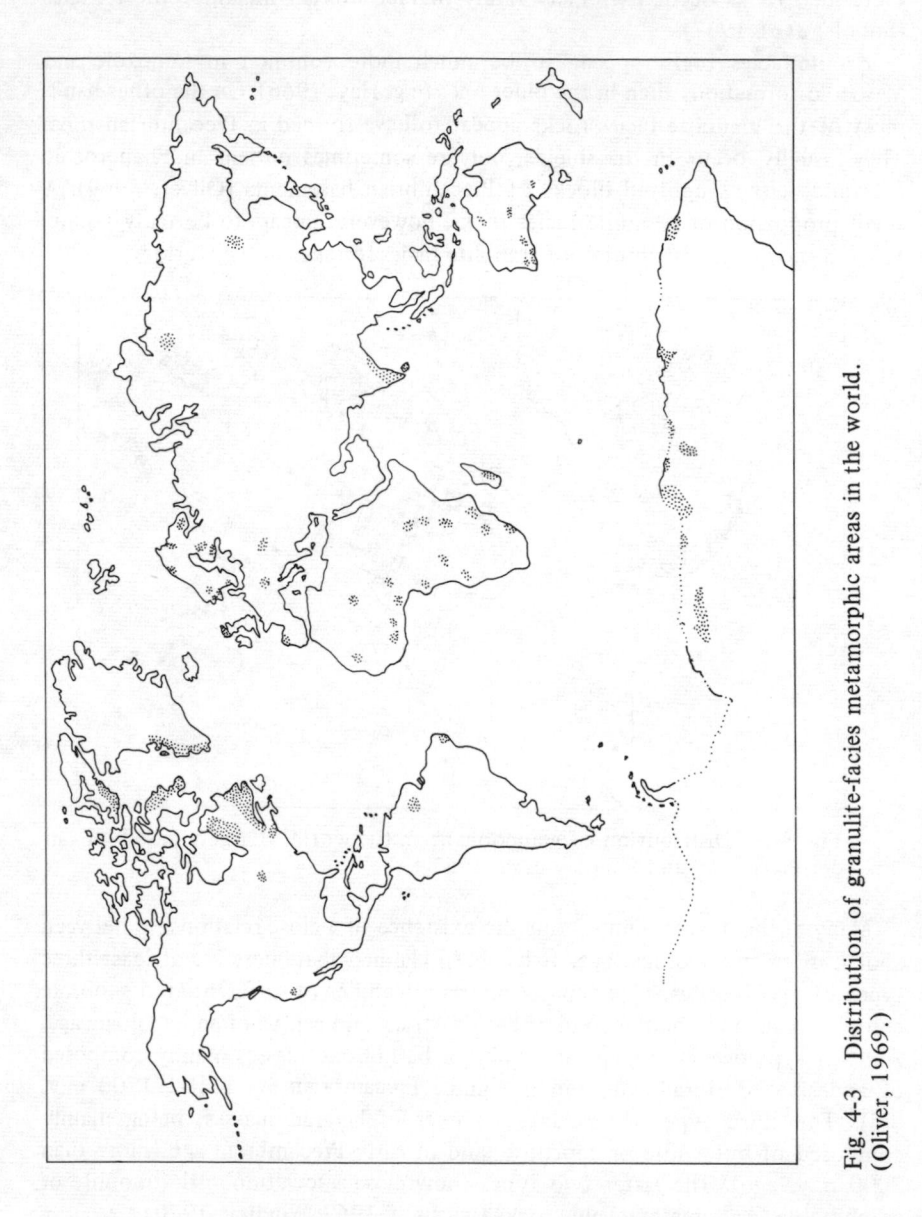

Fig. 4-3   Distribution of granulite-facies metamorphic areas in the world. (Oliver, 1969.)

occur, for example, in the Precambrian Svecofennian terranes, Palaeozoic terranes of eastern Australia, the Hercynian terranes of western Europe, and some Mesozoic terranes in Japan (table 4-5).

Apparent or Real?

It is an unsettled question whether these distribution rules have any real significance.

Many geologists have doubted the validity of the idea that glaucophane-schist facies rocks are confined to younger metamorphic terranes. Some have considered that since Precambrian terranes are very extensive, large areas of glaucophane-schist facies rocks may remain unnoticed. Others have suggested that glaucophane schists formed in Precambrian time may have been lost in later geologic time.

Zwart (1967a) has attempted to explain the rarity of glaucophane in old geologic ages as follows: Glaucophane-schist facies metamorphism takes place in the Alpinotype orogenic belts as already shown in table 4-1. Rapid and large uplift occurs characteristically in such belts, resulting in removal of all the low-temperature part (containing glaucophane) by erosion. Another possibility is that glaucophane may have been replaced by minerals of the greenschist facies, since such changes are common in high-pressure metamorphics.

The first explanation leads to the possibility that only glaucophane-schist facies rocks were lost from the Precambrian terranes, and rocks of high-temperature facies of the high-pressure type may be widely exposed unnoticed there. At present, we have no factual data to support or negate any of these possibilities.

Phanerozoic high-pressure metamorphic belts containing glaucophane schists appear to form along trenches in island arcs and continental margins, usually as the ocean-side members of paired belts. If such belts were formed in Precambrian time, and if a later Precambrian or Phanerozoic orogeny produced new paired belts roughly along the same zone but slightly on the ocean side of the older belts, the low-pressure metamorphism and granite emplacement of the new pair might take place on the same zone as the older high-pressure belt, which thus could be lost by the new recrystallization. If continental growth produces successively younger paired metamorphic belts slightly on the ocean side of the older ones, the high-pressure belts of the older pairs might be successively destroyed, and only the youngest pair might be observed undestroyed. This would give the appearance that all the older phases of metamorphism were of the low-pressure type.

Progressive Dehydration as a Possible Cause of the Distribution Rules

If the observed rules concerning the distribution of metamorphic facies through geologic time are tentatively assumed to be real, what could be the possible causes for them?

Table 4-5 *Baric types of Regional metamorphism and geologic ages*

	Precambrian	Palaeozoic	Mesozoic-Cenozoic
Low-pressure metamorphism	Svecofennides Karelides Canada (partly) Australia N.E. China	Hercynides Appalachians (partly) Eastern and South Australia Hida belt (Japan) Pichilemu series (Chile)	Ryoke-Abukuma belt (Japan) Hidaka belt (Japan)
Medium-pressure metamorphism	Canada (partly)	Caledonides Appalachians (partly)	North American Cordillera (partly)
High-pressure metamorphism	Anglesey (Wales)	Kiyama (Japan) Sangun belt (Japan) Curepto series (Chile) Penjina Range (NW Kamchatka)	Alps Franciscan group (California) Sanbagawa belt (Japan) Kamuikotan belt (Japan) New Caledonia Central Kamchatka

First, a large part of the rules appears to result from increasing dehydration with geologic age. Rocks of the zeolite- and glaucophane-schist facies are generally higher in H_2O content than those of other facies, whereas rocks of the granulite facies are generally the lowest in H_2O content among regional metamorphic rocks. Therefore, the abundance of the former in younger ages and of the latter in Precambrian time may mean a tendency for increasing dehydration with age. This trend can be noticed even when we focus our attention on zeolites and feldspars. Hay (1966) has emphasized the correlation between zeolites in sedimentary and metasedimentary rocks and their geologic age. The sum of phillipsite, clinoptilolite, erionite and chabazite appears to decrease, whereas analcime appears to increase with age within Cenozoic time. Analcime appears to decrease with age through Palaeozoic time, whereas albite and K-feldspar increase from Cenozoic to Precambrian. Such variations clearly indicate advancing dehydration (Miyashiro and Shido, 1970).

A deeper level of the orogenic belt would tend to be exposed by advancing denudation with age, leading to a generally higher temperature of metamorphism and a higher degree of dehydration on the exposed level. This is a possible, but not very convincing explanation.

Another possible process is the following: when a geosynclinal pile is subjected to regional metamorphism for the first time, pelitic and psammitic rocks contain abundant H_2O between and within mineral grains. Progressive rise of temperature would cause progressive dehydration of minerals. In the initial stages, metamorphic recrystallization would proceed in the presence of an intergranular aqueous fluid. A zeolite or glaucophane-schist facies zone would form under such a condition. If the duration of metamorphic recrystallization is prolonged, the H_2O in the rock complexes would be gradually expelled upward, and the condition would become drier. The zone of the zeolite and glaucophane-schist facies would tend gradually to disappear. Bryhni, Green, Heier and Fyfe (1970) pointed out that the onset of partial melting would also greatly contribute to dehydration.

Recent accumulation of radiometric age data has shown that many regional metamorphic terranes were subjected to two or more phases of metamorphism. With increasing age, the possibility of repeated metamorphism would also increase. In particular, most Precambrian rocks may be polymetamorphic. Repeated metamorphism should be similar in effect to prolonged recrystallization, and would result in greatly advanced dehydration (Winkler, 1967. p. 139).

Progressive metamorphic reactions in hydrous sediments are strongly endothermic. A large part of the heat transferred should be consumed in such reactions, resulting in a relatively small rise of temperature, which may favour the formation of rocks of high-pressure series. On the other hand, in repeated metamorphism, endothermic reactions would not be as intense, and the temperature rise may be more rapid, promoting the formation of lower pressure series rocks, especially in the Precambrian.

If this is true, the Precambrian shields must be generally drier than younger orogenic belts. Hyndman and Hyndman (1968) have pointed out that there is a layer of high electrical conductivity in the lower crust of young orogenic belts, and the high conductivity is probably due to the saturation of the rocks with H_2O, whereas such a layer is absent in shield regions.

Changes in the Thickness and the Motion of Plates as Possible Causes of the Distribution Rules

If the oceanic plates have become thicker, the descending slabs have become steeper, and the movements of the slabs have become more rapid in younger geologic time, the occurrence of high-pressure metamorphism should have become more common (Ernst, 1972a; Miyashiro, 1972b). Thus, we may assume that plates were thin and/or moved slowly in Precambrian and early Palaeozoic time. Under such conditions, metamorphism of the low- and medium-pressure types should be common, and occasionally atypical glaucophane schists could form.

In late Palaeozoic and early Mesozoic time, the plates may have become thicker and/or may have come to move more rapidly, with resultant common formation of high-pressure metamorphic rocks and paired metamorphic belts. The beginning of the present cycle of continental drift in early Mesozoic time may be another result of such a change in the plates.

Magnetic anomalies suggest that plate motion was more rapid in the Pacific than in other oceans at least in late Mesozoic and later time (Heirtzler *et al.* 1968). It is conceivable that similar rapid motion took place in early Mesozoic and late Palaeozoic time as well, leading to the common formation of high-pressure metamorphic rocks in the circum-Pacific regions.

Runcorn (1965) has suggested that a great change in the pattern of currents in mantle convection took place in early Mesozoic time, and as a consequence the present cycle of continental drift began. This change in the pattern of currents may have resulted also in an increase in the velocity of plate motion, particularly in the Pacific region and may have caused the common formation of glaucophane schists and paired metamorphic belts.

4-4 DIVERSITY IN LOW-TEMPERATURE METAMORPHISM

Observed Diversity. The metamorphic rocks formed at low temperatures, including the zeolite, prehnite–pumpellyite, glaucophane-schist and greenschist facies, show a greater mineralogical diversity than those formed at higher temperatures. Though P_s and P_{CO_2} are factors influencing mineral paragenesis at any temperature, their effects appear to be more evident at low temperatures.

Table 4-6 shows a synopsis of the observed diversity. In some regions, an unrecrystallized area grades directly into a greenschist-facies zone. In others, a

Table 4-6 *Classification of the low-temperature parts of metamorphic facies series*

Baric type	Facies series	Examples
Low-pressure	(a) Unrecrystallized zone . → greenschist facies zone	Ryoke metamorphic belt, Japan (Katada, 1965)
	(b) Unrecrystallized zone → zeolite facies zone → greenschist facies zone	Probably Mogami area, Japan (Utada, 1965)
	(c) Unrecrystallized zone → zeolite facies zone → prehnite-pumpellyite facies zone → greenschist facies zone	Tanzawa Mountains, Japan (Seki, *et al.* 1969*a*) Probably Akaishi Mountains, Japan (Matsuda and Kuriyagawa, 1965)
Medium-pressure	(a) Unrecrystallized zone → greenschist facies zone	Part of the Scottish Highlands (Wiseman, 1934) Part of the northern Appalachians, U.S.A. (Zen, 1960)
	(*b*) Unrecrystallized zone → zeolite facies zone → prehnite-pumpellyite facies zone → greenschist facies zone	Central Kii Peninsula, Japan (Seki *et al.* 1971) Probably part of the South Island New Zealand (Coombs *et al.* 1959)
High-pressure	(*a*) Unrecrystallized zone → glaucophane-schist facies zone	Bessi area, Japan (Banno, 1964) Kanto Mountains, Japan (Seki, 1958, 1960)
	(*b*) Unrecrystallized zone → prehnite-pumpellyite facies zone → glaucophane-schist facies zone	Katsuyama area, Japan (Hashimoto, 1968)

zeolite facies zone occurs in between. In still others, an unrecrystallized area grades into a prehnite–pumpellyite facies zone, which in turn grades into a glaucophane-schist facies zone.

A part of the diversity could be due to a difference in the threshold temperature of recrystallization, which would decrease with the increased duration of metamorphic recrystallization. Other possible factors will be discussed below.

Effect of P_{CO_2}. Higher P_{CO_2} tends to decompose Ca-bearing silicates with the resultant formation of carbonates. This greatly contributes to the mineralogical diversity of low-temperature metamorphic rocks.

The role of P_{CO_2} and P_{H_2O} in the formation of zeolites and a zeolite facies zone has been discussed particularly by Zen (1961*b*). The following compositional relations hold for Ca-zeolites:

heulandite + CO_2 =

    calcite + pyrophyllite (or kaolinite) + quartz + H_2O,

laumontite + CO_2 =

    calcite = pyrophyllite (or kaolinite) + H_2O.

A high P_{H_2O} should promote the formation of Ca-zeolites, whereas a high P_{CO_2} should tend to decompose them with the resultant formation of calcite and an aluminous mineral. It follows that under a high P_{CO_2}, an unrecrystallized area may directly grade into a greenschist facies zone, and a decrease in P_{CO_2} may produce a zone of the zeolite facies between these two areas.

Lawsonite, prehnite and pumpellyite, which are characteristic of the glaucophane-schist and prehnite–pumpellyite facies, are rich in Ca. They should be unstable under high P_{CO_2}.

In the greenschist facies, actinolite and epidote are common Ca-bearing silicates. These minerals are abundant in particular in greenschists recrystallized at low P_{CO_2}. With increasing P_{CO_2}, actinolite first and then epidote decompose to produce calcite and dolomite (or ankerite) together with Ca-free ferromagnesian minerals such as chlorite and stilpnomelane (§ 8A-6).

In the Sanbagawa and other low-temperature metamorphic terranes of Japan, Seki (1965) has observed that the frequency of occurrence of calcite in metabasites differs from area to area within the same metamorphic belt, that Ca-bearing silicates (pumpellyite, epidote and actinolite) are not common in areas where calcite is widespread in metabasites, and that the frequency of occurrence of calcite tends to decrease with increasing temperature of metamorphism. This suggests that P_{CO_2} differs from area to area within the same metamorphic belt, resulting in a difference in the degree of carbonate formation.

Effect of P_{H_2O}. Even if an aqueous fluid may exist during metamorphism at low temperatures, models (A), (B) and (C) described in § 2-3 give widely different P_{H_2O} under the same P_s. Higher P_{H_2O} should tend to form a greater proportion of hydrated minerals such as zeolites and chlorites.

In some regions, metamorphic recrystallization immediately follows the sedimentation of the geosynclinal pile, whereas in others recrystallization occurs long after the completion of sedimentation. Conceivably, the abundance of H_2O in a newly formed sediment pile may cause a higher P_{H_2O}, and hence may tend to form zeolite facies rocks.

Relation to Geothermal Gradient. Zeolite facies zones form under a wide range of geothermal gradients, that is, in metamorphic terranes of the low-, medium- and high-pressure types. However, the paragenetic relations of zeolites appear to differ in different baric types (§ 6A-1).

Zeolite facies areas, representing a relatively high geothermal gradient, show a great variety of zeolites, including mordenite, stilbite, yugawaralite, wairakite, heulandite, laumontite and analcime; e.g. the Tanzawa Mountains (Seki *et al.* 1969*a*). The present topographic surface in such areas represents a shallow depth of low-pressure regional metamorphic belts, or of contact- or hydrothermally metamorphosed areas. In contrast, zeolite facies areas, representing a low geothermal gradient, have only a very limited variety of zeolites such as heulandite, laumontite and analcime; e.g. the Taringatura area (Coombs *et al.* 1959). Such areas are parts of regional metamorphic terranes of the high-pressure type or close to it.

The physical factors directly controlling the difference in zeolite paragenesis between different baric types are not clearly known. Under the condition $P_s = P_{H_2O}$, the stability fields of some zeolites could be limited to a low-pressure range so that their occurrence is confined to metamorphism of the low-pressure type. It is conceivable also that the behaviour of H_2O may differ in different baric types. The threshold temperature of recrystallization would differ in different types.

4-5 NATURE OF CONTACT METAMORPHISM

Baric Types

It has been widely believed that contact metamorphism takes place under lower rock-pressures than does regional metamorphism. This idea is partly incorrect, however, since there is a great diversity in the *P–T* conditions of contact metamorphism.

Among the Al_2SiO_5 polymorphs, andalusite is common in contact aureoles, but sillimanite also occurs in many cases. Kyanite has been found occurring together with andalusite and sillimanite in a few aureoles. Thus, we may classify contact aureoles in terms of Al_2SiO_5 minerals into three categories as follows:

(*a*) Aureoles containing andalusite as the only Al_2SiO_5 mineral; e.g. the Oslo area, Norway (Goldschmidt, 1911).

(*b*) Aureoles containing andalusite in a low-temperature part and sillimanite in a high; e.g. the Arisu area, Japan (Seki, 1957, 1961*b*).

(*c*) Aureoles containing all three polymorphs of Al_2SiO_5; e.g. the Kwoiek area, British Columbia (Hollister, 1969*a, b*).

Polymorphs of Al_2SiO_5 may form by metastable crystallization. This possibility will be discussed in §§ 7B-13 and 10-3. Whether the Al_2SiO_5 minerals crystallized as stable or as metastable forms, it is probable that class (*c*) represents a relatively high rock-pressure in a transitional state between low- and medium-pressure metamorphism. Class (*b*) includes contact-metamorphic aureoles with a facies series more or less similar to that of low-pressure regional metamorphism. As was shown in fig. 2-1, andalusite is stable up to a higher temperature under lower pressures. The inner aureole of some examples of class (*a*) appears to have reached a very high temperature, and yet not to have produced sillimanite, suggesting that the pressure was very low. There would be no regional metamorphism corresponding to such aureoles. However, in cases where the maximum temperature attained is low, andalusite will form throughout the aureole even under a pressure comparable to that of low-pressure regional metamorphism. Therefore, class (*a*) probably represents a considerable range of pressure.

In chapter 10, well documented examples of contact aureoles will be described on the basis of this classification.

Syn- and late-tectonic plutons associated with medium-pressure regional metamorphism may give contact effects of the medium-pressure series. In most of such cases, however, it is not possible to distinguish contact from regional effects.

Metamorphic Facies

Amphibolite, Pyroxene-Hornfels, and Granulite Facies. The definitions of individual metamorphic facies are based on the externally controlled conditions and paragenetic relations of minerals and not on geologic relations of rocks. If the paragenetic relations in a contact aureole are virtually the same as those in some areas of regional metamorphism, the same name of metamorphic facies may well be applied to both.

The amphibolite facies is probably the most prevalent in contact aureoles. Many contact aureoles around large granitic masses in orogenic belts show amphibolite-facies mineral assemblages in all of the recrystallized zones. The paragenetic relations in such amphibolite facies are similar to those of the amphibolite facies in the low-pressure regional metamorphism. Though such aureoles are common, they have aroused little interest among petrologists in recent years.

In some aureoles, however, the outer zone belongs to the amphibolite facies, whereas the inner zone belongs to the pyroxene–hornfels facies. Such an aureole has been found around some masses of gabbroic and dioritic rocks, as exemplified by the diorite of the Comrie area in Scotland (Tilley, 1924) as well as around alkali plutonic masses (e.g. syenites and alkali granites) of the Oslo area (Goldschmidt, 1911).

In some andalusite–sillimanite-bearing contact aureoles, the innermost aureole, containing orthopyroxene, would belong to the granulite facies. This is the case in particular with contact metamorphism caused by syn- or late-tectonic plutons in low-pressure regional metamorphic terranes. Since schists and not hornfelses usually form in such cases, the outer limit of the contact aureoles is not clear. It appears to be common, however, that such contact metamorphism produces a metamorphic facies series very similar to, but reaching a higher temperature than, that of the associated regional metamorphism, as exemplified by the Nakoso aureole in the central Abukuma Plateau, where an ortho-pyroxene-bearing zone is formed in aureoles surrounding gabbros but not granites (Shido, 1958).

Low-Temperature Part of Contact Aureoles. The area on the outer side of an amphibolite facies zone is commonly too poorly recrystallized to be assigned to any metamorphic facies. However, some aureoles around late-tectonic plutons show the same facies as the surrounding regional metamorphism of the low-pressure type. In such a case, a zone of the greenschist facies occurs on the outer side of the amphibolite facies zone (e.g. Shido, 1958, for Nakoso), though distinction of the zone from the regional metamorphic terrane is difficult.

In typical contact aureoles composed of hornfelses, however, a zone of the greenschist facies is very rare. Instead, a zone characterized by metabasites with the actinolite + calcic plagioclase assemblage occurs commonly just on the outer side of an amphibolite facies zone. The actinolite in this case is colourless or very pale green with low refractive indices. The plagioclase is labradorite or andesine. Such an assemblage was observed in the Iritono aureole (Shido, 1958) and Arisu area (Seki, 1961b), both in Japan, and in the Sierra Nevada (Loomis, 1966), but has not been observed in any case of regional metamorphism.

The amphibolite is mainly composed of hornblende and a relatively calcic plagioclase. A decrease in temperature in this case is accompanied by replace-ment of the hornblende by actinolite, whereas the calcic plagioclase remains stable.

If H_2O and CO_2 are mobile, the actinolite + calcic plagioclase assemblage is virtually isochemical with common greenschist assemblages such as chlorite + epidote + albite + quartz and chlorite + calcite + albite + quartz, as is clear from the equations:

$$9 \text{ actinolite} + 114 \text{ anorthite} + 64 \ H_2O$$

$$= 10 \text{ chlorite} + 66 \text{ epidote} + 77 \text{ quartz,}$$

$$9 \text{ actinolite} + 15 \text{ anorthite} + 31 \ H_2O + 33 \ CO_2$$

$$= 10 \text{ chlorite} + 33 \text{ calcite} + 77 \text{ quartz.}$$

Here, the formula of chlorite is taken as $Mg_{4.5}Al_3Si_{2.5}O_{10}(OH)_8$. Therefore, the occurrence of the actinolite + calcic plagioclase assemblage would depend mainly on external conditions. If so, this assemblage is entitled to characterize a metamorphic facies. Thus, the metamorphic facies characterized by the actino-lite + calcic plagioclase assemblage was provisionally called the actinolite-calcic plagioclase hornfels facies (Miyashiro, 1961a, p. 307). Since available data are not ample, however, we will avoid the term facies and use the name *actinolite-calcic plagioclase zone* in this book.

In the first of the two preceding equations, the total volume of the solid phases is much smaller on the right-hand side. In the second, the relation is reversed. Therefore, the actinolite + calcic plagioclase assemblage could form under a low rock-pressure if P_{CO_2} is relatively low, but could not form under a relatively high P_{CO_2} even if the rock-pressure may be low. Since epidote usually contains much Fe_2O_3, the oxidation condition would also have some effect.

No Al_2SiO_5 minerals were found in the Iritono and Sierra Nevada aureoles. In the Arisu aureole, there is a wide medium-temperature zone containing andalusite and a narrow highest temperature zone containing sillimanite. The narrowness of the sillimanite zone suggests the possibility that the rock-pressure in this aureole may have been lower than that in ordinary low-pressure regional metamorphism. It is conceivable that all these aureoles represent rock-pressures lower than that of low-pressure regional metamorphism in combination with a low P_{CO_2}.

In the Iritono Aureole, the actinolite-calcic plagioclase zone represents the lowest temperature of recrystallization, whereas in the Arisu aureole, the actinolite-calcic plagioclase zone is followed on the outer side by a zone of the greenschist facies. Thus, this area could represent a state intermediate between those of the Iritono-like aureole and the low-pressure regional metamorphic area where a zone of the greenschist facies is well developed.

As will be summarized in table 8b-1, the composition of plagioclase at the temperature at which hornblende substitutes for actinolite in metabasites, tends to be more calcic with decreasing rock-pressure. The generation of the actinolite + calcic plagioclase assemblage in some contact aureoles may be regarded as a result of this trend.

Migration of Materials

Many contact aureoles contain a few unusual rocks that suffered intense metasomatism. It is known, however, that the majority of the hornfelses in such aureoles approximately preserve their pre-metamorphic compositions. This was first shown in the second half of the nineteenth century by Th. Kjerulf in the contact aureoles of the Oslo area (Barth, 1962, p. 262) and by H. Rosenbusch (1877) in the aureole around the Barr-Andlau granite in Alsace. The only marked change is dehydration and decarbonation.

However, the original pelites and psammites, for example, usually show a considerable range of composition, and hence it is difficult to prove the strict constancy of composition in metamorphism.

Oxygen isotope studies give a clue to this problem. Shieh and Taylor (1969) measured the $^{18}O/^{16}O$ ratios of metapelites in contact aureoles of the Santa Rosa Range and other places. The ratio in metapelites is similar to that in associated unmetamorphosed pelites and is much higher than that in associated plutons. In other words, there is virtually no change in $^{18}O/^{16}O$ of the metapelites due to the influence of the granitic plutons, except for a narrow zone, one metre thick or so, in direct contact with the plutons. This could mean that the outward migration of H_2O from a pluton into the aureole was negligible. This is in marked contrast with regional metamorphism, where the $^{18}O/^{16}O$ ratio of metamorphic rocks changes over a very wide region (§ 3-1). In general, the introduction of materials into contact aureoles would not be intense.

In high temperature zones, particularly in the pyroxene–hornfels facies, partial melting of pelitic and psammitic hornfelses could occur. Melts of broadly granitic compositions would form, and the removal of such melts would produce diverse solid residues of unusual chemical compositions. Highly desilicated hornfelses rich in corundum and/or spinel have been found in high-temperature contact aureoles of the Comrie area (Tilley, 1924), the Cashel–Lough Wheelaun area (Leake and Skirrow, 1960), the Belhelvie area (Stewart, 1946) and others. Such rocks may have been derived from very unusual sediments, but may conceivably represent solid residues (§ 10-4).

Processes of partial melting of metapelites and resultant formation of desilicated rocks have been traced in some cases of pyrometamorphism (§ 10-5).

Chapter 5

Diagrammatic Representation of Mineral Parageneses

5-1 MINERALOGICAL PHASE RULES AND COMPOSITION-PARA-GENESIS DIAGRAMS

When a group of rocks is subjected to a definite set of externally controlled conditions, the rocks should show variations in mineral composition corresponding to their chemical variations. In equilibrium, there should be a definite relationship between the chemical and the mineral compositions of rocks. In this case, the term chemical composition refers only to components that are not mobile. If a rock is open to H_2O, for example, its H_2O content varies with the external conditions.

The relationship between the chemical and the mineral compositions should vary with external conditions. A definite relationship should hold under definite values or small ranges of external conditions. Since such ranges of external conditions represent a metamorphic facies, each metamorphic facies is characterized by a definite relationship between the chemical and the mineral compositions. Thus the study of such a relationship is an essential part of the theory of metamorphic facies.

A considerable number of methods for the diagrammatic representation of such a relationship have been proposed. Here, we will begin with a simplified account of the connection of such diagrams with the phase rule of Willard Gibbs.

According to the phase rule,

$$F = c + 2 - p \geqslant 0$$

where F = number of degrees of freedom, c = number of independent components necessary to define the compositions of all phases in the system, and p = number of phases. The number of degrees of freedom means the maximum number of intensive thermodynamic variables which we may independently vary without changing the number of phases in the system in equilibrium.

Closed System

We now assume that a rock undergoing metamorphism is a *closed system* with no exchange of substance with the environment. The temperature T and rock-pressure P_s of the system are controlled by the external conditions. In other words, these two intensive variables depend on the external conditions. We assume that a mineral assemblage commonly found over a metamorphic area is stable in certain ranges of temperature and rock-pressure. Mineral assemblages stable only at a definite temperature and/or a definite rock-pressure would not be widespread. For a system to be in equilibrium in certain ranges of temperature and rock-pressure, the number of the degrees of freedom must be two or more. Thus,

$$F = c + 2 - p \geqslant 2,$$

therefore, $c \geqslant p$.

If there is no fluid phase in the system, p represents the number of minerals. If there is a fluid phase, $p - 1$ is equal to the number of minerals. Thus, the number of minerals that could coexist in stable equilibrium in a closed system for certain ranges of temperature and rock-pressure is equal to or less than the number of independent components. This is called Goldschmidt's *mineralogical phase rule* (Goldschmidt, 1911, 1912b).

If such rocks are composed of three components, the number of minerals in the commonly observed assemblages should be 3, 2, or 1. This means that when the mineral compositions of such rocks are represented in a triangular diagram with components [A], [B] and [C] at the apexes as shown in fig. 5-1, the paragenetic relations can be shown by dividing the diagram into a series of smaller triangles. In this figure A, B, C, D, E, and F represent minerals with definite compositions, and the diagram is divided into smaller triangles ADF, DEF, DBE, . . . A point within or on a side of the smaller triangles represents the coexistence of three or two phases respectively. A corner of the smaller triangles represents the presence of only one phase. Such a diagram is generally called a *composition-paragenesis diagram.*

Most metamorphic minerals, however, are solid solutions with considerable ranges of composition. Their compositions are represented by areas in the triangular diagram as shown by A, B, C, D, E, and F in fig. 5-2. Within these areas, $c = 3$ and $p = 1$, and hence $F = c + 2 - p = 4$. Two out of the four degrees of freedom refer to temperature and rock-pressure. The remaining two correspond to the divariant variation in composition. When two phases coexist, $c = 3$ and $p = 2$, and hence $F = 3$. Thus the compositional variation is univariant. In other words, the composition of each of two coexisting minerals is on a curve in the diagram. Such a curve represents the outline of the composition field of a mineral. In fig. 5-2, a point on the curve for a mineral is connected by a straight line to the point for another mineral in equilibrium. When three phases coexist, the composition of each of them is fixed, and the paragenesis is represented by a definite smaller triangle in the diagram.

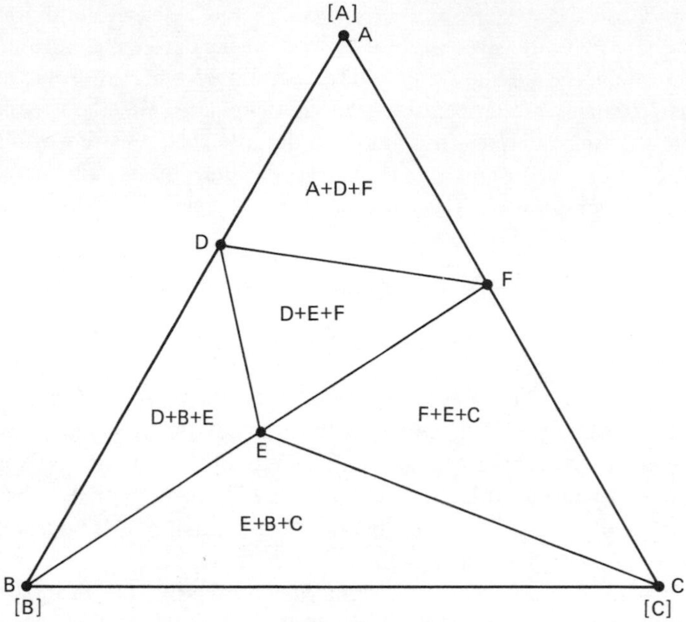

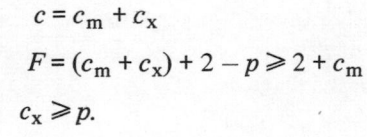

Fig. 5-1 Composition-paragenesis diagram of the three-component closed system [A]-[B]-[C] with minerals A, B, C, D, E and F of definite compositions. The same relation holds for a system which has fixed components [A], [B], [C] and is open for other components.

Open System

We assume that a rock undergoing metamorphism is open as regards mobile components (e.g. H_2O and CO_2). These components are here assumed to have extremely great mobility and hence are called *perfectly mobile components* after Korzhinskii (1959). The components that are not mobile are called *fixed components.* It is assumed for theoretical simplification that all the components of a rock undergoing metamorphism can be assigned as either perfectly mobile or fixed components. The temperature, the rock-pressure and chemical potentials of perfectly mobile components of the system are controlled by external conditions. The state of the system varies with the external conditions.

If c_m = number of the perfectly mobile components and c_x = number of the fixed components, the number of the degrees of freedom of the system must be equal to or greater than $2 + c_m$ for equilibrium to hold over certain ranges of externally controlled variables.

Hence,

$$c = c_m + c_x$$

$$F = (c_m + c_x) + 2 - p \geqslant 2 + c_m$$

$$c_x \geqslant p.$$

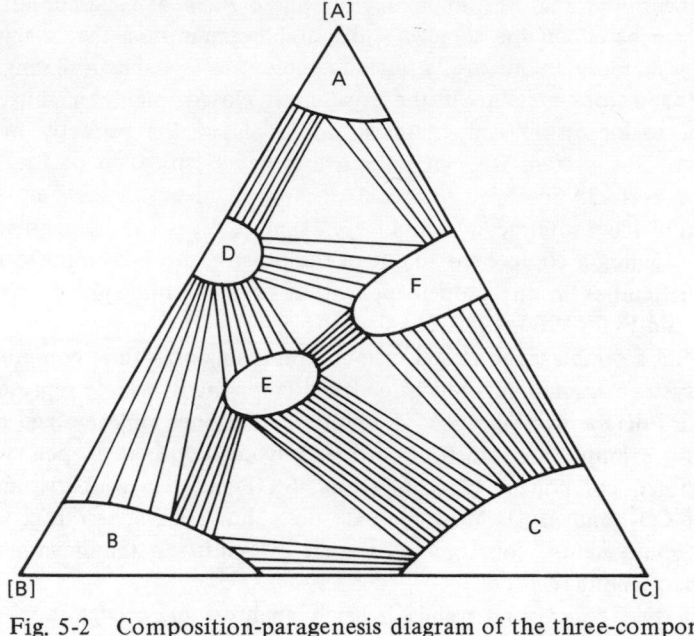

[A]

[B]   [C]

Fig. 5-2   Composition-paragenesis diagram of the three-component closed system [A]-[B]-[C] with solid-solution minerals A, B, C, D, E and F. The same relation holds for a system which has fixed components [A], [B], and [C] and is open for other components.

Thus, the number of minerals that could coexist in stable equilibrium in an open system for certain ranges of temperature, rock-pressure and chemical potentials of perfectly mobile components is equal to or less than the number of the fixed components. This is called Korzhinskii's mineralogical phase rule (Korzhinskii, 1936, 1959, 1965).

The state of a system is determined by the external conditions and the proportions of the fixed components in the system. Hence, the paragenetic relations for an open system with three fixed components can be represented by a triangular diagram with the fixed components at the apexes. We assume that an open system contains three fixed components [A], [B] and [C], as shown in fig. 5-1. According to Korzhinskii's phase rule, the number of minerals in commonly observed mineral assemblages should be 3, 2 or 1. If such minerals have fixed compositions as shown by points A, B, C, D, E and F in the figure, the triangular diagram is to be divided into smaller triangles to represent paragenetic relations. If the minerals are solid solutions, relations will be as shown in fig. 5-2. Thus, the composition-paragenesis diagram for an open system is apparently the same as that for a closed system, except that only fixed components are taken at the apexes. This is an important consequence of the mineralogical phase rules.

It is to be noted that the mineralogical phase rules of Goldschmidt and Korzhinskii are based on the somewhat disputable assumption that a mineral assemblage commonly found over a metamorphic area is stable over ranges of temperature and rock-pressure if the stystem is closed, and over ranges of temperature, rock-pressure and chemical potentials of the perfectly mobile components if the system is open. Moreover, the classification of the components into perfectly mobile and fixed ones is an idealization. Many components should have intermediate mobility. In most cases, we have no sound criteria for assigning a component to one or the other group. Accordingly, there are many difficulties in the proper application of these principles to natural rocks (Weill and Fyfe, 1964, 1967; Mueller, 1967).

The mineral assemblages in closed systems containing only three components or in open systems containing only three fixed components can be represented in triangular diagrams. Siliceous dolomitic limestones undergoing metamorphism, for example, are sometimes successfully dealt with as an open system with three fixed components CaO, MgO and SiO_2 and two perfectly mobile components CO_2 and H_2O. Most natural rocks, however, appear to contain more fixed components. This imposes a great difficulty on the diagrammatic analysis of paragenetic relations.

There are two contrasting trends in such analysis: (a) a large number of components and a wide range of composition in natural rocks may be represented in a diagram, though rigorous paragenetic analysis is not possible; (b) a more rigorous paragenetic analysis may be attempted, though the range of composition of the rocks to be treated cannot be wide. Eskola's ACF and $A'KF$ diagrams are examples of the former method whereas the AFM diagram proposed by Thompson is an example of the latter.

In the case of fig. 5-2, if a rock contains three minerals, the compositions of all the three are definite under definite external conditions, and should vary with external conditions but not with the variable bulk chemical composition of the rock. The effect of the variable bulk chemical composition on solid solution minerals could be eliminated by using such a rock. For such a reason, rocks containing a greater number of phases are generally more useful in the investigation of the relationship between the compositions of minerals and P-T conditions.

5-2 ESKOLA'S ACF AND $A'KF$ DIAGRAMS

The major components of ordinary metamorphic rocks are SiO_2, TiO_2, Al_2O_3, Fe_2O_3, FeO, MgO, MnO, CaO, Na_2O, K_2O, H_2O and P_2O_5. Some of these, however, are not important. It is the Al_2O_3 : CaO : (FeO + MgO) : K_2O ratios that have the dominant effects on the mineral composition of metamorphic rocks. The ACF and $A'KF$ diagrams proposed by Eskola (1915) are based on this empirical fact. These triangular diagrams are intended to focus our attention on the ratios of Al_2O_3 : CaO : (FeO + MgO) and of Al_2O_3 : K_2O : (FeO + MgO) respectively. We will examine the construction of these diagrams with reference to the mineralogical phase rule.

ACF Diagram

Metamorphic rocks usually contain small amounts of magnetite, hematite, ilmenite, sphene, rutile, apatite and sulphides. The *ACF* and *A'KF* diagrams, however, are intended for representing paragenetic relations among more abundant minerals. Accordingly, these accessory minerals are disregarded. TiO_2, P_2O_5 and S, being contained in them, are also disregarded.

ACF and *A'KF* diagrams are used only for rocks containing quartz (or very rarely another stable form of SiO_2). Such rocks are called rocks with *excess SiO_2*. Quartz is virtually pure SiO_2. An increase or a decrease of SiO_2 content of such rocks causes only an increase or a decrease in quartz without any effects on the paragenetic relations of minerals. Even if the component SiO_2 and the phase quartz are disregarded simultaneously, the number of the degrees of freedom $F = c + 2 - p$ remains unchanged. Hence, this disregard of SiO_2 is justified from the viewpoint of the phase rule.

When a group of rocks have been metamorphosed as closed systems, a component which is always present as a particular phase in the rocks under consideration as SiO_2 is in the present case, is called an *excess component* (Korzhinskii, 1959, p. 67). If the rocks have been metamorphosed as open systems, an excess component means a fixed component that is always present as a particular phase, either on its own or in combination with any amounts of perfectly mobile components.

Eskola disregarded the component H_2O on the basis of his statement that H_2O is always present in excess during metamorphism. Probably he meant that an intergranular fluid existed during metamorphism. Indeed, if an intergranular fluid phase, with virtually pure H_2O composition, at the same pressure as the surrounding solid phases, exists as in model (A) of § 2-3, we may disregard the component H_2O and the fluid phase in our diagram, just as we did SiO_2 and quartz. When we assume a rock undergoing metamorphism to be open for H_2O, H_2O will not be taken as a component at the apexes in the composition-paragenesis diagram, and hence we come to a diagram which is the same in effect as Eskola's.

The Na_2O present in metamorphic rocks is usually mainly contained in the albite molecule except in the glaucophane-schist and eclogite facies. In order to reduce the number of components, Eskola disregarded the component albite, though this is not justified from the phase rule.

Eskola added Fe_2O_3 to Al_2O_3 on the assumption that they play similar roles in solid-solution minerals. He added MgO and MnO to FeO for an analogous reason. Nowadays we know that the replacement of Al^{3+} by Fe^{3+} and of Fe^{2+} by Mg^{2+} and Mn^{2+} does not justify the grouping of these components from the viewpoint of the phase rule. These atoms have distinctive properties, and can substitute for one another in different ways in minerals. They should be treated as different components in more rigorous paragenetic analysis.

This procedure, however, is a necessary evil in order to reduce the number of compositional variables. Thus, we have the following four components: Al_2O_3 + Fe_2O_3, CaO, FeO + MgO + MnO, and K_2O. For the purpose of making triangular diagrams, we have to choose three components out of the four.

Here, we choose Al_2O_3 + Fe_2O_3, CaO and FeO + MgO + MnO. Since K_2O is disregarded in this case, K-feldspar cannot be represented in the diagram. As was explained above, albite has also been disregarded. Therefore, correction of the Al_2O_3 content is necessary. The molecular amount of Al_2O_3 in the alkali feldspars is the same as that of $(Na_2O + K_2O)$. Thus, we have the following three groups of components:

$$A = Al_2O_3 + Fe_2O_3 - (Na_2O + K_2O)$$

$$C = CaO$$

$$F = FeO + MgO + MnO.$$

The amounts of these three groups of components are calculated on a molecular basis. A triangular diagram having the three on the apexes is called an *ACF* diagram, as exemplified by figs. 5-3 and 5-4.

The micas and hornblendes contain considerable amounts of K_2O and Na_2O. Strictly speaking, they cannot be represented on the *ACF* diagram. However, they are customarily shown.

Calcite may be shown or disregarded in the diagram. If we regard a rock as being open for CO_2, the formation of calcite depends on the chemical potential of CO_2.

The merit of the *ACF* diagram is that all the common metamorphic rocks including pelitic, psammitic, calcareous, and basic ones can be plotted to show their mutual relations and broad paragenetic features. As is clear from the above, however, the *ACF* diagram is based on a number of procedures that are not justifiable from the phase rule. Accordingly, it is not a rigorous composition-paragenesis diagram. The *ACF* diagram cannot be divided into smaller triangles rigorously showing parageneses. In many cases, four or five minerals on the diagram occur in the same rock. *ACF* diagrams in the literature, however, are usually divided into smaller triangles indicating parageneses. This might be for aesthetic reasons rather than for the truth. We follow this custom in this book.

Solid-solution minerals should appear properly as lines or areas in the *ACF* diagram. However, as the diagram is not rigorous, very accurate representation of

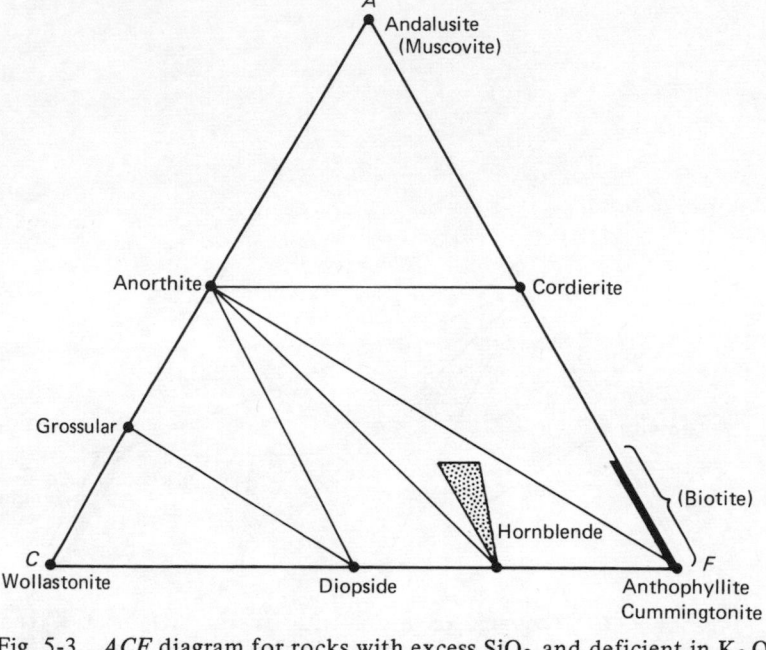

Fig. 5-3   *ACF* diagram for rocks with excess SiO_2 and deficient in K_2O in the low-pressure amphibolite facies of the Orijärvi area, Finland. Usually K-feldspar is virtually the only K_2O-bearing mineral in rocks with excess SiO_2 near the *C* apex. Near the *C* apex, therefore, only the rocks virtually devoid of K_2O fall into Eskola's category of 'rocks deficient in K_2O'.

the solid-solution range on the diagram is of little value. In this book, most minerals are schematically shown by dots.

Moreover, the *ACF* diagram has little significance for the glaucophane-schist facies where transformations in Na-bearing minerals play an important role.

The composition fields of some rock types in the *ACF* diagram are shown in fig. 5-5.

Excess and Deficiency in K_2O
In the epidote-amphibolite facies and the low-temperature part of the amphibolite facies, muscovite and biotite are stable minerals. Since K-feldspar has a higher content of K_2O than micas, it reacts with K-free minerals on the *AF* side of the *ACF* diagram such as andalusite, kyanite, cordierite, almandine and anthophyllite to form micas. Accordingly, only rocks very poor in K_2O could contain such K-free minerals. With increasing K_2O, K-free minerals are gradually transformed into micas. After complete transformation, a further increase in K_2O causes the appearance of K-feldspar. This relation was found by Eskola in a paragenetic study of the amphibolite-facies rocks in the Orijärvi area.

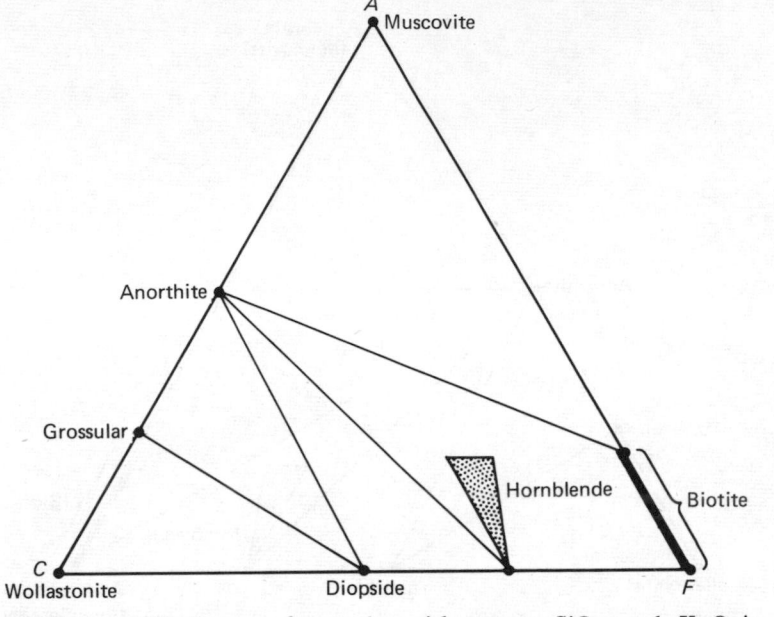

Fig. 5-4 *ACF* diagram for rocks with excess SiO_2 and K_2O in the low-pressure amphibolite facies of the Orijärvi area, Finland.

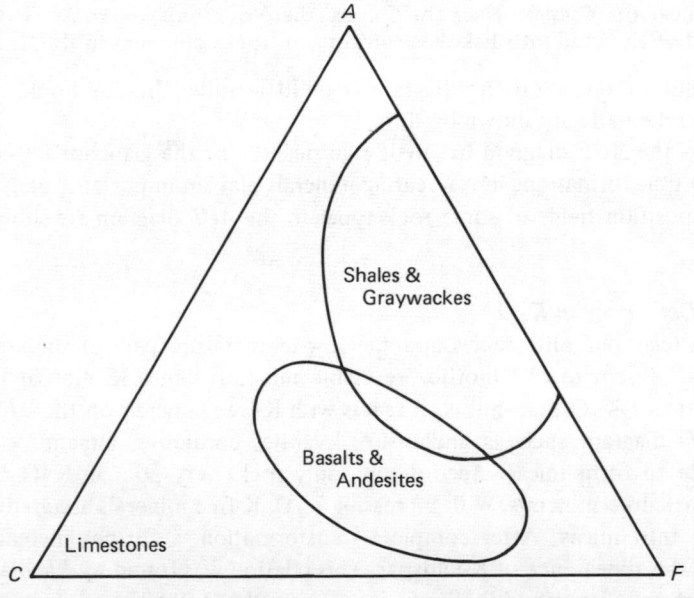

Fig. 5-5 Composition fields of some rock types in the *ACF* diagram.

Since this variation is related to the K_2O content, it cannot be represented in the *ACF* diagram. Eskola attempted to show it qualitatively by using two *ACF* diagrams: one for rocks *deficient in K_2O* (fig. 5-3) and the other for rocks *with excess K_2O* (fig. 5-4) for the same metamorphic facies.

Here, 'rocks dificient in K_2O' means rocks containing a mineral or minerals that cannot coexist with K-feldspar. 'Rocks with excess K_2O' means rocks not containing such a mineral or minerals. This classification has significance only for the rocks plotting on and near the *AF* side of the diagram. In the rocks plotting near the *C* apex, practically all the K_2O is contained in K-feldspar, so long as SiO_2 is in excess.

At higher temperatures (i.e. the highest amphibolite facies and the granulite facies), muscovite and then biotite become unstable, and hence K-feldspar can coexist with minerals on the *AF* side.

A'KF Diagram for Metapelites

Pelitic rocks are usually low in CaO, containing $Al_2O_3 > Na_2O + K_2O + CaO$ (molecular basis). Such rocks are said to have *excess Al_2O_3*. Their Al_2O_3 contents are higher than the amount necessary to combine with all the Na_2O, K_2O and CaO present to form feldspars. They are plotted within the triangle *A*–anorthite–*F* of the *ACF* diagram (fig. 5-4).

In pelitic rocks of the amphibolite and higher facies, the CaO is contained mainly in the anorthite molecule. The effects of K_2O on the minerals on the *AF* side of the *ACF* diagram can be clarified by examining the Al_2O_3 remaining after the formation of Na-, K- and Ca-feldspars. (In the epidote-amphibolite and greenschist facies, epidote would be formed instead of the anorthite molecule. Since epidote and anorthite have somewhat different compositions, this must have some effect to other minerals. However, it might well be neglected.)

For rocks with excess Al_2O_3, we can use the *A'KF* diagram with the following groups of components (molecular basis) on the apexes:

$$A' = Al_2O_3 + Fe_2O_3 - (Na_2O + K_2O + CaO)$$

$$K = K_2O$$

$$F = FeO + MgO + MnO.$$

Eskola named such a diagram an *AKF* diagram. However, the meaning of *A* in this case differs from that of *A* in the *ACF* diagram. To avoid possible confusion, the name of the *A'KF* diagram is used in this book after Winkler (1967).

In the *A'KF* diagram, SiO_2 is an excess component and H_2O may be regarded as a perfectly mobile component (fig. 5-6). The construction of the *A'KF* diagram also involves procedures unjustifiable from the viewpoint of the phase rule. Hence, four phases on the diagram may stably coexist.

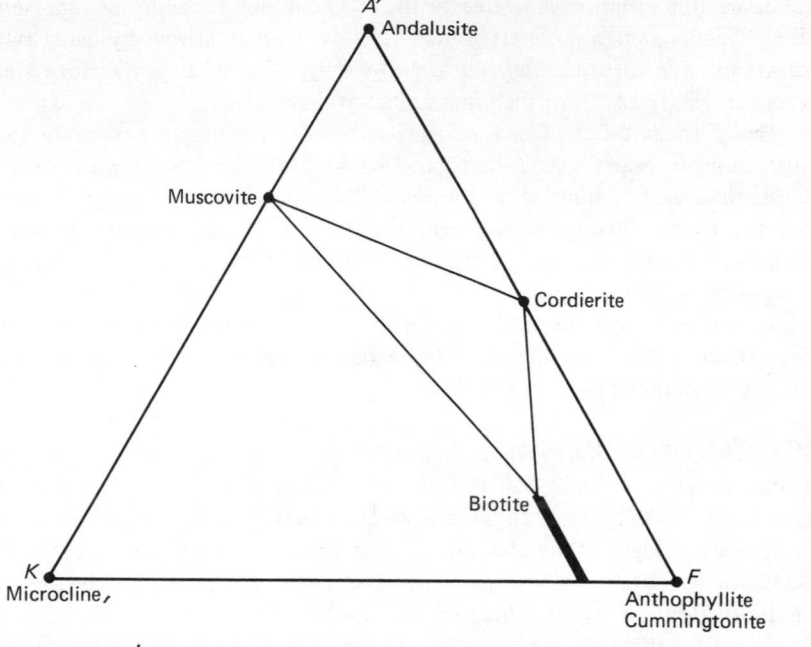

Fig. 5-6 A'KF diagram for pelitic rocks with excess silica in the low-pressure amphibolite facies of the Orijärvi area, Finland.

5-3 *AFM* DIAGRAMS FOR METAPELITES

In the *ACF* and A'*KF* diagrams, FeO and MgO are jointly treated as *F*. Among the minerals of metapelites, however, garnet, chloritoid, and staurolite are high in FeO/MgO, whereas cordierite is low. The FeO/MgO ratio of rocks has a great influence on the mineral assemblage. Hence, a triangular diagram with FeO and MgO at different apexes is helpful in paragenetic analysis.

Most minerals in metapelites fall on and near the $A'F$ side of the $A'KF$ diagram. Accordingly, paragenetic relations could be represented to a considerable extent by a triangular diagram with A', FeO and MgO at the apexes, or the like, as was used, for example, by Korzhinskii (1959, fig. 49) and Chinner (1962, 1967).

Thompson AFM Diagram.

Thompson (1957) has proposed a more rigorous method of representation. Metapelites are regarded as being composed of six components SiO_2, Al_2O_3, MgO, FeO, K_2O and H_2O by neglecting minor ones such as TiO_2, Fe_2O_3, MnO, CaO, and Na_2O. He regards the rocks as being open for H_2O, and hence H_2O does not appear in the composition-paragenesis diagram. Most metapelites contain quartz, and his discussion is confined to quartz-bearing rocks. Accordingly, SiO_2 is an excess component as defined in the preceding section, and may be neglected in the diagram.

Thus, only four main components Al_2O_3, MgO, FeO and K_2O remain in our graphical representation, and the composition of a metapelites is approximately represented by a point in a tetrahedron as shown in fig. 5-7.

Muscovite is very widespread in metapelites. Thompson confined his discussion to muscovite-bearing rocks, and hence this mineral need not be expressed in the composition-paragenesis diagram. The muscovite composition has been assumed to be $KAl_3Si_3O_{10}(OH)_2 = KAl_3O_5 . 3SiO_2 . H_2O$, and all the other minerals within the tetrahedron are projected from the muscovite point KAl_3O_5 onto the Al_2O_3–FeO–MgO plane, as shown in fig. 5-7. The maximum

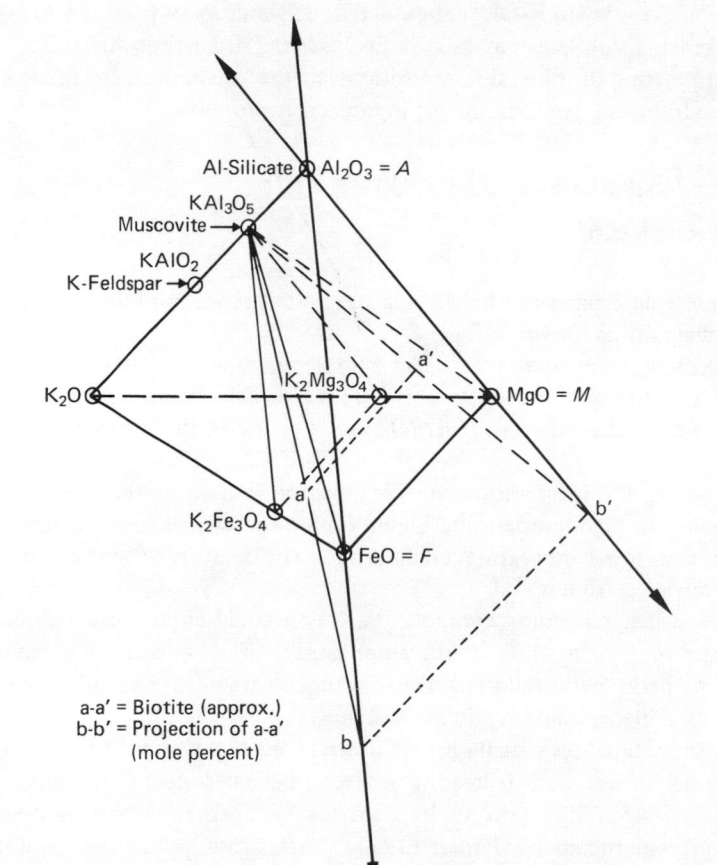

Fig. 5-7   Thompson's projection through the idealized muscovite composition onto the plane determined by Al_2O_3, FeO and MgO.

number of phases is four according to the mineralogical phase rule. Excepting muscovite, the remaining three phases can be strictly represented on a triangular diagram on the Al_2O_3–FeO–MgO plane.

Compositions in the sub-tetrahedron Al_2O_3–FeO–MgO–KAl_3O_5 will project onto the triangle Al_2O_3–FeO–MgO, whereas those in the volume KAL_3O_5–FeO–MgO–$K_2Mg_3O_4$–$K_2Fe_3O_4$ will project onto the extension of the plane Al_2O_3–FeO–MgO beyond the line FeO–MgO. Compositions in the sub-tetrahedron K_2O–$K_2Fe_3O_4$–$K_2Mg_3O_4$–KAl_3O_5 will project onto the extension of the plane Al_2O_3–FeO–MgO beyond Al_2O_3. However, the last-mentioned projection beyond Al_2O_3 is usually unimportant, since each possible assemblage is usually represented in the other parts of projection. Hence, this part of projection may be neglected in ordinary cases. The tie lines from almandine, cordierite or biotite toward K-feldspar extend downwards away from the Al_2O_3 apex. In this sense, K-feldspar may be regarded as being situated downwards.

The actual plotting of mineral compositions on the diagram can be made by calculating the following two parameters in molecular proportions:

$$(Al_2O_3 - 3\,K_2O)/(Al_2O_3 - 3K_2O + FeO + MgO)$$

$$MgO/(FeO + MgO).$$

These parameters determine the height and horizontal position of the projected point on the diagram, as shown in fig. 5-8.

In the coexisting minerals, the Fe^{2+}/Mg ratio increases in the following order: cordierite → chlorite → biotite → staurolite (and chloritoid) → almandine (Thompson, 1957; Albee, 1968). This relation is of use in the construction of the diagram.

In this diagram, the ideal muscovite composition is used as the projection point. Muscovites in high-temperature metamorphic rocks have such a composition, but those in low-temperature metamorphic rocks are now known to be considerably deviated from it.

Among the neglected minor components, Na_2O could enter muscovite and K-feldspar, and CaO and MnO could enter garnet. This should cause some modification of paragenetic relations. Generation of garnet, for example, would be promoted by a higher content of CaO and MnO.

Muscovite in metapelites usually breaks down in the high-temperature part of the amphibolite facies, and instead K-feldspar becomes nearly ubiquitous. Another type of *AFM* diagrams to be used for metapelites with quartz and K-feldspar was constructed by Barker (1961). In this case, points representing minerals and rocks situated within the Al_2O_3–FeO–MgO–K_2O tetrahedron are projected onto the Al_2O_3–FeO–MgO plane from the projection point of the K-feldspar composition.

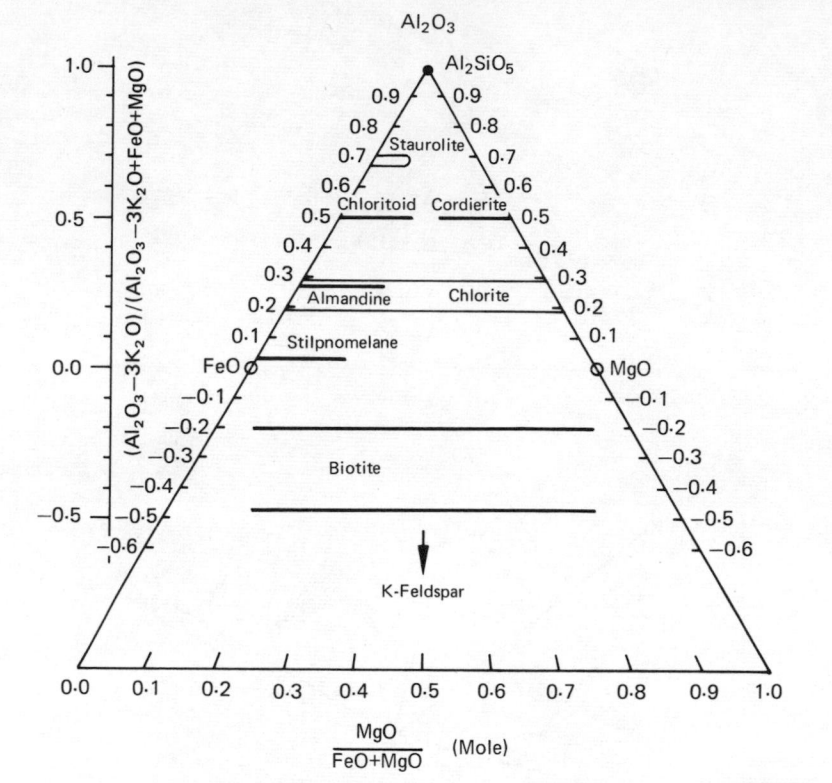

Fig. 5-8   Thompson *AFM* diagram showing possible phases in equilibrium
with quartz and muscovite.

5-4   SCHREINEMAKERS BUNDLE

So far in this chapter, we have discussed diagrammatic representation of
variations of mineral parageneses with chemical composition. Now we intend to
give a brief general discussion on the variation of mineral parageneses in relation
to temperature and pressure.

If a system composed of c components contains $c + 2$ phases of fixed
composition, the phase rule tells us that the number of the degrees of freedom
(F) is zero. In a rectangular diagram with temperature, and pressure as the
coordinate axes such as fig. 5-9, the assemblage with $c + 2$ phases is represented
by a definite point (invariant point). If any one of the phases is removed, the
system comes to have $c + 1$ phases and hence $F = 1$, and should be stable on a
curve (univariant curve) in the diagram. There are $c + 2$ different ways in the
removal of a phase. Thus, in general, $c + 2$ curves exist, which should radiate
from the invariant point representing the coexistence of $c + 2$ phases. Such a

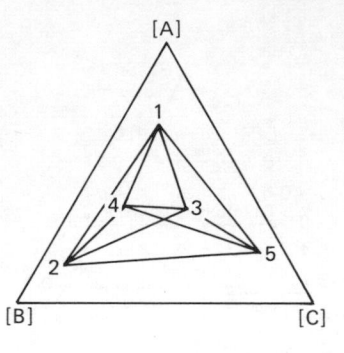

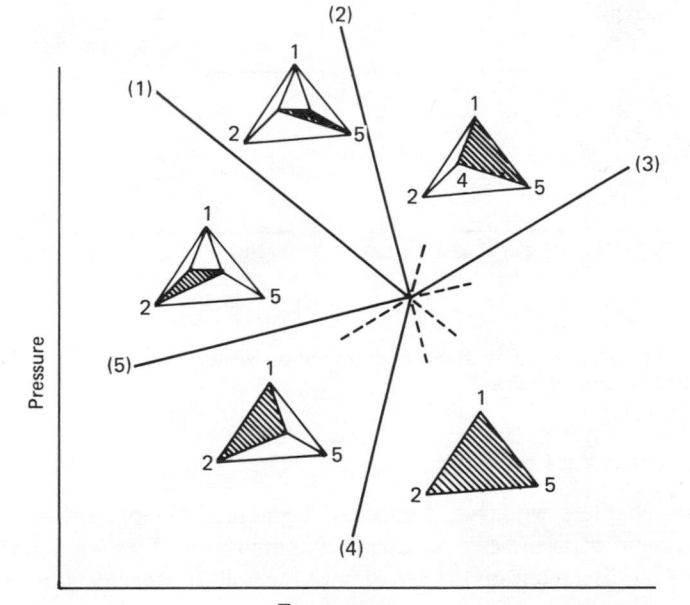

Fig. 5-9   Schreinemakers bundle for a three-component system [A]-[B]-[C] with five possible phases 1, 2, 3, 4 and 5 in the compositional relation illustrated in the large triangular diagram at the top. Five univariant curves radiate from the invariant point representing the coexistence of all five phases. Full lines represent stable equilibria, whereas broken lines represent metastable ones. Univariant curve (1), for example, represents the four-phase assemblage that is obtained by removing phase 1 from the five phases. The paragenetic relations for each divariant field are shown by a small triangular diagram in which a shaded area gives the phase assemblage characteristic of the field.

group of curves around an invariant point has been called by the name *'Schreinemakers bundle'*. Each of the univariant curves is divided into two parts by the invariant point. One part represents stable and the other metastable equilibrium. Each curve may be regarded as corresponding to a chemical equation among the phases coexisting on it.

These curves divide the temperature–pressure diagram into $c + 2$ fields (divariant fields), in each of which c or less phases can coexist in equilibrium. Within each field, mineral assemblages change with their chemical composition according to the rule characteristic to the field. Such a rule can be represented by a diagram as illustrated in fig. 5-9.

As a simple example, we take the one-component system Al_2SiO_5, in which a three-phase point (triple point) is fixed in a temperature–pressure diagram as was shown in fig. 2-1. Three curves representing two-phase equilibria radiate from the point. Only one phase is stable in each of the fields bounded by these curves.

The shapes and positions of individual univariant curves may be known by synthetic experiments, or can be calculated if enough thermodynamic data are available. The slope of the curves can be calculated by the Clausius–Clapeyron equation, if the entropy and volume changes are known. Even when the shapes, positions and slopes of individual curves are not known, there is a general rule governing the sequence of univariant curves around the invariant point. It was found by F. A. H. Schreinemakers in the 1910s. The sequence depends only on the chemical formulae of the relevant phases. Fig. 5-9 was drawn in accordance with it. For a detailed account, refer to Zen (1966).

An example of the use of the Schreinemakers rule was shown in § 3-3.

II   Progressive Metamorphism

Outline

The recent advance in petrographic study throughout the world has clarified the great diversity of metamorphic terranes. A cursory survey of the diversity was given in § § 3-3 and 3-4. Part II treats detailed classifications and descriptions of well-documented examples of progressive metamorphic terranes.

The nature of the diversity is revealed by examination of progressive metamorphic reactions, which vary not only with T, P_s, P_{H_2O} and P_{CO_2} but also with chemical composition of the dominant metamorphic rocks.

Thus, the backbone of Part II is composed of chapters 7, 8, and 9, which are allotted respectively to the progressive regional metamorphism of pelitic, basic and calcareous rocks. Chapters 7 and 8 are divided into two sub-chapters: A and B. In each case, sub-chapter A treats a classification and description of pertinent metamorphic terranes in order to show the extent of their diversity, whereas sub-chapter B treats the progressive metamorphic reactions, except for those of the zeolite and prehnite-pumpellyite facies, in order to clarify the nature and cause of the diversity.

The diversity is greatly enhanced by the presence or absence of zeolite and prehnite-pumpellyite facies zones in the low-temperature part. Moreover, zeolite facies zones themselves show diversity. These problems are treated in chapters 6A and 6B prior to the above-mentioned backbone of Part II.

Chapter 10 treats the diversity of progressive contact metamorphism.

As a summary of all these descriptions, three generalized series of metamorphic facies will be formulated and described in chapter 11. A rigorous definition of metamorphic facies will also be discussed.

Eclogite does not occur as a usual member of progressive metamorphic zones. Therefore, the problem of eclogites and the eclogite facies will be discussed in chapter 12 at the end of Part II.

Chapter 6A

Zeolite and Prehnite–Pumpellyite Facies Metamorphism: Its Diversity

6A-1 CLASSIFICATION OF BURIAL METAMORPHIC TERRANES IN TERMS OF GEOTHERMAL GRADIENTS

Burial metamorphism, usually in the zeolite and prehnite–pumpellyite facies, takes place in the lowest temperature part of regional and ocean-floor metamorphic terranes, though occasionally it results in apparently independent small metamorphic areas. Rocks of the zeolite and prehnite–pumpellyite facies in all geologic settings will be treated together in chapters 6A and 6B for the following three reasons: (a) Comparison of burial metamorphism in different geologic settings is instructive; (b) The limited variation in the chemical composition of rocks which are extensively recrystallized under burial metamorphism means that they can hardly be treated in terms of the compositional classification which will be used in our treatment of regional metamorphic rocks in chapters 7, 8 and 9; (c) Many regional metamorphic terranes have no burial metamorphic part, and hence burial metamorphism may well be treated separately for convenience.

Recrystallization is usually incomplete in the zeolite and prehnite–pumpellyite facies. Pyroclastic rocks and lavas as well as graywackes containing abundant pyroclastic materials are susceptible to recrystallization at low temperatures, whereas other rocks associated with them may show little recrystallization. The temperature of recrystallization is too low to produce anhydrous Al_2SiO_5 minerals. However, the geothermal gradients and baric types may be inferred from the estimated depth and temperature of recrystallization.

Recent petrographic studies, particularly in Japan, have revealed that the paragenetic relations in the zeolite facies vary considerably from area to area (§ 4-4). It appears that the paragenetic relations are partly related to the geothermal gradient. The terranes with higher geothermal gradients tend to produce a greater number of zeolites including mordenite, stilbite, yugawaralite, wairakite, heulandite, laumontite and analcime. On the other hand, the terranes with lower geothermal gradients produce only the last three of the above-mentioned zeolites, or even only laumontite.

Thus, metamorphic terranes with zeolite and prehnite–pumpellyite facies zones will be reviewed below on the basis of a classification in terms of geothermal gradients, or baric types.

6A-2 ZEOLITE AND PREHNITE-PUMPELLYITE FACIES TERRANES OF THE LOW-PRESSURE TYPE

Many of the low-pressure regional metamorphic terranes, such as the Ryoke belt of Japan, have no rocks belonging to the zeolite and the prehnite–pumpellyite facies, so far as we are aware. Their lowest temperature zone is in the greenschist facies. On the other hand, Coombs, Horodyski and Naylor (1970) have found a wide zone belonging to the prehnite–pumpellyite facies in northeastern Maine. This zone probably represents the lowest temperature part of the low-pressure facies series in the northern Appalachian Mountains.

Three well-documented areas of zeolite facies rocks in Japan will be reviewed below. The geothermal gradient appears to have been very great in all of them. Hence, probably they belong to the low-pressure type, though Al_2SiO_5 minerals do not occur.

Akaishi Mountains, Japan

Matsuda and Kuriyagawa (1965) have described a metamorphosed Mesozoic and Tertiary terrane, about 40 km wide, with abundant volcanic and pyroclastic rocks ranging from the zeolite through the prehnite–pumpellyite to the greenschist facies in the eastern Akaishi Mountains to the west of Mt Fuji in central Japan (figs. 15-8 and 15-9). The metamorphism is of the burial type on a regional scale showing no relationship to plutonic masses. The temperature of metamorphism generally increases westward (fig. 6A-1).

The zeolites identified in metamorphosed basalts, dolerites and andesites were mordenite, heulandite (clinoptilolite), laumontite, analcime and thomsonite. Iron saponite and celadonite occur in the zeolite facies zone, whereas prehnite begins to occur in a higher temperature part of the same zone and persists to the prehnite–pumpellyite facies zone. Relict minerals are abundant not only in the zeolite but also in the prehnite–pumpellyite facies zone.

The isograds cross formational boundaries obliquely. The boundary between the zeolite and the prehnite–pumpellyite facies zone was probably at a depth of 2–4 km during the metamorphism. The temperature that prevailed on the boundary would be around 300–350 °C. Accordingly, the average geothermal gradient should have been 75–175 °C km^{-1}. Even if we make a liberal allowance for error in our estimation, the metamorphism is probably of the low-pressure type.

The rocks of the zeolite facies have no noticeably preferred orientation, whereas those in the prehnite–pumpellyite facies occasionally show preferred

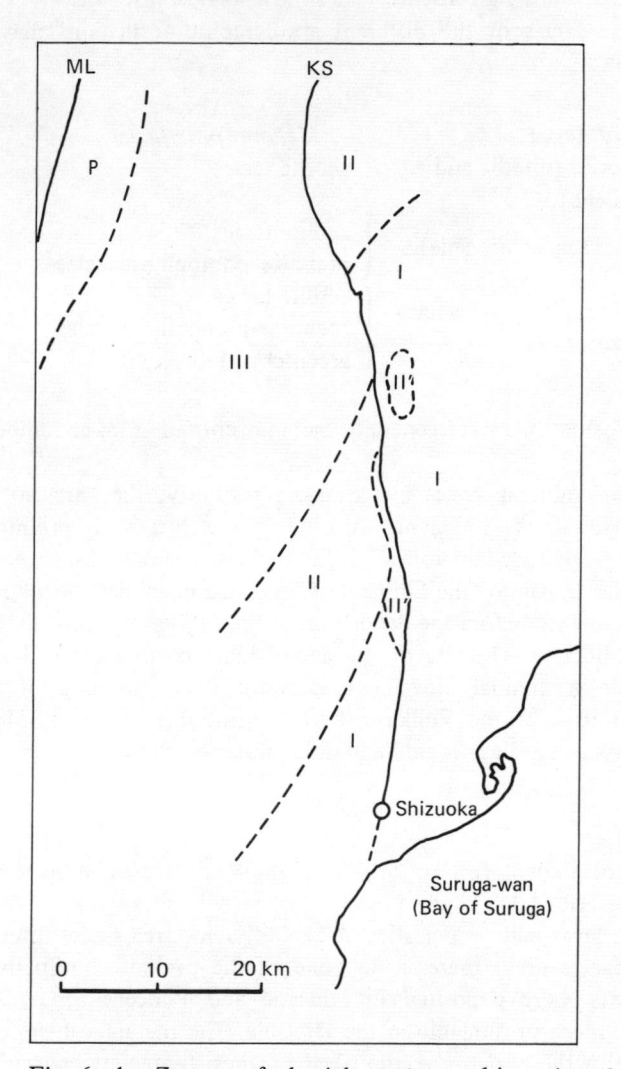

Fig. 6A-1 Zones of burial metamorphism in the eastern Akaishi Mountains. (Matsuda and Kuriyagawa, 1965.)

    I: Zeolite-facies zone.

    II: Prehnite-pumpellyite facies zone.

    II': A zone similar to II but more advanced in recrystallization.

    III: Greenschist facies zone.

P: Metamorphosed Palaeozoic formations; ML: Median Tectonic Line; and KS: Kobuchizawa-Shizuoka thrust.

orientation. Many rocks in the greenschist facies show weak schistosity.

The metamorphic facies of the different stratigraphic formations may be summarized as follows.

Formation (age)	Metamorphic facies
Fujikawa Series (middle and late Miocene)	zeolite facies
Misaka Series (early Miocene)	{ zeolite facies prehnite–pumpellyite facies
Shimanto Group (early Tertiary to Mesozoic)	{ zeolite facies prehnite–pumpellyite facies greenschist facies

Thus, the younger formations were generally metamorphosed at lower temperatures than the older.

Though the three mineral zones are arranged regularly, they are not in progressive relations. Rock fragments of the greenschist and prehnite–pumpellyite facies are included in rocks of the Fujikawa Series in the zeolite facies zone. Therefore, a part of the Shimanto Group had been metamorphosed and exposed on the surface before the deposition of the Fujikawa Series, that is, before the middle Miocene. The site of the depositional basin and the burial metamorphism made a gradual movement eastward from the area of the Shimanto Group to that of the Fujikawa Series during Tertiary time. Metamorphism in the Fujikawa Series should be middle Miocene or later.

Mogami Area, Japan

The metamorphism of a small Tertiary terrane of the Mogami area, in northeast Honshu has been described by Utada (1965). The area is about 6 km across, containing abundant lavas and pyroclastics. A zeolite facies area grades directly into a greenschist facies area; there is no zone of the prehnite–pumpellyite facies. The sediments were deposited in Miocene and Pliocene times. The metamorphism took place presumably in the Pliocene. The metamorphism as a whole appears to be of the burial type, though the highest temperature zone has small intrusions of granite, rhyolite, diorite and dolerite.

The terrane was divided into the following five zones in order of increasing depth (fig. 6A-2).

Zone I (unrecrystallized zone). Recrystallization is negligible. Even volcanic glass remains fresh. Interstitial spaces remain mostly open, though sometimes filled with opal.

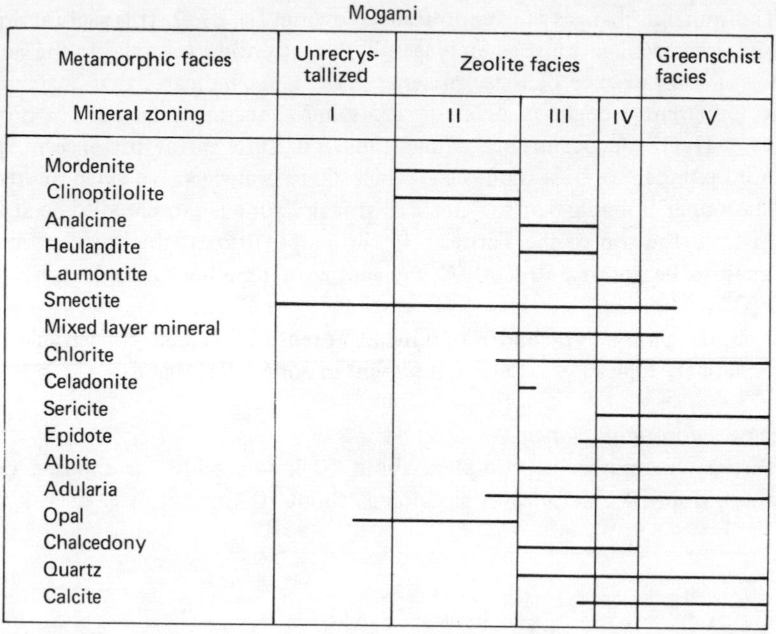

Mogami

Metamorphic facies	Unrecrystallized	Zeolite facies		Greenschist facies
Mineral zoning	I	II	III	IV	V

Mordenite
Clinoptilolite
Analcime
Heulandite
Laumontite
Smectite
Mixed layer mineral
Chlorite
Celadonite
Sericite
Epidote
Albite
Adularia
Opal
Chalcedony
Quartz
Calcite

Fig. 6A-2   Progressive mineral changes in pyroclastic rocks in the Mogami area, Japan. (Utada, 1965.)

Zone II (mordenite–clinoptilolite zone). Mordenite and clinoptilolite together with smectite replace glass. Interstitial spaces of pumice and vitric tuffs and of clastic rocks are commonly filled with opal and zeolites. Augite and hornblende are frequently replaced by smectite (saponite), but most plagioclase remains unchanged.

Zone III (analcime–heulandite zone). The opal, mordenite and clinoptilolite disappear, whereas analcime, heulandite, albite, adularia and chlorite begin to appear. Volcanic glass is mostly replaced by analcime and/or heulandite associated with clay minerals and sometimes by adularia. Vesicles in pumice are filled with analcime or heulandite associated with quartz. Augite and hornblende are replaced by chlorite, while plagioclase by albite and heulandite. Epidote begins to occur.

Zone IV (laumontite layer). This is actually a thin layer, about 2 m thick. It contains laumontite together with albite, adularia and quartz.

Zone V (chlorite–albite–epidote zone). Greenschist-facies mineral assemblages with chlorite, albite, epidote and calcite occur.

The mineral changes are synoptically shown in fig. 6A-2. It is said that quartz or opal is present in most assemblages, but no attention was paid to the possible effect of the presence of these minerals. The figure indicates that analcime and albite apparently begin to occur at the same zone. Since analcime + quartz = albite + H_2O, the occurrence of analcime and albite in the presence of quartz cannot be understood. It is not clear where there is always associated quartz.

The upper boundary of the greenschist facies zone is estimated to lie at about 3 km from the top of the Tertiary. If the temperature at the boundary may be assumed to be about 350–400 °C, the geothermal gradient must have been over $100\,^\circ\mathrm{C}\,\mathrm{km}^{-1}$.

Prehnite, pumpellyite and actinolite are absent in this area, conceivably owing to a relatively high P_{CO_2}. Calcite is present in zones III, IV and V.

Tanzawa Mountains, Japan

A Tertiary metamorphic complex, about 10 km in width, is exposed on the southern slope of the Tanzawa Mountains, about 70 km southwest of Tokyo and

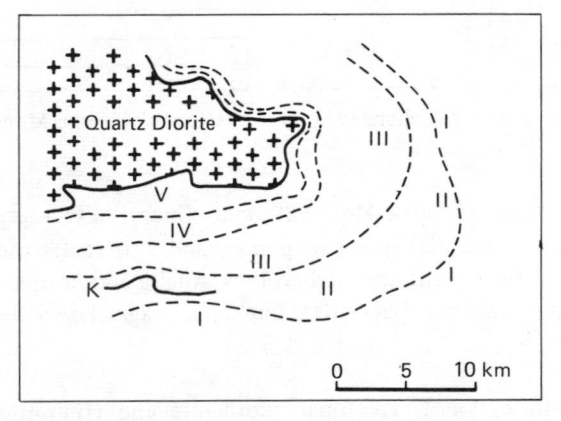

Fig. 6A-3 Zones of metamorphism in the Tanzawa Mountains, Japan. I-V represent Zones I-V respectively in fig. 6A-4. Line K is the boundary between the Tanzawa Group on the north and the Ashigara Group on the south. (Seki *et al.* 1969*a*.)

halfway between Tokyo and the Akaishi Mountains (fig 6A-3). The metamorphism ranges from the zeolite to the amphibolite facies. The high-temperature parts of the complex are strongly deformed and schistose, and even partly overturned, and have also been influenced by thermal effects of the associated mass of quartz diorite. On the other hand, the low-temperature rocks are non-schistose and are a product of burial metamorphism. The original rocks were mostly submarine pyroclastics and lavas of basaltic and andesitic compositions that formed in early Miocene to early Pliocene time. Seki *et al.* (1969*a*) have divided the terrane into the following five zones in order of increasing temperature of metamorphism (fig. 6A-4).

Zone I (low zeolite facies). Both sedimentary and mafic igneous rocks preserve their original texture and show no schistosity. Only the fine-grained matrix and groundmass are recrystallized. The recrystallized part is characterized by clinoptilolite, heulandite, stilbite and mordenite as well as by mixed-layer smectite–vermiculite and vermiculite–chlorite. Chlorite proper does not occur.

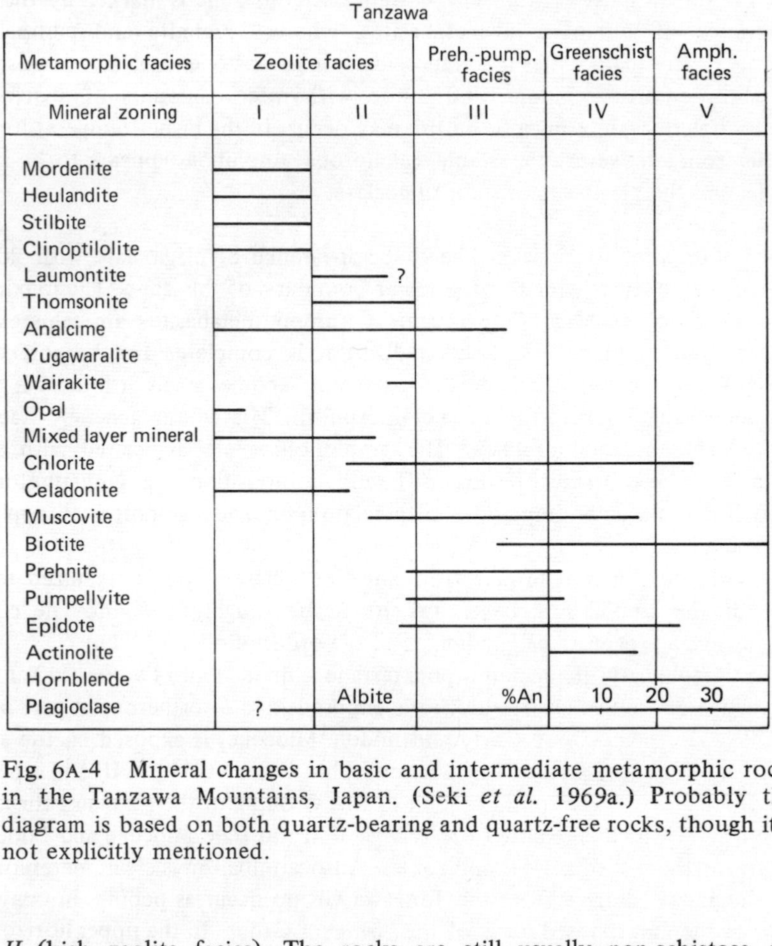

Tanzawa

Metamorphic facies	Zeolite facies		Preh.-pump. facies	Greenschist facies	Amph. facies
Mineral zoning	I	II	III	IV	V
Mordenite					
Heulandite					
Stilbite					
Clinoptilolite					
Laumontite		— ?			
Thomsonite					
Analcime					
Yugawaralite					
Wairakite					
Opal					
Mixed layer mineral					
Chlorite					
Celadonite					
Muscovite					
Biotite					
Prehnite					
Pumpellyite					
Epidote					
Actinolite					
Hornblende					
Plagioclase	?	Albite	%An	10 20 30	

Fig. 6A-4 Mineral changes in basic and intermediate metamorphic rocks in the Tanzawa Mountains, Japan. (Seki *et al.* 1969a.) Probably this diagram is based on both quartz-bearing and quartz-free rocks, though it is not explicitly mentioned.

Zone II (high zeolite facies). The rocks are still usually non-schistose and incompletely recrystallized. This zone is characterized by the occurrence of the laumontite–quartz assemblage. A mixed-layer clay mineral is common in the low-temperature half of this zone, whereas chlorite is common instead in the high-temperature half. Wairakite and yugawaralite were rarely found to occur in association with quartz in the higher half.

Zone III (prehnite–pumpellyite facies). The original textures are still preserved, though some rocks show weak schistosity. This zone is characterized by the occurrence of prehnite and pumpellyite. Chlorite, epidote, and albite also occur.

Zone IV (greenschist facies). The entrance to this zone is marked by the first appearance of actinolite in metabasites, whereas prehnite and pumpellyite disappear. The metabasites are greenschists or greenstones mainly composed of actinolite, chlorite, epidote, and albite with minor amounts of quartz and sphene. Calcite, white mica or biotite may occur. In the higher temperature part of this zone, however, the stable calciferous amphibole appears to be hornblende, and the plagioclase is sodic oligoclase.

Zone V (amphibolite facies). The first appearance of plagioclase with 20 per cent An has been regarded as the lower boundary of this zone. The maximum An content observed is 73 per cent. Common metabasites are schistose or non-schistose amphibolites. Recrystallization is complete. The hornblende is blue–green in the lowest temperature part, and becomes green or brownish green with increasing temperature of metamorphism. The colour generally deepens with advancing metamorphism. The amphibolites are associated with small amounts of metabasites having different compositions: a cummingtonite–plagioclase rock, an anthophyllite–plagioclase rock, and a grandite–salite–plagioclase rock.

The distance on the surface from the line where albite is stabilized to the onset of the amphibolite facies appears to be roughly 5–8 km. The corresponding average geothermal gradient is of the order of 40–60 °C km^{-1}.

Stratigraphically, the metamorphic terrane is divided into two units: Tanzawa and Ashigara Groups. Quartz diorite is exposed in the northernmost part of the area. The Tanzawa Group (early and middle Miocene) is exposed on the south side of it, and successively forms metamorphic zones V, IV and III and a part of zone II (fig. 6A-3). The southern limit of this Group is at the Kannawa thrust, to the south of which is exposed the Ashigara Group (late Miocene and Pliocene) that forms the rest of zone II and zone I. Autocannibalism is clear. Metamorphic rock fragments derived from the Tanzawa Group occur as pebbles in conglomerates of metamorphosed parts of the Ashigara Group. In the upper horizons of the Ashigara, quartz diorite also occurs as pebbles (Seki *et al.* 1969a, pp. 6–11).

It follows that the submarine volcanic deposition of the Tanzawa Group in early and middle Miocene time was probably simultaneous with an early major phase of metamorphism, which formed a progressive series from a low-temperature part up to zone V. Then, the quartz diorite mass was intruded, which resulted in some modifications in the highest temperature part. Later uplift and erosion had exposed the metamorphic and plutonic complex by late Miocene time. A new depositional basin was formed to the south of the uplifted mountains, and many fragments of the exposed rocks were transported southward and deposited, leading to the formation of the Ashigara Group. Metamorphism continued and was spatially extended into the newly formed sediments of the Ashigara. Thus, a part of zone II and zone I formed there in late Miocene and Pliocene time.

6A-3   ZEOLITE DISTRIBUTION IN ACTIVE GEOTHERMAL FIELDS

In many areas that have hot springs and fumaroles, wells have been drilled to provide steam and hot water, or rarely carbon dioxide. Mineralogical investigations of drill cores and cuttings gave unusual opportunities for correlating mineral assemblages with temperatures actually measured and other conditions. Comparison of such mineral assemblages with those in burial or regional metamorphism helps us understand the nature of metamorphic assemblages. The geothermal gradients in such areas are still higher than those in the low-pressure metamorphic areas described in the preceding section.

In active geothermal fields, zeolites occur in the following generalized order with increasing depth (in the presence of excess silica) : mordenite (and heulandite) → laumontite (and yugawaralite) → wairakite. This is very similar to the zeolite sequence in low-pressure metamorphic terranes, though a zone characterized by prehnite and pumpellyite has not been noticed.

Wairakei, New Zealand
A well-documented example of active geothermal fields is the Wairakei area in North Island of New Zealand (Steiner, 1953, 1955; Coombs *et al.* 1959). The following zones were observed in hydrothermally altered rhyolite tuffs and breccias, from the surface downwards:

(a) *Acid leached zone.* The characteristic minerals are kaolinite, alunite and opal.
(b) *Montmorillonite zone.* Rhyolite glass is altered to montmorillonitic clays.
(c) *Upper zeolite zone* (roughly 150–200 °C). Mordenite (ptilolite) is characteristic, filling vesicles in pumice. It is sometimes accompanied by minor heulandite.
(d) *Lower zeolite zone* (roughly 200–230 °C). Wairakite replaces andesine, and also fills veins and cavities by direct crystallization from solution.

Laumontite was found roughly in a zone transitional between the upper and lower zeolite zones (roughly 195–220°C).

(e) *Feldspathization zone* (roughly 230–250 °C). An upper part is characterized by albitization and a lower one by adularia replacing plagioclase. Rarely prehnite occurs, and calcite is locally precipitated.

(f) *Hydromica zone.* Primary plagioclase is replaced by a hydromica as a result of leaching of alkalis by rising carbon dioxide solutions.

The hydrothermal solutions are rich in silica which precipitates at all levels. The temperature and depth of each zone differ in different wells. The depth of the upper and lower zeolite zones ranges roughly from 90 to 400 m. The geothermal gradient is roughly 200–1000 °C km^{-1}.

The hot solutions in the deepest zone tapped have a high concentration of K, which is progressively lowered by the formation of adularia. In contrast, the concentrations of Na and Ca are increased until some of these ions are precipitated in the zone of zeolite formation. In the deeper zones, the water contains sufficient CO_2 to make it slightly acid with respect to pure water at the same temperature. Towards the surface, the CO_2 is lost from the water, which hence becomes slightly alkaline.

In Wairakei, the pressure of the aqueous fluid measured in fissures is often close to that of the water column, that is, much lower than the rock-pressure. It corresponds to model (c) discussed in § 2-3.

Onikobe, Japan
Seki (1966, 1969) and Seki, *et al.* (1969b) studied drill cores from the Onikobe active geothermal area in northeast Japan. The following zones were noticed in descending order.

(a) *Native sulphur zone.*

(b) *Smectite zone.*

(c) *Mordenite zone.* Analcime and albite also occur.

(d) *Laumontite zone.* Yugawaralite, analcime, and albite occur in some rocks.

(e) *Wairakite zone.* Albite also occurs.

(f) *Albite zone.*

Quartz is present in all rocks. The depth and temperature of each zone differ in different drill holes. Laumontite occurs in the range of 80–240 m and 80–170 °C. The geothermal gradient is roughly 300–1000 °C km^{-1}.

Geothermal Fields Unfavourable to Zeolite Formation
The composition of the hot water ascending from depths differs in different fields, and should have important effects on the formation of zeolites and other

minerals. In some geothermal fields, there is scarcely any formation of zeolites, probably owing to a higher concentration of CO_2 and other chemical characteristics of the water (§4-4).

For example in the North Island of New Zealand, there is another well-documented geothermal field in the Waiotapu area. Though the area is near Wairakei, the observed mineral sequence differs greatly. In terms of the alteration of plagioclase in acidic volcanics, three zones were noticed in descending order as follows: (a) zone of sulphuric acid alteration, characterized by alunite, (b) zone of potassium silicate alteration characterized by the formation of hydromica and K-feldspar, and (c) zone of albitization associated with formation of calcite. Wairakite occurs occasionally in the zone of albitization. No other zeolites occur (Steiner, 1963).

No zeolites were found in the geothermal field of the Salton Sea area, California, which is characterized by solutions containing abundant CO_2 (§ 1-5).

6A-4   ZEOLITE AND PREHNITE-PUMPELLYITE FACIES TERRANES OF THE MEDIUM-PRESSURE TYPE

Coomb's zeolite facies terrane of the Taringatura area, New Zealand, had a much lower geothermal gradient than the areas discussed above. The difference in the varieties of zeolites observed is partly related to this difference in geothermal gradient. Mordenite, stilbite, yugawaralite and wairakite were not found, whereas heulandite, laumontite and analcime do occur. This terrane is a part of the New Zealand geosyncline, other parts of which contain lawsonite and crossite. Hence, the zeolite facies metamorphism of this area is probably of the medium-pressure type, close to the high-pressure type.

The Sanbagawa belt of Japan is largely of the high-pressure type. However, no glaucophane occurs in the central Kii Peninsula, where instead zeolite facies rocks occur. Therefore, the central Kii terrane also is probably of the medium-pressure type, relatively close to the high-pressure type. Laumontite was the only zeolite found there.

Judging from the incompatibility of glaucophane and zeolites in these cases, perhaps a zeolite facies zone hardly form in high-pressure metamorphic areas, and at least in typical high-pressure ones. (Relevant data concerning zeolites in the Franciscan Formation and a prehnite–pumpellyite facies zone in high-pressure terranes will be reviewed in § 6A-5.)

Taringatura Area, New Zealand
This area is situated near the southern end of South Island (Coombs, 1954; Coombs et al. 1959). The rocks form a part of the sediment pile of the New Zealand geosyncline, being Triassic in age and about 10 km thick. They are

mainly composed of tuffs and volcanic graywackes of the andesite–dacite–rhyolite group.

In the upper part of the stratigraphic section, recrystallization is poor. However, glass in tuff beds has been completely replaced by heulandite or less commonly by analcime (fig. 6A-5). Both zeolites can coexist with quartz and fine-grained phyllosilicates. Igneous plagioclase remains unchanged. In successively lower horizons, however, plagioclase is increasingly replaced by pseudomorphs of dusty albite with sericite inclusions. Simultaneously with the

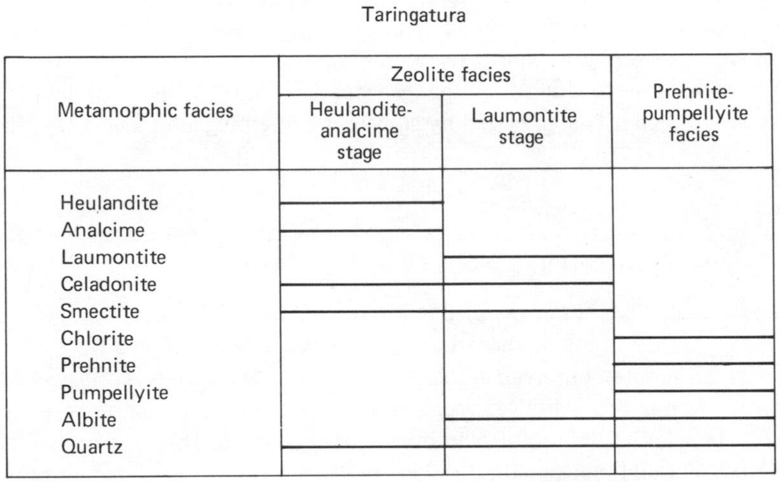

Taringatura

Metamorphic facies	Zeolite facies		Prehnite-pumpellyite facies
	Heulandite-analcime stage	Laumontite stage	
Heulandite	▬		
Analcime	▬		
Laumontite		▬	
Celadonite	▬▬▬		
Smectite	▬		
Chlorite			▬
Prehnite			▬
Pumpellyite			▬
Albite		▬▬	
Quartz	▬▬▬▬▬▬		

Fig. 6A-5 Progressive mineral changes of volcanic graywackes and tuffs in the Taringatura area, New Zealand. (Coombs, 1954; Coombs *et al.* 1959.)

replacement, analcime and heulandite disappear, and laumontite associated with quartz begins to occur. Adularia also occurs. Toward the base of the section, pumpellyite and prehnite appear.

Thus, with regard to recrystallized mineral assemblages, the following three stages in order of increasing temperature were noticed.

(*a*) Zeolite facies; heulandite–analcime stage (upper 5 km):
Quartz + heulandite + smectite
Quartz + heulandite + celadonite
Quartz + analcime + celadonite
(*b*) Zeolite facies; laumontite stage:
Quartz + laumontite + albite + celadonite + adularia
Quartz + albite + smectite + adularia

(c) Prehnite–pumpellyite facies:
   Quartz + albite + pumpellyite + adularia
   Albite + chlorite + pumpellyite
   Quartz + albite + chlorite + prehnite

If the post-Triassic cover that has been removed may be assumed to have been a few km thick, the reaction analcime + quartz = albite + H_2O at the boundary between the heulandite–analcime and the laumonite stage appears to have taken place at a depth of about 8 km. If we may regard this reaction as taking place at about 180°C, the goethermal gradient is estimated at about 22°C km^{-1}.

Canterbury Region, New Zealand

Another wide terrane of graywackes which occurs in Canterbury, South Island, has been recrystallized at a grade to produce prehnite and pumpellyite (Coombs et al. 1959). This is probably in the prehnite–pumpellyite facies, and is a part of the accumulations in the New Zealand geosyncline. Toward the west, it appears to grade into a greenschist facies zone, which grades into an epidote–amphibolite facies zone, which in turn grades into an amphibolite facies zone. This group of schists is called the Alpine schists.

In the transitional part between the prehnite–pumpellyite facies and the greenschist facies, prehnite disappears first, and then pumpellyite does. The first appearance of actinolite takes place roughly at the same temperature as the disappearance of prehnite.

No intrusive igneous rocks are present. Penetrative deformation was intense in the zone of the greenschist and higher facies, but was weak in the prehnite–pumpellyite facies zone.

Central Kii Peninsula, Japan

Though the Sanbagawa belt is generally of the high-pressure type, an area in the central part of the Kii Peninsula (fig. 15-8) within the belt has no glaucophane, nor lawsonite, nor jadeite. There has developed, instead, the following series of metamorphic facies: zeolite (laumontite) facies → prehnite–pumpellyite facies → greenschist facies (Seki et al. 1971).

This area is made up of late Palaeozoic and Jurassic sediments with interbedded mafic volcanics. The degree and temperature of recrystallization increase northward over a distance of about 40 km. There are no plutonic rocks influencing the temperature distribution. Metaclastic rocks show only slight mineralogical changes, being graphite-white mica-chlorite-albite-quartz schist with porphyroblastic development of albite in the highest temperature part. On the other hand, metavolcanic rocks show marked progressive changes (Fig. 6A-6). On the basis of their mineral assemblages, the following zones have been delineated from south to north:

Central Kii Peninsula

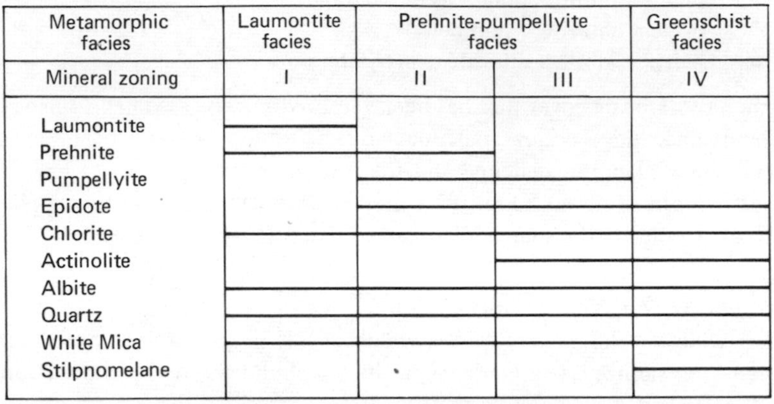

Metamorphic facies	Laumontite facies	Prehnite-pumpellyite facies		Greenschist facies
Mineral zoning	I	II	III	IV
Laumontite	▬			
Prehnite	▬	▬		
Pumpellyite		▬	▬	
Epidote		▬	▬	▬
Chlorite	▬	▬	▬	▬
Actinolite			▬	▬
Albite	▬	▬	▬	▬
Quartz	▬	▬	▬	▬
White Mica	▬	▬	▬	▬
Stilpnomelane				▬

Fig. 6A-6   Progressive mineral changes in the central Kii Peninsula in the Sanbagawa metamorphic belt. (Seki *et al.* 1971.)

(a) *Zone of laumontite.* The characteristic metamorphic assemblage in metavolcanics is: quartz + albite + laumontite + chlorite. No zeolites other than laumontite have been found.

(b) *Zone of prehnite-pumpellyite facies.* Neither laumontite nor actinolite occurs. The characteristic assemblage is: quartz + albite + prehnite + pumpellyite + epidote + chlorite.

(c) *Transitional zone with both pumpellyite and actinolite.* Prehnite disappears and actinolite begins to appear at the threshold to this zone. Since pumpellyite is still present, this zone is regarded as belonging to the prehnite–pumpellyite facies, and not to the greenschist facies. The characteristic assemblage is: quartz + albite + pumpellyite + epidote + actinolite + chlorite.

(d) *Zone of greenschist facies.* Pumpellyite disappears and the following greenschist-facies assemblage appears: quartz + albite + epidote + actinolite + chlorite + stilpnomelane.

The whole rock Rb-Sr age of metamorphic rocks of this area is about 110 m.y. (Yamaguchi and Yanagi, 1970).

6A-5   RELEVANT   DATA   FROM   HIGH-PRESSURE   METAMORPHIC TERRANES

A zone of the prehnite–pumpellyite facies occurs in the lowest temperature part of the high-pressure metamorphic terrane in the Katsuyama area, Japan, but no zone of the zeolite facies has been observed. The significance of laumontite in the Franciscan metamorphics is not clear.

Katsuyama Area, Japan

The Sangun metamorphic belt of Japan contains sporadically glaucophane-bearing areas but no zeolite facies areas, though a thorough search has not been made.

In the Katsuyama area (fig. 15-8) within the belt in western Honshu, Hashimoto (1968) found a zone of the prehnite–pumpellyite facies which grades with increasing temperature into a zone of the glaucophane-schist facies. Prehnite is confined to the former zone, whereas pumpellyite occurs in both. The glaucophane-schist facies zone in turn grades with increasing temperature into a greenschist facies zone with actinolite but without pumpellyite.

Franciscan Formation, California

In the Panoche Pass and Pacheco Pass areas of the Franciscan Formation, California (Ernst, 1965; Ernst et al. 1970), a well-recrystallized zone having glaucophane, lawsonite, jadeite and aragonite, grades into an area where recrystallization is less advanced, and pumpellyite occurs besides albite, calcite, actinolite, chlorite and stilpnomelane. This might be regarded as a transition from the typical glaucophane-schist facies toward the prehnite–pumpellyite facies (fig. 7A-15).

In the western part of the Franciscan terrane in California, laumontite is widespread. The relation of this part to the glaucophane-schist facies area to the east has not been clarified (Bailey et al. 1964). Apparently, the laumontite-bearing rocks in the west, the pumpellyite-bearing rocks in the central part, and the typical glaucophane-schist facies rocks in the east form a progressive metamorphic facies series. However, it is possible that the laumontite-bearing rocks in the west were deposited and metamorphosed at a later time than the typical glaucophane-schist facies rocks in the east (fig. 13-3).

6A-6   OTHER AREAS

Aragonite-bearing Prehnite-Pumpellyite Facies Area in Washington State

The occurrence of aragonite in a burial metamorphic terrane is of some interest, since aragonite is stable only under high rock-pressures. Vance (1968) described it in the San Juan Islands, west of Bellingham in the State of Washington. The terrane is mainly composed of Upper Palaeozoic and Lower Mesozoic pyroclastic and sedimentary rocks, which were subjected to recrystallization in the prehnite–pumpellyite facies in Mesozoic time. Metabasalts and meta-andesites show the mineral assemblage chlorite + albite + prehnite + pumpellyite. Prehnite is more abundant than pumpellyite. Epidote was rarely found.

The limestone lenses in this terrane are composed of aragonite and calcite. The largest lens measures about 200 m in length. Vance considers from structural data that the rock-pressure of the prehnite–pumpellyite facies terrane was probably lower than the experimentally determined stability field of aragonite. He doubted the validity of the use of aragonite as a criterion of high rock-pressure. Newton, Goldsmith and Smith (1969) have supported this argument by demonstrating that deformed calcite can be transformed into aragonite at a lower pressure than undeformed one (see § 7B-19).

Quartz-free Zeolitic Rocks in Eastern Iceland

Quartz occurs in most rocks of the petrographically well-documented areas of the zeolite facies. Our knowledge on quartz-free zeolite assemblages is very meager.

A variety of quartz-free zeolitic rocks was described by Walker (1960) in the Tertiary basaltic lava plateau of eastern Iceland. The lava piles are roughly 5 km thick, and were subjected to rise of temperature sufficient to cause weak recrystallization in the zeolite facies. In the recrystallized olivine basalt lavas, silica minerals are absent, even in amygdales. These rocks contain such zeolites as gismondine, thomsonite, chabazite, scolecite and phillipsite, which all are stable probably only in the absence of free silica (§ 6B-1).

Zones based not only on the kinds, but also on the relative abundance, of zeolites have been distinguished.

(a) *Unzeolitized basalt zone* (about 200 m thick).

(b) *Slightly zeolitized zone.* Chabazite and thomsonite are common, and levyne, phillipsite, and gismondine also occur in amygdales.

(c) *Analcime-heulandite zone* (about 150 m thick). Analcime, mesolite, chabazite, thomsonite, levyne, phillipsite, stilbite and heulandite occur in amygdales.

(d) *Scolecite–laumontite zone.* This extends from a maximum altitude of 760 m down to sea level. Zeolitization is most intense here. Zeolites occur not only in amygdales but also in interspaces between or replacing igneous minerals in ordinary parts of rocks. The zeolites include mesolite, scolecite, chabazite, thomsonite, analcime, stilbite, heulandite and levyne, and in the lower part laumontite.

A strongly hydrated zeolite, gismondine, occurs only in the uppermost zone, whereas relatively weakly hydrated zeolites, scolecite and laumontite, occur in the lowest zone. The zeolite zones are probably approximately parallel to the original top of the lava pile and clearly cut across the stratification of the lavas, and zeolitization should have taken place at a time long subsequent to the effusion and cooling of individual lavas.

Quartz-free Zeolite Assemblages in the Mid-Atlantic Ridge near 24°N

A large amount of quartz-free metabasalts of the zeolite facies was dredged from a transverse fracture zone near 24°N on the Mid-Atlantic Ridge (Miyashiro *et al.* 1971). These metabasalts were probably recrystallized at some depth and were caught and brought up by serpentinite masses rising along the fracture zone.

Natrolite, thomsonite, analcime, chabazite, laumontite and stilbite were identified, among which analcime is the most widespread. Natrolite and thomsonite occur due to the absence of quartz in these rocks. Natrolite, being deficient in silica, does not coexist with stilbite rich in silica. The zeolite assemblages are considered to be close to chemical equilibrium. Laumontite and chabazite, differing only in H_2O content, occur. This suggests that the dredge hauls contain metabasalts recrystallized over a considerable range of temperature.

The original basalts were probably abyssal tholeiites. Recrystallization under zeolite facies conditions was accompanied by introduction of Na_2O and H_2O. The Na_2O content varies sympathetically with the H_2O content.

The mafic minerals stable in zeolite facies metabasalts are smectite and mixed-layer minerals made up of chlorite, smectite and vermiculite components. Chlorite proper does not occur.

Chapter 6B

Zeolite and Prehnite - Pumpellyite Facies Metamorphism: Progressive Mineral Changes

6B-1   PROGRESSIVE CHANGES IN ZEOLITE ASSEMBLAGES

Though considerable amounts of petrographic and synthetic data are available on zeolites, they have not given a clear picture of the progressive changes in zeolite assemblages. It is regrettable that most existing descriptions are of individual zeolites and not of zeolite-bearing assemblages.

Miyashiro and Shido (1970) have therefore attempted to formulate a theoretical scheme for the possible stability relations of zeolites on the basis of a few basic assumptions and the available petrographic data, as discussed below. Since necessary petrographic data are not always available, the scheme is not complete, and will probably have to be modified in the light of future progress in the study of zeolite assemblages.

It is assumed in this scheme that zeolites in metamorphic rocks form in equilibrium with a fluid phase composed of virtually pure H_2O, and that the pressure of the fluid phase is equal to that acting on the solid phases. In some cases, however, a fluid phase may contain a large proportion of CO_2 (Zen, 1961b), may be highly saline (Hay, 1966), or may not exist. The present scheme is not applicable to such cases. We further assume that a rise in temperature always tends to form more strongly dehydrated zeolite assemblages.

The zeolites in general are a large group of hydrated alumino-silicates of alkalies and alkaline earths. Some of them contain a large amount of Ba, Sr, or K. However, most are poor in these elements, and may be approximately regarded as being composed of the following four components: An (= $CaAl_2Si_2O_8$), Ne (= $Na_2Al_2Si_2O_8$), SiO_2 and H_2O. In zeolite structures, Ca^{2+} can usually be replaced by $2Na^+$. Only the zeolites composed of these four components will be treated here. Such zeolites will be divided into two groups: Ca-zeolites composed of An, SiO_2 and H_2O and Na-zeolites composed of Ne, SiO_2 and H_2O. This division is artificial, since natural zeolites always contain both Ca and Na. We assume, however, that the main features of the stability relations of zeolites are preserved in this simplification.

According to the phase rule, the maximum number of stable phases coexisting over a range of temperature and of pressure in a three-component system is three. Since we have assumed the existence of an aqueous fluid phase, the maximum number of the coexisting mineral phases is only two. Thus, the stable assemblages should be represented by smaller triangles having an apex at the H_2O corner of the triangular diagram $An-SiO_2-H_2O$, or $Ne-SiO_2-H_2O$. The position of the two mineral phases in stable assemblages should move away from the H_2O corner with progressive dehydration.

Ca-Zeolite Assemblages

The composition relations of Ca-zeolites are shown in fig. 6B-1. Possible steps of dehydration are shown in fig. 6B-2. Brief comments will be made on the individual steps.

Step (a). The Ca-zeolite assemblages with the highest possible H_2O contents are quartz-stilbite, stilbite-chabazite and chabazite-gismondine, as shown in fig. 6B-2a. These assemblages are thought to represent the lowest temperature of recrystallization. Stilbite occurs in the lowest temperature part of the Tanzawa area (fig. 6A-4). Chabazite and gismondine occur in the low-temperature part of basaltic piles of eastern Iceland (Walker, 1960).

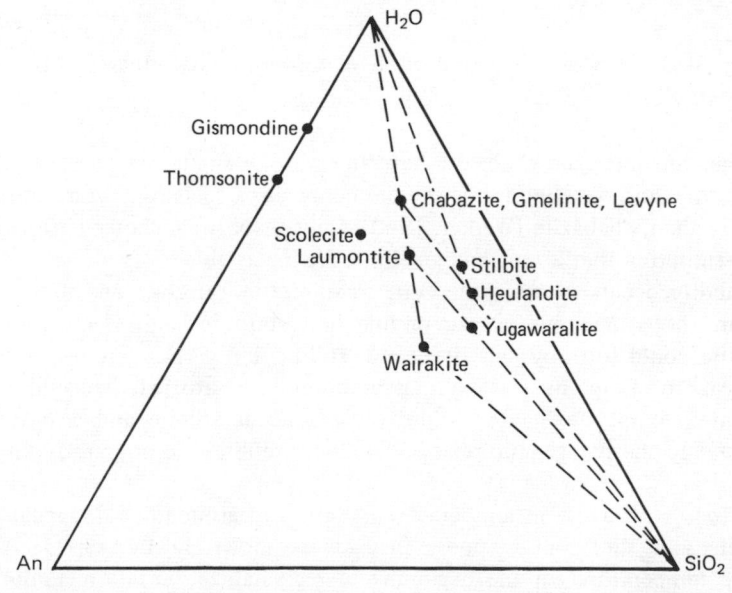

Fig. 6B-1   Composition relations of Ca-zeolites. An = $CaAl_2Si_2O_8$.

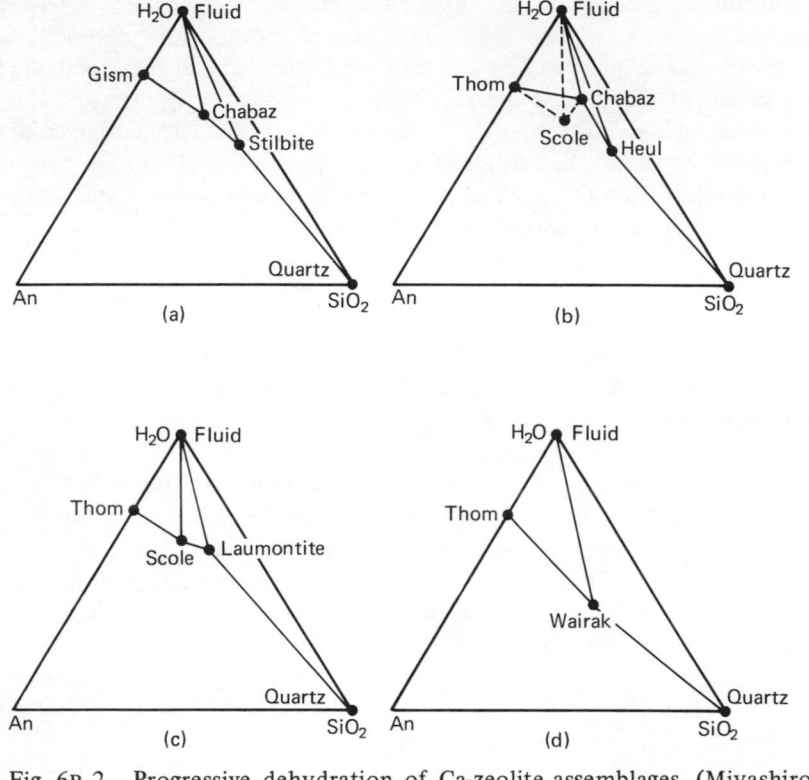

Fig. 6B-2   Progressive dehydration of Ca-zeolite assemblages. (Miyashiro and Shido, 1970.)

Levyne, gmelinite and chabazite have the same or similar chemical composition. Levyne and gmelinite occur commonly in association with, but less frequently than, chabazite (Walker, 1960). Only chabazite is shown in fig. 6B-2*a* on the assumption that levyne and gmelinite are metastable.

Heulandite occurs in the same zone as stilbite in Tanzawa, and thomsonite occurs in the same zone as gismondine in eastern Iceland. Heulandite and thomsonite could form by dehydration of stilbite and gismondine respectively, as is clear from fig. 6B-1. If our assumptions are satisfied, heulandite and thomsonite cannot be stable in the same zone as stilbite and gismondine. Unfortunately the paragenetic relations of these zeolites are not clearly known.

Step (b). With a rise in temperature, stilbite and gismondine disappear, and heulandite and thomsonite appear instead, as shown in fig. 6B-2*b*. If the threshold temperature for the beginning of recrystallization in a metamorphic terrane lies at this step, stilbite and gismondine would not form at all, as in

Taringatura (fig. 6A-5) and Mogami (fig. 6A-2). In the higher temperature part, scolecite may begin to form, as shown by broken lines in the figure.

Step (c). With a further rise of temperature, chabazite and heulandite become unstable, and laumontite appears instead, as shown in fig. 6B-2c. The laumontite + quartz assemblage is stable in the higher temperature part of the zeolite facies zone in Taringatura (fig. 6A-5) and Tanzawa (fig. 6A-4). Scolecite and thomsonite should occur in quartz-free rocks, though pertinent petrographic data are not available.

Step (d). Wairakite occurs in the highest temperature part of the zeolite facies zone of Tanzawa (fig. 6A-4). If our basic assumptions are satisfied, laumontite cannot be stable in this part. Wairakite is the only zeolite in drill cores from the deepest part of the active geothermal fields of Wairakei in New Zealand and Onikobe in Japan.

The zones observed in individual areas may be correlated roughly with the above steps as summarized below.

Taringatura (Coombs *et al.* 1959)	(b) (c)
Tanzawa Mountains (Seki *et al.* 1969*a*)	(a) (b) (c) (d)
Eastern Iceland (Walker, 1960)	(a) (b) (c)
Wairakei geothermal field (Coombs *et al.* 1959)	(b) (c) (d)
Onikobe geothermal field (Seki *et al.* 1969*b*)	(c) (d)

Na-Zeolite Assemblages

A similar line of discussion has been put forward for Na-zeolites by Miyashiro and Shido (1970). The composition relations and four possible steps are shown in figs. 6B-3 and 6B-4 respectively.

The analcime + quartz assemblage in the heulandite–analcime stage of Taringatura (fig. 6A-5) corresponds to step (c$'$), and the occurrence of albite in the laumontite stage to step (d$'$) in fig. 6B-4.

6B-2 EXPERIMENTAL STUDIES ON THE STABILITIES OF ZEOLITES

Many authors have carried out synthetic experiments on zeolites. However, most experiments show metastable phase relations. Charges in starting materials and in time duration give a great variation in the results (e.g. Coombs *et al.* 1959; Koizumi and Roy, 1960). Only a few experimental studies giving important results will be reviewed here.

Synthetic analcime shows a wide range of solid solution from the albite-like composition $NaAlSi_3O_8 \cdot 1 \cdot 5H_2O$, through the idealized analcime composition

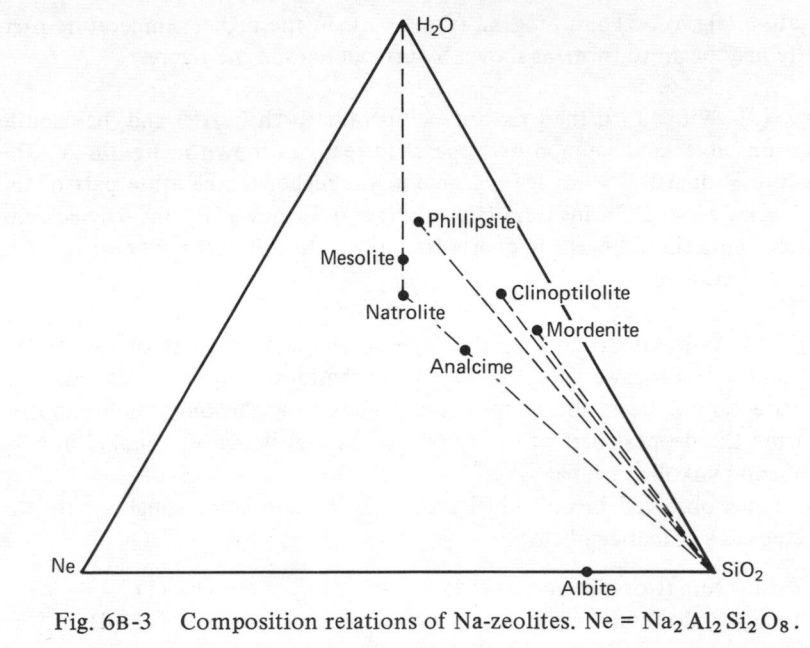

Fig. 6B-3   Composition relations of Na-zeolites. Ne = $Na_2 Al_2 Si_2 O_8$.

$NaAlSi_2 O_6.H_2 O$, to the natrolite-like composition $NaAlSi_{1.5}O_5.0.75H_2 O$ (Saha, 1961).

Natural analcime appears to range from Na : Al : Si = 1:1:2.7 to 1:1:2 composition. Analcimes formed by reaction of siliceous volcanic glass with water commonly have a 1:1:2.5 or more siliceous composition. They may be metastable, having formed in environments with an unusually high chemical potential of silica. Analcimes formed in environments free from siliceous volcanic glass and quartz usually have a composition around 1:1:2 (Coombs and Whetten, 1967).

The equilibrium curve for the reaction analcime + quartz = albite + $H_2 O$ corresponds to the boundary between step (c') and (d') in fig. 6B-4. According to field evidence in Taringatura, this agrees approximately with the boundary between steps (b) and (c) in fig. 6B-2. The equilibrium curve for the reaction has a negative slope above 0.4 kbar, as shown in fig. 6B-5 (Campbell and Fyfe, 1965; Thompson, 1971).

The equilibrium curve for the breakdown reactions of laumontite and wairakite, determined by Liou (1970), are also shown in the same figure. These correspond to the boundary between steps (c) and (d) and the upper limit of step (d) respectively in fig. 6B-2. Liou (1971b) gives a breakdown curve of stilbite into laumontite + quartz.

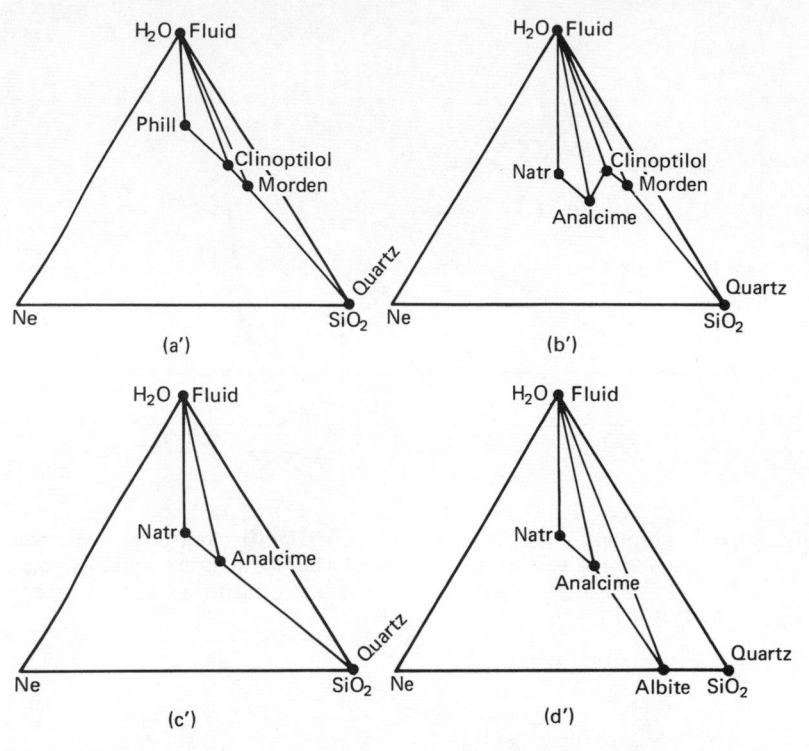

Fig. 6B-4 Progressive dehydration of Na-zeolite assemblages. (Miyashiro and Shido, 1970.)

In § 2-4, when the equilibrium curves of dehydration reactions were discussed, it was pointed out that they usually have a very small positive slope at pressures below 0·1 kbar, show a marked bend in the pressure range 0·1–3·0 kbar, and become very steep at pressures above 3 kbar. At pressures above 10 kbar, an aqueous fluid is so highly compressed that equilibrium curves may come to have a negative slope. The volume changes of the solid phases in dehydration reactions of certain zeolites are greater than those in other dehydration reactions. This tends to cause some zeolite dehydration curves to turn to a negative slope at a relatively low pressure. The curve for the reaction analcime + quartz = albite + H_2O is an example. On the other hand, the breakdown curve of laumontite has a positive slope.

A possible configuration of equilibrium curves for Ca-zeolites based on their volume relations is shown schematically in fig. 6B-6. The high-pressure parts of the stability fields of Ca-zeolites are cut off in natural rocks by the fields of lawsonite, pumpellyite, prehnite and epidote, as shown for example in figs. 6B-5 and 7B-14.

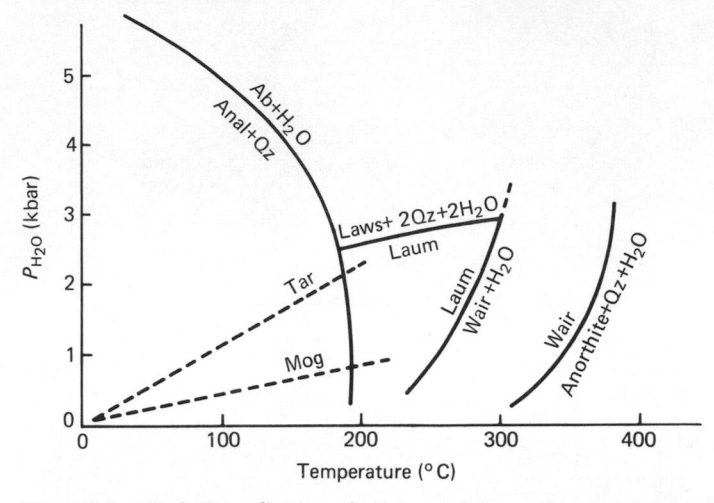

Fig. 6B-5 Stability fields of the analcime + quartz assemblage and laumontite (Campbell and Fyfe, 1965; Liou, 1970). Broken lines 'Tar' and 'Mog' indicate possible geothermal curves of the Taringatura and Mogami areas. The high-pressure stability limit of laumontite is shown after Thompson (1970*b*).

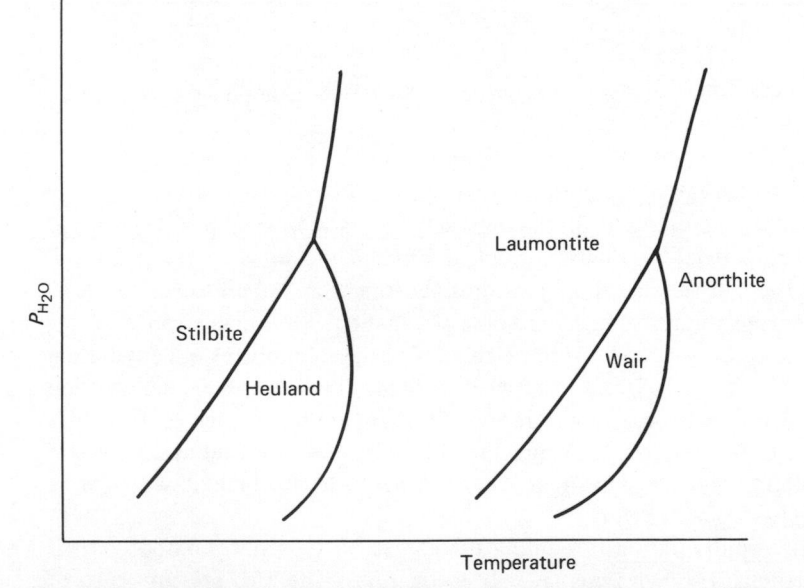

Fig. 6B-6 Possible form of phase boundaries between Ca-zeolites in the presence of an aqueous fluid. (Modified from Coombs *et al.* 1959.) 'Heuland' and 'Wair' mean heulandite and wairakite. The high-temperature and low-pressure part of this figure is more precisely shown in fig. 6B-5.

6B-3   SMECTITE, ILLITE AND MIXED-LAYER CLAY MINERALS

In rocks of the low greenschist facies, chlorite and muscovite are the most common phyllosilicates. These two minerals continue to be stable down to the prehnite–pumpellyite facies. In the zeolite facies, chlorite disappears, whereas smectite, illite and various members of mixed-layer clay minerals appear.

The name smectite (montmorillonoids) here means minerals of the montmorillonite group, including montmorillonite, nontronite and saponite. The name illite means mica-type clay minerals, including white mica (sericite) and celadonite (glauconite).

Identification of these minerals requires special techniques. A conventional procedure of sample preparation for the X-ray method is as follows: a rock is crushed into a fine powder, which is dispersed in water and left to settle for several hours. Fine-grained clay minerals remain suspended in the upper part of the water column. This suspension is then settled out by the centrifuge. To make the identification of smectite and vermiculite certain, the following pre-treatment before X-ray study is strongly recommended. A part of the sample is treated with MgCl solution and another treated with KCl or NH_4NO_3 solution to obtain Mg^{2+}- and K^+- or NH_4^+-saturated clays. When treated with NH_4NO_3, several hours of boiling in about 1N solution is recommended to assure the saturation of a certain type of vermiculite with NH_4NO_3. After these treatments, the samples are washed repeatedly with distilled water and concentrated with the centrifuge. They are then transferred onto slide glasses and dried in air. The mounted samples are glyceroled (i.e. immersed in glycerol for more than several hours) or heated at 550–600 °C for about 1 hour. The changes in angles and relative intensities of their basal reflections are clues for identification (table 6B-1).

The refractive indices of smectite and vermiculite vary according to the immersion medium. When measured values are reported, the liquid in which they were determined should be recorded. Illites and chlorites do not show such a variation.

Present-day marine sediments contain smectite, kaolinite, illite, and chlorite. Smectite is abundant where volcanic ash has been an important source material. Kaolinite tends to be abundant near shore. At least some of illite in sediments seems to be detrital. Recent K–Ar dating of deep-sea illitic clays has given various geologic ages which suggest an airborne derivation from continental regions. Discrete chlorite does not occur in zeolite facies rocks, but does occur in rocks recrystallized at higher temperatures. Therefore, chlorite in present-day sediments is probably of detrital origin. Chlorite is abundant in sediments on ocean floors at high latitudes, because the mineral contained in exposed continental metamorphic rocks is transported and deposited without chemical weathering owing to the cold climate (Griffin, Windom and Goldberg, 1968).

Table 6B-1 *Some characteristics used in identification of low-temperature phyllosilic*

Treatment {	Untreated			M
	Dried in air	Heated at 300°C	Heated at 550°C	Dried in air
Smectite	12·4-15·4	9·5-9·8	9·5-9·8	15·4
				$I(001) > I(002)$
Vermiculite	12·6-15	9 (quickly rehyd. in air)	9	14·2-14·4 $I(001) \gg I(002)$
Chlorite	14·0-14·3 $I(001), (003) <$ $I(002), (004)$	X	13·8-14·1 $I(001) \gg I(002)$ etc.	X
Swelling chlorite	14·0-14·3 $I(001), (003) <$ $I(002), (004)$	X	13·8-14·1 $I(001) \gg I(002)$ etc.	X
Illite	10	X	X	X
Kaolinite	7·15	X	X	X
Halloysite	7·2	X	Collapse	X

X = No change as compared with sample untreated and dried in air.
'etc.' = (002) and other higher order reflections.

Thus, smectite, kaolinite and illite appear to be stable in unmetamorphosed pelites. Slight recrystallization under zeolite facies conditions would begin to form mixed-layer minerals with a chlorite component. Then, with a further increase in temperature, chlorite as a discrete mineral forms in the prehnite–pumpellyite and the glaucophane-schist facies.

In a few sedimentary piles it has been observed that smectite is abundant in the uppermost layer, 1–2 km thick, and it is joined by mixed-layer smectite-illite at a deeper level, until smectite ultimately disappears at a depth of a few kilometres (Burst, 1959; Powers, 1959). However, the chemical environment would also have an important effect on their stability. The relationship between physical and chemical controls has not been analysed.

6B-4 PREHNITE AND PUMPELLYITE

These two minerals are hydrosilicates of Ca and Al, though pumpellyite contains considerable amounts of Mg, Fe^{2+} and Fe^{3+} also.

Prehnite $\qquad$ $Ca_2Al_3Si_3O_{10}(OH)_2$

Pumpellyite $\qquad$ $Ca_4(Mg, Fe^{2+})(Al, Fe^{3+})_5Si_6O_{23}(OH)_3 \cdot 2H_2O$

They are chemically similar to wairakite, laumontite, chabazite, scolecite,

| ing (Å) | | | | Optic properties | |
| ated | | K$^+$- or NH$_4^+$-saturated | | | |
nmersed in water	Glyceroled	Dried in air	Glyceroled	$\gamma - \alpha$	Sign of elongation
> 18	17·8	12·4	17·8	0·03-0·04	+
< 14·8	14·3-14·5	10·4-10·8	10·4-10·8	0·02-0·03	+
X	X	X	X	0·000-0·007	±
?	16-18	?	?		±
X	X	X	X	0·03-0·04	+
X	X	X	X	0·006-0·007	+
X	X	X	X	0	?

lawsonite, epidote, zoisite and anorthite. In progressive metamorphism, prehnite and pumpellyite would form by breakdown of some zeolites, and with further increase in temperature would be decomposed to produce epidote, and in some cases of glaucophanitic metamorphism lawsonite. The high-temperature stability limit of prehnite is at about 400 °C for 2-5 kbar P_{H_2O} (Liou, 1971a).

Prehnite occurs in low-pressure as well as in high-pressure series of regional metamorphic rocks. However, it hardly occurs in typical lawsonite-bearing high-pressure metamorphic rocks.

Pumpellyite also occurs in low-pressure as well as in high-pressure series of regional metamorphic rocks. It is associated with lawsonite in some areas.

Hashimoto (1968, p. 132) has shown that the refractive indices of pumpellyite tend to decrease with increasing metamorphic temperature, probably owing to a decrease of the Fe^{3+} content as a result of progressive reduction.

6B-5   OPAL, CHALCEDONY AND QUARTZ

Opal is a common mineral in the low-temperature zones of active geothermal fields. In high-temperature zones, quartz occurs instead, probably owing to a higher rate of reaction leading to the formation of stable minerals.

Opal was reported as occurring in a low-temperature part of the zeolite facies zones in Mogami (Utada, 1965) and Tanzawa (Seki *et al.* 1969*a*), which both have a high geothermal gradient. Chalcedony occurs in a high-temperature part of the zeolite facies zones of Mogami. Quartz is the only silica mineral in the greenschist facies zones of these areas.

Zeolite facies regions with smaller geothermal gradients such as Taringatura and central Kii have only quartz, and not opal or chalcedony. Deposition of opal would indicate an unusually high chemical potential of SiO_2. This may induce the metastable formation of highly siliceous minerals, such as mordenite.

Chapter 7A

Metapelites: Diversity in Progressive Regional Metamorphism

7A-1 NATURE OF THE DIVERSITY

In this chapter an attempt will be made to systematize the diverse features of regional metamorphism of pelitic rocks all over the world, largely on the basis of baric types. Many features can be explained in this way (table 7A-1). Among the metamorphic terranes of each type, however, there are further variations, the causes of which have not always been clearly understood.

For example, certain high-pressure metamorphic terranes have aragonite and the jadeite + quartz assemblage, whereas others do not. This difference may be ascribed to a difference in rock-pressure within the category of the high-pressure type (§ 3-3). On the other hand, certain low-pressure terranes have abundant chloritoid, staurolite and/or almandine, whereas others do not. Such differences should be due to a combination of T, P_s, P_{H_2O}, P_{CO_2}, and the chemical composition of the predominant rocks. In some cases, one of these factors (e.g. P_s) may have played a dominant role. This could be determined only by a rigorous paragenetic analysis.

Albee (1965a), Hoschek (1969), Hess (1969) and Hensen (1971) have made theoretical attempts to coordinate observed mineral assemblages of metapelites in terms of temperature and pressure (fig. 3-9). Future progress in this direction will enable us to speak of minor variations of temperature and pressure and of relationships between controlling factors.

7A-2 LOW-PRESSURE METAPELITES IN THE RYOKE BELT AND ABUKUMA PLATEAU, JAPAN

Regional metamorphic rocks of the low-pressure type are widely exposed in the Ryoke belt running along the Southwest Japan Arc. Metapelites ranging from slates to gneisses are abundant. Andalusite and sillimanite occur in a lower and a higher temperature part respectively.

The metamorphic terrane of the Abukuma Plateau could represent a northeastern extension of the Ryoke belt. The rock-pressure of metamorphism there, however, appears to have been slightly higher than that in the Ryoke belt, since

Table 7A-1 *Baric types of regional metamorphism and characteristic minerals of metapelites*

Baric type	Glaucophane-schist facies	Greenschist facies	Epidote-amphibolite and amphibolite facies	Granulite facies
Low-pressure		Andalusite, spessartite-almandine, chloritoid, chlorite	Andalusite, sillimanite, cordierite, staurolite, Mn-rich almandine	Orthopyroxene, sillimanite, cordierite, almandine
Medium-pressure		Mn-rich almandine, chloritoid, chlorite, (stilpnomelane)	Kyanite, sillimanite, staurolite, almandine, chlorite	Orthopyroxene, sillimanite, pyrope-almandine
High-pressure	Lawsonite, jadeite, aragonite, chloritoid, chlorite, (stilpnomelane)	Chloritoid, chlorite	Kyanite, staurolite, almandine	

in addition to andalusite and sillimanite, the occasional occurrence of kyanite has been noted.

For the geologic relations of the Ryoke belt and Abukuma Plateau, refer to chapter 15, and especially § 15-7.

Three areas within these terranes will be reviewed below.

Shiojiri-Takato Area in the Ryoke Belt

This area is at the eastern end of the Ryoke belt (fig. 15-8). The progressive mineral changes in the western part of this area have already been summarized in fig. 3-4 (Oki, 1961*a, b*; Katada, 1965), whereas those in metapelites and metapsammites in the eastern part of the area will be reviewed here on the basis of recent studies by Ono (1969*a, b*; 1970).

In this region, the sporadic occurrence of recrystallized biotite begins roughly at the same temperature as the beginning of recrystallization, and the chlorite zone in the proper sense is lacking. The area under consideration has been divided into three zones, here called I, II and III in order of increasing temperature, as shown in fig. 7A-1. The temperature increases towards the southeast over a distance of 50 km, until the terrane is cut by a large fault, called the Median Tectonic Line. The widespread metamorphism is evidently of a regional type. The distribution of granitic masses is independent of the distribution of the temperature of metamorphism.

The progressive mineral changes are summarized in fig. 7A-2. Metapelites are dominant and metabasites are scarce. Hence, the metamorphic facies names shown in this figure are tentative. The individual zones show the following characteristics (Ono, 1969*b*).

Zone I (chlorite–biotite zone). Chlorite is abundant in the metapelites of this zone, but a small amount of brown biotite ($\beta = 1.635$-1.644) also occurs. Recrystallization is incomplete, and the common assemblages of recrystallized minerals are as follows:

1. Quartz + plagioclase + chlorite + muscovite + sphene + ilmenite + graphite,
2. Quartz + albite + sodic plagioclase + chlorite + muscovite + biotite + sphene + ilmenite + graphite.

In conventional geologic mapping, these rocks have been regarded as unmetamorphosed slates. Most metapelites of this zone contain two plagioclases with 7 per cent An and 14 per cent An. Ono has ascribed this to the effect of the peristerite solvus. The metapelites do not contain K-feldspar, but some metapsammites do, though it is not clear whether the mineral is stable or relict. Muscovite is very fine-grained and sometimes greenish, suggesting a phengitic composition.

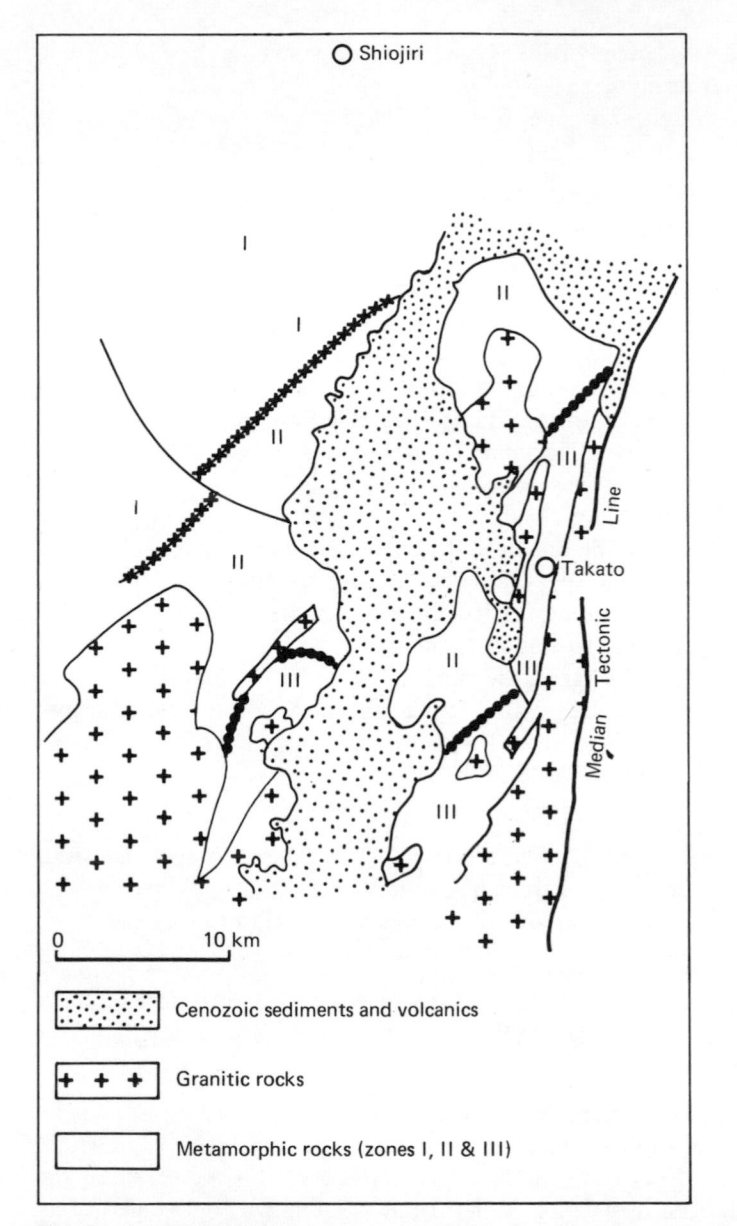

Fig. 7A-1 Zones of progressive regional metamorphism in the Shiojiri-Takato area of the Ryoke belt, Japan (Ono, 1969*a*, *b*). The location of this area is shown in fig. 15-8.

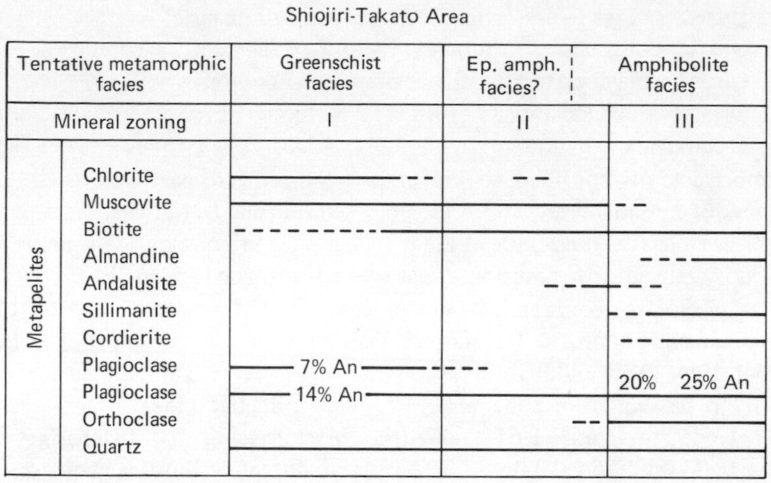

Shiojiri-Takato Area

Tentative metamorphic facies		Greenschist facies	Ep. amph. facies?	Amphibolite facies
Mineral zoning		I	II	III
Metapelites	Chlorite	———————— – –	– –	
	Muscovite	————————————		– –
	Biotite	– – – – – – – – – – ————		
	Almandine			– – –
	Andalusite		– –	– –
	Sillimanite			⊢ – – ————
	Cordierite			– – ————
	Plagioclase	——— 7% An ————	– – – –	
	Plagioclase	——— 14% An ———		20% 25% An
	Orthoclase		– –	
	Quartz	————		

Fig. 7A-2   Progressive mineral changes in the shiojiri-Takato area (Ono, 1969*b*).

Zone II (*biotite–andalusite zone*). This zone is characterized by the disappearance of chlorite. Biotite becomes abundant. Andalusite occurs in the high-temperature part of this zone. The observed mineral assemblages are:

1. Quartz + sodic plagioclase + muscovite + biotite
2. Quartz + oligoclase + orthoclase + muscovite + biotite
3. Quartz + oligoclase + orthoclase + muscovite + biotite + andalusite.

Assemblages (2) and (3) are confined to the higher-temperature part of this zone. The rocks are slates and schists.

Plagioclase usually contains 14–20 per cent An. In the lower temperature part of this zone, the metapelites do not contain K-feldspar but some metapsammites do. In the highest temperature part, the metapelites commonly contain a small amount of orthoclase, which shows monoclinic symmetry in X-ray powder pattern, and can coexist with andalusite. Ono considered that this orthoclase is produced by decomposition of the phengite component in muscovite. Biotite is brown in the low-temperature part, but becomes slightly reddish brown in the high-temperature part.

Zone III (*sillimanite zone*). The entrance to this zone is marked by the appearance of sillimanite. Though andalusite still occurs in some rocks, it is regarded as a relict mineral. The observed assemblages are:

1. Quartz + oligoclase + orthoclase + muscovite + biotite + andalusite + sillimanite.

2. Quartz + oligoclase + orthoclase + biotite + sillimanite,
3. Quartz + oligoclase + orthoclase + biotite + sillimanite + cordierite,
4. Quartz + plagioclase + biotite + garnet + cordierite,
5. Quartz + oligoclase + orthoclase + biotite + garnet + sillimanite + cordierite.

Assemblage (5) is confined to the highest temperature part of this zone. The amounts of muscovite are around 25 per cent in zone II, but become 8 per cent or less in zone III. Muscovite in zone III has been interpreted as probably being formed by retrogressive reaction. Most rocks of this zone are gneisses.

The plagioclase contains 20–30 per cent An. Orthoclase ($2 V = 60$–$65°$) is widespread and abundant. The muscovites are low in iron content ($FeO + Fe_2O_3 = 0·6$–$1·2$ per cent) and high in Al_2O_3 (33–34 per cent). The biotites are reddish brown. In passing from zone II to III, their refractive index β increases from $1·637$–$1·645$ to $1·645$–$1·652$ and the basal spacing $d(001)$ changes from $10·06$–$10·11$ to $10·04$–$10·08$ Å. The Fe_2O_3 content of biotites decreases from 7–2 per cent in zones I and II to $1·0$–$0·0$ per cent in zone III. The TiO_2 content increases from 2–3 per cent in zones I and II to 3–4 per cent in zone III.

Andalusite in zone II is confined to rocks exceptionally high in Al_2O_3 just as in other parts of the Ryoke belt, whereas sillimanite in zone III occurs in rocks of a much wider range of composition. Therefore, sillimanite forms partly by the transformation of andalusite, but largely by the breakdown of muscovite at the threshold to zone III by a reaction such as muscovite + quartz = sillimanite + orthoclase + H_2O. Orthoclase forms as a by-product. Usually the Na_2O/K_2O ratio of orthoclase is higher than that of the associated muscovite, and hence the resultant orthoclase absorbs Na_2O from the coexisting plagioclase, leading to an increase of the An percentage of plagioclase in zone III, as described above.

Almandine (24–28 per cent FeO, 3–6 per cent MnO) and cordierite occur in a high-temperature part of zone III in association with orthoclase. In the highest temperature part, sillimanite decreases and, instead, cordierite increases, possibly by some reaction such as sillimanite + biotite + quartz = cordierite + orthoclase + H_2O.

The H_2O content of the metapelites has decreased and Fe_2O_3 has been reduced in progressive metamorphism. There was no metasomatism on a regional scale.

Granitic Rocks and Pegmatites. The granitic rocks in this area are mostly granodiorite or quartz diorite. They occur in all zones ranging from unmetamorphosed rocks to sillimanite gneisses, and show no relationship to the distribution of the temperature of the regional metamorphism. They form discordant intrusions with hornfelsic aureoles. The age of intrusion would be much less than that of the regional metamorphism. The K-Ar and Rb-Sr ages of the granites are around 60–70 million years.

Pegmatitic bodies containing K-feldspar (orthoclase) occur in zone III, and are especially abundant in its high-temperature part. The thinner bodies, usually containing biotite, garnet, cordierite and sillimanite as well as quartz, plagioclase (oligoclase) and orthoclase, have similar mineral assemblages to the surrounding gneissic rocks. These bodies appear to be approximately in chemical equilibrium with the surrounding gneisses, and probably have been formed by some local migration of materials. However, it is doubtful whether the mechanism is partial melting and squeezing, since their orthoclase content is commonly as high as 50–65 per cent.

On the other hand, pegmatitic bodies more than 50 cm thick are commonly discordant to the surrounding rocks and show a distinctly different mineral composition from them. These bodies appear to be certainly of melt origin (Ono, 1970).

Tukuba Area

The Tukuba (Tsukuba) area is about 60 km northeast of Tokyo, and could represent an isolated exposure of buried metamorphic complexes connecting the Ryoke belt with the Abukuma Plateau. It is a classical area studied by Sugi (1930), who claimed that large-scale injection of granitic magma took place into gneisses. Since then, the idea of large-scale migration of materials in the Ryoke metamorphic complexes either by mechanical injection of magma or by chemical diffusion, has been supported by most Japanese geologists.

The metamorphic complex is mainly composed of metapelites. Andalusite occurs in the medium-temperature part, whereas sillimanite, cordierite and rarely garnet occur in the high-temperature part. Uno (1961) has shown that the metapelites show no systematic changes in chemical composition with increasing metamorphism, except for a decrease of H_2O content, in spite of their transformation from biotite slates to sillimanite gneisses. There is no evidence for large-scale migration of materials.

Central Abukuma Plateau

In the central Abukuma Plateau, 150–190 km NNE of Tokyo (fig. 15-8), the unmetamorphosed area and the transitional zone into the metamorphosed area are covered by Tertiary sediments. Miyashiro (1958) and Shido (1958) have divided the exposed metamorphic terrane into three zones A, B, and C in terms of progressive mineral changes in metabasites, which are especially abundant in the eastern half of the area (figs. 3-5 and 8A-1).

Zone A (greenschist facies). Biotite-chlorite-quartz schist is common.

Zone B (amphibolite facies). Biotite-plagioclase-quartz schist is common. Andalusite occurs rarely.

Zone C (amphibolite facies). Biotite–K feldspar–plagioclase–quartz gneiss with some garnet and/or sillimanite is common. Andalusite occurs in the lower temperature part of this zone, and sillimanite in the higher part. The reaction between muscovite and quartz to produce sillimanite and K-feldspar appears to begin in the middle of this zone.

Highly aluminous metapelites are rare. The transformation temperature from andalusite to sillimanite lies within the low-temperature part of zone C.

Recently, the rare occurrence of staurolite and kyanite has been discovered, and it has been suggested that these formed by a metamorphic event older than that now under consideration (Hara *et al.* 1969). However, the evidence for this is trifling. Staurolite is common in many low-pressure regional metamorphic terranes of the world as reviewed below, and does not need a peculiar sequence of events for its formation. Kyanite is extremely rare here, being confined to a zone close to the isograd representing the transformation of andalusite to sillimanite. The *P–T* curve at some stage in the metamorphism could have passed through, or situated in the close vicinity of, the triple point of the Al_2SiO_5 system. This situation is similar to that in the southern Abukuma Plateau (Tagiri, 1971) and the Mount Lofty Range in Australia.

7A-3 LOW-PRESSURE METAPELITES IN AUSTRALIA

Cooma and Wantabadgery Areas, New South Wales.
Low-pressure regional metamorphic rocks, mostly of pelitic composition, are widely exposed in the Paleozoic orogenic belts of eastern Australia (Vallance, 1967). Detailed studies were made in the Cooma area (Joplin, 1942, 1943; Pidgeon and Compton, 1965) and in the Wantabadgery–Tumbarumba area (Vallance, 1953, 1960, 1967), both in New South Wales. Abundant granitic rocks are exposed.

These two areas are near the southeast coast of Australia and are similar to each other in character. The following sequence of progressive metamorphic zones was found in the metapelites:

1. *Chlorite zone,* containing slates and phyllites with chlorite and muscovite. The Cooma and Wantabadgery areas differ from the Ryoke belt of Japan in having a wide chlorite zone, which however may not belong to the same metamorphic event and facies series as the higher temperature zones, as mention in § 4-1.
2. *Biotite zone*, containing schists with biotite and muscovite.
3. *Andalusite–cordierite zone.*
4. *High-temperature zone*, containing andalusite, sillimanite, cordierite and K-feldspar.

Garnet and staurolite are very rare in these areas.

Mount Lofty Ranges, South Australia.

The early Palaeozoic orogenic belt in the Mount Lofty Ranges to the east of Adelaide, South Australia, has a low-pressure metamorphic belt which somewhat resembles a medium-pressure one (Offler and Fleming, 1968; A. J. R. White, 1966; White, Compston and Kleeman, 1967). The temperature of metamorphism increases toward the axial zone over a distance of about 40 km. At least four mineral zones have been distinguished in pelitic schists:

1. *Chlorite zone*
2. *Biotite zone.*
3. *Andalusite–staurolite zone.* Andalusite begins to occur before staurolite does in one part of the zone, and the reverse sequence has been observed in another part.
4. *Sillimanite zone.* In the higher temperature part of this zone, sillimanite can coexist with orthoclase.

Kyanite was found to occur sporadically within the andalusite-staurolite and sillimanite zones. The *P–T* curve at some stage of metamorphism appears to have been situated in the vicinity of the triple point of the Al_2SiO_5 system. Granitic and migmatitic rocks occur in the axial zone.

East Kimberley Area.

Gemuts (1965) has published a preliminary petrographic note on a Precambrian metamorphic complex composed of pelitic, basic and calcareous rocks and ranging from the greenschist to the low granulite facies in the East Kimberley area near the northeastern corner of Western Australia. The temperature of metamorphism tends to increase northwards over a distance of 150 km. Though all the three polymorphs of Al_2SiO_5 occur in metapelites, there is a tendency for kyanite, andalusite and sillimanite to occur in this order, each in turn being the most abundant form as the temperature rises. Hence, the geothermal curve would pass the triple point on the low-pressure side. Chloritoid, staurolite and cordierite occur in the low-, medium- and high-temperature parts respectively.

A brief general account of metamorphic belts of Australia will be given in § 16-2.

7A-4   LOW-PRESSURE METAPELITES IN FRANCE AND SPAIN

Regional metamorphic rocks of the low-pressure type are widespread in the Hercynian terranes of France and Spain. Andalusite and sillimanite occur there. Staurolite also occurs in some areas. Small parts of the terranes may belong to the medium-pressure type, as suggested by the occurrence of kyanite (Capdevila, 1968). There are associated granitic masses, especially common in high-temperature zones in the metamorphic terranes. Three areas will be reviewed below.

Aracena Area, Southwest Spain

Metamorphic rocks ranging from the greenschist to the granulite facies occur in the Aracena area, southwest Spain, near the border with Portugal (Bard, 1969). No staurolite was found.

Silurian and older formations exposed there are composed of nearly equal amounts of pelitic, basic volcanic and calcareous rocks with subordinate acidic volcanics. The metamorphic belt has a thermal axis (fig. 7A-3).

The progressive mineral changes in metapelites, metabasites and limestones are correlated in fig. 7A-4. The lowest temperature zone of the area is characterized by the pelitic assemblage chlorite + muscovite + quartz possibly with occasional pyrophyllite. With a rise of temperature, chloritoid and andalusite begin to occur in some rocks, and then biotite. With a further rise in temperature, cordierite and sillimanite begin to occur commonly in association with K-feldspar and muscovite. The highest temperature zone is characterized by the decomposition of muscovite, the change of microcline to orthoclase, and the common occurrence of almandine. Some rocks in this zone contain orthopyroxene, but biotite still remains.

The distance from the biotite isograd to the thermal axis is 3–10 km. Granitic rocks are distributed in all zones, but granodiorites with cordierite and garnet or cordierite and orthopyroxene occur in and near the highest temperature zone.

Guitard and Raguin (1958) and Fonteilles and Guitard (1968) have described a low-pressure metamorphic complex without staurolite in the Agly area, eastern Pyrenees. The highest temperature zone in this area has reached the granulite facies.

Bosost Area, Pyrenees

The Bosost area in the central Pyrenees was studied by Zwart (1962, 1963). Early and middle Palaeozoic pelites have undergone deformation and metamorphism. Four successive phases of deformation occurred throughout a large part of the Pyrenees. Each of them is characterized by folds having a rather constant attitute for both axial planes and axes. The first phase formed the major structure of this and surrounding areas with E–W trending fold axes. The deformation phases were separated by periods of tectonic inactivity (fig. 7A-5).

The Palaeozoic rocks of the region are widely metamorphosed at the chlorite grade. In addition, the following four mineral zones indicating progressive rise of temperature were distinguished:

    I. Biotite zone,
   II. Staurolite–andalusite–cordierite zone,
  III Andalusite–cordierite zone,
   IV. Cordierite–sillimanite zone.

Zone IV forms the high-temperature core of this area, around which zone III and then zone II and zone I form shells. Zone I is absent in the southern part of

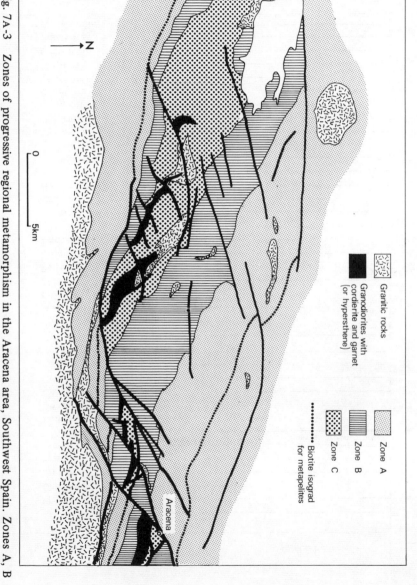
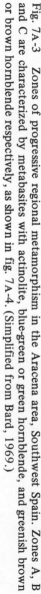

Fig. 7A-3   Zones of progressive regional metamorphism in the Aracena area, Southwest Spain. Zones A, B and C are characterized by metabasites with actinolite, blue-green or green hornblende, and greenish brown or brown hornblende respectively, as shown in fig. 7A-4. (Simplified from Bard, 1969.)

Granitic rocks

Granodiorites with cordierite and garnet (or hypersthene)

Zone A

Zone B

Zone C

Biotite isograd for metapelites

Aracena

N

0   5km

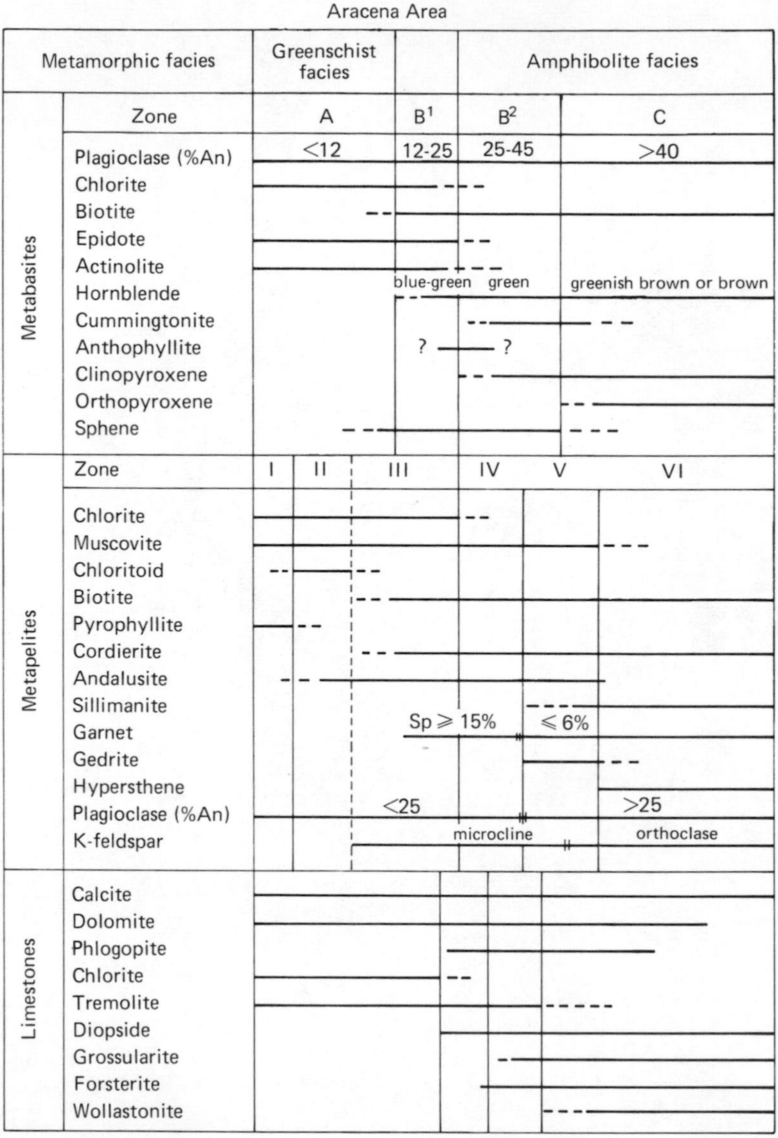

Aracena Area

Metamorphic facies		Greenschist facies		Amphibolite facies		
Zone		A	B¹	B²		C
	Plagioclase (%An)	<12	12-25	25-45		>40
Metabasites	Chlorite					
	Biotite					
	Epidote					
	Actinolite					
	Hornblende		blue-green	green		greenish brown or brown
	Cummingtonite					
	Anthophyllite		?	?		
	Clinopyroxene					
	Orthopyroxene					
	Sphene					
	Zone	I	II	III	IV V	VI
Metapelites	Chlorite					
	Muscovite					
	Chloritoid					
	Biotite					
	Pyrophyllite					
	Cordierite					
	Andalusite					
	Sillimanite					
	Garnet			Sp ≥ 15%	≤ 6%	
	Gedrite					
	Hypersthene					
	Plagioclase (%An)			<25		>25
	K-feldspar				microcline	orthoclase
Limestones	Calcite					
	Dolomite					
	Phlogopite					
	Chlorite					
	Tremolite					
	Diopside					
	Grossularite					
	Forsterite					
	Wollastonite					

Fig. 7A-4. Progressive mineral changes in the Aracena area, southwest Spain (Bard, 1969).

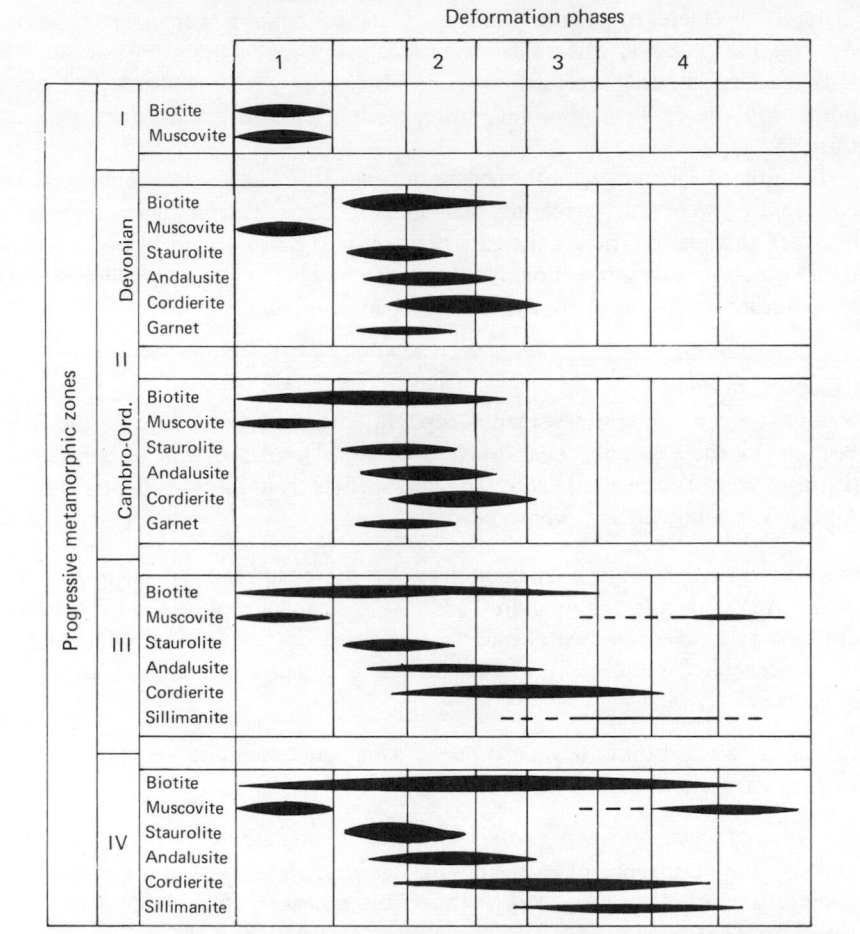

Fig. 7A-5 Relations between mineral formation and deformation in metapelites of the Bosost area, Central Pyrenees. The four phases of deformation are separated by periods of tectonic inactivity (Zwart, 1963).

the area, where zone II is in direct contact with chlorite-grade phyllites. The metamorphic zones cut across the structural and stratigraphic boundaries, and zone IV does not occur in the lowest rocks stratigraphically, which are exposed in the area.

By combined structural and petrologic analyses, Zwart has concluded that the four zones represent a succession in time. First the rocks passed through the stage of zone I, then part of the rocks went through the stage of zone II, next most of these rocks passed through the stage of zone III, and finally some through the stage of zone IV. The depth of formation of zone II was estimated at 3·5–4·0 km.

It is considered that the biotite schists of zone I were formed during deformation phase 1, when schistosity was produced in the infrastructure and slaty cleavage in the suprastructure. In the static period following phase 1, porphyroblasts of staurolite, andalusite, cordierite, biotite and garnet began to form.

Granitic rocks and pegmatites occur in zones II, III and especially in zone IV. No granites were influenced by deformation phase 1, and hence they must postdate this phase. The granites are presumed to have formed by metasomatic transformation of metasediments. Pegmatite began to form in phase 2. The temperature was not high enough to cause partial melting.

Cévennes, France

Metapelites containing widespread pyrophyllite have been described on the River Beaume in the Cévennes near the southeastern border of the Massif Central (Palm, 1958; Tobschall, 1969). The metapelites contain 16-20 per cent of Al_2O_3. The following zones have been observed.

Ia. *Chlorite zone.* This is characterized by the occurrence of albite (0-8 per cent An), quartz, muscovite chlorite and clinozoisite. Pyrophyllite optically resembles muscovite, but was identified by the X-ray powder method on concentrated samples. It was estimated to constitute 2-5 per cent of the rock.

Ib. *Biotite zone.* Biotite begins to occur. This zone resembles the preceding one in other respects. Pyrophyllite is widespread.

II. *Zone of manganiferous garnet.* This was distinguished from the preceding zone by the occurrence of manganiferous garnet, an analysis of which showed a composition of $Alm_{50}Sp_{39}Py_{11}$. The occurrence of garnet in this zone may have been controlled mainly by a compositional variation of the metapelites.

III. *Andalusite zone.* A further increase in temperature of metamorphism caused decomposition of pyrophyllite to produce andalusite. In addition to andalusite, oligoclase (12-18 per cent An), quartz, muscovite, biotite and garnet occur. Primary chlorite and clinozoisite are no longer present, and part of the andalusite is assumed to have formed by the reaction of chlorite with muscovite.

7A-5   LOW-PRESSURE METAPELITES IN NORTH AMERICA

Northern Appalachians

In the northern Appalachians, a wide region including southeastern New York, Connecticut, Massachusetts, and Vermont was subjected to medium-pressure regional metamorphism in Palaeozoic time, whereas to the east and northeast, another region including northern and eastern New Hampshire and Maine was subjected to low-pressure regional metamorphism (§ 13-3). The boundary between the two regions is shown in fig. 7A-6. It is possible that the

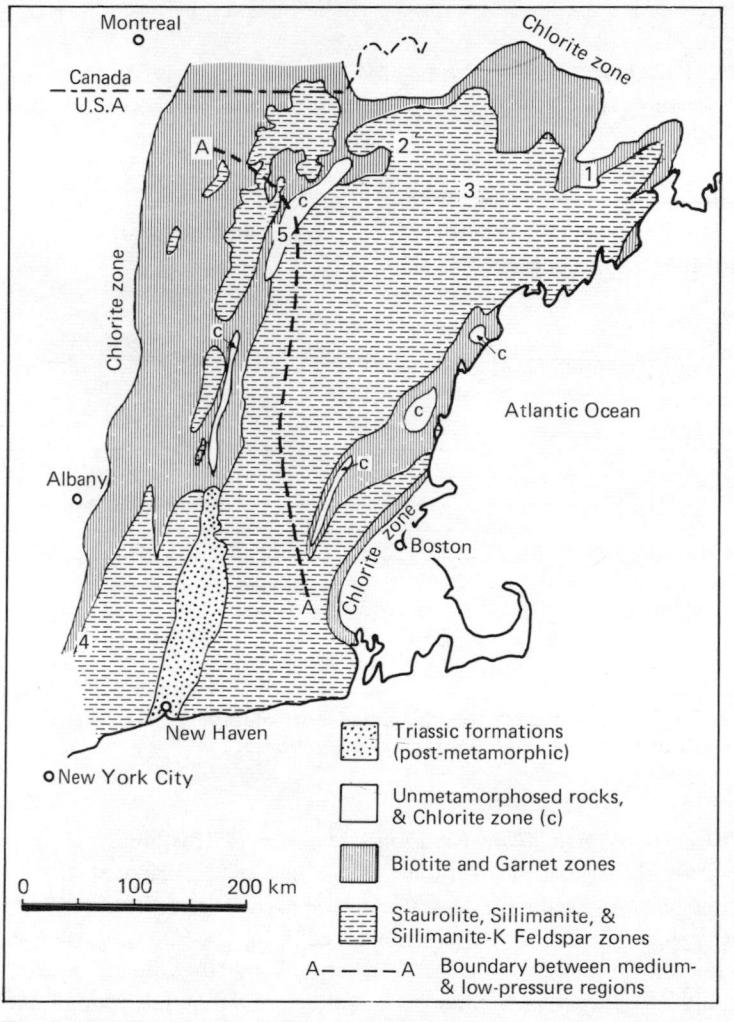

Fig. 7A-6 Palaeozoic regional metamorphic zones in the northern Appalachians. Granitic rocks are omitted. The regions on the southwest and the northeast side of the line A-A are the medium-pressure and the low-pressure regions respectively. Locations: 1, Waterville, 2, Errol, 3, Bryant, 4, Dutchess County, 5, Woodsville. (Modified from Warner, Doyle and Hussey, 1967; Doll, 1961 and Thompson and Norton, 1968.)

rock-pressure of metamorphism tends to become lower toward the northeast in the latter region.

Prehnite–pumpellyite facies rocks occur in the northeast corner of the metamorphic region in Maine (§ 6A-2) and along the west margin of the region in New York State (E-an Zen and Yotaro Seki, pers. comm. 1970).

Two areas in the metamorphic region will be reviewed below.

Waterville–Vassalboro Area, Central Maine. This area is distant from the medium-pressure region (fig. 7A-6). Andalusite and staurolite are common (Osberg, 1968, 1971).

The progressive mineral changes are summarized in fig. 7A-7. The almandine is probably more or less manganiferous. Cordierite begins to occur at a higher temperature than andalusite but at a lower temperature than sillimanite. Staurolite does not persist into the sillimanite zone.

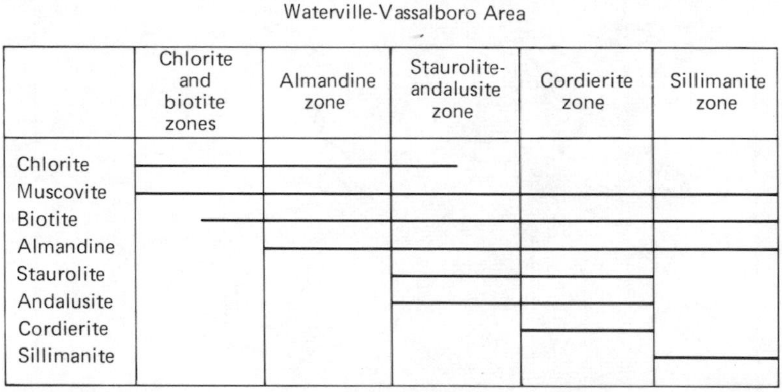

Waterville-Vassalboro Area

	Chlorite and biotite zones	Almandine zone	Staurolite-andalusite zone	Cordierite zone	Sillimanite zone
Chlorite					
Muscovite					
Biotite					
Almandine					
Staurolite					
Andalusite					
Cordierite					
Sillimanite					

Fig. 7A-7 Progressive mineral changes in metapelites of the Waterville-Vassalboro area, central Maine. (Osberg, 1968.)

Errol–Bryant Pond Area, New Hampshire and Maine. This area is closer to the medium-pressure region of Vermont than the preceding area (fig. 7A-6). Metapelites recrystallized at relatively low temperatures have been studied in the Errol quadrangle (Green, 1963), as summarized in fig. 7A-8.

The lowest temperature part is represented by the biotite zone, where chlorite and muscovite are common. The aluminous minerals pyrophyllite and paragonite were looked for without success. Only manganiferous beds contain pyralspite garnet, probably rich in MnO.

With a slight increase in temperature, almandine begins to occur. With a further rise of temperature, staurolite and andalusite begin to occur. Chlorite appears to be still stable at this stage.

The highest metamorphic temperature in the Errol quadrangle is represented by the sillimanite zone. Sillimanite occurs in fine needles or prismatic crystals usually enclosed in muscovite, biotite, chlorite, garnet or quartz. There is no evidence for direct polymorphic transformation of andalusite to sillimanite.

A sillimanite-bearing area, east of the Errol quadrangle, has been divided into the lower and upper sillimanite zones (Guidotti, 1968, 1970). The former is characterized by the coexistence of staurolite with sillimanite, whereas the latter has no staurolite. Within the lower sillimanite zone, staurolite has been partially replaced by muscovite from the crystal margins. In the higher sillimanite zone,

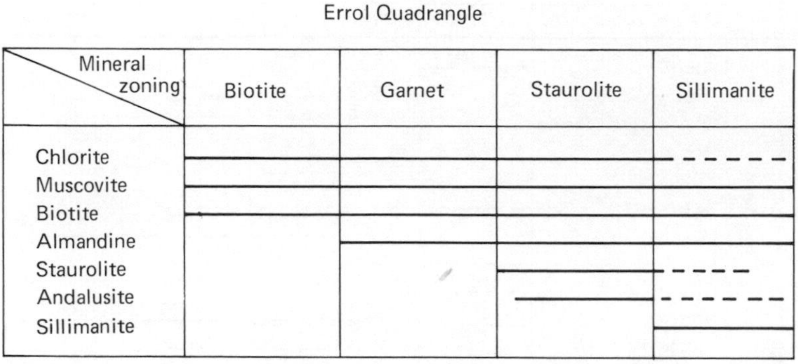

Errol Quadrangle

Mineral zoning	Biotite	Garnet	Staurolite	Sillimanite
Chlorite				- - - - -
Muscovite				
Biotite				
Almandine				
Staurolite				- - - -
Andalusite				- - - - -
Sillimanite				

Fig. 7A-8 Progressive mineral changes in metapelites of the Errol Quadrangle, New Hampshire and Maine. (Green, 1963.)

aggregates of muscovite in pseudomorphs after staurolite tend to recrystallize into single large plates, up to 2 cm across and commonly lying at high angles to the schistosity. It is claimed that the muscovite has formed by a progressive reaction and not by a retrogressive one.

The mineral parageneses in a sillimanite-bearing area have been analysed in the Bryant Pond quadrangle (Guidotti, 1963; Evans and Guidotti, 1966). Here, the pelitic gneisses are very coarse-grained, being mainly composed of quartz, oligoclase, orthoclase, biotite, muscovite, garnet and sillimanite. Quartz and biotite are present in all the rocks. The so-called sillimanite-K feldspar isograd can be drawn on the geologic map. Sillimanite and orthoclase occur individually on the low-temperature side of this isograd, while they can coexist on the high-temperature side.

Northern Michigan

The middle Precambrian (Huronian) formations in northern Michigan on the south shore of Lake Superior have undergone low-pressure metamorphism. James (1955) has studied the area with special reference to the mineral changes of iron formations. The metamorphism took place in pre-Keweenawan time.

The metamorphic terrane has been divided into five zones characterized by chlorite, biotite, almandine, staurolite and sillimanite (fig. 7A-9). A large part of the Huronian is in the chlorite zone defined by the assemblages in the metapelites. There are four thermal domes in which the metamorphism increases up to the sillimanite grade. These are not elongated in shape nor aligned in a belt.

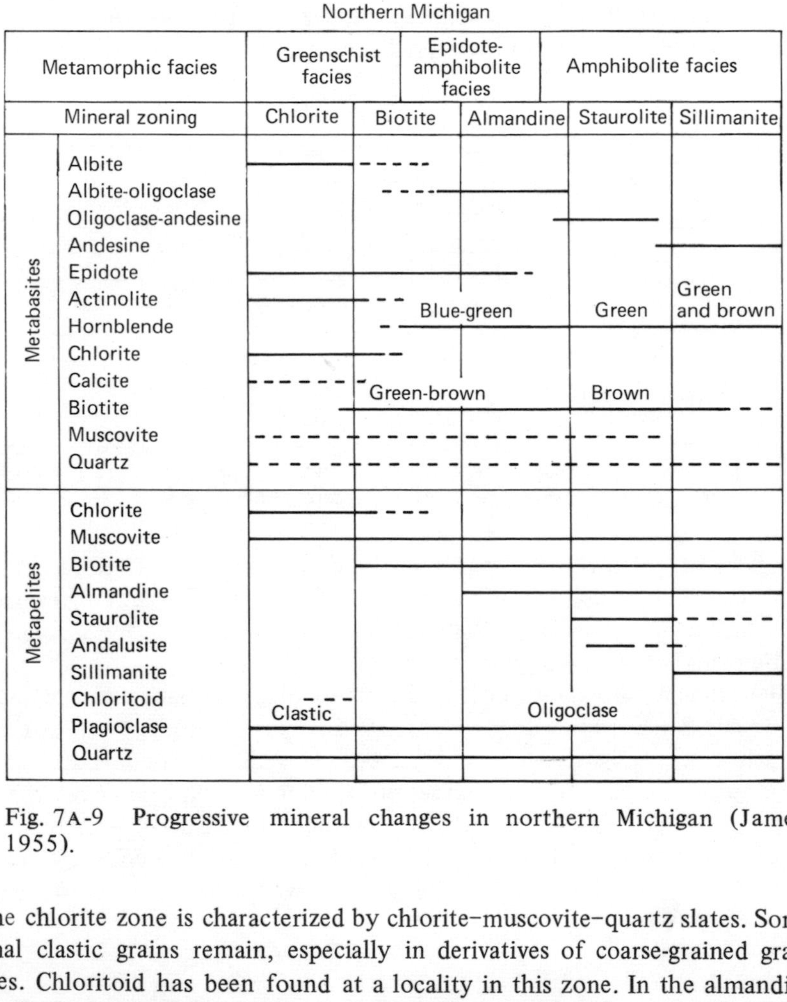

Fig. 7A-9 Progressive mineral changes in northern Michigan (James, 1955).

The chlorite zone is characterized by chlorite–muscovite–quartz slates. Some original clastic grains remain, especially in derivatives of coarse-grained gray-wackes. Chloritoid has been found at a locality in this zone. In the almandine zone, the textural reconstruction is almost complete. Though a staurolite zone was set up in this area, staurolite is very rare. Andalusite occurs in association with it. The graywackes show almost no change in outward appearance with increasing metamorphism. Under the microscope, however, they are seen to be recrystallized to an interlocking mosaic of quartz and oligoclase. The sillimanite zone is confined to the central part of some thermal domes.

The associated iron formations were metamorphosed to produce minnesotaite, stilpnomelane, grunerite, blue–green hornblende and garnet.

7A-6   MEDIUM-PRESSURE   METAPELITES   IN   THE   SCOTTISH   HIGHLANDS

As shown in fig. 7A-10, the Dalradian and Moine metamorphic terranes of the Scottish Highlands are divided into two regions. The major part shows a medium-pressure facies series, including the classical area studied by Barrow, whereas a smaller southeastern part shows a low-pressure series. The two regions are commonly called the *Barrovian and Buchan regions* respectively, as will be discussed in § 14.3.

The progressive metamorphism of pelitic rocks in the Barrovian region will be described below on the basis of Harker's (1932) and later work. A synopsis of mineral changes has been given in fig. 3-2.

Chlorite zone. Common pelitic rocks in the low-temperature part of the chlorite zone are slates mainly composed of muscovite, chlorite and quartz in varying proportions. Sometimes albite and graphite-like matter are also present. Quartz may preserve the outlines of original detrital grains or may have lost them by recrystallization. With a slight increase in temperature, the grain sizes increase, leading to the formation of chlorite–muscovite–albite–quartz phyllite or schist. In some rocks, albite occurs as porphyroblasts.

The muscovites are usually phengite. The chlorites are usually ripidolite. Paragonite was found to occur only sporadically in metapelites (McNamara, 1965). Stilpnomelane occurs rarely in metapsammites. Pyrophyllite has not been found (Chinner, 1967).

Biotite zone. Typical pelitic rocks in this zone are biotite–chlorite–muscovite–albite–quartz schist. The amount of biotite may be very small. Many pelitic rocks are still rich in chlorite and are devoid of biotite. Chloritoid occurs in rocks high in Al_2O_3 and FeO.

Almandine zone (garnet zone). With an increase of temperature, almandine begins to occur. The almandines on the almandine isograd contain about 26 per cent FeO, 4–5 per cent MnO, 1–2 per cent MgO and 6–8 per cent CaO (Atherton, 1964). Almandine sometimes occurs in porphyroblasts with numerous inclusions of quartz and other minerals, and shows marked zonal structure having a decreasing MnO content toward the margin (Atherton, 1968).

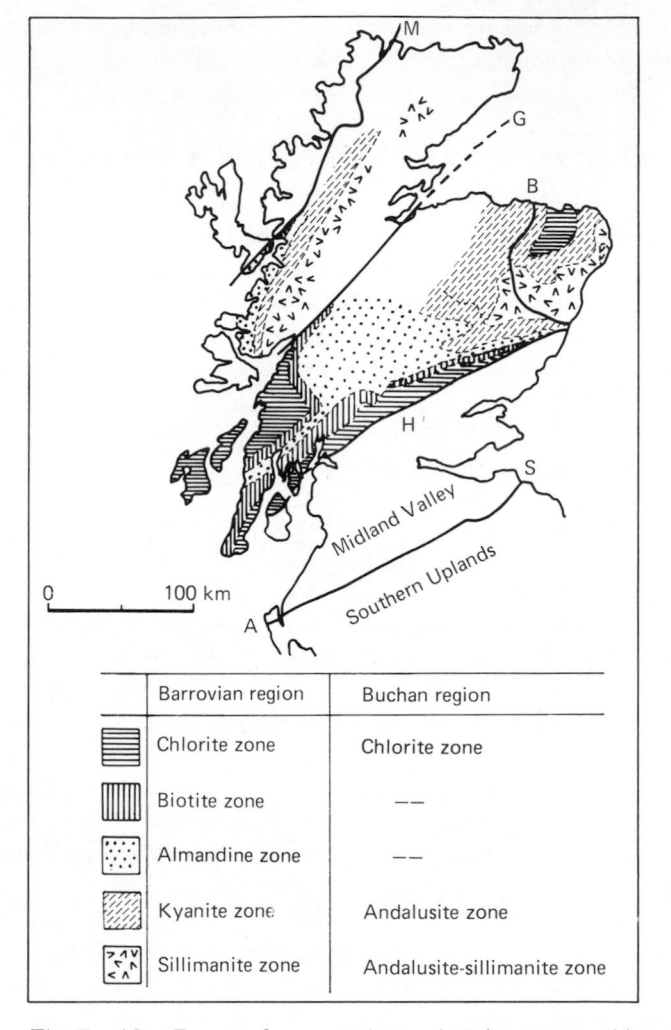

	Barrovian region	Buchan region
	Chlorite zone	Chlorite zone
	Biotite zone	——
	Almandine zone	——
	Kyanite zone	Andalusite zone
	Sillimanite zone	Andalusite-sillimanite zone

Fig. 7A-10 Zones of progressive regional metamorphism in the Scottish Highlands. The Barrovian and the Buchan region are to the west and to the east of the line B respectively. The Moine thrust (M) and Highland Boundary fault (H) represent the northwestern and the southeastern limit of the metamorphic regions. The Great Glen fault (G) is a strike-slip fault with a great sinistral displacement. S represents the Southern Upland fault and A the Ballantrae area. (Modified from Johnson, 1963 and Chinner, 1966*a*.)

Typical pelitic rocks of this zone are almandine–biotite–muscovite–albite–quartz schists commonly with compositional layering. Chlorite may still be present. Chloritoid occurs in some rocks. Almandine persists into higher zones. The FeO and MgO contents of almandine tend to increase, whereas the MnO and CaO tend to decrease with rising temperature (Sturt, 1962; Atherton, 1968).

Staurolite and kyanite zones. With a further increase of temperature, staurolite and kyanite begin to occur, commonly with a porphyroblastic habit. Barrow (1893) showed that staurolite begins to form at a slightly lower temperature than kyanite. Hence, the staurolite zone was distinguished from the kyanite. Staurolite persists into the kyanite zone. The significance of the staurolite and kyanite zones in progressive metamorphism, however, is still very obscure.

Staurolite and kyanite are high in Al_2O_3. If such aluminous hydrous minerals as pyrophyllite, chloritoid or paragonite were common in pelitic rocks of the almandine zone, the breakdown of these minerals should produce staurolite and kyanite. In the low-temperature parts of this area, however, pyrophyllite has not been found, and paragonite is rare. Chloritoid occurs only in a region near the east coast. Accordingly, the nature of the reactions which produce staurolite and kyanite is not known with certainty in most cases (Chinner, 1967). The occurrence of the staurolite and kyanite zones, or the separation of the former from the latter may largely depend on variations in the chemical composition of the predominant pelitic rocks rather than mineral reactions with rising temperature.

Chinner (1967) suggested the possibility of the migration of K on a regional scale to produce the staurolite and kyanite zones.

Typical rocks in the staurolite and kyanite zones are staurolite–almandine–biotite–muscovite –plagioclase–quartz schists and kyanite-almandine–biotite–muscovite–plagioclase–quartz schists respectively.

Sillimanite zone. The sillimanite isograd represents the boundary of the stability fields of kyanite and sillimanite. However, the real mechanism of sillimanite generation does not appear to be the direct transformation of kyanite to sillimanite (§ 7B-13).

In the highest temperature part of the sillimanite zone, muscovite reacts with quartz to produce sillimanite and K-feldspar. This marks the first appearance of K-feldspar in metapelites in this region. The frequency and amount of sillimanite increase remarkably. Typical pelitic rocks in this zone are sillimanite–almandine –biotite–muscovite–K-feldspar–plagioclase–quartz gneisses.

Most metapelites in the Scottish Highlands seem to contain a small amount of graphite and other carbonaceous matter, which should keep such rocks in relatively reduced states during progressive metamorphism. Some pelitic rocks, however, are devoid of such substances, and as a result show highly variable oxidation states. Chinner (1960) described a group of such pelitic gneisses from lowest temperature part of the sillimanite zone. With increase in the degree of oxidation, biotite reacts with garnet to produce muscovite and iron oxides, as may be schematically shown by the following equation:

$$Fe^{2+} \text{ eastonite (biotite)} + \text{almandine} + \text{oxygen}$$

$$= \text{muscovite} + \text{iron oxides} + \text{quartz.}$$

With such an increase in oxidation, the biotite changes in colour from reddish brown to green, and becomes higher in MgO/FeO ratio while the garnet becomes higher in MnO/FeO. The less-oxidized rocks contain graphite as well as ilmenite and magnetite, whereas more strongly oxidized rocks contain no graphite and do contain magnetite and haematite.

The variation in the oxidation state of the graphite-free gneisses appears to result mainly from the compositional differences of pre-metamorphic sedimentary rocks.

7A-7 MEDIUM-PRESSURE METAPELITES IN THE NORTHERN APPALACHIANS

In the southwestern parts of the northern Appalachians, the pelitic metamorphic rocks commonly contain kyanite and staurolite (fig. 7A-6). Barth (1936) made a pioneer petrologic study of such rocks in Dutchess County in southeastern New York State. He described a sequence from a muscovite slate zone, through a kyanite–almandine–biotite schist zone to a sillimanite–almandine–biotite gneiss zone. Almandine, staurolite and kyanite begin to appear at nearly the same temperature, and therefore the zones characterized by each of these minerals were not distinguished.

In Massachusetts, Vermont and New Hampshire, a large number of areas have been mapped petrographically by workers from universities in New England. Zones of chlorite, biotite, garnet (almandine), staurolite–kyanite, sillimanite, and sillimanite–K feldspar have been delineated (Thompson and Norton, 1968; Albee, 1968).

The chlorite zone is wide and stilpnomelane occurs in some slates of the zone (Zen, 1960). In the biotite zone, chloritoid occurs in highly aluminous rocks, and staurolite and kyanite occur in almost the same zone. In the highly aluminous rocks, kyanite appears to precede staurolite. Much sillimanite forms by reactions other than the phase transformation of kyanite. K-feldspar begins to coexist with sillimanite at a higher temperature than that of the first formation of sillimanite; hence the sillimanite and the sillimanite–K-feldspar zone.

The metamorphic facies corresponding to these zones were determined from the mineral assemblages of the associated metabasites. The chlorite and biotite zones have been assigned to the greenschist facies, the garnet zone to the epidote–amphibolite facies, the staurolite–kyanite and sillimanite zones to the amphibolite facies, and the sillimanite–K-feldspar zone to the lower part of the granulite facies.

The quartz diorite, quartz monzonite and granite of the New Hampshire plutonic series occur in the high-temperature parts of the terrane, and appear to be genetically related to the regional metamorphism.

7A-8   HIGH-PRESSURE METAPELITES IN THE ALPS

Orogeny and metamorphism appear to have taken place repeatedly at widely different times in the Alps. Alpine, Hercynian and other older metamorphic complexes are exposed there. Older metamorphic rocks are usually modified by more recent metamorphic events.

Alpine metamorphism is most clearly observed in post-Hercynian rocks. In these, the lowest temperature zone which is exposed in the northernmost part of the metamorphic terrane in the Swiss Alps is characterized by the occurrence of prehnite, pumpellyite and stilpnomelane. In a wider zone representing higher temperatures and exposed on the south side, glaucophane and chloritoid occur in rocks of appropriate compositions (Niggli and Niggli, 1965; Niggli, 1970; van der Plas, 1959). These rocks may belong to a series of the prehnite–pumpellyite and the glaucophane-schist facies. Apparently another series of metamorphic rocks up to the amphibolite facies (sillimanite zone) is exposed in the Simplon-Ticino region, about 100 km wide in an east–west direction, in the southernmost part of the Pennine Alps (fig. 7A-11). This region, probably belonging to the medium-pressure type, would have formed by temperature rise at a stage later than that of the recrystallization in the prehnite-pumpellyite and the glaucophane-schist facies.

Frey (1969a, b) investigated the Upper Triassic (Keuper) red-bed formation, the metamorphism of which increases generally southward across Switzerland. The unmetamorphosed pelites exposed in the northern areas contain 1 Md-illite, 1 Md mixed-layer smectite–illite and haematite. On entering the Helvetic zone to the south, effects of very slight metamorphism appear as follows. The mixed-layer smectite–illite decomposes to produce illite, chlorite and quartz, and then 1 Md-illite changes to $2M_1$ phengite. A mixed-layer paragonite–phengite appears.

Between the Aar and Gotthard Massifs (fig. 7A-11), the intensity of metamorphism increases to produce paragonite and chloritoid. On the south side of the massifs, i.e. in the Pennine nappes, staurolite, kyanite and almandine begin to appear, though paragonite and chloritoid still occur.

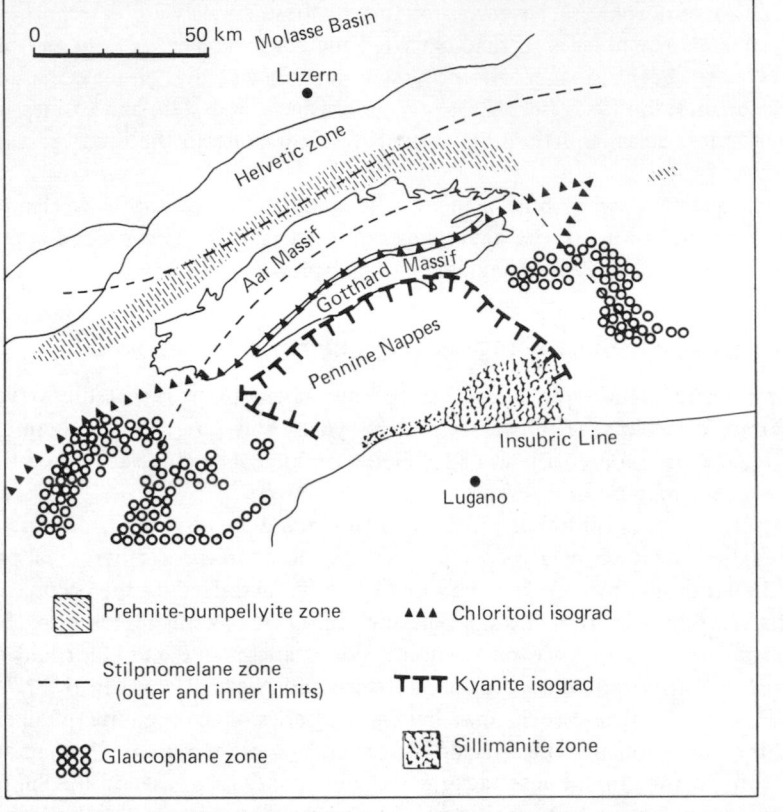

Fig. 7A-11   Distribution of metamorphic minerals in the Swiss Alps. (Based on Niggli and Niggli, 1965 and Niggli, 1970.)

Chatterjee (1961) investigated metapelites in the Pennine nappes to the south of the Aar massif. The An content of plagioclase coexisting with epidote increases toward the southeast. The following zones were distinguished.

Zone Ia (chlorite zone). Ordinary metapelites are mainly composed of chlorite, muscovite, epidote, albite and quartz, sometimes with calcite or chloritoid.

Zone Ib (biotite zone). Ordinary metapelites are composed of chlorite, muscovite, biotite, epidote, albite and quartz. The plagioclase coexisting with epidote is still albite.

Zone IIa. The plagioclase coexisting with epidote is oligoclase in this zone. Ordinary metapelites are composed of chlorite, muscovite, biotite, almandine, chloritoid, epidote, plagioclase and quartz.

Zone IIb. The plagioclase coexisting with epidote is andesine in this zone. The metapelites commonly contain staurolite and/or kyanite, and rarely sillimanite, together with muscovite, biotite, almandine, calcite, epidote and quartz.

The boundaries between these zones cross the stratigraphic horizons and tectonic structures. The temperature rise which caused the progressive metamorphic recrystallization appears to have taken place at a later time than the large-scale folding and thrusting movements. Almost all the porphyroblasts grew after the penetrative movement had died out.

7A-9   HIGH-PRESSURE METAPELITES IN THE SANBAGAWA BELT, JAPAN

The Sanbagawa metamorphic belt of Japan generally belongs to the high-pressure type, though glaucophane, lawsonite and jadeite are not very widespread nor abundant. The temperature was not high enough to produce biotite except in the Bessi and Iimori areas. Detailed petrographic studies have been made in many places, but our knowledge on the poorly recrystallized, low-temperature part of the belt is still insufficient. Fig. 15-9 shows the widespread presence of a zone of prehnite–pumpellyite facies rocks on the low-temperature side of a zone of rocks of the greenschist and glaucophane-schist facies, though it has not been well established (§ 15-8).

The characteristics of metamorphism appear to differ in different parts of the belt. For example, facies series with the jadeite + quartz assemblage were found only in the eastern half of the belt. An exceptional area with well-developed zeolite facies rocks in central Kii Peninsula has already been reviewed in § 6A-4.

It is to be noted that glaucophane occurs in metabasites, but not in metapelites. Two areas are reviewed below (fig. 15-8).

Bessi-Ino Area, Southwest Japan

Banno (1964) has shown that a virtually unmetamorphosed Upper Palaeozoic terrane in southern Sikoku grades northwards into a zone of the glaucophane-schist facies, which in turn grades into a relatively narrow zone of the epidote–amphibolite facies, the highest temperature part of the Sanbagawa belt, in the vicinity of Bessi (fig. 15-9). The distance from the southern end to the highest temperature part is about 40 km (fig. 7A-12).

The common metapelites in the glaucophane-schist facies are chlorite–muscovite–calcite–epidote–graphite–albite–quartz schists. Magnetite and manganiferous garnet occur in some rocks (fig. 3-6).

On entering the zone of the epidote–amphibolite facies, biotite and ilmenite begin to occur and garnet becomes more common. Thus, typical pelitic rocks are garnet–chlorite–muscovite–calcite–epidote–graphite–albite–quartz schists with or without biotite. In the high-temperature part of the epidote–amphibolite

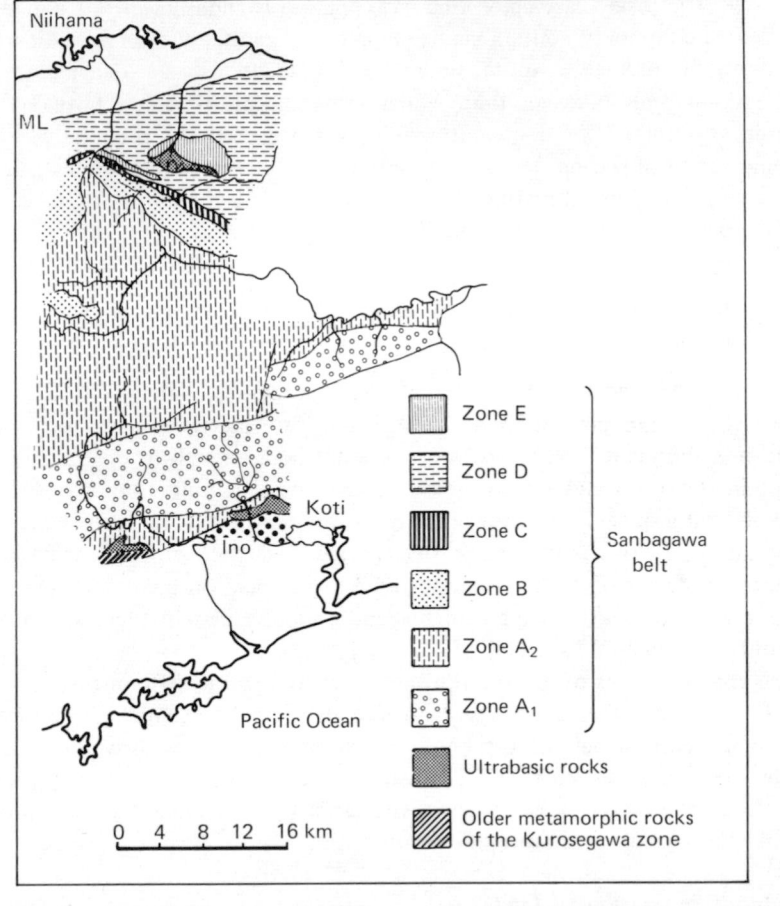

Fig. 7A-12   Zones of progressive regional metamorphism in the San-bagawa terrane of the Bessi-Ino area, Japan (Banno, 1964). The blank areas at the southern and northern ends represent the Pacific Ocean and the Seto Inland Sea respectively. ML, Median Tectonic Line. Refer to figs. 15-8 and 17-1. For explanation of the zones shown in the key see fig. 3-6 and § 8A-8.

facies, biotite becomes more common, and chlorite almost disappears in metapelites. Typical rocks there are biotite–garnet–muscovite–calcite–epidote–graphite–albite–quartz schists.

Lawsonite and jadeite are virtually absent in this area. Glaucophane occurs in metabasites but not in metapelites. Paragonite has not been found in the metapelites. Probably the pelitic rocks are not aluminous enough to produce paragonite.

The highly silicic metasediments (probably metacherts) of this area show a more diversified mineralogy. They contain stilpnomelane, garnet, piemontite, riebeckite, magnesioriebeckite, magnesioarfvedsonite or aegirine in addition to quartz, albite, chlorite, muscovite, epidote and opaque minerals.

Kanto Mountains, Central Japan

Seki (1958, 1960, 1961a) has mapped metamorphic zones in the Kanto Mountains, several tens of kilometres northwest of Tokyo, which represents the eastern end of the exposed part of the Sanbagawa belt. The metamorphism in this area has produced lawsonite and a definite zone characterized by the jadeite + quartz assemblage.

An unrecrystallized Palaeozoic terrane in this area grades into a zone of glaucophane-schist facies, which with rising temperature grades into a zone of greenschist facies. The change of mineral assemblages with temperature is slight (fig. 3-7). The typical pelitic rocks are chlorite–muscovite–albite–quartz schists in all zones except for zone IV, where lawsonite and jadeite occur. Stilpnomelane occurs in all zones. Garnet and piemontite occur in the high-temperature part.

The jadeite is so fine-grained that microscopic identification is very difficult. Concentration by heavy liquids and identification by the X-ray powder method were successfully used on a large number of rocks.

7A-10   HIGH-PRESSURE METACLASTICS IN THE CALIFORNIA COAST RANGES

Coleman and Lee's Nomenclature

The Franciscan complex in the Coast Ranges of California represents eugeosynclinal sedimentary accumulations deposited during Late Jurassic to Late Cretaceous time. It consists mainly of graywacke with moderate amounts of shale and mafic volcanics. Radiolarian cherts and serpentinites are commonly associated with them. Parts of the Franciscan have been subjected to feeble or intense recrystallization. Structural analysis has not been successful, because the metamorphic terrane is strongly deformed and cut by many faults, and moreover lacks suitable marker beds.

According to the degree of recrystallization, Coleman and Lee (1963) divided the Franciscan metamorphics in the Cazadero area into four 'types' as follows:

Type I: Unrecrystallized rocks,
Type II: Incipiently recrystallized, non-schistose rocks,
Type III: Well-recrystallized, schistose metamorphic rocks,
Type IV: Well-recrystallized, coarse-grained metamorphic rocks (schistose or non-schistose).

This nomenclature would be applicable to most or all of the Franciscan rocks.

The type IV rocks occur commonly as large isolated blocks resting directly upon and within less intensely metamorphosed terranes of types II and III. The origin of these enigmatic blocks is not clear. Coleman and Lee considered that they are concentrated in a band which is roughly concordant with some of the major faulting, and that they appear to have been tectonically transported upwards from greater depths. Coleman and Lanphere (1971) have suggested that these blocks are fragments of the crystalline basement on which the Franciscan rocks were deposited.

There is a gradational relation between areas of types I and II and between those of types II and III. The sequence of types I→II→III represents progressive metamorphism, or at least progressive increase in the degree of recrystallization. A large part of the eastern half of the Franciscan metamorphic terranes is in the glaucophane-schist facies. The rocks of this region could be regarded as the most typical glaucophane schist facies terrane in the world (e.g. Ernst *et al*. 1970).

Two areas where there is a progressive metamorphic sequence in metaclastics will be reviewed below.

Panoche Pass Area, Central California
This area, being about 180 km southeast of San Francisco, is in the Diablo Range (fig. 13-3). Metamorphosed graywackes are predominant, though some metacherts and metabasites occur in association. Ernst (1965) has shown that metabasites in a part of the terrane have pumpellyite-bearing greenschist-like mineral assemblages, while those in the rest of the terrane show glaucophane-schist facies mineral assemblages (fig. 3-8). The former may be considered type II rocks in Coleman and Lee's nomenclature, and the latter type III rocks.

The type II metagraywackes and most of the type III metagraywackes do not contain jadeitic pyroxene. They are usually incompletely recrystallized, being composed of clastic grains of quartz, sodic plagioclase and rock fragments set in a mesostasis with chlorite, white mica and stilpnomelane. Lawsonite occurs within the plagioclase. Both calcite and aragonite have been found.

The rest of the type III metagraywackes contain jadeitic pyroxene. Such rocks occur in a few separate areas, 1–2 km across, surrounded by an area of the preceding group of rocks (fig. 7A-13). They are thoroughly recrystallized. Glaucophane occurs in more than half of the rocks. Jadeitic pyroxene occurs as radial sprays and fibrous intergrowths with lawsonite and glaucophane. It contains about 70–80 per cent of the jadeite component. Chlorite, white mica and stilpnomelane occur in these metagraywackes as in those of the preceding group. The primary carbonate is aragonite.

Thus, the most notable progressive reaction in the metagraywackes is the formation of jadeitic pyroxene from albite. The isograd corresponding to this reaction has been called the jadeitic pyroxene + lawsonite isograd (Ernst, 1965).

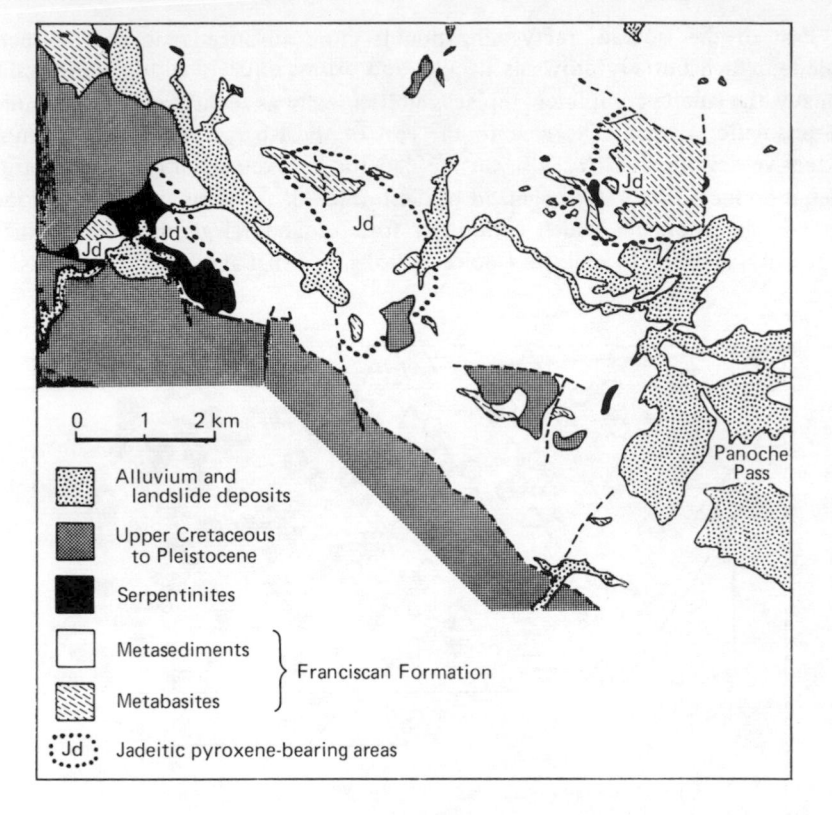

0   1   2 km

Alluvium and
landslide deposits

Upper Cretaceous
to Pleistocene

Serpentinites

Metasediments ⎫
                  ⎬ Franciscan Formation
Metabasites   ⎭

Jd   Jadeitic pyroxene-bearing areas

Fig. 7A-13   Zones of metamorphism in the Franciscan terrane of the Panoche Pass area in the Diablo Range, California. (Ernst, 1965.) The location of the Diablo Range is shown in fig. 13-3.

Pacheco Pass Area, Central California

Pacheco Pass, being about 130 km southeast of San Francisco, is also in the Diablo Range. The Franciscan rocks of this area are metamorphosed in the typical glaucophane-schist facies. The terrane is composed mainly of metagraywackes.

The jadeitic pyroxene + lawsonite isograd runs in a NE–SW direction through the centre of this area (fig. 7A-14). The rocks on the west and east sides of the isograd belong respectively to types II and III. West of the isograd, jadeite is virtually absent, and albite is abundant. Slight recrystallization of the fine-grained mesostasis of the graywackes produces white mica, chlorite, albite and quartz. With increasing recrystallization, lawsonite grows as prismatic or tabular crystals (fig. 7A-15).

East of the isograd, recrystallization is more advanced and has produced jadeite, which initially grows as needles and prisms entirely within albite grains. Finally the jadeite completely replaces albite usually as slightly radiating bundles of prismatic crystals. The area to the east of the isograd represents the most extensive development (*c.* 200 km^2) of jadeitic pyroxene- and aragonite-bearing metamorphic rocks yet recognized in California. Many of jadeitized sandstones contain glaucophane, which commonly forms subhedral grains in the matrix, especially adjacent to chlorite (McKee, 1962*a, b*; Ernst and Seki, 1967; Ernst *et*

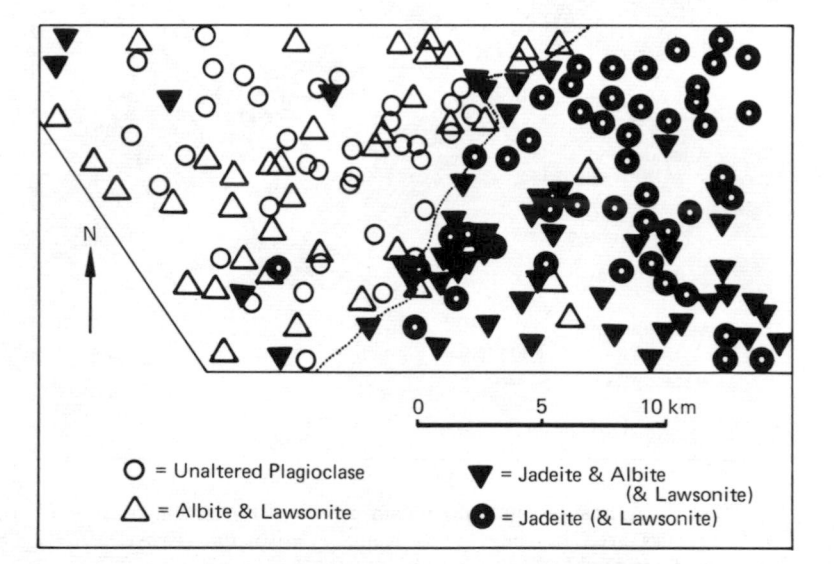

Fig. 7A-14   Zones of metamorphism in the Franciscan terrane of the Pacheco Pass area in the Diablo Range, California. (McKee, 1962*a*.)

al. 1970). Stilpnomelane occurs on both sides of the isograd. Associated metacherts may contain stilpnomelane, garnet, piemontite, riebeckite and deerite.

Type IV rocks occur as many scattered outcrops and blocks. These rocks show the following mineral assemblages:

Glaucophane + garnet (+ epidote),
Glaucophane + albite + epidote (+ chlorite),
Glaucophane + lawsonite + epidote (+ chlorite + white mica),
Glaucophane + lawsonite + jadeite,
Actinolite + chlorite (+ talc),
Clinopyroxene + garnet (= eclogite).

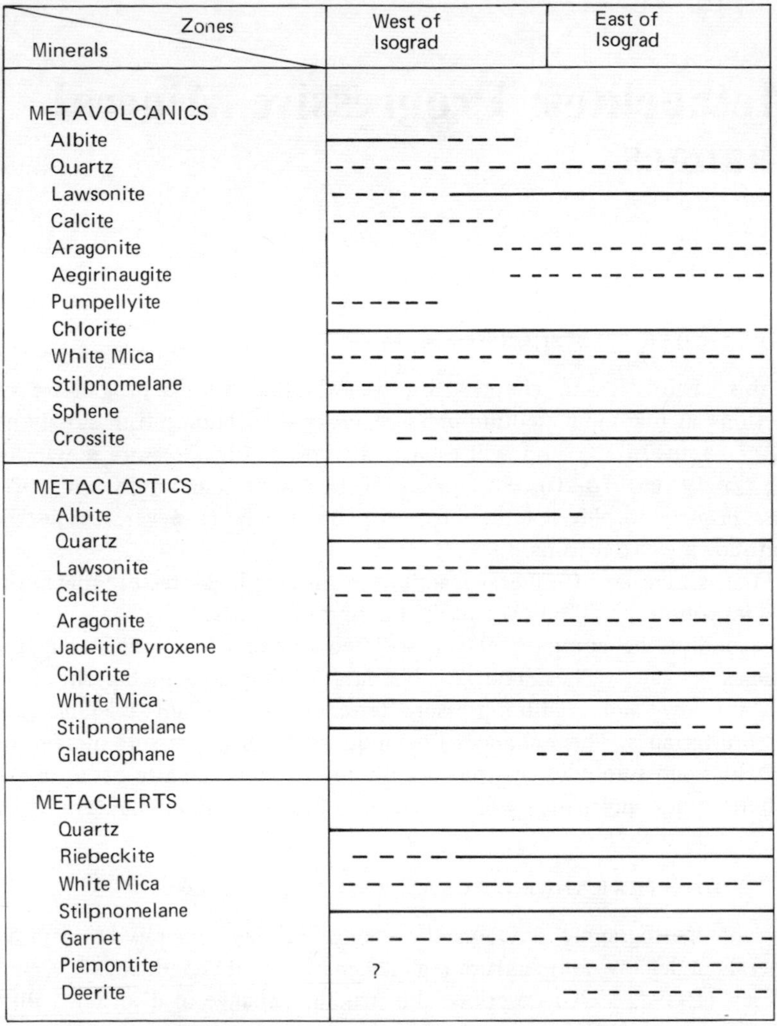

Minerals \ Zones	West of Isograd	East of Isograd
METAVOLCANICS		
Albite		
Quartz		
Lawsonite		
Calcite		
Aragonite		
Aegirinaugite		
Pumpellyite		
Chlorite		
White Mica		
Stilpnomelane		
Sphene		
Crossite		
METACLASTICS		
Albite		
Quartz		
Lawsonite		
Calcite		
Aragonite		
Jadeitic Pyroxene		
Chlorite		
White Mica		
Stilpnomelane		
Glaucophane		
METACHERTS		
Quartz		
Riebeckite		
White Mica		
Stilpnomelane		
Garnet		
Piemontite	?	
Deerite		

Fig. 7A-15   Progressive mineral changes in the Pacheco Pass area, California. (Ernst and Seki, 1967; Ernst *et al.* 1970.)

Quartz may be present in any of these assemblages. Pumpellyite, stilpnomelane, rutile and sphene may occur. The distribution of these rocks shows no clear relation to the above-mentioned regular increase in metamorphic recrystalliz-ation nor to the jadeitic pyroxene + lawsonite isograd.

Chapter 7B

Metapelites: Progressive Mineral Changes

7B-1 ORDER OF DISCUSSIONS

In this chapter as in chapter 8B, we will first discuss progressive mineral reactions in low- and medium-pressure regional metamorphism simultaneously (§§ 7B-1 to 7B-17), and will then treat those in high-pressure metamorphism (§§ 7B-19 and 7B-20). This order of treatment unnaturally cuts off high-pressure metamorphic reactions from the other ones. However, it is justified for the following two reasons.

First, a number of mineral reactions involved in high-pressure metamorphism have no counterparts in low- and medium-pressure ones.

Secondly, the common clastic sediments observed in typical high-pressure terranes are graywackes. Petrographic data on aluminous metapelites are abundant for low- and medium-pressure terranes, but are not available for high-pressure terranes. This enhances the uniqueness of high-pressure metamorphism.

Oxides and sulphides of iron, organic matter, and graphite occur in rocks of all baric types, and hence will be discussed near the end of chapters 7B and 8B.

7B-2 DISAPPEARANCE OF CLAY MINERALS

Available petrographic data on the change of clay minerals to metamorphic minerals under low-temperature regional metamorphic conditions is very scarce.

Frey (1969a, 1970) described the structural change of disordered illite into phengite, and the breakdown of a mixed-layer smectite-illite to produce phengite and chlorite.

Kaolinite could change to pyrophyllite by dehydration in progressive metamorphism. McNamara (1965) has stated that although clay minerals are usually absent in the chlorite-zone schists in the Dalradian Series, kaolinite occurs rarely in rocks rich in Al_2O_3 and poor in FeO and MgO. However, Chinner (1967) has suggested that kaolinite occurs in the area only as an alteration product.

Pyrophyllite has not been found in the Dalradian. In a detailed petrographic study of the chlorite-zone and biotite-zone metapelites in Vermont, Zen (1960) looked for pyrophyllite without success. It can only occur very rarely, if at all.

In the Cévennes, on the other hand, Tobschall (1969) has found that pyrophyllite is widespread in phyllites and schists of the chlorite and biotite zones, and is decomposed to produce andalusite with a rise in the temperature of metamorphism. The occurrence of pyrophyllite depends mainly on the chemical composition of the predominant metasediments (§7A-4).

The transformation of kaolinite to pyrophyllite is represented by the equation:

$$Al_2Si_2O_5(OH)_4 + 2SiO_2 = Al_2Si_4O_{10}(OH)_2 + H_2O. \tag{7B-1}$$

$$\underset{\text{kaolinite}}{} \quad \underset{\text{quartz}}{} \quad \underset{\text{pyrophyllite}}{}$$

This reaction was experimentally investigated by A. B. Thompson (1970a), as shown in fig. 7B-1.

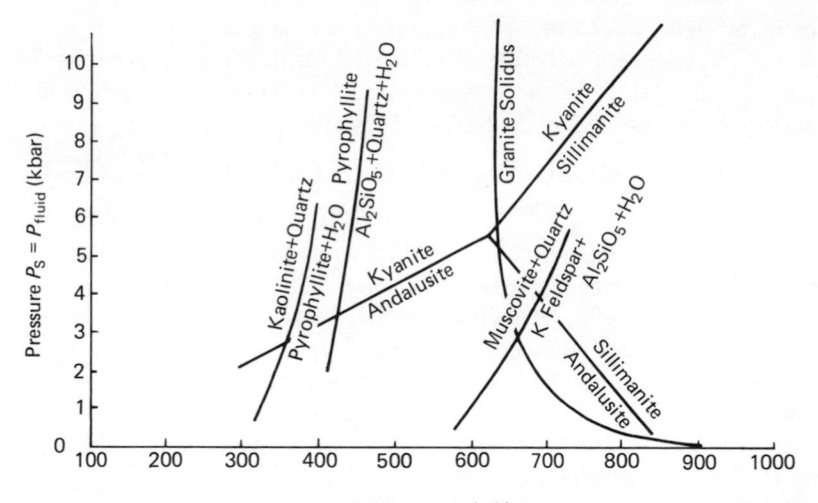

Fig. 7B-1   Experimentally determined equilibrium curves relevant to metamorphism of pelitic rocks. The dehydration curves and the granite solidus have been determined in the presence of an aqueous fluid.

7B-3   PARAGONITE

Though paragonite is the sodium analogue of muscovite, it behaves as if it were a more aluminous mineral than the latter because of its limited stability field. In low-temperature metapelites with an Al_2O_3 content just in excess of that required to form alkali feldspar, muscovite is formed. If the Al_2O_3 content is greater, paragonite also forms. Muscovite, including illite and phengite, is stable from the zeolite to the middle amphibolite facies, whereas paragonite is stable only over a fraction of this temperature range. The decomposition of paragonite at a moderate temperature should result in the formation of another highly aliminous mineral such as kyanite.

Paragonite had been regarded as a rare mineral on earth prior to 1954, when Eugster and Yoder discovered its common occurrence in metapelites in Vermont. Paragonite is usually mistaken for muscovite on cursory microscopic observation.

Low- and High-Temperature Limits

In contrast to the widespread occurrence of muscovite (illite) in unmetamorphosed soils and shales, the occurrence of paragonite has been extremely rarely reported. Under ordinary sedimentary and lowest metamorphic conditions, paragonite seems to be unstable. Zen (1960) has noted that paragonite is absent in chlorite-zone metapelites (slates), and begins to occur in chloritoid-bearing phyllite of the biotite zone in the Castleton area, Vermont. He has ascribed this relation to the existence of a low temperature stability limit for paragonite, as suggested by the following equation:

$$NaAlSi_3O_8 + Al_2Si_2O_5(OH)_4 = NaAl_3Si_3O_{10}(OH)_2 + 2\ SiO_2 + H_2O.$$

| albite | kaolinite | paragonite | quartz | (7B-2) |

The albite + kaolinite assemblage on the low-temperature side of this equation was observed, for example, by Coombs (1954, p. 81).

Frey (1969a, 1969b, 1970) found a mixed-layer paragonite-phengite in slates from the Glarus Alps, and considered that it represents an intermediate step in the transformation of Na-bearing illite or montmorillonite to paragonite in metamorphism.

The equilibrium curve for the breakdown of paragonite into albite, corundum and H_2O lies at temperatures about $100\ °C$ lower than that for the breakdown of muscovite (Chatterjee, 1970).

Paragonite and Pyrophyllite

In the medium-pressure terrane of the northern Appalachians, paragonite is widespread in metapelites ranging from the biotite to the kyanite zone, usually in association with quartz and muscovite and rarely with pyrophyllite. Paragonite occurs sporadically in metapelites of the Dalradian of western Scotland (Chinner, 1967). The mineral was found in many mica schists and glaucophane schists in the Alps (Harder, 1956).

Thus, the paragonite + quartz assemblage is stable over a wide $P-T$ field of medium- and high-pressure regional metamorphism. The composition relation of this assemblage to pyrophyllite is shown by the equation:

$$NaAl_3Si_3O_{10}(OH)_2 + 4\ SiO_2 = NaAlSi_3O_8 + Al_2Si_4O_{10}(OH)_2. \quad (7B-3)$$

| paragonite | quartz | albite | pyrophyllite |

The assemblage albite + pyrophyllite was found in the chlorite and biotite zones of the low-pressure regional metamorphism in the Cévennes (Tobschall, 1969). There, this assemblage appears to be stable up to the temperature of breakdown of pyrophyllite to form andalusite.

The existing data suggest that entropy and volume increase as reaction (7B-3) proceeds. Hence, the right-hand side should represent higher temperatures and lower pressures than the left. The occurrence of the albite + pyrophyllite assemblage in low-pressure metamorphism may be due to this. The plagioclase associated with epidote tends to have a higher An content in metamorphism under lower rock-pressures (Table 8B-1), and this may promote the stabilization of the right-hand side of equation (7B-3).

7B-4   MUSCOVITE

Illite and Muscovite
Illite is a general term for the mica-structure clay minerals including both dioctahedral and trioctahedral types. Though illite is abundant in unmetamorphosed and weakly metamorphosed pelites, its nature and stability are not clear. Illite in unmetamorphosed sediments commonly appears to be detrital in origin and unstable in many or most cases. The occurrence of celadonite has been commonly reported in zeolite-facies rocks.

Sericite or muscovite is said to occur in the high zeolite and higher facies (Utada, 1965; Seki *et al.* 1969*a*). In the greenschist, glaucophane-schist and higher facies with which we are now dealing, muscovite is very widespread in metapelites.

Metamorphic muscovites usually show a slight deficiency of alkali ions as compared with the idealized formula $KAl_3Si_3O_{10}(OH)_2$. A very small part (up to about 5 per cent) of the alkali sites may be occupied by H_3O^+ or H_2O.

Phengite
Complete substitution of $(Mg, Fe^{2+})Si$ for $AlAl$ in the idealized muscovite formula, accompanied by limited substitution of Fe^{3+} for Al, gives the formula $K(Mg, Fe^{2+})(Al,Fe^{3+})Si_4O_{10}(OH)_2$, which is that of the mineral celadonite. This is a distinct mineral species occurring in sedimentary and hydrothermal environments.

Metamorphic muscovites are solid-solutions between idealized muscovite and celadonite. The observed compositions fall in the range of 0–50 mol per cent celadonite. The muscovites near the idealized formula may be called muscovite proper, and those rich in celadonite component are called phengite. The term muscovite has been and will be used in this book in a wide sense so as to include muscovite proper, phengite and sericite.

Muscovites in metapelites of the amphibolite facies are usually close to the idealized muscovite formula in composition. On the other hand, Lambert (1959), Ernst (1963c), Butler (1967) and others have established that muscovites from metapelites of the epidote–amphibolite, greenschist and glaucophane-schist facies are usually phengites.

Phengite appears to be stable in the low-temperature part of low-, medium- and high-pressure regional metamorphism. However, abundant data are available only for the latter two types. The regular compositional variation of muscovite with temperature or metamorphic facies is shown in fig. 7B-2. The idealized

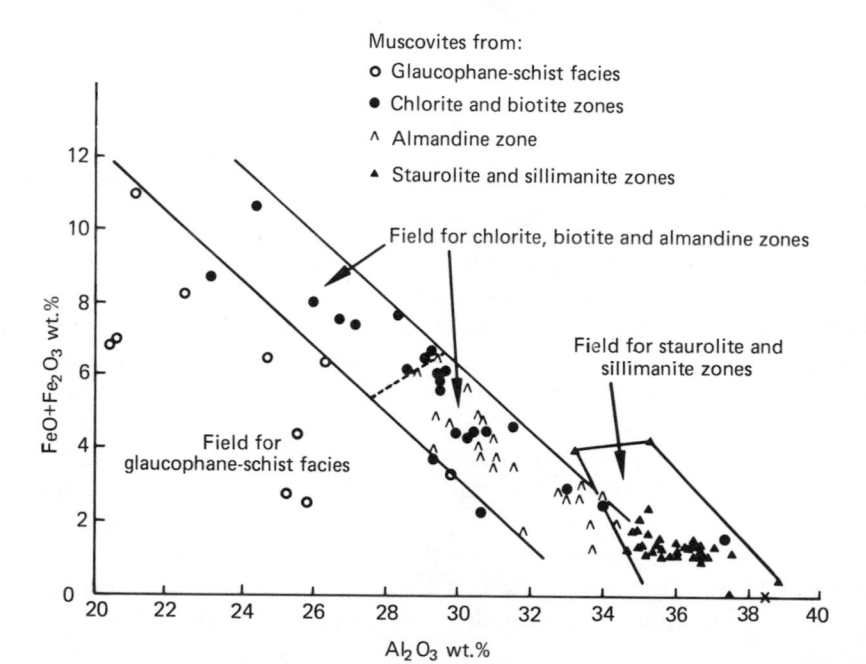

Fig. 7B-2 Composition of muscovites from metapelites. The cross on the abscissa at 38·4 per cent Al_2O_3 represents the idealized muscovite composition. Muscovites from the glaucophane-schist facies tend to have higher $MgO/(FeO + Fe_2O_3)$ ratios than chlorite and biotite zone muscovites. (Compiled from Butler, 1967; Guidotti, 1969 and others.)

muscovite composition has 38·4 per cent Al_2O_3 and 0 per cent $(FeO + Fe_2O_3)$. Muscovites from the staurolite and higher zones have compositions close to it. At a lower temperature, muscovite compositions tend to show greater departures from the idealized formula towards celadonite (Lambert, 1959; Butler, 1967; Guidotti, 1969).

The content of the celadonite molecule in muscovite changes also with variation in mineral assemblage at a definite temperature. Among the muscovites from the almandine zone of the Scottish Moine schists, for example, those in rocks with the muscovite + plagioclase + K-feldspar + quartz assemblage are systematically lower in Al_2O_3 and Na_2O than those in rocks with the muscovite + plagioclase + quartz assemblage (without K-feldspar). Guidotti (1969) has suggested that muscovite in the muscovite + plagioclase + K-feldspar + quartz assemblage should give the highest celadonite contents and be a good indicator of metamorphic temperature.

Muscovites near the idealized composition crystallize in the $2M_1$ structural type, whereas phengites occur usually in the $2M_1$ and rarely in the 3T type. Muscovites near the idealized composition have $\gamma = 1 \cdot 598 - 1 \cdot 605$ and $2V(-) = 36 - 40°$. Phengites usually have $\gamma = 1 \cdot 600 - 1 \cdot 626$ and $2V(-) = 28 - 35°$ with a pale greenish or pinkish colour. Thus, the compositional variation can be detected to some extent by the measurement of refractive indices (e.g. Miyashiro, 1958, p. 254) or of $2V$ (e.g. Iwasaki, 1963, p. 77).

Velde (1965) has made a synthetic study on the stability relations of phengite. In the presence of an aqueous fluid, the composition range of muscovite expands to phengite with increasing pressure and decreasing temperature. This is consistent with its observed modes of occurrence. The breakdown of phengite produces muscovite + biotite + K-feldspar + quartz. The equilibrium curve for the breakdown reaction of phengite with a composition of 70 per cent muscovite + 30 per cent (Mg, Al)-celadonite lies at 4 kbar at $400°C$.

Muscovite–Paragonite Relations

There is a wide miscibility gap between muscovite and paragonite. Eugster and Yoder (1955) and Iiyama (1964) made synthetic experiments to determine the limit of the gap. It becomes smaller with rising temperature, as shown schematically in fig. 7B-3, thus indicating a possibility of its use as a geologic thermometer.

The basal spacing of muscovite is around $9 \cdot 9 - 10 \cdot 0$ Å, whereas that of paragonite is around $9 \cdot 6 - 9 \cdot 7$ Å. Thus, two micas can be easily distinguished by the X-ray powder method. The d_{002} values of such coexisting micas are a function of the miscibility gap, and hence could be used as a geologic thermometer. Zen and Albee (1964) have shown the validity of this idea, as shown in fig. 7B-4.

After the decomposition of paragonite near the boundary between the kyanite and sillimanite zones, the maximum value of $Na/(Na + K)$ in muscovite rapidly decreases. After muscovite begins to decompose to sillimanite and K-feldspar, the $Na/(Na + K)$ ratio of muscovite appears to show a further decrease. According to Evans and Guidotti (1966), the maximum value of the atomic ratio $Na/(Na + K)$ in muscovite in the sillimanite zone of western Maine is $0 \cdot 18$ before the beginning of muscovite decomposition, and $0 \cdot 09$ after.

Natural muscovite and paragonite are not simple binary solutions. Muscovite may be phengitic. Paragonite commonly shows a small extent of the NaSi $\rightleftharpoons$

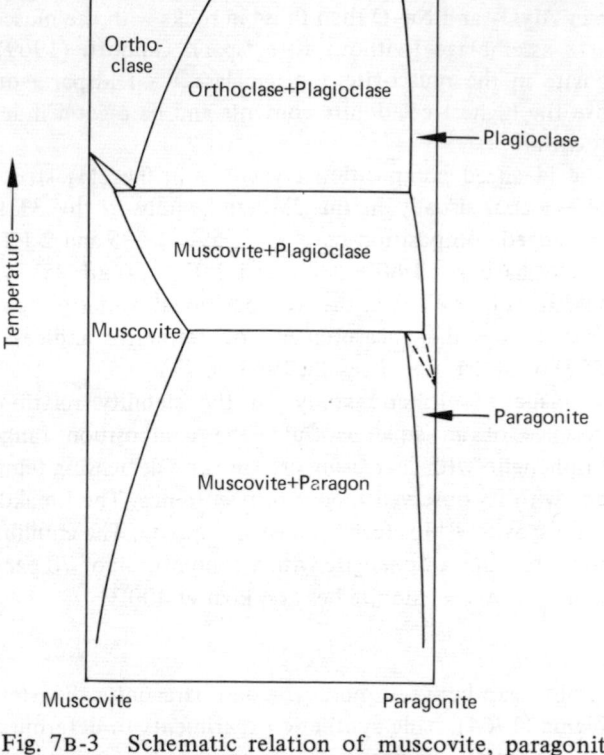

Fig. 7B-3 Schematic relation of muscovite, paragonite and feldspars in the presence of quartz. An Al_2SiO_5 mineral is present in all the feldspar-containing fields. (Evans and Guidotti, 1966.)

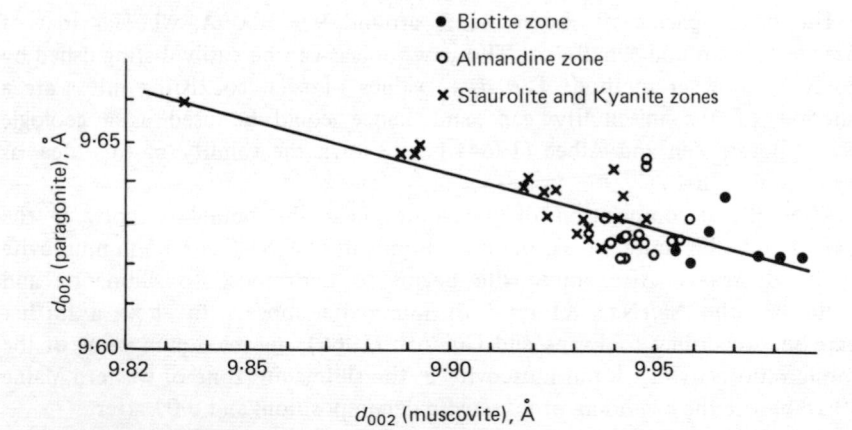

Fig. 7B-4 Basal spacing of coexisting muscovite and paragonite in relation to metamorphic zones. (Zen and Albee, 1964.)

CaAl replacement. These should cause a change in the miscibility gap and deviation of d_{002} from the value in the idealized case.

7B-5 CHLORITOID

Like paragonite, chloritoid (Fe^{2+}, Mg, Mn)$Al_2SiO_5(OH)_2$ occurs characteristically in highly aluminous low-temperature metapelites. It has been found in some contact aureoles and low-pressure regional metamorphic terranes in association with andalusite and/or cordierite as well as in the low-temperature zones of the Barrovian sequence and the Swiss Alps and other high-pressure metamorphic terranes. Thus, the range of rock-pressure for the formation of chloritoid is very wide. Chloritoid has monoclinic and triclinic forms.

Chloritoid occurs only in rocks with an Al_2O_3 content in excess of the quantity present in micas and epidote. The rocks must also have high Fe^{2+}/Mg ratios (Halferdahl, 1961). Only a small portion of metapelites fulfill these conditions. Chloritoids usually have $Fe^{2+}/(Mg + Fe^{2+})$ ratios above 0·6. Some chloritoids have a considerable MnO content.

It is not clear what are the low-temperature equivalents of chloritoid. Possible reactions to form chloritoid with rising temperature are as follows:

5 haematite + chlorite

$$= 5 \text{ magnetite} + 2 \text{ chloritoid} + 2 \text{ quartz} + 4 H_2O \qquad (7\text{B-4})$$

$$5 \text{ kaolinite} + \text{chlorite} = 7 \text{ chloritoid} + 7 \text{ quartz} + 2 H_2O \qquad (7\text{B-5})$$

5 pyrophyllite + chlorite

$$= 7 \text{ chloritoid} + 17 \text{ quartz} + 4 H_2O \qquad (7\text{B-6})$$

5 paragonite + chlorite + 3 quartz

$$= 7 \text{ chloritoid} + 5 \text{ albite} + 4 H_2O. \qquad (7\text{B-7})$$

These reactions would be greatly influenced by the Fe^{2+}/Mg ratio of the rocks. The pyrophyllite + chlorite assemblage is widespread but chloritoid is not present in the Cévennes (Tobschall, 1969).

Chloritoid occurs in the low-temperature parts (probably from the chlorite to the lower staurolite zone) of the Dalradian metamorphics in the Scottish Highlands. The localities with chloritoid are within a narrow belt about 40 km long extending from the east coast toward the southwest. Paragonite has not been found in this belt. Further to the southwest, chloritoid has not been found, and instead paragonite has been found at many localities. Equation (7B-7) may represent the differing stable assemblages in these northeastern and southwestern areas (Chinner, 1967).

The breakdown of chloritoid with rising temperature usually gives rise to staurolite. In some cases of low-pressure metamorphism, however, it may produce cordierite. Since chloritoid is a hydrous mineral high in FeO, its breakdown reactions depend not only on the rock-pressure, but also on P_{H_2O} and P_{O_2}. Halferdahl (1961), Hoschek (1967), Ganguly (1968), Ganguly and Newton (1968), and Richardson (1968) have made experimental investigations on the breakdown reactions of pure Fe-chloritoid. According to these studies, chloritoid by itself could be stable over all the pressure range ($P_S = P_{H_2O}$) from near zero to the highest investigated value (about 20 kbar) and up to about 500–700 °C depending on P_{O_2}. In natural rocks, reactions with associated minerals could decrease the breakdown temperature.

7B-6 CHLORITE

Chlorites may be regarded as solid solutions composed of the following two components and their iron analogues as shown in fig. 7B-5.

antigorite	$Mg_6Si_4O_{10}(OH)_8$
amesite	$Mg_4Al_4Si_2O_{10}(OH)_8$

Thus, the main substitution is $Si(Mg, Fe^{2+}) \rightleftharpoons Al\, Al$. The structures of natural chlorite are made up by the stacking of unit layers, about 14 Å thick. In hydrothermal experiments, not only chlorite with the 14 Å structure but also another mineral of similar composition with a 7 Å structure is synthesized. Such 7 Å minerals are called septechlorites. The natural minerals antigorite and amesite are septechlorites. All the 'chlorites' in pelitic and basic schists so far examined have been found to have a 14 Å structure (Banno, 1964; McNamara, 1965).

The range of the $Si(Mg, Fe^{2+}) \rightleftharpoons Al\, Al$ substitution in chlorites from metapelites, however, is very limited. The atomic proportion of Si on the basis of 18(O, OH), or on the anhydrous basis of 14O, is usually in the range of 2·4–2·9 as shown in fig. 7B-5. Thus, an idealized formula for such chlorites is $(Mg, Fe^{2+})_{4.5}Al_3Si_{2.5}O_{10}(OH)_8$.

Metamorphic chlorites with Si contents in the range described above, and with a wide variation in the Fe^{2+}/Mg ratio, belong mostly to ripidolite in Hey's (1954) classification. The $Fe^{2+}/(Fe^{2+} + Mg)$ ratio of chlorites in metapelites is usually in the range 0·2–0·8 (Albee, 1962). The chlorites of metapelites tend to show a decrease in the Fe^{2+}/Mg ratio with rising metamorphic temperature. This tendency is noticeable in the Barrovian zones as well as in high-pressure terranes (fig. 7B-5). It has been detected, for example, by a decrease of the refractive indices with increasing metamorphic temperature (Banno, 1964, p. 259–60).

Muscovite (including phengite) has a lower Fe^{2+}/Mg ratio than the associated chlorite, and hence its breakdown causes a decrease of this ratio in chlorite. Biotite and garnet have higher Fe^{2+}/Mg ratios than the associated chlorite, and hence their formation also results in a decrease of the ratio in chlorite (e.g. Ernst, 1964; Atherton, 1968).

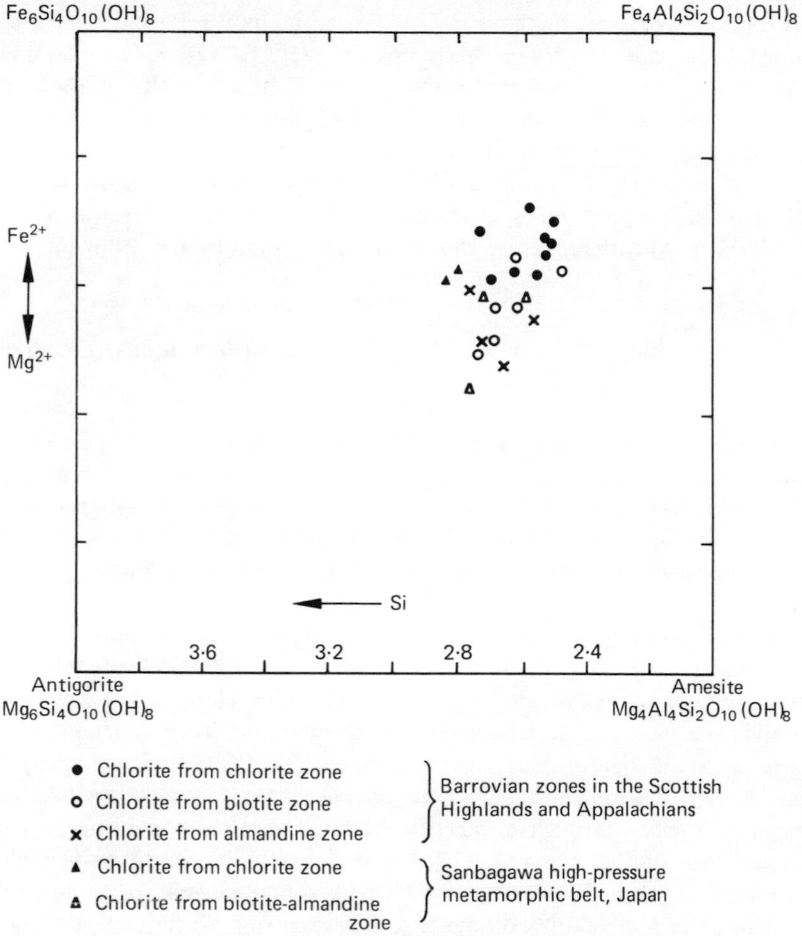

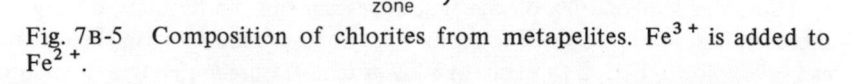

Fig. 7B-5 Composition of chlorites from metapelites. Fe^{3+} is added to Fe^{2+}.

$Fe_6Si_4O_{10}(OH)_8$ · · · $Fe_4Al_4Si_2O_{10}(OH)_8$

Fe^{2+} ↕ Mg^{2+}

← Si

3·6   3·2   2·8   2·4

Antigorite
$Mg_6Si_4O_{10}(OH)_8$

Amesite
$Mg_4Al_4Si_2O_{10}(OH)_8$

● Chlorite from chlorite zone
○ Chlorite from biotite zone   } Barrovian zones in the Scottish Highlands and Appalachians
✗ Chlorite from almandine zone

▲ Chlorite from chlorite zone   } Sanbagawa high-pressure metamorphic belt, Japan
△ Chlorite from biotite-almandine zone

Mg-chlorite by itself is stable up to about 800 °C in the presence of an aqueous fluid (Fawcett and Yoder, 1966). In metamorphic rocks, however, it decomposes usually at a much lower temperature by reaction with associated minerals such as quartz and muscovite. In metapelites and metapsammites, chlorite is always associated with quartz. The chlorite + quartz assemblage is said to be stable up to about 650 °C, above which chlorite reacts with quartz to produce cordierite + talc at low pressures, yoderite + talc + quartz at high pressures, and kyanite + talc at very high pressures (Schreyer, 1968).

7B-7 FORMATION OF BIOTITE

In some low-pressure metamorphic terranes, the biotite isograd for metapelites roughly coincides with the beginning of recrystallization, whereas in other low-pressure terranes and the Barrovian series there is a chlorite zone on the low-temperature side of the isograd. Most of the high-pressure terranes have not reached a temperature high enough to form biotite.

Prior to the middle 1950s, a false idea was popular that muscovite was nearly constant in composition. Accordingly, the formation of biotite was regarded as resulting in Al-enrichment of chlorite as represented by the following relation:

6 muscovite + 3 antigorite component of chlorite

= 6 biotite + 3 amesite component of chlorite + 14 quartz + $8H_2O$

$$(7B-8)$$

However, this relation is at variance with recent mineralogical data. The chlorites in metapelites do not become richer in Al_2O_3 with rising temperature (fig. 7B-5). As stated before, muscovites in low-temperature metapelites have a wide range of solid solution. The biotite-producing reactions were discussed from the new viewpoint by Ernst (1963c), Velde (1965), Mather (1970), and Brown (1971).

Figure 7B-6 represents three progressive stages of biotite formation. In stage (a), biotite forms in rocks to the right of the microcline–chlorite join. Then, a reaction such as microcline + chlorite → biotite + white mica + quartz + H_2O would take place in a particular range of temperature, above which biotite forms in rocks to the right of microcline–phengite–chlorite join as shown in (b). With a further rise of temperature, the composition range of muscovite solid solutions becomes smaller through a reaction such as phengite + chlorite → idealized muscovite + biotite + quartz + H_2O. Consequently, biotite forms in rocks of a composition range that becomes wider with rising temperature, as shown in (c).

Thus, the temperature of the first appearance of biotite depends largely on the chemical composition of rocks in the terranes. In the Scottish Highlands, for example, biotite begins to occur at a lower temperature in metagraywackes than in metapelites (Mather, 1970). Since the metapelites of an area usually show a considerable variation in composition, a biotite isograd cannot appear as a sharp line on a map. The progressive metamorphism of pelitic rocks usually causes partial reduction of Fe^{3+} to Fe^{2+}. The Fe^{2+} thus produced is incorporated in biotite and other minerals.

It has been established that the Fe^{2+}/Mg ratio of muscovite and that of the coexisting biotite in metapelites show a regular and sympathetic variation (Evans and Guidotti, 1966; Butler, 1967). This indicates that the coexisting muscovite and biotite are usually in close approach to chemical equilibrium with each other.

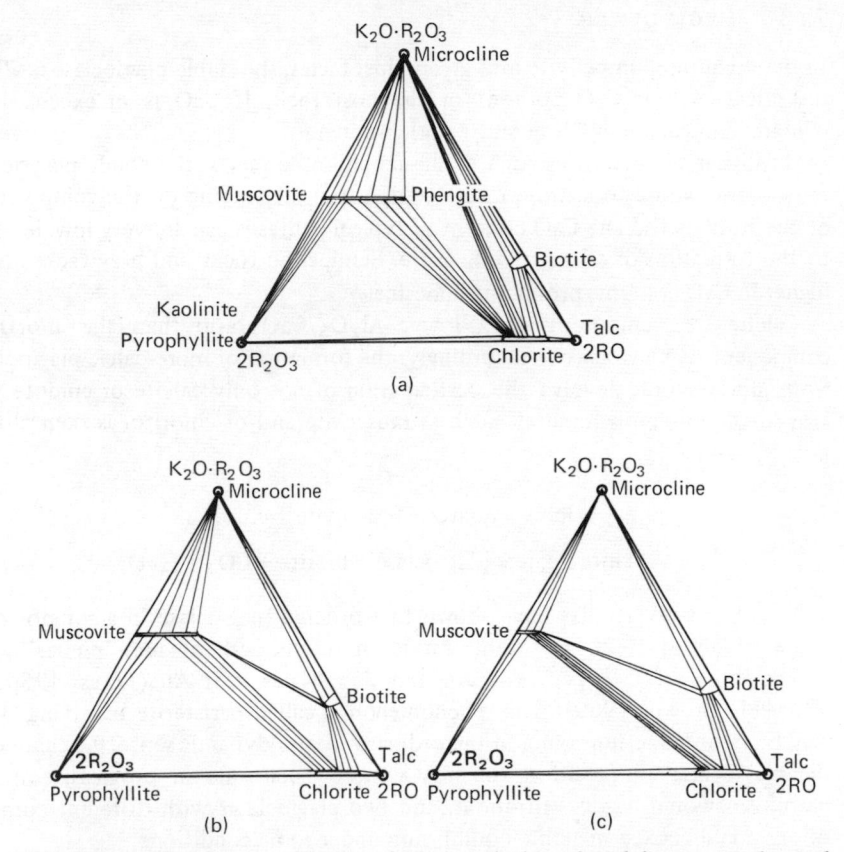

Fig. 7B-6 Progressive changes of phase relations involving muscovite and biotite in metapelites.

Muscovite and chlorite have lower Fe^{2+}/Mg ratios than the associated biotite. Hence, the progress of biotite-producing reactions would tend to cause a decrease in the Fe^{2+}/Mg ratio of biotite. A further rise in temperature would tend to produce almandine in medium- and high-pressure metamorphism. This would promote a further decrease of the Fe^{2+}/Mg ratio in biotite. If cordierite forms in low-pressure metapelites, however, the Fe^{2+}/Mg ratio of the associated biotite should increase.

The compositional change of biotite in metapelites with rising temperature has been discussed by a number of authors. Decrease of Fe^{2+}/Mg ratio and increase of TiO_2 content have been widely observed (Miyashiro, 1958; Lambert, 1959; Engel and Engel, 1960; Oki, 1961b; Binns, 1969).

7B-8 PLAGIOCLASE

In the prehnite–pumpellyite and greenschist facies, the stable plagioclase is albite regardless of the CaO content of the host rock. If CaO is in excess, it is contained in calcite, epidote and/or other minerals.

At higher temperatures such as the amphibolite facies, the stable plagioclase ranges from albite to a more calcic composition depending on the composition of the host rocks. The CaO content of typical pelites is usually very low, leading to the formation of albite or oligoclase. Semipelitic rocks and graywackes may higher in CaO content, producing andesine.

Calcite and epidote have a lower Al_2O_3/CaO ratio than the anorthite component of plagioclase. Accordingly, the formation of more calcic plagioclase from albite should involve the participation of not only calcite or epidote but also some aluminous minerals such as muscovite and/or chlorite, as exemplified by:

$$\text{albite + calcite + quartz + muscovite + chlorite}$$

$$\rightleftharpoons \text{more calcic plagioclase + biotite} + CO_2 + H_2O. \qquad (7\text{B-9})$$

Detailed X-ray studies have shown that ordered plagioclase in a composition range of about 2–20 per cent An is an intergrowth of two 'phases' with compositions of 0–1 per cent An and 25–28 per cent An (Laves, 1954, p. 409–411; Brown, 1960). This phenomenon is called peristerite unmixing. If it may be regarded as unmixing in the ordinary thermodynamic sense, the change of the plagioclase composition towards a more calcic one in progressive metamorphism would be discontinuous, and two plagioclases with different compositions could coexist in stable equilibrium under some conditions.

An apparently discontinuous change of plagioclase composition from about 5 or 10 per cent to about 20 per cent An with increasing metamorphism has been reported from some areas, and this may be a result of the peristerite unmixing (Lyons, 1955; Brown, 1962; Wenk, 1962; Crawford, 1966). According to Crawford (1966), plagioclase in semipelites of the chlorite zone in Vermont contains usually less than 1·0 per cent An, and that of the lower biotite zone contains a few to about 5 per cent An. In the higher part of the biotite zone and in the lower part of the almandine zone, two plagioclases coexist, one with a few or several per cent of An and the other with about 20–30 per cent An. The sodic plagioclase disappears in the staurolite zone, where the temperature of metamorphism may have been above that of the top of the peristerite solvus. In the Alpine schists of New Zealand, the observed relation is similar, but the sodic member of the two coexisting plagioclases contains less than 2 per cent An. When two plagioclases coexist in the same rock, the more calcic one may occur as a rim around the more sodic core. The boundary between them is sharp.

The coexistence of plagioclases with different compositions was reported from various other metamorphic rocks, including intermediate and calcic plagioclases in granulite facies rocks (e.g. Tsuji, 1967; Ono, 1969b; Mall and Singh, 1972). In any of these descriptions, however, sufficient proof for equilibrium of the coexisting phases has not been given.

7B-9 CORDIERITE

Cordierite and almandine are common in metapelites in low- and medium-pressure regional metamorphic terranes, respectively. They are fairly similar in chemical composition: i.e. cordierite (Me, Fe)$_2$Al$_4$Si$_5$O$_{18}$ and almandine (Fe, Mg)$_3$Al$_2$Si$_3$O$_{12}$. They begin to occur on the high-temperature side of the biotite isograd, usually at a considerable distance from the biotite isograd. However, metamorphic cordierites usually have an atomic ratio Fe^{2+}/(Mg + Fe^{2+}) lower than 0·6, whereas almandines have a higher ratio and may contain significant amounts of MnO and CaO.

Cordierite is a low-pressure mineral with a relatively low density. Figure 7B-7 shows that the high pressure stability limit of Mg-cordierite is close to the boundary between the sillimanite and kyanite fields (Schreyer, 1968; Seifert and Schreyer, 1970). The association of cordierite with sillimanite or andalusite is common in metapelites, whereas that of cordierite with kyanite is very rare. Hietanen (1956) described a rock showing the coexistence of cordierite, andalusite, sillimanite and kyanite. The high pressure stability limit of Fe-cordierite, being about 3·5 kbar, is much lower than that of Mg-cordierite (Richardson, 1968). Cordierite shows structural variations and hydration (Miyashiro, 1957a; Iiyama, 1960; Schreyer, 1964–6; Gibbs, 1966).

Almandine is much more widespread in medium-pressure metamorphic terranes than in low-pressure ones. However, almandine by itself is not a high-pressure mineral. It is stable on its own composition over the entire pressure range on the low temperature side of a line representing the high temperature limit, which varies markedly with oxidation condition (Hsu, 1968). Hensen (1971) has clearly shown that cordierite is stable in many low-pressure assemblages, whereas almandine is stable in many high-pressure ones.

A petrogenetic grid for cordierite and related minerals characteristic of metapelites is shown in fig. 3-9 (Hess, 1969 and pers. comm. 1970). It suggests that at low pressures cordierite becomes stable at about 450–500 °C in rocks with very low Fe^{2+}/Mg ratios and in association with quartz, muscovite, biotite and chlorite. In rocks with higher Fe^{2+}/Mg ratios, garnet and staurolite occur instead. At 2–3 kbar, chlorite reacts with andalusite with a rise of temperature to form cordierite and staurolite. The composition field of the rocks which form cordierite becomes much wider by this reaction. At 4 kbar, on the other hand,

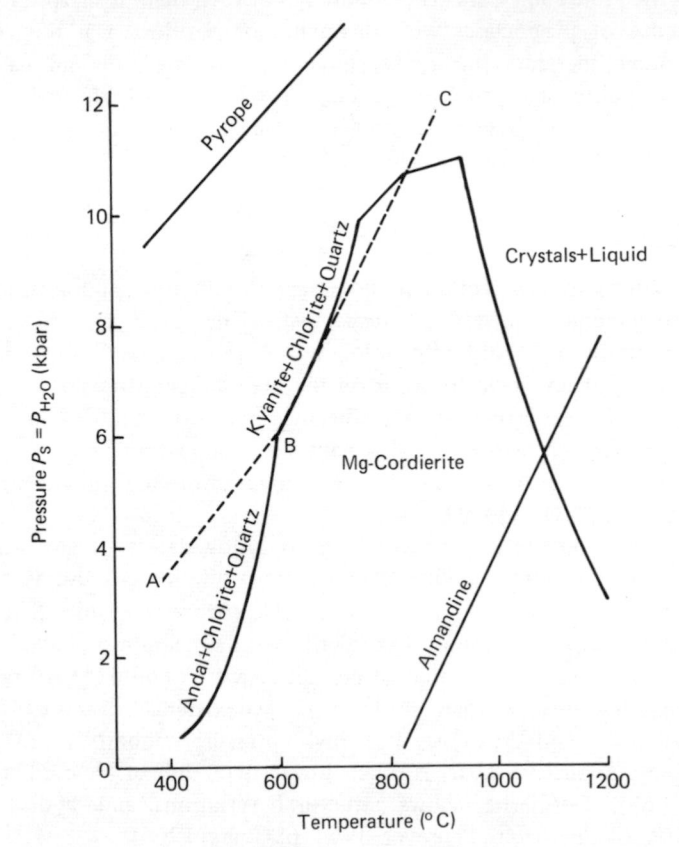

Fig. 7B-7   Stability fields of Mg-cordierite, almandine and pyrope, each by itself (Seifert and Schreyer, 1970; Hsu, 1968: Boyd and England, 1959). The stability field of almandine is for the iron-magnetite and iron-wüstite buffers. Lines AB and BC represent the field boundaries between andalusite and kyanite, and between sillimanite and kyanite respectively.

chlorite reacts with garnet with a rise of temperature to form staurolite and biotite. Generally speaking, the composition field of the rocks which form cordierite becomes narrower with increasing pressure.

In this grid, the composition of each solid-solution mineral is assumed to be constant in order to simplify the calculations. Chinner (1962) has discussed possible paragenetic relations of metapelites with special reference to the variation of the composition ranges of solid solution minerals. The schematic *AFM* diagram of fig. 7B-8a represents an assumption that under a very low rock-pressure, a complete Fe–Mg solid-solution series of cordierite is stable, whereas garnet can occur only in rocks with unusually high Fe^{2+}/Mg ratios and

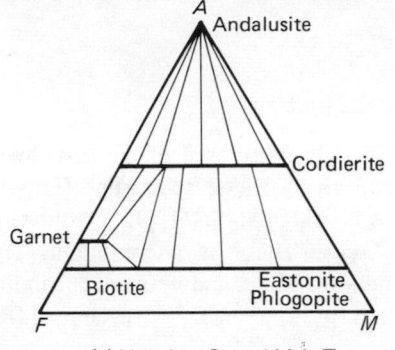

(a) Very low P_S and high T

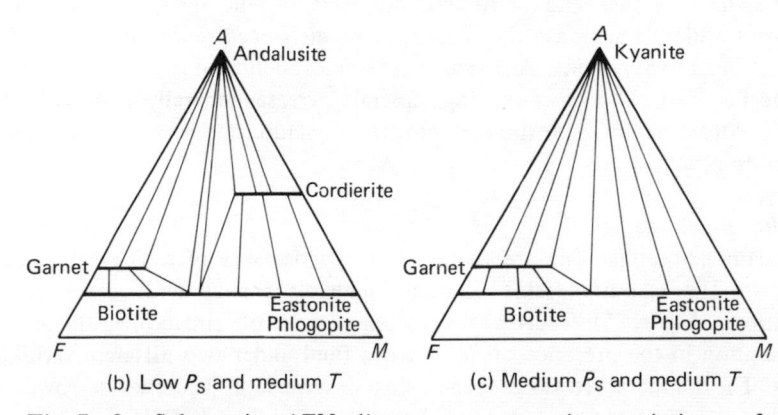

(b) Low P_S and medium T (c) Medium P_S and medium T

Fig. 7B-8 Schematic *AFM* diagrams representing variations of paragenetic relations of garnet and cordierite in metapelites with temperature and pressure. $A = Al_2O_3 - (K_2O + Na_2O)$, $F = FeO$, $M = MgO$. (Modified from Chinner, 1962.)

relatively low Al_2O_3/(FeO + MgO) ratios. This would correspond to high-temperature, low-pressure contact aureoles.

With increasing rock-pressure and decreasing temperature, the Fe^{2+}/Mg range of cordierite becomes narrower, resulting in the enlargement of the rock composition field favourable for the formation of garnet (fig. 7B-8b). The cordierite-garnet join would be replaced by the andalusite-biotite join, which conceivably corresponds to the medium-temperature zones of low-pressure regional metamorphism.

With a further rise in pressure, the Fe^{2+}/Mg range of cordierite dwindles and then the mineral becomes physically unstable (fig. 7B-8c). The composition field of metapelites which form garnet would become wider. This situation would correspond to a medium-temperature zone of the medium-pressure regional metamorphism.

7B-10 PYRALSPITE GARNETS

Pyralspite garnets are mainly composed of the first three of the following five common components of the garnets: spessartine ($Mn_3Al_2Si_3O_{12}$), almandine ($Fe_3Al_2Si_3O_{12}$), pyrope ($Mg_3Al_2Si_3O_{12}$), grossular ($Ca_3Al_2Si_3O_{12}$), and andradite ($Ca_3Fe_2Si_3O_{12}$). There is a continuous series of solid solution between spessartine and almandine, and between almandine and pyrope, but not between spessartine and pyrope owing to a large difference in ionic radius between Mn^{2+} and Mg.

Pyralspites are widespread in metapelites and metapsammites in medium- and high-pressure regional metamorphic areas. Though they occur in some rocks of low-pressure regional and some contact metamorphic terranes, higher rock-pressures evidently increase the frequency of their occurrence and their average amount in ordinary rocks. Almandine is the most common garnet in such rocks.

The Fe^{2+}/Mg ratio of coexisting minerals increases generally in the following order: cordierite $\rightarrow$ chlorite $\rightarrow$ biotite $\rightarrow$ staurolite $\rightarrow$ almandine garnet (Thompson, 1957).

Stability Relations

Spessartine and almandine are stable under a wide range of rock-pressure down to zero. The stability fields of these minerals change with the oxidation conditions. Figure 7B-9 gives the stability field of almandine on its own composition in the presence of an aqueous fluid under two different oxidation states. The temperature range of stability of almandine becomes narrower and disappears with increasing P_{O_2} as shown in fig. 7B-10. Relatively reducing conditions are essential for the formation of the mineral (Hsu, 1968).

In synthetic experiments on the spessartine composition with excess H_2O, hydrated and unhydrated spessartines form respectively below and above about 600 °C. The low temperature limit of stability of spessartine is much lower than that of almandine and is nearly independent of P_{O_2} within the P_{O_2} range commonly realized in metamorphism. Manganese changes its oxidation state less readily than iron (Hsu, 1968).

Pyrope is stable only at pressures above 10 kbar at 400 °C as shown in fig. 7B-7 (Boyd and England, 1959). Garnets rich in the pyrope component are not stable in regional metamorphism except for the granulite and eclogite facies.

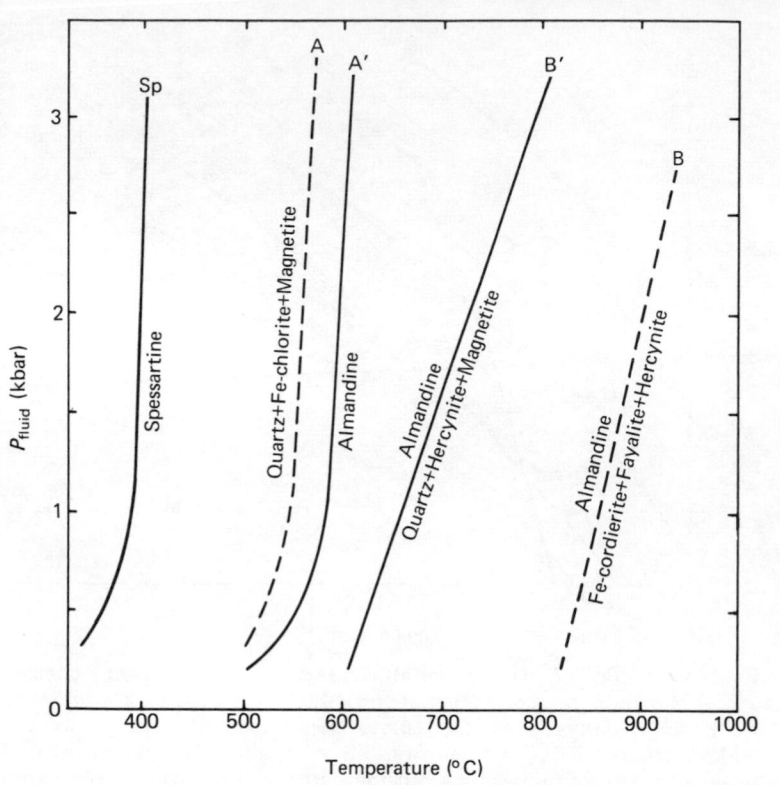

Fig. 7B-9   Stability relations for almandine bulk composition in the presence of an aqueous fluid. Curves A and B are with the iron-magnetite and iron-wüstite buffers, whereas curves A′ and B′ are for a higher P_{O_2} with the fayalite-magnetite-quartz buffer. Curves A and A′ have the same mineral assemblages on their low-temperature side as well as on their high-temperature one. Curve Sp represents the low-temperature limit of the stability field of spessartine on its own composition in the presence of an aqueous fluid. (Hsu, 1968.)

Compositional Changes in Progressive Metamorphism and Zoning

The progressive compositional change of garnet has been investigated in many areas. In the high-pressure terrane of the Bessi-Ino area, the MnO content of garnet in metapelites is as high as 9–13 per cent in the lowest temperature part up to the biotite isograd, and rapidly decreases to about 3 per cent with further rise in temperature (Banno, 1964).

In the Barrovian region of the Scottish Highlands, Sturt (1962) and Atherton (1968) have studied the compositional variation of pyralspites. On the almandine isograd, almandine contains about 5–6 per cent MnO and 6–9 per cent CaO. With rise in temperature, the MnO and CaO contents decrease rapidly, whereas the FeO and MgO contents increase. Almandine in metapelites of the high-

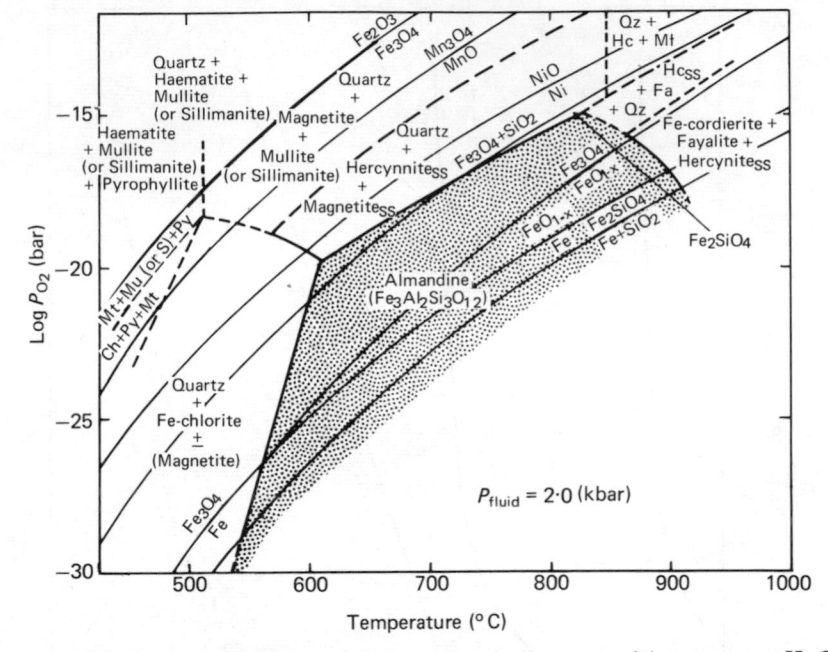

Fig. 7B-10 P_{O_2}-T diagram for almandine bulk composition + excess H_2O at 2 kbar fluid pressure (Hsu, 1968). Field boundaries are presented as thick lines. Oxygen buffer curves are presented as thin solid lines. *Abbreviations:* CH = Fe-chlorite, Fa = fayalite, Hc = hercynite, Mt = magnetite, Mu = mullite, Py = pyrophyllite, Qz = quartz, S = sillimanite.

temperature part of the almandine zone commonly contain about 0·6 per cent MnO, 28 per cent FeO, 3–4 per cent MgO and several per cent CaO. In both the Bessi-Ino area and Barrovian region, the almandine isograd lies very close to the threshold to the epidote-amphibolite facies zone.

Müller and Schneider (1971) re-investigated the compositional variation of garnets in metapelites of the Stavanger area, Norway, which had formerly been described by Goldschmidt (1921). The garnets of the chlorite zone showed 14–5 per cent MnO. The MnO content decreases with rising temperature down to values around 1 per cent in the amphibolite facies. The garnets in metamorphosed arkoses and pyroclastics did not show such a regular variation in composition.

A remarkable decrease of the MnO content of metapelite garnets was found in the low-pressure metamorphic terrane of the central Abukuma Plateau (Miyashiro, 1953b, 1958). Here, the garnet of a biotite-bearing greenschist facies schist gave 19·74 per cent MnO. With a rise of temperature, the MnO content decreases to about 16 per cent at the beginning of the amphibolite facies and to about 3–5 per cent in the high-temperature part of the same facies.

Thus, the trend of compositional variation in garnet with rising temperature is similar in all the baric types: The MnO content decreases and the FeO content increases. With further rise in temperature, the MgO content increases.

Garnets in metapelites are strongly zoned with decreasing MnO and CaO contents and an increasing MgO content toward the rim (Atherton, 1968, Brown, 1969). This trend is the same as that of the above-summarized progressive changes.

Garnet-Chlorite and Garnet-Biotite Relations

When garnet begins to occur at a relatively low temperature, the necessary materials would be supplied mainly by gradual decomposition of chlorite. The $(Mg + Fe^{2+} + Mn^{2+})/Al$ ratio of metapelite chlorites is roughly in the range of $1\cdot2-2\cdot5$ and that of pyralspite is similar, being $1\cdot5$ or a little less. Thus, the formation of garnet from chlorite in metapelites would be approximately represented by the following sliding equilibrium:

$$2 (Mg, Fe, Mn)_{4.5}Al_3Si_{2.5}O_{10}(OH)_8 + 4 SiO_2$$

chlorite   quartz

$$= 3 (Mg, Fe, Mn)_3Al_2Si_3O_{12} + 8 H_2O. \qquad (7B-10)$$

garnet

After chlorite is used up, garnet continues to coexist with biotite in metapelites. Garnets always have much greater Mn/Fe^{2+} and Fe^{2+}/Mg ratios than the associated chlorites and biotites. In chlorite, biotite and most other ferromagnesian minerals, Mg^{2+}, Fe^{2+} and Mn^{2+} ions are in six-fold coordination, whereas in garnet they are in eight-fold coordination together with Ca^{2+}. This exceptional structural feature of garnet should control the equilibrium distribution of Mg, Fe and Mn between garnet and associated ferromagnesian minerals. The sites in eight-fold coordination in garnet are so large that Ca^{2+} and Mn^{2+} ions can be properly accommodated, but small ions of Mg^{2+} would be concentrated in associated ferromagnesian minerals (Miyashiro, 1953b; Zermann, 1962).

A number of workers calculated the Mn–Fe and Mn–(Fe, Mg) distribution coefficients between garnet and chlorite, and between garnet and biotite. The values obtained showed irregular variations without a clear relationship to the temperature of metamorphism. This irregularity would come largely from the strong zoning of garnets.

On the other hand, the Fe—Mg distribution constant between garnet and biotite, here denoted as K, shows a regular decrease with increasing temperature of metamorphism (e.g. Albee, 1965b; Lyons and Morse, 1970):

$$Mg(\text{in biotite}) + Fe^{2+}(\text{in garnet})$$

$$= Fe^{2+}(\text{in biotite}) + Mg(\text{in garnet}) \qquad (7B-11)$$

$$\frac{(\text{Fe/Mg}) \text{ in garnet}}{(\text{Fe/Mg}) \text{ in biotite}} = K \qquad (7\text{B-}12)$$

At a temperature within the amphibolite facies, the reaction of muscovite and biotite may begin to produce almandine, as follows:

$$KAl_3Si_3O_{10}(OH)_2 + K(Mg, Fe)_3AlSi_3O_{10}(OH)_2 + 3\ SiO_2$$

$$\text{muscovite} \qquad\qquad \text{biotite} \qquad\qquad \text{quartz}$$

$$= (Mg, Fe)_3Al_2Si_3O_{12} + 2\ KAlSi_3O_8 + 2\ H_2O. \qquad (7\text{B-}13)$$

$$\text{almandine} \qquad\qquad \text{K-feldspar}$$

Garnets have much higher Mn/Fe^{2+} and Fe^{2+}/Mg ratios than coexisting biotites. Above this temperature garnet can coexist with K-feldspar.

Complete Fractionation Model for the Progressive Decrease of the MnO Content of Garnet

The progressive decrease of the MnO content in garnet will be discussed below with the help of two models representing two extreme cases (Miyashiro and Shido, 1973). Hollister (1966) and Atherton (1968) have demonstrated that the variation in the MnO content of some zoned garnets can be semiquantitatively explained under the following assumptions: (*a*) There is no diffusion in garnet. Only the outermost layer of garnet crystals is in equilibrium with the surroundings, and once crystallized the garnet is immediately removed from further reaction with the system. (*b*) Diffusion in the surroundings is complete. (*c*) The fractionation factor of MnO between the outermost layer of garnet crystals and the surroundings is constant. These assumptions give a complete fractionation model for garnet growth.

In this case, MnO is highly concentrated in the early formed garnet, which is then removed from the active diffusional system, with a resultant decrease in the MnO content of the system. Hence, the subsequently formed garnet becomes rapidly poorer in MnO. The following relation holds:

$$x = SC(1 - g)^{S-1}. \qquad (7\text{B-}14)$$

where
- x = weight per cent of MnO in the outermost layer of garnet,
- g = weight fraction of all the already crystallized garnet in all the ferromagnesian minerals of the rock,
- S = (weight per cent of MnO in a garnet rim)/(average weight per cent of MnO in the associated ferromagnesian minerals), and
- C = average weight per cent of MnO in all the ferromagnesian minerals of the rock (including garnet).

Here, the 'rock' should be understood to mean materials in the volume within the effective range of diffusion around the garnet.

If S remains constant, the average MnO content (X_{av}) of zoned garnet crystals is calculated from the above equation as follows:

$$X_{av} = \frac{C}{g} \{1 - (1-g)^S\} = C\{S - \tfrac{1}{2}g[S/(S-1)] + \ldots\} \qquad (7\text{B-}15)$$

Therefore, when $g \to 0$, $X_{av} \to CS$. For $S > 10$, the value of $\{1 - (1-g)^S\}$ is close to unity in the range of $g > 0.2$; hence the following approximate relation holds for this range:

$$X_{av} = \frac{C}{g}. \qquad (7\text{B-}16)$$

The change of X_{av} with g is illustrated in fig. 7B-11.

The actual growth of garnet may take place with a practically constant value of S, or may take place with changing S with rising temperature.

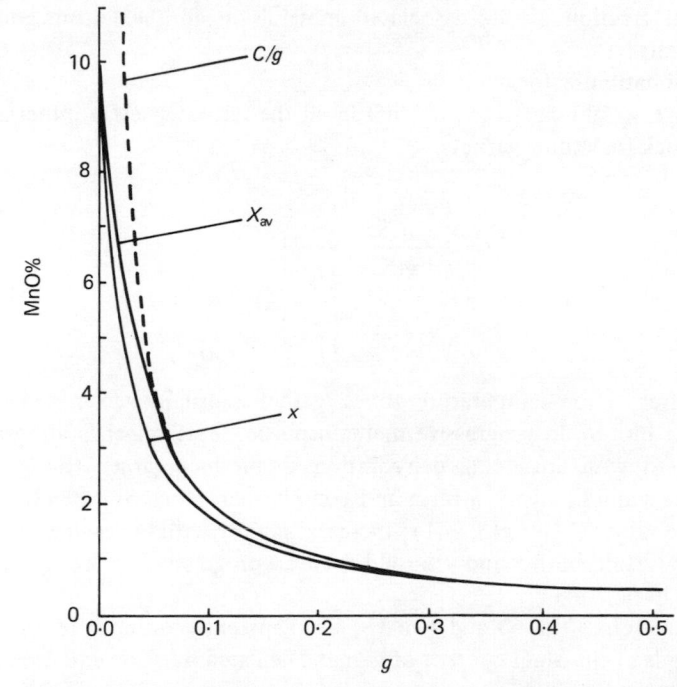

Fig. 7B-11   The MnO contents (X_{av} and x) of garnet as a function of the weight fraction of garnet (g) for complete fractionation and complete equilibrium models, respectively (Miyashiro and Shido, 1973). It is assumed that $C = 0.2$, and $S = S' = 50$.

Complete Equilibrium Model for the Progressive Decrease of the MnO Content of Garnet

As an alternative case, let us consider a garnet-bearing metapelite in which garnet crystallizes in complete equilibrium. In this model, the garnet crystals are assumed to be homogeneous and to have been in equilibrium with the associated minerals during their growth.

The following relations hold:

$$\left. \begin{array}{c} g + a = 1.00 \\[2mm] \dfrac{x}{y} = S' \\[2mm] gx + ay = C. \end{array} \right\} \tag{7B-17}$$

where

- x = weight per cent of MnO in garnet,
- y = average weight per cent of MnO in the associated ferromagnesian minerals,
- g = weight fraction of garnet in all the ferromagnesian minerals,
- a = weight fraction of the associated minerals in all the ferromagnesian minerals,
- S' = fractionation factor, and
- C = average weight per cent of MnO in all the ferromagnesian minerals of the rock (including garnet).

Hence, we have

$$\left. \begin{array}{c} x = \dfrac{CS'}{1 + g(S' - 1)} \\[4mm] y = \dfrac{C}{1 + g(S' - 1)}. \end{array} \right\} \tag{7B-18}$$

In metapelites of low-temperature zones, garnet is usually accompanied by chlorite and/or biotite. In progressive metamorphism, these minerals are gradually decomposed with advancing dehydration to produce garnet, though the actual relations should be very diverse and complicated. Since S' is much larger than unity, the value of $\{1 + g(S' - 1)\}$ increases rapidly with increasing amount of garnet. As a result, both x and y should decrease progressively. The change of x is illustrated in fig. 7B-11.

Though equations (7B-15) and (7B-18) are apparently quite different, the values and trends of the MnO content of garnet calculated by these equations are similar, as shown in fig. 7B-11.

Thus, the progressive decrease of the MnO content of garnet with rising temperature could be mainly a result of the gradual increase in the amount of this mineral (Miyashiro and Shido, 1973). The high values of the fractiona tion factors (S and S') are essential to this decrease. Change in their values with temperature may enhance the progressive compositional change of garnet. However, such a change of fractionation factor is not necessary to effect the progressive decrease of MnO in both models. The effective removal of the previously crystallized garnet from the active diffusional system also is not a necessary condition for the progressive decrease of MnO.

Effect of Rock-Pressure on the MnO Content of Garnet

It appears that the frequency of occurrence and the amount of garnet in metapelites tend to increase with increasing rock-pressure, P_s. This is partly due to the higher density of garnet in comparison with other ferromagnesian silicates. Higher P_{H_2O} in high-pressure metamorphism may counteract the effect of high P_s.

The increase of garnet with rock-pressure should have the same effect on the MnO content of the mineral as the increase of garnet with temperature. We may expect that the MnO content of garnet tends to be lower in metamorphism under high rock-pressure, if not greatly disturbed by the effect of P_{H_2O}.

Effect of Bulk Chemical Composition on the MnO Content of Garnet

In both models, the MnO content of garnet varies in proportion to the MnO content of the host rock. It is likely that in most metamorphic terranes the observed apparent progressive compositional change of garnet is enhanced by a regular variation of the bulk chemical composition of the garnet-bearing rocks selected for investigation. Usually, the bulk chemical composition of metapelites exposed in an area should vary within some range. Garnet should be more easily formed in some rocks than in others. In low-temperature zones, garnet could form only in rocks with favorable compositions, presumably those with higher MnO and CaO contents than those of typical pelites. Garnets formed in such rocks could be expected to have higher MnO and CaO contents, at least partly owing to the effect of the bulk chemical composition as indicated by equations (7B-15) and (7B-18). On the other hand, in high-temperature zones, most or all metapelites may contain garnet. If so, the average MnO and CaO contents of the garnet-bearing metapelites should become lower in high-temperature zones. Thus, apparently randomly selected garnet-bearing metapelites may actually have statistically decreasing MnO contents with increase in the temperature of recrystallization.

The progressive increase of MgO content in garnet can be explained in a similar way. The conversion of biotite to garnet causes an increase of the MgO content of garnet. In this case, the MgO fractionation factor = (MgO per cent in garnet)/(MgO per cent in biotite) increases regularly with temperature, and this should enhance the increase of the MgO content of garnet (Miyashiro and Shido, 1973).

7B-11 STAUROLITE

In the Scottish Highlands, staurolite begins to occur at a higher temperature than almandine but at a lower temperature than kyanite. In the northern Appalachians, however, kyanite begins to occur nearly at the same temperature as, or even at a lower temperature than, staurolite (Thompson and Norton, 1968). There are a number of possible reactions leading to the formation of staurolite and of kyanite. Each of them may be valid for rocks of limited compositions and physical conditions.

Staurolite forms over a wide range of rock-pressure. It occurs in the high-pressure regional metamorphic terrane of the Alps, in the medium-pressure ones of various regions, and in the low-pressure ones of the Pyrenees and the Appalachians. It was found in some contact aureoles also.

In the presence of an aqueous fluid and under P_{O_2} of the fayalite-magnetite-quartz buffer, the staurolite + quartz assemblage is stable roughly between 530 and 700 °C above 2 kbar, below which cordierite-bearing assemblages form instead (Richardson, 1968).

Though there is some doubt about the chemical formula of staurolite, it resembles chloritoid in that they are both FeO-rich aluminous minerals and could be stable over a wide range of pressure. Natural occurrences and synthetic experiments suggest that in progressive metamorphism of aluminous pelites, chloritoid breaks down roughly at the beginning of the amphibolite facies to produce staurolite-bearing assemblages. However, the actual reactions for staurolite formation appear to depend on the associated minerals, P_{H_2O} and P_{O_2}. Reactions to produce staurolite from chloritoid in metapelites would be such as:

$$31 \text{ chloritoid} + 5 \text{ muscovite} + \text{quartz} = 8 \text{ staurolite} + 5 \text{ biotite} + 27 \text{ H}_2\text{O}. \tag{7B-19}$$

Hoschek (1967) has pointed out that the composition range of staurolite-bearing rocks is considerably wider than that of the chloritoid-bearing ones. Staurolite in some metapelites, therefore, should form by a reaction not involving chloritoid. Possible reactions to form staurolite in chloritoid-free rocks would be such as:

$$31 \text{ chloritoid} + 5 \text{ muscovite}$$
$$= 26 \text{ staurolite} + 55 \text{ biotite} + 20 \text{ quartz} + 173 \text{ H}_2\text{O}. \tag{7B-20}$$

A reaction of this type was found to take place at 540 °C under 4 kbar and at 565 °C under 7 kbar in the presence of an aqueous fluid and a fayalite-magnetite-quartz buffer (Hoschek, 1969).

Some crystals of staurolite in contact metamorphic rocks have been found to be composed of sectors with slightly different MgO, Al_2O_3 and TiO_2 contents. The compositions within any one sector are relatively uniform. The compositional difference results from growth in different crystallographic directions (Hollister and Bence, 1967; Hollister, 1970). It is not known whether such a difference is common in staurolites of regional metamorphic origin.

7B-12   FORMATION OF ANDALUSITE AND KYANITE

In low-pressure regional metamorphism, andalusite begins to occur usually in a low-temperature part of the amphibolite facies, if highly aluminous rocks are present. In medium-pressure regional metamorphism, kyanite begins to occur at the kyanite isograd usually in the lowest amphibolite facies. In the northern Appalachians and Scottish Highlands (figs. 7A-6 and 7A-10), a low-pressure metamorphic region grades into a medium-pressure one. The andalusite-bearing zone in the low-pressure region lies alongside the kyanite-bearing zone in the medium-pressure one. The reactions which form the Al_2SiO_5 minerals may be similar in both regions apart from the polymorphism.

However, a number of different reactions could lead to the generation of the Al_2SiO_5 minerals, probably at different temperatures. The meaning of the andalusite and kyanite isograds, therefore, is not clear until the relevant reaction is specified.

Possible reactions are as follows:

$$Al_2Si_4O_{10}(OH)_2 = Al_2SiO_5 + 3\ SiO_2 + H_2O \qquad \text{(7B-21)}$$
$$\text{pyrophyllite} \qquad\qquad\qquad \text{quartz}$$

3 chlorite + 7 muscovite + quartz

$$= 13\ Al_2SiO_5 + 7\ biotite + 18\ H_2O \qquad \text{(7B-22)}$$

6 staurolite + 4 muscovite + 7 quartz

$$= 31\ Al_2SiO_5 + 4\ biotite + 3\ H_2O \qquad \text{(7B-23)}$$

6 staurolite + 11 quartz = 23 Al_2SiO_5 + 4 almandine + 3 H_2O.   (7B-24)

Andalusite and kyanite are usually associated with quartz in metapelites. Accordingly, whatever reactions may form them, the $P\text{-}T$ condition on the andalusite and the kyanite isograd should be within the stability field of the Al_2SiO_2 + quartz assemblage. The low temperature limit for this assemblage is given by the equilibrium curve for equation (7B-21).

Experimental investigations of the breakdown of pyrophyllite have been reported by several authors in the last twenty years. Most of the early authors gave equilibrium temperatures between 500–580°C in the presence of an aqueous fluid at 2 kbar. However, a recent experiment by Kerrick (1968) gave much lower temperatures: 410 °C at 1·8 kbar and 430 °C at 3·9 kbar, as shown in Fig. 7B-1.

In some regions, staurolite disappears in the upper part of the andalusite or the kyanite zone (e.g. Francis, 1956; Atherton, 1965; Osberg, 1968), whereas in others staurolite persists to a lower part of the sillimanite zone (Thompson, 1957; Green, 1963; Guidotti, 1968). The difference may be a result of different rock-pressures as shown in fig. 3-9, if we may neglect the effects of possible differences in P_{H_2O} and P_{O_2}.

Chinner (1965) has shown in Glen Clova of the Scottish Highlands that staurolite of the staurolite zone has a wider range of Fe^{2+}/Mg than that of the kyanite zone. Hence, relatively Mg-rich staurolite of the staurolite zone would become unstable in the kyanite zone, giving rise to a less Mg-rich staurolite and kyanite with some compositional adjustment by the associated micas.

7B-13   PHASE RELATIONS OF ANDALUSITE, KYANITE AND SILLIMANITE

The stability of Al_2SiO_5 minerals depends mainly on the rock-pressure and temperature. Differences in entropy and free energy between the three polymorphs are very small, however. It follows that the phase transformations between them might be somewhat influenced by structural disorder or slight replacement by other components. In reality Fe^{3+} and some other ions can replace a small part of the Al^{3+} in these minerals. The replacement by Fe^{3+} appears to be promoted by an oxidizing environment (Chinner, Smith and Knowles, 1969). Possible effects of such replacement on the stability relations have been experimentally demonstrated (Althaus, 1969).

Despite these reservations, we are impressed by the existence of a clear regularity in the modes of occurrence of the Al_2SiO_5 polymorphs. It strongly suggests that the main features of their modes of occurrence are controlled by their stability relations. Miyashiro (1949) proposed a model of an inverted Y form for the phase diagram of Al_2SiO_5, which since 1963 has been experimentally proved by a number of researchers.

Recent experiments by Althaus (1967, 1969) and Richardson et al. (1969) indicate that the triple point is around 600 °C (± about 50 °C) and 6 kbar (± about 0·5 kbar). The slope of the andalusite-kyanite field boundary is about 1·0 kbar/100 °C. The triple point is on the high-temperature side of the equilibrium curve for the decomposition reaction of pyrophyllite (7B-21) in the presence of an aqueous fluid (fig. 7B-1). Accordingly, even in the presence of an intergranular aqueous fluid, andalusite and kyanite could form by dehydration of pyrophyllite or other reactions as was mentioned in the preceding section, and with a further rise of temperature, andalusite and kyanite would transform into sillimanite.

In detailed examination, however, the coexistence of two polymorphs of Al_2SiO_5 within the same rock is not rare, and even the coexistence of all the three polymorphs was reported from several areas in the world (e.g. Hietanen, 1956; Pitcher, 1965; Chinner, 1966b; White et al. 1967). Since the free energy differences between the polymorphs are very small, a polymorph may form metastably near the transformation temperature. Once formed a polymorph would persist in the $P-T$ field where another polymorph is stable.

In regional metamorphic terranes, it is rare that sillimanite forms by the direct phase transformation of andalusite or kyanite. It forms more commonly as small needles (fibrolite) independent of the pre-existing andalusite and kyanite. In the sillimanite zone of the Scottish Highlands, Chinner (1961) ascribed the common formation of sillimanite needles on biotite flakes to ease of nucleation. The material for the growth of sillimanite was thought to have been derived from kyanite undergoing decomposition within the same rock. Sillimanite sometimes forms in lenticular or vein-forming aggregates in gneisses, suggesting chemical migration during metamorphism.

7B-14   MICROCLINE TO ORTHOCLASE

Microcline occurs commonly in metamorphosed psammites and semipelites of the greenschist to the amphibolite facies. It is not so common in typical metapelites, in which higher contents of Al_2O_3, MgO and FeO tend to eliminate this mineral in the greenschist and epidote-amphibolite facies. Above the temperatures of the beginning of muscovite breakdown in the amphibolite facies, however, K-feldspar becomes very widespread in metapelites.

The transformation from microcline to orthoclase has been observed in the middle or high amphibolite facies in progressive metamorphism. The continuous change of symmetry is readily noticed by the X-ray powder method. Correspondingly, the optic angle $2V$ becomes smaller continuously. Some grains show a zonal structure with a rim having a smaller optic angle than the core (Heier, 1957, 1961; Shido, 1958).

However, microcline is known to occur in some high amphibolite and granulite facies areas (e.g. Eskola, 1952). Presumably, the orthoclase that had formed at the maximum of metamorphic temperature was later transformed retrogressively to microcline. In western Maine, K-feldspar in calc-silicate rocks was found to be microcline in an area where K-feldspar in pelitic gneisses is orthoclase (Evans and Guidotti, 1966). In Broken Hill, Australia, K-feldspar in quartzo–feldspathic gneisses was found to be microcline in a zone where K-feldspar in pelitic gneisses is orthoclase (Binns, 1964). These would be due to a difference in composition of K-feldspar in different rocks, or to selective retrogressive changes.

7B-15 BREAKDOWN OF MUSCOVITE

In a part of the sillimanite zone, muscovite reacts with quartz as shown schematically by the equation:

$$\text{muscovite} + \text{quartz} = \text{orthoclase} + \text{sillimanite} + H_2O. \qquad (7\text{B-}25)$$

Muscovite and orthoclase contain some Na, and hence the real reaction is more complicated. A closer approach to the reaction would be:

$$\text{muscovite} + \text{quartz} + \text{plagioclase}$$

$$= \text{orthoclase} + \text{sillimanite} + \text{more calcic plagioclase} + H_2O. \quad (7\text{B-}26)$$

The isograd corresponding to this reaction is usually called the sillimanite–K-feldspar (or orthoclase) isograd.

The accurate location of the sillimanite–K-feldspar isograd on the map is difficult, because the above reaction takes place over a considerable range of temperature, and moreover muscovite readily forms retrogressively. Evans and Guidotti (1966) have made a detailed study of this reaction in western Maine.

In the northern Appalachians, the sillimanite–K-feldspar isograd coincides roughly with the boundary between the amphibolite and granulite facies (Thompson and Norton, 1968). However, in the Ryoke belt in Japan, the isograd appears to be within the amohibolite facies (Ono, 1969b).

The reaction muscovite + quartz = K-feldspar + Al_2SiO_5 + H_2O has been experimentally investigated by Evans (1965) and by Althaus and Nitsch (Winkler, 1967, p. 74). The result of the latter workers is shown in fig. 7B-1.

If quartz is absent, the breakdown reaction of muscovite would be such as muscovite = K-feldspar + corundum + H_2O. As was discussed in §2-5, this reaction should take place at higher temperature than equation (7B-25). The equilibrium curve for this reaction was experimentally investigated by Yoder and Eugster (1955), Evans (1965) and Velde (1966).

7B-16 PARTIAL MELTING

The temperatures for the beginning of melting of granites and pelites in the presence of an aqueous fluid were determined by Wyllie and Tuttle (1961), Yoder and Tilley (1962), Luth, Jahns and Tuttle (1964), Boettcher and Wyllie (1968), and Winkler (1967). They lie between 620 and 700 °C for most of the pressure range within the crust, as shown in figs. 1-1 and 7B-1. It follows that if an aqueous fluid is present between mineral grains, pelitic and psammitic rocks should undergo partial melting in most of the sillimanite zone (amphibolite and granulite facies), and even in parts of the andalusite-bearing and kyanite-bearing zones.

In the sillimanite zone of regional metamorphism, pockets and veins of pegmatite as well as granitic masses occur very commonly. Some or all of them may be crystallization products of liquids that formed by partial melting. Migmatites may be formed by such melts (von Platen, 1965; Lundgren, 1966; Winkler, 1967, pp. 192–224; Thompson and Norton, 1968).

In the northwest Adirondacks, Engel and Engel (1958, 1960) have found that quartz–feldspar–biotite–muscovite gneisses derived from semi-psammitic rocks are in the amphibolite facies at a distance of about 55 km from the Adirondack plutonic complex. The metamorphism advances to the lower granulite facies near the complex, thus causing the decomposition of muscovite and the formation of garnet. At this high temperature stage, partial melting appears to have taken place. Silicic liquids so formed would have been removed, leaving solid residues lower in SiO_2 and K_2O and in Fe^{3+}/Fe^{2+} ratio, and higher in Al_2O_3, FeO, MgO and CaO.

Some granulite-facies terranes were shown to be less silicic than the associated lower-temperature ones (Ramberg, 1951; Lambert and Heier, 1968). Most granulite facies terranes may be composed of solid residues of partial melting (chapter 18).

Some solid residues of partial melting may have rather unusual chemical compositions. Grant (1968) has discussed the possibility of the generation of cordierite–anthophyllite rocks as solid residues of the partial melting of graywackes.

7B-17   BREAKDOWN OF BIOTITE

Biotite is stable in metapelites up to a higher temperature than muscovite. As long as the lower temperature limit of the granulite facies is defined by the formation of two pyroxenes in metabasites, biotite in the associated metapelites persists into the granulite facies. Ultimately biotite decomposes, but such high-temperature metamorphism is rare on earth.

The reaction for the breakdown of biotite may be shown schematically as follows:

$$K(Mg, Fe)_3 AlSi_3 O_{10}(OH)_2 + 3\ SiO_2 = 3(Mg, Fe)SiO_3 + KAlSi_3O_8 + H_2O$$
$$\text{biotite} \qquad \text{quartz} \quad \text{orthopyroxene} \quad \text{K-feldspar} \qquad (7B\text{-}27)$$

Heier (1960) described a progressive metamorphic sequence of metasediments from the amphibolite to the granulite facies on the Island of Langöy in northern Norway. The rocks are so well recrystallized that they look as though they were igneous. However, interbedded graphite schist and limestone reveal their original sedimentary nature. Metasedimentary gneisses of the amphibolite facies are composed of quartz, microcline–perthite, oligoclase, biotite, green hornblende and sphene. In the zone of the granulite facies, not only hornblende but also biotite is decomposed. Thus, the rocks there are composed of quartz, ortho-clase–perthite, oligoclase (or andesine), orthopyroxene and clinopyroxene and opaque minerals.

In the absence of quartz, the breakdown of biotite needs an exceptionally high temperature. The equilibrium curve for breakdown reaction: phlogopite = forsterite + leucite + $KAlSiO_4$ + H_2O has been determined by Yoder and Eugster (1954). It lies above 1000 °C in the presence of an aqueous fluid at a pressure higher than 1 kbar.

The stability relations of iron-bearing biotite have been investigated by Eugster and Wones (1962) and Wones and Eugster (1965). They are strongly influenced by P_{O_2}. Their breakdown temperature tends to decrease with in-creasing Fe^{+2}/Mg ratio.

7B-18 *AFM* DIAGRAMS FOR PROGRESSIVE METAMORPHISM OF PELITES

In order to summarize the progressive changes in paragenetic relations of metapelites, two series of Thompson *AFM* diagrams, one for low-pressure and the other for medium-pressure metamorphism, are shown in figs. 7B-12 and 7B-13. Since the actually observed characteristics of regional metamorphism are very diversified, these diagrams are merely intended to show a generalized, schematic model for progressive mineral changes.

7B-19 LAWSONITE AND ARAGONITE

Reactions to form the three most characteristic minerals of metaclastics in high-pressure terranes, i.e. lawsonite, aragonite and jadeite, will be examined in this and the succeeding section. Glaucophane is not discussed in this chapter inasmuch as it does not occur in typical metapelites.

The occurrence of lawsonite is practically confined to rocks of the glauco-phane-schist facies. The calculated and synthetic curves representing phase relations of lawsonite are shown in fig. 7B-14. Since lawsonite is a hydrous Ca-bearing mineral, its stability is controlled largely by P_{H_2O} and P_{CO_2} (§ 4-4). Hence, it cannot be a good indicator of rock-pressure.

$CaCO_3$ appears to have at least three stable polymorphs, calcite I (ordinary calcite), calcite II, and aragonite, with a triple point at about 480 °C and 10·0 kbar, as shown in fig. 7B-14 (Boettcher and Wyllie, 1967; Goldsmith and Newton, 1969). The calcite I–aragonite transition pressure at 100 °C is said to be about 4·35 kbar (Crawford and Fyfe, 1964). Hence, aragonite becomes stable at a considerably lower pressure than the pure jadeite + quartz assemblage. It is conceivable that in many areas the aragonite which had been formed during a phase of high-pressure metamorphism, was later transformed back to calcite.

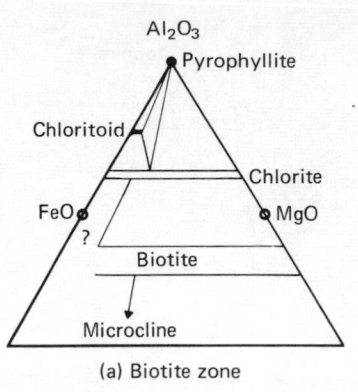

(a) Biotite zone

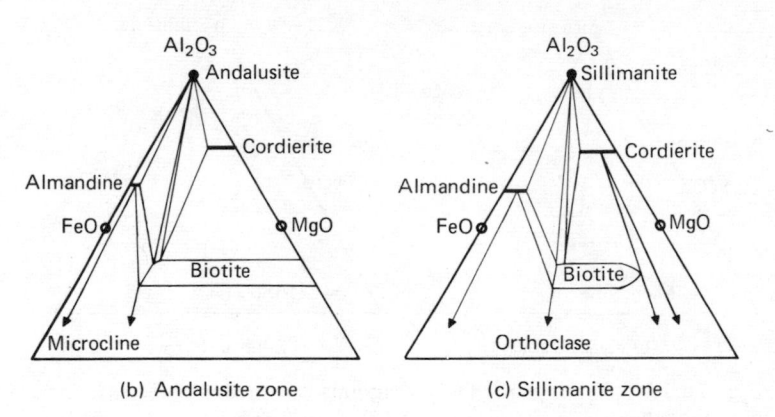

(b) Andalusite zone   (c) Sillimanite zone

Fig. 7B-12   Thompson *AFM* diagrams for low-pressure progressive meta-morphism.

Calcite is widespread in low-temperature metamorphic rocks of the low-, medium- and high-pressure types. In some high-pressure metamorphic rocks of the Franciscan formation in California, however, aragonite was found to occur (Coleman and Lee, 1962; McKee, 1962*b*; Ernst, 1965). Such aragonite is characteristically associated with jadeite and lawsonite, and is regarded as formed at high rock-pressures.

Vance (1968) has found the occurrence of aragonite limestones in a prehnite–pumpellyite facies terrane in Washington State (6A-6). It may or may not indicate that aragonite commonly forms metastably at a lower pressure than its stability field. Newton *et al.* (1969) have pointed out that deformed calcite would store a considerable amount of plastic strain energy and hence may become unstable relative to aragonite at a much lower pressure than undeformed

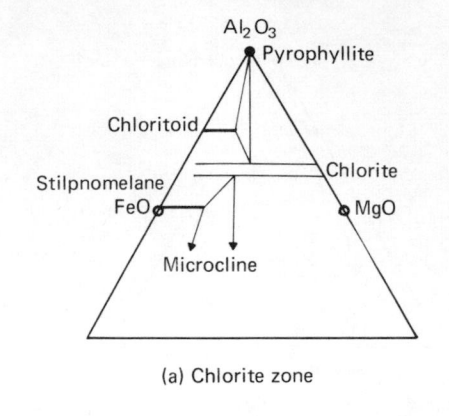

(a) Chlorite zone

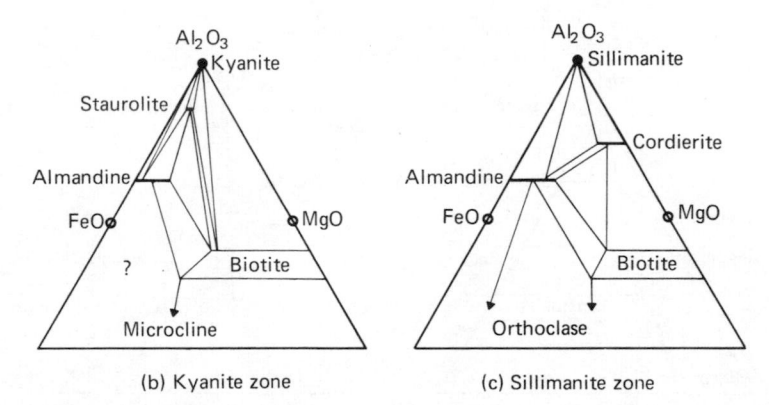

(b) Kyanite zone   (c) Sillimanite zone

Fig. 7B-13   Thompson *AFM* diagrams for medium pressure progressive metamorphism.

calcite. At low temperatures, aragonite growth is a faster process than strain recovery of calcite. Thin section observations suggest that the aragonite has selectively replaced highly strained calcite crystals. The Washington aragonite would have formed from deformed calcite at relatively low pressures.

7B-20   FORMATION OF JADEITE

Idealized reactions to form pure jadeite $NaAlSi_2O_6$ are given by equations:

$$albite = jadeite + quartz \tag{7B-28}$$

$$albite + nepheline = 2\ jadeite. \tag{7B-29}$$

The stability fields of jadeite in the presence and the absence of quartz are shown in figs. 2-1 and 7B-14. Jadeite occurs as a major component in quartz-free

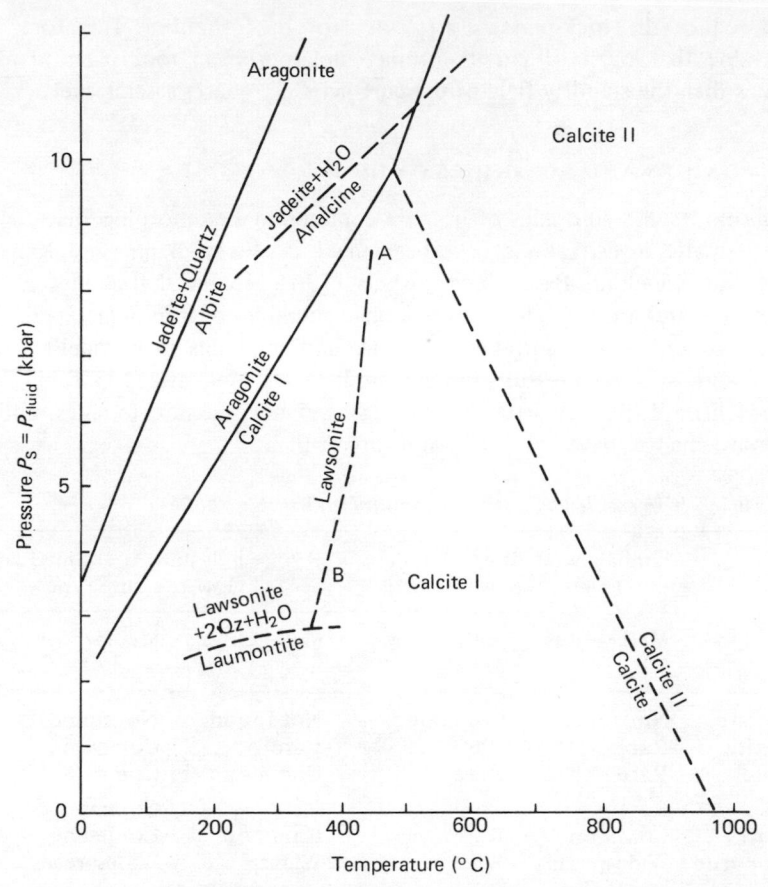

Fig. 7B-14 Equilibrium curves for reactions relevant to high-pressure metamorphism. The phase relations of $CaCO_3$ are after Boettcher and Wyllie (1967) and Crawford and Fyfe (1964). The high-pressure limits of analcime and laumontite are after Newton and Kennedy (1968) and Thompson (1970b), respectively. The high temperature limit of lawsonite (curve A-B) is after Newton and Kennedy (1963).

inclusions within serpentinite, and also as a constituent of metaclastics usually in association with quartz, both in high-pressure terranes.

Coleman and Clark (1968) have emphasized that jadeites occurring in serpentinite are commonly of pure jadeite composition, while jadeites in quartzose metamorphic rocks always contain more than 10 per cent of other pyroxene components (diopside, hedenbergite and aegirine). Hence, the jadeite-forming reactions in metamorphic rocks would be of the type: sodic plagioclase + minor ferromagnesian components (in chlorite for instance) → jadeitic pyroxene + quartz + lawsonite (Ernst, 1965). The impure composition of jadeite should reduce the rock-pressure necessary for its formation. Therefore, it is conceivable that most of jadeite-forming metamorphism took place at lower pressures than the stability field of the pure jadeite + quartz assemblage.

7B-21   OXIDES AND SULPHIDES OF IRON

Though oxides and sulphides of iron are common in metamorphic rocks, only a few systematic investigations have been made on them. Banno and Kanehira (1961) and Kanehira, Banno and Nishida (1964) found that oxides of iron (except ilmenite) are much less common than sulphides of iron in metapelites.

The frequency of occurrence of oxides and sulphides in metapelites from high-pressure and low-pressure terranes in Japan is summarized in Table 7B-1. There is little difference between the high and low-pressure terranes, and no progressive changes have been noticed in any of them.

Table 7B-1   *Iron oxides and sulphides in metamorphic rocks*

	Sanbagawa belt and Omi area (high-pressure type)		Ryoke belt and Abukuma Plateau (low-pressure type)	
	Metapelites	Metabasites	Metapelites	Metabasites
Haematate	Not found	Common	Not found	Not found
Magnetite	Rare	Rare	Rare	Common
Ilmenite	Rare	Rare	Common	Common
Pyrite	Rare	Common	Rare	Common
Pyrrhotite	Common	Rare	Common	Common
Chalcopyrite	Widespread	Widespread	Widespread	Widespread

Note: After Banno and Kanehira (1961) and Kanehira *et al.* (1964).

7B-22   ORGANIC MATERIAL AND GRAPHITE

Ordinary pelitic sediments contain a considerable amount of organic material, which may be either dominantly coaly or dominantly bituminous, or a mixture of both. In progressive metamorphism, such organic materials suffer a series of decomposition reactions, ultimately leading to the generation of well-crystallized graphite. They exert a reducing effect on associated mineral assemblages (§2-7).

The progressive increase in the crystallinity of such materials was investigated in particular by Landis (1971). In zeolite facies rocks, such materials are virtually amorphous, or show only a rudimentary graphitic structure. Their crystallinity increases with rise in temperature. The ordered graphite structure is realized in the epidote–amphibolite and the amphibolite facies.

Izawa (1968) separated such materials from unmetamorphosed and metamorphosed sediments for chemical investigation. They were shown to resemble various ranks of coal in composition and X-ray diffraction pattern. Materials from metapelites of the glaucophane-schist and greenschist facies, giving C = 84–94 per cent and H = 1·8–1·0 per cent in complete analysis, were similar to semianthracite and anthracite, whereas those from epidote-amphibolite facies rocks, giving C = 93–96 per cent and H = 0·7–0·4 per cent, were found to be similar to anthracite in composition.

The metamorphism of coal seams has rarely been investigated in relation to the surrounding rocks. Coal seams in zeolite facies areas are composed of either bituminous coal or semianthracite. The rank of coal becomes higher with rise in temperature of metamorphism of the surrounding rocks (Kisch, 1966).

Chapter 8A

Metabasites: Diversity in Progressive Regional Metamorphism

8A-1   MINERALOGICAL SENSITIVITY OF METABASITES

Metabasites are mineralogically sensitive to variations in temperature and pressure. This is the reason why Eskola (1920, 1939) based his definitions of individual metamorphic facies on the mineral assemblages of metabasites. The main characteristics are synoptically shown in table 8A-1. The mineralogical sensitivity is especially great under high rock-pressures, as exemplified by the formation of glaucophane schist and eclogite.

Metabasites derived from pyroclastic rocks are more easily recrystallized and show more marked schistosity and more complex variation in composition than those derived from lavas and minor intrusives. The former are commonly mixtures of volcanic with some sedimentary materials, and more susceptible to pre-metamorphic chemical changes including weathering and hydrothermal action. Some altered rocks could produce very unusual mineral assemblages on metamorphism (Vallance, 1967).

8A-2   LOW-PRESSURE METABASITES IN THE RYOKE BELT AND ABUKUMA PLATEAU, JAPAN

The Ryoke metamorphic belt of Japan has abundant metapelites and metapsammites, but few metabasites. Recently, however, Katada (1965) has been able to follow progressive mineral changes of metabasites in the Shiojiri area of the belt. On the other hand, the Higo metamorphic complex and Abukuma Plateau, the possible western and the eastern extension of the belt respectively, have abundant metabasites. Two areas will be reviewed.

Shiojiri Area (Northern Kiso) in the Ryoke Belt
The basaltic rocks in this area are more easily recrystallized than the clastic sediments at low temperatures. The mineral changes in this area (Katada, 1965) have been summarized in fig. 3-4. A sketch map of metamorphic zones is shown in fig. 7A-1.

Table 8A-1 *Baric types of regional metamorphism and characteristic minerals of metabasites*

Baric type	Prehnite-pumpellyite and glaucophane-schist facies	Greenschist facies	Epidote-amphibolite and amphibolite facies	Granulite facies
Low-pressure		Actinolite, chlorite, epidote, muscovite, biotite	Hornblende, clinopyroxene, biotite	Orthopyroxene, clinopyroxene
Intermediate-pressure		Actinolite, chlorite, epidote, stilpnomelane, muscovite, biotite	Hornblende, epidote, clinopyroxene, biotite	Orthopyroxene, clinopyroxene
High-pressure	Glaucophane, actinolite, chlorite, lawsonite, epidote, pumpellyite, stilpnomelane, muscovite	Actinolite, chlorite, epidote, stilpnomelane, muscovite	Hornblende, barroisite, clinopyroxene, epidote, muscovite, biotite	

In the lowest temperature zone, the original plagioclase is largely preserved. The recrystallized minerals are quartz, chlorite, calcite, and muscovite with subordinate albite, epidote and green biotite.

With a slight rise of temperature, recrystallization becomes complete, and actinolite appears. The plagioclase becomes more calcic than 10 per cent An, leading to the formation of the actinolite + oligoclase assemblage. Brown biotite becomes common. This zone belongs to the greenschist or a closely related facies. (The plagioclase is more calcic than in the typical greenschist facies.)

With a further rise in temperature, pale blue-green hornblende appears (zone I*c* in fig. 3-4). When it is in contact with actinolite, the boundary is sharp. The associated plagioclase is oligoclase or andesine, indicating the onset of the amphibolite facies.

Advancing metamorphism then begins to form green hornblende and andesine (zone II*a*). Actinolite disappears and calcic clinopyroxene may appear.

In this area, the zones of the greenschist and a closely related facies (zones I*a,b*) thus grade through a narrow transitional zone (I*c*) into zones of the amphibolite facies (zones II*a,b*, and III). In the adjacent Shiojiri–Takato area, however, Ono (1969*b*) found a few epidote amphibolites in association with actinolite greenschists in a zone intermediate between the zones of the greenschist and of the amphibolite facies. Since metabasites are very scarce in the latter area, it is not clear whether it means the existence of a narrow zone of the epidote–amphibolite facies.

Central Abukuma Plateau

The metamorphic terranes of the central Abukuma Plateau (fig. 8A-1) have abundant metabasites, as was mentioned in §7A-2. The temperature of metamorphism increases generally westward. The metabasites range from the greenschist to the high amphibolite facies, as summarized in fig. 3-5.

The exposed metamorphic terrane has been divided into three zones A, B and C on the basis of progressive changes in calcic amphibolite in metabasites (Miyashiro, 1958). Zone A is characterized by actinolite, zone B by blue–green hornblende, and zone C by green or brown hornblende (without a bluish tint). The changes in optic constants and chemical composition of the amphiboles will be described in §8B-4.

Common metabasites are as follows:

Zone A (*greenschist facies*). Chlorite-epidote-actinolite schist.
Zone B and C (*amphibolite facies*). Hornblende-plagioclase schist or amphibolite.

In a zone transitional between zones A and B, actinolite coexists with blue–green hornblende (§8B-4). The plagioclase in zone A is usually albite. In

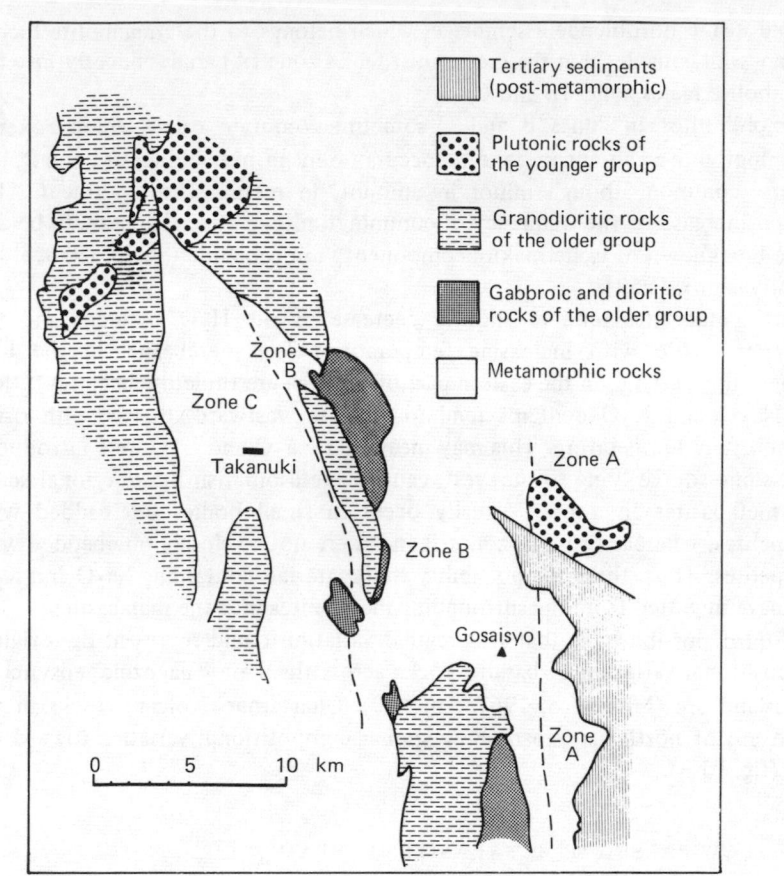

Fig. 8A-1   Zones of progressive regional metamorphism in the central Abukuma Plateau. The transitional area from an unmetamorphosed area to zone A is covered by Tertiary sediments. Plutonic rocks of the older group have given rise to some thermal effects in the surrounding schists especially in the eastern part of the region, whereas those of the younger group produced a hornfelsic contact aureole. (Miyashiro, 1958.)

zones B and C, it is oligoclase, andesine and labradorite. Almandine garnet does not occur in ordinary metabasites.

In metabasites in the area of Barrovian zones of the Scottish Highlands, hornblende (blue–green) begins to occur at the almandine isograd, where the associated plagioclase is still albite. Hence, the albite + epidote + hornblende assemblage characteristic of the epidote–amphibolite facies is produced. In the central Abukuma Plateau, the plagioclase of metabasites at the onset of zone B contains about 20–30 per cent An, thus forming the calcic oligoclase or andesine

(+ epidote) + hornblende assemblage, which belongs to the amphibolite facies. We may well consider that the greenschist facies zone (A) grades directly into the amphibolite facies zones (B and C).

Amphibolites in zones B and C sometimes contain calcic clinopyroxene. Cummingtonite is absent in zone A, occurs rarely in metabasites of zone B, but is very common, though minor in amount, in metabasites of zone C. The marked increase of the frequency of cummingtonite in zone C appears to be due to the breakdown of tschermakite component in hornblende (Shido, 1958; Shido and Miyashiro, 1959).

The metabasites tend to show a decrease of the H_2O^+ content and the Fe^{3+}/Fe^{2+} ratio with increasing temperature. The metabasites in the low-temperature part i.e. in the eastern part of the area are tholeiitic in composition. The Na_2O and K_2O contents tend to increase westward, that is, with rising metamorphic temperature. This may mean that Na_2O and K_2O were introduced from some source lying to the west, causing metasomatism on a regional scale. The metabasites in zone C usually occur in small bodies interbedded with metapelites, whereas those in zones B and A are not so closely interbedded with metapelites. Thus, there is a possibility that materials containing Na_2O and K_2O may have migrated from the surrounding metapelites into the metabasites.

A third possibility is that this regular variation could represent the original compositional variation of basaltic rocks across the Late Palaeozoic geosyncline and island arc (Miyashiro, 1967b, p. 440). Quarternary volcanic rocks in the island arc of northeast Japan show similar compositional variation toward the west (fig. 15-3).

8A-3 LOW-PRESSURE METABASITES IN SPAIN

Fabriès (1963, 1968) studied metabasites interbedded with metapelites in northeast Séville Province, Spain. The Hercynian metamorphic complex exposed there is in the low-pressure amphibolite facies, being divided into three zones:

Zone I. Greenish blue hornblende with an acicular habit occurs in the metabasites.

Zone II. Green hornblende with an acicular or prismatic habit occurs in the metabasites. Andalusite occurs in the associated metapelites.

Zone III. Greenish brown hornblende with a prismatic or dumpy habit occurs in the metabasites. Sillimanite occurs in the associated metapelites.

The amphiboles of zones I and II are high in tschermakite component, while those of zone III have higher contents of Na in vacant sites.

Bard (1969, 1970) described a progressive series of metabasites ranging from the greenschist to the low granulite facies in the Hercynian terrane of the Aracena area (figs. 7A-3 and 7A-4).

The lowest temperature zone contains actinolite greenschists. With a rise in temperature, blue–green hornblende with a high tschermakite content begins to occur in association with actinolite. A further rise causes disappearance of actinolite and the change of hornblende from blue–green to green colour. At this stage, the associated plagioclase contains 25–45 per cent An. Then cummingtonite and clinopyroxene appear. In the highest temperature zone, cummingtonite and sphene disappear and orthopyroxene appears. Green–brown and brown hornblendes and biotite are still stable (§8B-4).

8A-4 LOW-PRESSURE METABASITES IN MICHIGAN

The middle Precambrian metamorphic terrane of northern Michigan contains basaltic volcanics, dolerites, and gabbroic dikes and sills (James, 1955). All these rocks were subjected to metamorphism along with the surrounding sedimentary ones. The area has been divided into the chlorite, biotite, almandine, staurolite, and sillimanite zones in terms of mineral changes in the metapelites (fig. 7A-9).

The chlorite zone metabasites show textural relics, but their mineralogical reconstitution is virtually complete. The common rocks are actinolite–chlorite–epidote–albite rocks. In the biotite zone, blue-green hornblende appears. In the garnet zone, the common metabasites are epidote-amphibolites with blue–green hornblende and sodic plagioclase (about 15 per cent An). Epidote, biotite and quartz are common. Garnet is extremely rare. In the staurolite zone, epidote disappears, and the common metabasites are amphibolites with green hornblende and andesine. In the sillimanite zone, some amphibolites contain brown hornblende. Generally, the metabasites of this area lack garnet.

8A-5 MEDIUM-PRESSURE METABASITES IN THE SCOTTISH HIGHLANDS AND NORWAY

Metabasites in the Scottish Highlands may be classified into two groups: Green Beds and epidiorites. The former represent metamorphosed pyroclastic deposits of broadly basaltic composition occasionally mixed with some clastic sedimentary materials. Epidiorite is an old field term used by British geologists, in particular in the Scottish Highlands, which represents metamorphosed doleritic intrusions cutting through metasediments. The following description will be made with reference to the Barrovian zones (see §3-1 and fig. 7A-10).

Green Beds of the Scottish Highlands
The Green Beds in the chlorite zone are usually composed of chlorite–albite–epidote schist with or without calcite and quartz. Very small amounts of white

mica and magnetite are present occasionally. In the middle of the chlorite zone, green–brown biotite and actinolite begin to occur in some rocks. In the biotite zone, biotite in Green Beds becomes brown in colour.

With rising temperature, chlorite and epidote disappear and almandine may appear. Thus, in the almandine zone for metapelites, the rocks of the Green Beds become hornblende–oligoclase schist with or without biotite or almandine. The Green Beds in the kyanite and sillimanite zones are composed of coarse-grained hornblende–andesine schists with or without almandine (Phillips, 1930; Harker, 1932).

Epidiorites of the Scottish Highlands

A detailed description has been given by Wiseman (1934). The epidiorites from the chlorite and biotite zones as indicated by the metapelites are actinolite-chlorite-epidote–albite felses or schists with or without biotite. Actinolite is absent in some rocks. Quartz or calcite are present in some rocks. Some epidiorites preserve remnants of the porphyritic and ophitic textures of the original rocks, but most are schistose. Actinolite occurs as slender needles commonly intergrown with chlorite, or as big crystals pseudomorphous after igneous augites. Rarely, original augites remain surrounded by actinolite. Biotites of a green and a brown colour occur in small amounts in epidiorites of both chlorite and biotite zones. White mica occurs in some epidiorites from the chlorite and biotite zones, but not from the higher zones. Stilpnomelane occurs very rarely. These epidiorites belong to the greenschist facies (fig. 3-2).

On entering the almandine zone of the metapelites, the epidiorites show two marked mineral changes: the appearance of blue-green hornblende and of almandine. Near the beginning of the zone, blue–green hornblende occurs sometimes in parallel growth with actinolite and sometimes as a rim around an actinolite core.

With a further rise in temperature, actinolite disappears. The blue–green hornblendes in almandine-bearing epidiorites from the almandine zone have $\beta = 1 \cdot 663 - 1 \cdot 685$, whereas those in almandine-free ones from the same zone have $\beta = 1 \cdot 639 - 1 \cdot 675$. Rocks with higher FeO/MgO ratios tend to form almandine. The mineral may be present in one part of an outcrop of epidiorite and absent in another. Quartz and biotite are present in some rocks and the biotites are commonly brown but sometimes green. The plagioclase is still albite. Epidote is widespread, but chlorite is almost absent. Thus, typical epidiorites from the lower temperature part of the almandine zone are blue-green hornblende-epidote–albite schists with or without almandine. In ordinary nomenclature, they are called epidote amphibolites (fig. 3-2).

With a little rise in temperature, however, epidote reacts with albite to produce more calcic plagioclase. Thus, the highest temperature part of the almanine zone contains epidiorites with andesine. Plagioclases may show zoning with increasing An content towards the rim in some cases and with decreasing An in others.

In the kyanite and sillimanite zones, the epidiorites are usually mainly composed of green or brown hornblende and plagioclase (usually andesine). Brown hornblende appears to occur in the highest temperature part of the sillimanite zone. Typical epidiorites in the kyanite and sillimanite zones are thus coarse-grained amphibolites with or without almandine or clinopyroxene. In the vicinity of syn-metamorphic granites, metasomatic effects on the epidiorites are noticed.

The K_2O content of the epidiorites ranges from 'trace' to 1·2 per cent whereas that of the Green Beds ranges from 0·6 to 2·0 per cent. The higher contents of the latter may be due to the clay admixture in the original materials. Alternatively, some glass which was contained in original pyroclastics, may have absorbed K_2O from sea water during palagonitization, as observed on the present-day ocean floors.

Sulitjelma Area, Norway

The Sulitjelma area on the boundary between Norway and Sweden was made a classical metamorphic terrane by Vogt's (1927) study, though a part of his geologic interpretation was in error (Mason, 1967). The metapelites range from the chlorite zone to the staurolite–kyanite zone (cf. Henley, 1970). The associated metabasites range from the amphibolite through the epidote–amphibolite to the greenschist facies (Mason, 1967). In the lowest temperature part of the greenschists, the CaO tends to be lost, resulting in the decomposition of actinolite and then of epidote to produce chlorite.

8A-6 MEDIUM-PRESSURE METABASITES IN THE NORTHERN APPALACHIANS

Thompson and Norton (1968) give the correlation of mineral zones in the metapelites with the mineral assemblages of interbedded metabasites in the northern Appalachians, as follows:

Metapelites	Metabasites
Biotite zone	Greenschist facies
Almandine zone	Epidote–amphibolite facies
Staurolite zone ⎱ Sillimanite zone ⎰	Amphibolite facies
Sillimanite-K feldspar zone	Low granulite facies.

Woodsville Quadrangle, Vermont and New Hampshire

Billings and White (1950) have described the progressive metamorphism of basic dikes that were intruded into a Palaeozoic geosynclinal pile in the Woodsville quadrangle, Vermont and New Hampshire (fig. 7A-6). The metabasites range from the greenschist to the amphibolite facies. The surrounding metapelites range from the chlorite to the sillimanite zone. The most interesting point of this study is the discovery of a great variety of metamorphic assemblages, which formed in response to variable P_{CO_2}, in the chlorite zone metabasites.

In the chlorite zone, some rocks in dikes preserve their original minerals and textures, but most have been completely recrystallized and become schistose. The completely recrystallized metabasites show the following four mineral assemblages (mineral names in decreasing order):

1. Actinolite+albite+epidote+chlorite+quartz+calcite,
2. Chlorite+albite+epidote+calcite+quartz,
3. Chlorite+albite+calcite+quartz+epidote,
4. Albite+chlorite+ankerite+quartz+epidote.

(Ankerite is a dolomite-structure mineral in which Mg is partly replaced by Fe and it shows porphyroblastic development.)

This sequence (1) to (4) indicates that actinolite was decomposed first and then epidote, with resultant formation first of chlorite, calcite and quartz, and then of ankerite. Probably this change took place with increasing P_{CO_2} under essentially the same temperature and rock-pressure. Ankerite becomes stable at a higher P_{CO_2} than calcite. In some large dikes, the carbonate content of the rocks increases toward the margin. Rocks with higher carbonate contents tend to show clear schistosity.

It is noted that schistose albite–epidote amphibolites occur in a higher temperature part of the chlorite zone and in the biotite zone (not in the almandine zone). Massive or weakly schistose amphibolites occur in the higher temperature part of the almandine zone and in the staurolite and sillimanite zones. The plagioclase is andesine or labradorite. Almandine is virtually absent in the metabasites of this area.

8A-7 METABASITES IN PASSAGE FROM THE AMPHIBOLITE TO THE GRANULITE FACIES

The progressive changes of metabasites from the amphibolite to the granulite facies in the Broken Hill area, Australia, and the northwest Adirondack Mountains in New York State, are well documented. In both cases, it is not clear whether they belong to the medium-pressure series or to the low-pressure, since no diagnostic minerals occur. The associated metasediments contain sillimanite but no andalusite nor kyanite. The metabasites of the two areas are similar to each other except that garnet and cummingtonite occur in the former but not in the latter.

Broken Hill, Australia

The Broken Hill area is near the western border of New South Wales. The older Precambrian Willyama complex has been metamorphosed under conditions of the amphibolite to the granulite facies (Binns, 1964, 1965*b*).

Three zones have been distinguished in terms of regular mineral changes in metabasites as follows:

Zone A. The metabasites are amphibolites with blue–green or green hornblende (low amphibolite facies). Garnet and epidote are present in some metabasites. The associated metapelites are sillimanite-bearing muscovite–biotite schists often containing garnet. The K-feldspar is orthoclase and is confined to sillimanite-free metapelites.

Zone B. The metabasites are amphibolites with brown or green-brown hornblende (high amphibolite facies). The colour change of the hornblende is ascribed to an increasing content of TiO_2. Almandine, cummingtonite and clinopyroxene occur in some metabasites. The associated metapelites are orthoclase-bearing silimanite gneisses, where muscovite is no longer stable. Garnet and cordierite occur in some metapelites.

Zone C. The metabasites are hornblende–orthopyroxene–clinopyroxene gneisses (low granulite facies). The associated metapelites are nearly the same as in the preceding zone.

Zones A, B and C are at least 11 km wide, 13 km at its widest, and more than 18 km wide respectively. There are not contemporaneous batholithic intrusions exposed in this region. Quartzo-feldspathic gneisses occur in all the zones. Some of them could have been derived from sedimentary rocks and others from igneous. The K-feldspar in quartzo–feldspathic gneisses is microcline in zone A and orthoclase in zones B and C.

The limestones in zone A contain wollastonite and diopside.

Northwest Adirondacks, New York State

Progressive metamorphism of Precambrian sedimentary and basic rocks has been followed along a belt 55 km long that extends across the Grenville Lowlands into the central massif of the Adirondack Mountains (Engel and Engel, 1958, 1960, 1962*a,b*; Engel, Engel and Havens, 1964).

The metabasites at the low-temperature end of the belt are amphibolites with the composition of saturated basalts, essentially composed of hornblende, andesine and quartz accompanied by minor biotite and sphene. With increasing temperature, the amounts of hornblende and quartz decrease, whereas that of plagioclase increases and sphene disappears. Clinopyroxene appears, and then with a further increase in temperature, orthopyroxene also appears. Thus, near the high-temperature end, the metabasites are hornblende-pyroxene granulites with olivine basaltic or noritic composition.

The chemical composition of the metabasites changes with increasing temperature: The contents of SiO_2, K_2O, Fe_2O_3 and H_2O decrease, whereas those of CaO and MgO increase. The associated metasediments show a similar trend of compositional change, including a decreasing SiO_2 content with increasing temperature. The compositional change in the metasediments has been ascribed to partial melting and removal of the resultant melt, whereas that of the metabasites has been ascribed to ionic diffusion or migration of an aqueous fluid, since the temperature is not considered high enough for partial melting.

The hornblendes are bluish-green in the low-temperature part, and become brownish with rising temperature. Compositionally, their contents of TiO_2, Na_2O and K_2O increase and those of MnO and $(OH^- + F + Cl)$ decrease.

In the Broken Hill area, wollastonite was found in calcareous rocks of the lower amphibolite facies zone. In the Adirondacks, on the other hand, the calcite + quartz assemblage is stable to the highest temperature zone.

8A-8 HIGH-PRESSURE METABASITES IN THE SANBAGAWA BELT, JAPAN

Metabasites are abundant in the Sanbagawa metamorphic belt of Japan. In the lowest temperature part of the belt, the basaltic volcanics preserve their original structure. They are mostly pyroclastics mixed with some lava flows. The rocks in higher temperature zones are well-recrystallized schists. Two well-documented areas will be reviewed below.

Bessi-Ino Area
The metamorphic terrane of this area has been divided into five progressive zones A, B, C, D and E (figs. 3-6 and 7A-12; §7A-9). The first two zones belong to the glaucophane-schist facies, and the last two to the epidote-amphibolite facies (Banno, 1964).

Zone A. The typical metabasites of this zone are pumpellyite–epidote–chlorite–albite–quartz schists and glaucophane–epidote–chlorite–albite–quartz schists. The mafic rocks in the low-temperature part of this zone are poorly recrystallized and are usually non-schistose, whereas those in the high-temperature part are considerably recrystallized and schistose. Pumpellyite-bearing metabasites are lower in Fe_2O_3 than pumpellyite-free ones. Jadeite was found in two rocks and lawsonite only in one. Prehnite, paragonite and laumontite were found only very rarely.

Zone B. The typical metabasites are chlorite–epidote–actinolite–albite–quartz schists with or without glaucophane. Glaucophane with $b = Z$ (i.e. crossite) is much more common than that with $b = Y$. Glaucophane-bearing metabasites tend to be higher in Fe_2O_3 content than glaucophane-free ones. Correspondingly, the epidotes in glaucophane-bearing metabasites tend to have a higher Fe_2O_3/Al_2O_3 ratio than those in glaucophane-free ones.

Zone C. This zone is transitional between the zones of the glaucophane-schist facies and the epidote–amphibolite facies. The most important feature of the metabasites is the disappearance of actinolite and appearance of barroisite. Glaucophane still occurs in some metabasites in coexistence with barroisite. Under the microscope the barroisite looks like blue–green hornblende. In this and succeeding zones, albite commonly shows porphyroblastic development.

Zone D. The typical rocks of this zone are blue–green hornblende-epidote-chlorite–albite–quartz schists with or without garnet. Not only blue–green hornblende but barroisite occurs, though the latter is rare. Some metabasites contain biotite.

Zone E. The typical rocks are epidote amphibolites with almandine or calcic clinopyroxene. Biotite, muscovite and quartz may occur. The hornblende is blue–green. Kyanite, omphacite, zoisite and paragonite have been found occasionally. Kyanite was found in a zoisite–kyanite–paragonite–hornblende–quartz rock.

Hematite occurs in the metabasites of all the zones, whereas ilmenite occurs only in those of zones D and E. Magnetite is very rare.

Kanto Mountains.
A synoptic description of this area has already been given in fig. 3-7 and §7A-9 (Seki, 1958, 1960, 1961a, 1965).

The recrystallization of basaltic rocks begins with the incipient formation of chlorite, albite and stilpnomelane (zone I in fig. 3-7). Then, actinolite, glaucophane, and pumpellyite begin to form (zone II). With a slight rise of temperature, epidote enters (zone III). At this stage, recrystallization is still incomplete, and foliation is very weak.

With a further rise of temperature, lawsonite and jadeitic pyroxene begin to occur (zone IV). These two minerals are confined to a narrow zone. Some metabasites in this zone show the coexistence of lawsonite, pumpellyite and epidote. With an additional increase of temperature, lawsonite and jadeitic pyroxene disappear (zone V). At this stage, recrystallization becomes fairly complete and the metabasites are usually schistose.

Glaucophane in a broad sense occurs in some of the metabasites ranging from nearly the stage of incipient recrystallization up to this zone. Glaucophane rimmed by actinolite and actinolite rimmed by glaucophane are common. Glaucophane with $b = Z$ occurs in all the above-mentioned range, whereas glaucophane with $b = Y$ is nearly confined to the lawsonite–jadeite zone. The latter is probably closer to the Mg–Al end member of glaucophane.

With an additional rise in temperature, glaucophane and pumpellyite disappear, and thus the metabasites come to have a mineral assemblage characteristic of greenschist (zone VI): chlorite + actinolite + epidote + albite. Some of the albites show porphyroblastic development. Some 'actinolites' of this zone may turn out to be barroisite.

8A-9 HIGH-PRESSURE METABASITES OF THE CALIFORNIA COAST RANGES

A great variety of metamorphosed pillow lavas and tuff breccias of broadly basaltic compositions are associated with metaclastics in the Franciscan group of the Coast Ranges of California. Two areas will be reviewed below.

Cazadero Area

Coleman and his co-workers studied the metabasites in the Cazadero area, about 100 km NW of San Francisco (Coleman and Lee, 1962, 1963; Lee, Coleman and Erd, 1963; Coleman et al, 1965; Coleman and Clark 1968; Coleman and Papike, 1968).

The poorly recrystallized metabasalts (type II in Coleman and Lee's nomenclature as commented upon in §7A-10) are non-schistose and resemble unmetamorphosed basalts in field appearance. Pillow structure is well preserved. Microscopic observation has revealed that features of the original subophitic and variolitic textures are preserved but mineralogical recrystallization is well advanced. Glaucophane, lawsonite, muscovite (1M), chlorite, omphacite, sphene and quartz are common constituents. Rarely, relict calcic plagioclase and clinopyroxene are present. Albite and paragonite have not been found.

With advancing recrystallization, the metabasalts become schistose (Coleman and Lee's type III). The pillow structure is considerably destroyed. Glaucophane is abundant. Muscovite $(2M_1)$, chlorite, lawsonite, pumpellyite, garnet, omphacite, aragonite, sphene and quartz are common. Pyrite is ubiquitous, whereas magnetite, hematite and ilmenite are absent.

Coarsely crystalline metabasalts (type IV) occur as isolated blocks in this area. In such rocks, epidote and rutile occur, whereas lawsonite and sphene dwindle.

The stable form of $CaCO_3$ is aragonite in well-recrystallized glaucophane schists (type III), whereas it is partly aragonite and partly calcite in poorly recrystallized metabasites (type II). Garnet does not occur in poorly recrystallized metabasalts, but does occur in well-recrystallized metabasalts of type III in which it contains spessartine, almandine and grossular components, and in coarsely crystalline metabasalts of type IV in which it contains almandine and grossular components. The pyroxenes of the metabasalts are aegirine-jadeite in poorly recrystallized rocks (type II), and are omphacite with only a small proportion of the acmite component in well-recrystallized rocks (type III).

Panoche Pass Area
Synoptic descriptions of this area (Ernst, 1965) were given in figs 3-8. and 7A-13 and §7A-10.

The metabasites in the poorly recrystallized part of the metamorphic terrane (Coleman and Lee's type II) sometimes preserve relict ophitic texture and relict plagioclase and augite. The plagioclase is heavily altered and ranges from labradorite (relict) to albite. Minute lawsonite tablets are present in many of the plagioclases. The augites are peripherally altered to aegirine or to an intergrowth of actinolite and chlorite. In some rocks, fine-grained plagioclase, stilpnomelane, chlorite and actinolite consitute an almost irresolvable mesostasis. Pumpellyite is present in most specimens, generally as irregular stringers and patches. Actinolite needles commonly exhibit a feathery periphery of crossite. Vesicles are filled with pumpellyite, chlorite, and rarely calcite and quartz, and/or albite. These rocks resemble greenschists insofar as chlorite, actinolite and albite are main products of recrystallization, but differ from them in containing lawsonite and pumpellyite instead of epidote.

On the other hand, metabasites in the completely recrystallized part of the metamorphic terrane (type III) are crossite schists mainly composed of crossite, lawsonite, and chlorite together with minor quartz, white mica, stilpnomelane, sphene and aragonite (partly altered into calcite). Actinolite is lacking, and albite has been found very rarely. A minor amount of aegirine is common. These rocks are in the typical glaucophane-schist facies.

Epidote–albite amphibolitic rocks, garnet–albite amphibolitic rocks and eclogites have been found in this area as well-recrystallized, coarse-grained rocks (Coleman and Lee's type IV). The amphibolitic rocks contain blue–green hornblende commonly rimmed by crossite.

Metabasites: Progressive Mineral Changes

8B-1 CHLORITE

The chlorites of metabasites show a limited range of the $Si(Mg,Fe^{2+}) \rightleftharpoons AlAl$ substitution, just like those of metapelites (§7B-6). The number of Si atoms on the basis of $18(O,OH)$ is in the range $2 \cdot 5 - 3 \cdot 0$. Hence, the same formula $(Mg,Fe)_{4 \cdot 5}Al_3Si_{2 \cdot 5}O_{10}(OH)_8$ may be used for both metabasite and metapelite chlorites. The $Fe^{2+}/(Mg + Fe^{2+})$ ratio of metabasite chlorites, however, is usually in the range $0 \cdot 1 - 0 \cdot 5$, tending to be lower than that of metapelite chlorites.

8B-2 EPIDOTE AND ZOISITE

Epidote and zoisite have the composition $HCa_2(Al,Fe^{3+})_3Si_3O_{13}$. The $Fe^{3+}/(Al + Fe^{3+})$ ratio is $0 \cdot 0 - 0 \cdot 2$ (mostly $0 \cdot 0 - 0 \cdot 1$) in zoisite (orthorhombic), and $0 \cdot 0 - 0 \cdot 4$ (mostly $0 \cdot 1 - 0 \cdot 4$) in epidote (monoclinic). Thus, epidote is usually higher in iron content than zoisite, though their composition ranges overlap. In the overlapping range, they are in a polymorphic relation. The stability relation between them, however, is not clear.

Epidote, including both clinozoisite and pistacite, is widespread in low-temperature metabasites. In many high-pressure terranes, it occurs in association with prehnite, pumpellyite or lawsonite, but persists to a higher temperature than any of them. In many medium- and low-pressure terranes such as the Scottish Highlands and the Ryoke belt, epidote is the predominant calcium-aluminium silicate in low temperature zones.

Most iron in epidote and zoisite is in the trivalent state. The stability of these minerals depends not only on temperature and rock-pressure but also on P_{H_2O} and P_{O_2}. With rising temperature, these minerals usually dwindle and finally disappear. This change involves dehydration and reduction. The liberated iron will mostly be reduced into the divalent state to enter other silicates. The liberated CaO and Al_2O_3 mostly enter plagioclase to make anorthite component.

The high-temperature stability limit of zoisite + quartz assemblage was experimentally determined by Newton (1966a). It is about 670 °C at 6 kbar in the presence of an aqueous fluid.

In progressive metamorphic terranes, zoned epidotes with decreasing iron content towards the rim are widespread. Probably this means that the ferric iron in epidote was partly reduced and transferred to other minerals with rising temperature. In relatively rare areas, a zonal structure with the reverse trend has been observed, suggesting retrogressive adjustment (Miyashiro and Seki, 1958).

In various parts of the Sanbagawa glaucophanitic belt, progressive enlargement of the composition field of epidote has been found (Miyashiro and Seki, 1958; Banno, 1964, p. 249). In the lowest temperature zones, the epidotes have a narrow range of composition near $Fe^{3+}/(Al + Fe^{3+}) = 0.3$. With rising temperature, the composition range is expanded in both directions, i.e. towards lower and higher $Fe^{3+}/(Al + Fe^{3+})$, though enlargement towards lower $Fe^{3+}/(Al + Fe^{3+})$ is more marked. In the contact aureole of the Arisu area, northern Japan, the composition range is expanded but also shifts toward a lower $Fe^{3+}/(Al + Fe^{3+})$ ratio with rising temperature (Seki, 1961b).

8B-3 PLAGIOCLASE

In the prehnite–pumpellyite, greenschist and epidote–amphibolite facies, prehnite, pumpellyite and epidote occur instead of the anorthite component in plagioclase, which is therefore albite. Intermediate and calcic plagioclase can form in the amphibolite and higher facies. The anorthite component usually forms at the expense of epidote. Gradual increase of the An content of plagioclase in metabasites with rising temperature has been clearly shown, for example, in the central Alps (Wenk and Keller, 1969) and the Tanzawa Mountains (Seki *et al*. 1969a). In the former area, however, there exists a compositional gap in the range of calcic albite and sodic oligoclase. For a discussion of this, refer to §7B-8.

The reaction of sodic plagioclase with epidote should be influenced by T, P_s, P_{H_2O}, P_{O_2} and by the minerals occurring in association with the albite and epidote. The rate of compositional variation of plagioclase with rising temperature should differ in different metamorphic terranes. Petrographic experience, however, has indicated that the rate of compositional variation is usually gentle up to 30 per cent An, above which it becomes very fast. This is the reason why a plagioclase composition of 30 per cent An in equilibrium with epidote is taken as the boundary between the epidote–amphibolite and amphibolite facies (Ramberg, 1952).

At present, a reliable scale relating temperature to the compositional change of plagioclase is not available. However, the change of the stable calcic amphibole in metabasites from actinolite to hornblende will be tentatively used here as a metamorphic grade reference against which plagioclase changes will be compared. At this grade, the composition of metabasite plagioclase is probably about 5 per cent An in the high-pressure terrane of the Sanbagawa belt and the Barrovian region of the Scottish Highlands, but is 20–30 per cent An in the low-pressure regional metamorphic terranes in Japan, as summarized in table 8B-1. While the reference point defined above does not represent a definite temperature, the relations described suggest the possibility that the formation of a more calcic plagioclase is promoted by lower rock-pressure.

Table 8B-1 *Composition of plagioclase associated with epidote or of most calcic plagioclase if epidote is not present at a metamorphic grade where predominant calcic amphibole changes from actinolite to hornblende in metabasites.*

Baric type	Area and author	Plagioclase (% An)
High-pressure (atypical)	Bessi (S. Banno, pers. comm. 1969)	5
Medium-pressure	Northern Appalachians (Billings and White, 1950)	Albite
	Scottish Highlands (Wiseman, 1934)	5
Low-pressure regional	Aracena (Bard, 1969)	20
	Ryoke (Katada, 1965)	30
	Central Abukuma (Miyashiro, 1958)	20-30
Low-pressure, small scale	Tanzawa (Seki *et al.* 1969*a*)	20
Low-pressure contact	Iritono (Shido, 1958)	Labradorite
	Arisu (Seki, 1961*b*)	Labradorite

Köhler (1941, 1949) has pointed out that the optical properties of plagioclase vary with its thermal history. Plagioclase in volcanic rocks has what are called high-temperature optics, whereas that in plutonic and metamorphic rocks usually has the different optic properties characteristic of low temperatures. The difference in optic angle between the two forms is great in relatively sodic plagioclase (Smith, 1958). There are continuous intermediate states between the high- and low-temperature optics. The effect of thermal history is also evident in the X-ray diffraction pattern (Smith and Yoder, 1956; Engel *et al.* 1964).

Though most metamorphic plagioclases show low-temperature optics, the granulite-facies metamorphic rocks of the Broken Hill area, Australia, were found to contain plagioclases showing intermediate optic and X-ray properties (Binns, 1965*b*).

Albites in low-temperature metamorphic rocks are mostly water-clear, untwinned and unzoned. With rising temperature, the frequency of twinning tends to increase (e.g. Iwasaki, 1963; Engel and Engel, 1962*a*). Gorai (1951) has pointed out that twinning on the albite and pericline laws occurs in both igneous and metamorphic rocks, whereas that on the Carlsbad and albite–Carlsbad laws is confined to igneous rocks. However, some observations at variance with this have been recorded (e.g. Binns, 1965*b*).

It is generally said that Carlsbad twinning forms at the time of crystal growth, whereas albite and pericline twinning can also form secondarily, for example, by deformation (Smith, 1962; Brown, 1962).

8B-4 CALCIC AND SUBCALCIC AMPHIBOLES

The general formula for the calcic and subcalcic amphiboles is:

$$(Ca, Na, K)_{2-3}(Mg, Fe^{2+}, Fe^{3+}, Al)_5 Si_6 (Si, Al)_2 O_{22} (OH)_2$$

where $Ca = 1\cdot5 - 2\cdot0$ in *calcic amphiboles*, and
$\quad\quad Ca = 1\cdot0 - 1\cdot5$ in *subcalcic amphiboles*,
$\quad\quad$ for 24 (O, OH) or 23O on anhydrous basis.

Comprehensive reviews on the chemical composition of these amphiboles have given by Hallimond (1943) and Leake (1968). A general review of their crystal chemistry, phase relations and occurrence has been given by Ernst (1968).

Calcic Amphiboles

To a first approximation, the composition of calcic amphiboles may be regarded as being controlled by the following three variables:

1. Degree of the Tschermak substitution: $(Mg, Fe^{2+})Si \rightleftharpoons AlAl$,
2. Total of Ca + Na + K. Since the calcic amphibole structure has large vacant sites that can accommodate Na^+ and K^+ ions, the total number of Ca^{2+} + $Na^+ + K^+$ ions is variable,
3. Degree of the $Mg \rightleftharpoons Fe^{2+}$ substitution.

The first two variables are shown in fig. 8B-1. Each point on the diagram still represents amphiboles whose composition may vary because of the $Mg \rightleftharpoons Fe^{2+}$ substitution.

Calcic amphiboles may be divided into two groups: the *actinolite group* and the *hornblendes*. The actinolite group comprises tremolites, actinolites and ferrotremolites (= ferroactinolites). The rest of the calcic amphiboles are here collectively called hornblendes. Fig. 8B-2 shows the curve for the frequency of occurrence of calcic and subcalcic amphiboles. There is a clear minimum in the curve at about $Si = 7\cdot2$ for 24(O + OH). Hence, the Si value of $7\cdot2$ is here regarded as the boundary between the actinolite group and the hornblendes.

The significance of this minimum is in dispute. It may be the result of a miscibility gap between the two groups of amphiboles under a limited range of

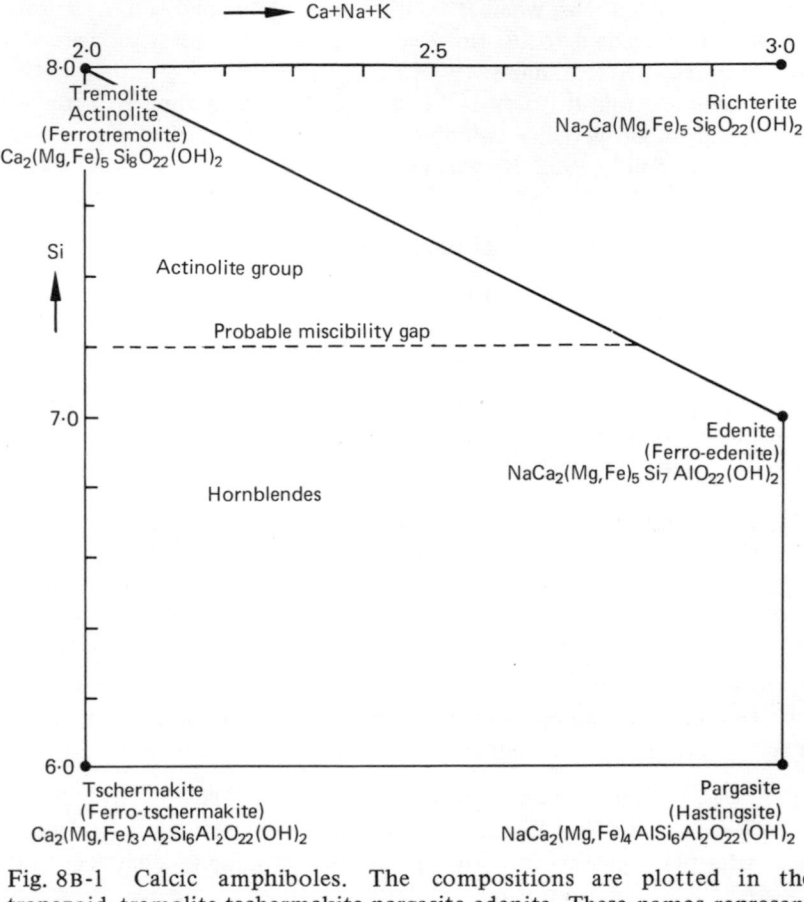

Fig. 8B-1   Calcic amphiboles. The compositions are plotted in the trapezoid tremolite-tschermakite-pargasite-edenite. These names represent Mg-end members. The names of Fe-end members are given in brackets below the corresponding Mg-end members.

metamorphic conditions (Shido, 1958; Shido and Miyashiro, 1959; Klein, 1969; Cooper and Lovering, 1970). The possible gap would not result in zero frequency for two reasons: (1) The gap may exist only under a limited range of conditions, especially at relatively low temperatures, and (2) there are a number of compositional variables, and the frequency diagram for Si does not reveal a gap within a two- or poly-dimensional composition space.

Modes of Occurrence.   In metabasites, actinolite does not form in the zeolite facies and the lower part of the prehnite-pumpellyite facies, but does form in the higher part of the latter facies and in the greenschist facies.

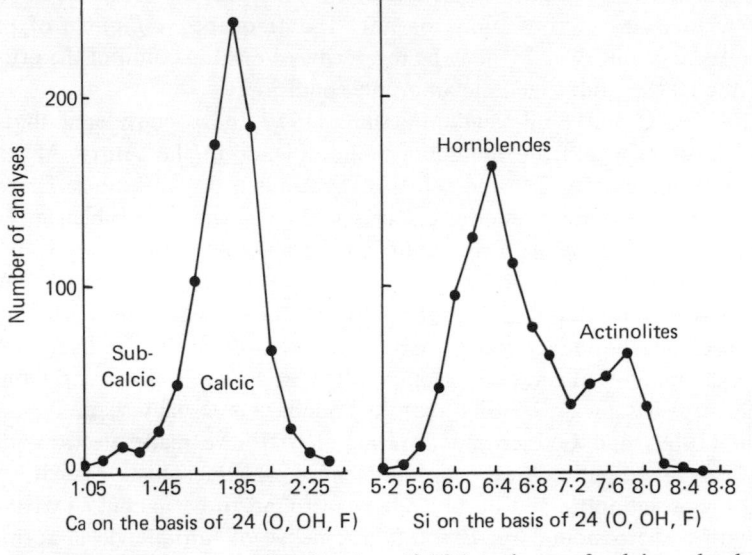

Fig. 8B-2 Frequency distribution of 936 analyses of calcic and sub-calcic amphiboles. (After Leake, 1968, with a correction in the Si diagram based on his personal communication, 1970.)

In greenschist facies terranes, the temperature of formation of actinolite would increase if P_{H_2O} and P_{CO_2} were high, thus delaying the formation of actinolite in progressive metamorphism. This may be illustrated schematically by the following equation:

$$5(Mg, Fe)_3Si_2O_5(OH)_4 + 6\,CaCO_3 + 14\,SiO_2$$

antigorite molecule   calcite   quartz
of chlorite

$$= 3\,Ca_2(Mg, Fe)_5Si_8O_{22}(OH)_2 + 6\,CO_2 + 7\,H_2O. \qquad (8B\text{-}1)$$

actinolite

Actinolite-free greenschists occur mixed with actinolite-bearing ones in some metamorphic terranes. In the equation given below, the left-hand side would be stabilized under a very high P_{CO_2}. This was observed in assemblage (4) of the Woodsville quadrangle given in §8A-6.

$$5\,Ca(Mg, Fe)(CO_3)_2 + 8\,SiO_2 + H_2O$$

ankerite   quartz

$$= Ca_2(Mg, Fe)_5Si_8O_{22}(OH)_2 + 3\,CaCO_3 + 7\,CO_2 \qquad (8B\text{-}2)$$

actinolite   calcite

In the epidote–amphibolite and amphibolite facies, the calcic amphiboles in ordinary metabasites are hornblendes. It has not been settled whether the compositional change from actinolite to hornblende in progressive metamorphism of metabasites is continuous or not. The frequency minimum of calcic amphiboles at about Si = 7·2 may be regarded as a manifestation of the possible compositional gap under such metamorphic conditions.

Shido (1958) and Shido and Miyashiro (1959) have considered that the compositional change from actinolite to hornblende in the central Abukuma Plateau and probably also in the Scottish Highlands is discontinuous. There is a transitional zone where actinolite coexists with blue–green hornblende. Hornblende occurs sometimes as a rim embracing an actinolite crystal, sometimes in parallel growth with actinolite, and sometimes as discrete crystals. When actinolite is in direct contact with hornblende, the boundary between them is sharp. When the boundary looks to be gradational under the ordinary microscope, examination of the thin section tilted in an appropriate angle on the universal stage has always revealed that the boundary is actually sharp.

Klein (1969) and Cooper and Lovering (1970) have made electron microprobe analysis of actinolites and hornblendes which coexist in such rocks. Homogeneous actinolite and hornblende were found to be in contact with each other with a sharp boundary. Exsolution lamellae of hornblende in actinolite and of actinolite in hornblende were found. These give a strong support for the existence of a miscibility gap at the relevant low temperatures. At higher temperatures, the gap may vanish.

Hornblendes in some metabasites not uncommonly have exsolution lamellae of cummingtonite. The lamellae are parallel to (001), (100) and ($\bar{1}$01) (Binns, 1965b; Jaffe, Robinson and Klein, 1968).

Colour Change. The colour of hornblende for Z shows a relatively regular change. In metabasites recrystallized at low temperatures (usually epidote-amphibolite facies or low amphibolite facies), the colour for Z is usually blue–green (greenish-blue or bluish-green). At higher temperatures (usually amphibolite facies), the colour for Z is green (without a bluish tint) and at still higher temperature it is brown. There can be complete gradation between these colours, and in some metamorphic terranes green and brown hornblendes are known to occur in rocks of different compositions within the same area. However, the general scheme of colour change for Z of blue–green → green → brown with rising temperature is widespread.

For this reason, actinolite and hornblendes with variable colours have been used as index minerals for zonal mapping of progressive metamorphic terrane in the Abukuma Plateau (Miyashiro, 1958; Shido, 1958), Aracena area (Bard, 1969) and Broken Hill (Binns, 1965b). Similar colour change was observed in the Ryoke belt (Katada, 1965), Séville province, Spain (Fabriès, 1968), Scottish Highlands (Wiseman, 1934), northern Michigan (James, 1955), northwest Adirondack, New York (Engel and Engel, 1962a, b), northern Appalachians

(Billings and White, 1950) and east Kimberley area, Australia (Gemuts, 1965). Blue–green hornblende tends to have a higher H_2O content and a higher Fe^{3+}/Fe^{2+} ratio than other hornblendes (Seitsaari, 1953). The brown colour has been ascribed to higher contents of TiO_2 (Binns, 1965b).

An approximate relationship between colour and optical constants is shown in fig. 8B-3.

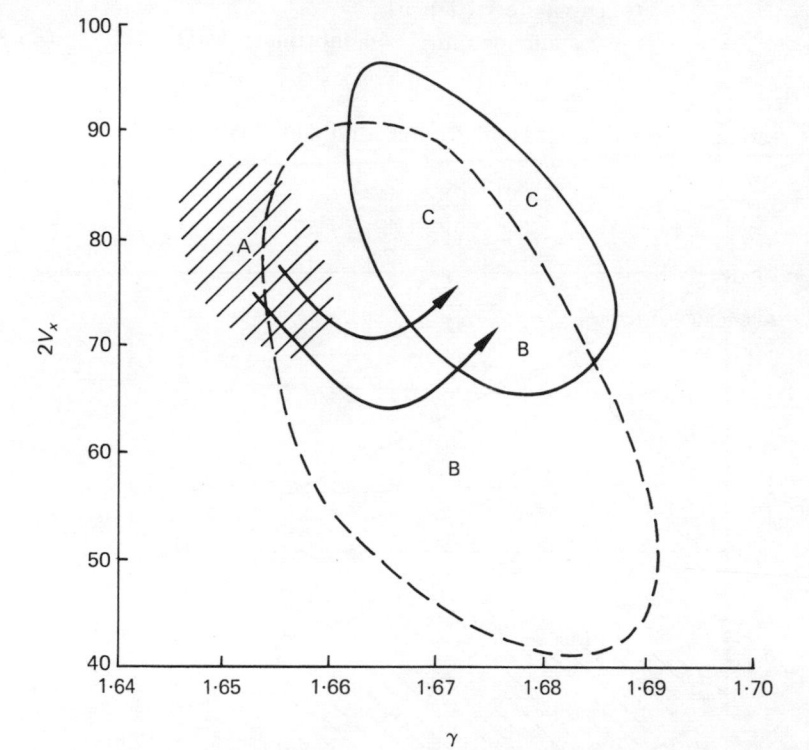

Fig. 8B-3   Optical constants of actinolites (A), Blue-green and green hornblendes (B) and greenish brown and brown hornblendes (C). The general trends of optical changes of calcic amphiboles with increasing temperature are shown by arrows. These trends are similar in high- and low-pressure metamorphic facies series.

Compositional Change. Since actinolite is poor and hornblende is rich in Al_2O_3, some authors have suggested that the total Al_2O_3 content or the tetrahedral Al content of such amphiboles increases with increasing temperature (e.g. Harry, 1950). However, the change is probably discontinuous as discussed above, and once hornblende becomes the predominant calcic amphibole, the Al_2O_3 and tetrahedral Al contents do not show a regular increase.

The compositional changes of calcic amphiboles with rising temperature in the central Abukuma Plateau (Shido, 1958; Shido and Miyashiro, 1959) and Broken Hill (Binns, 1965*b*, 1969) are shown as the fields A → B → C in fig. 8B-4. The blue–green hornblende is rich in the tschermakite component. With a rise in temperature, the component tends to react with quartz as follows:

$$\begin{aligned} &7 \text{ tschermakite} + 10 \text{ quartz} \\ &\quad = 3 \text{ cummingtonite} + 14 \text{ anorthite} + H_2O. \end{aligned} \qquad (8\text{B-}3)$$

This reaction tends to decrease the tetrahedral and octahedral Al. At the same

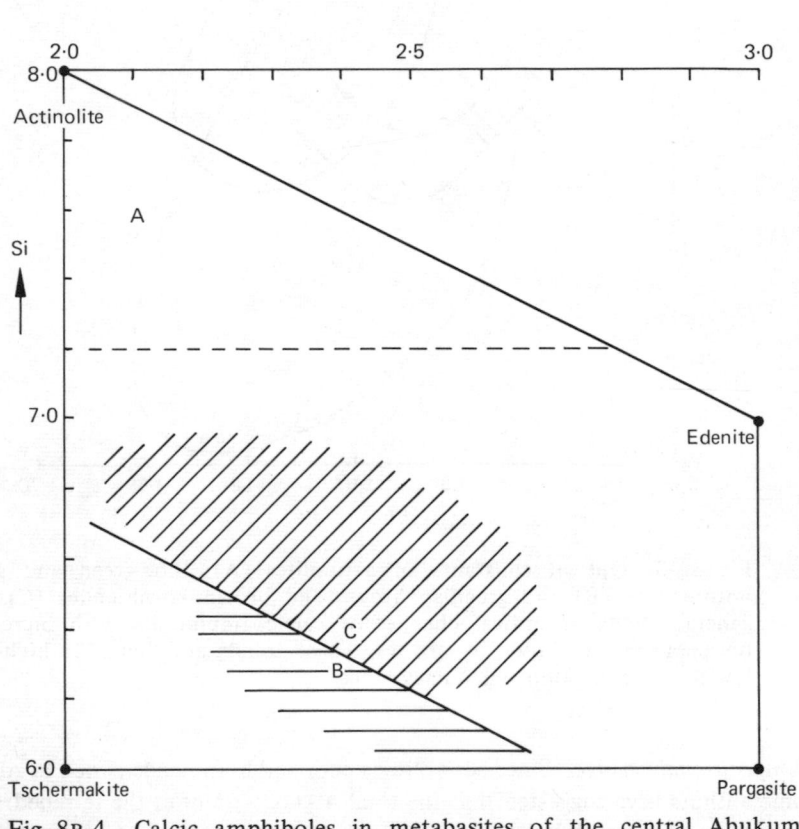

Fig. 8B-4 Calcic amphiboles in metabasites of the central Abukuma Plateau (Shido and Miyashiro, 1959) and Broken Hill (Binns, 1965*b*). A = actinolite field, B = blue-green hornblende field, C = field of green and brown hornblendes in the high amphibolite and granulite facies. Compare with fig. 8B-1.

time, the Na in vacant sites tends to increase by the replacement Si $\rightleftharpoons$ NaAl, resulting in a change toward the edenite or pargasite-hastingsite composition. The two reactions take place in the gradual transition from blue–green to green and brown hornblendes, and during a further increase in temperature into the granulite facies. Thus, the net results are a progressive decrease of octahedral Al and an increase of Na in vacant sites. It should be noted that the trend of compositional change differs between the transformation of actinolite → blue–green hornblende and that of blue–green hornblende → green and brown hornblende.

In the Haast Schist Group (Cooper and Lovering, 1970), amphibolite facies hornblendes show only slightly higher contents of (Ca + Na + K) than epidote–amphibolite facies ones.

In the Séville province (Fabriès, 1963, 1968) and Aracena area (Bard, 1969, 1970), a somewhat different trend of change has been noticed. In the transformation of blue–green to green and brown hornblendes, the Na in vacant sites increases. In fig. 8B-4, the field of such hornblendes moves simply to the right in the zone of Si = 6·2–7·0.

Leake (1965, 1968) and Binns (1969) have emphasized that the TiO_2 content tends to increase with metamorphic temperature.

The calcic amphiboles of metabasites in the Sanbagawa high-pressure terrane were found to have slightly higher SiO_2 + Al_2O_3 and Na_2O contents and lower TiO_2 and CaO contents than those showing a very similar mineral assemblage in medium- and low-pressure terranes (Ernst, 1972b).

Subcalcic amphiboles (Barroisites)

Na-rich subcalcic hornblendes, called barroisite, are rather common in high-pressure metamorphic terranes (Iwasaki, 1963; Banno, 1964) and in eclogites (Banno, 1964; Coleman *et al.* 1965; Binns, 1967).

The status of barroisite in progressive glaucophanitic metamorphism has been studied by Banno (1964) in the Bessi–Ino area. The amphiboles occurring in metabasites in three progressive stages are as follows:

Glaucophane-schist facies zone	glaucophane, actinolite,
Transitional zone	glaucophane, barroisite,
Epidote-amphibolite facies zone	barroisite, blue–green hornblende.

Glaucophane can coexist with actinolite or barroisite in the same rock. The compositional change from actinolite to barroisite appears to be continuous, and actinolite does not coexist with barroisite. The compositional variation from barroisite to hornblende also appears to be continuous.

8B-5   CUMMINGTONITE, ANTHOPHYLLITE AND GEDRITE

The chemical compositions of cummingtonite (monoclinic) and anthophyllite and gedrite (both orthorhombic) can be approximately represented by the formula:

$$(Mg, Fe, Al)_7 (Si, Al)_8 O_{22}(OH)_2.$$

Here, cummingtonite (including grunerite) has a low Al_2O_3 content and $Fe^{2+}/(Mg + Fe^{2+}) = 0\cdot3-1\cdot0$; anthophyllite has a low Al_2O_3 content and $Fe^{2+}/(Mg + Fe^{2+}) = 0\cdot0-0\cdot5$; gedrite has a high Al_2O_3 content and $Fe^{2+}/(Mg + Fe^{2+}) = 0\cdot4-1\cdot0$.

It appears that the composition fields of these three amphiboles are separated from one another by miscibility gaps (Stout, 1971, Robinson, Ross and Jaffe, 1971). However, the compositional difference between cummingtonite and anthophyllite for $Fe^{2+}/(Mg + Fe^{2+}) = $ about $0\cdot4$ is very slight.

Gedrite is apparently a relatively rare mineral and is similar in chemical composition to the assemblages: anthophyllite + cordierite, cummingtonite + cordierite, and almandine + cordierite.

Anthophyllite is a rare constituent of metabasites. However, it is abundant in some unusual rocks such as anthophyllite–cordierite rock (e.g. Eskola, 1914).

When anthophyllite coexists with cummingtonite, the former is always higher in Al_2O_3 than the latter. The $Fe^{2+}/(Mg + Fe^{2+})$ ratio of the former is sometimes higher and sometimes lower than that of the latter (Klein, 1968).

Cummingtonite is a relatively common but minor constituent of many low-pressure amphibolites. The frequency of occurrence appears to increase in the high amphibolite facies, where the breakdown of the tschermakite component of hornblende leads to the production of cummingtonite and anorthite component of plagioclase, as shown by equation (8B-3). Cummingtonite is not so common in amphibolites of medium- and high-pressure metamorphic terranes, where probably garnet forms instead (Shido, 1958, p. 211–12). There is a wide miscibility gap between cummingtonite and hornblende (e.g. Klein, 1968). Cummingtonite in some metamorphic rocks contains exsolution lamellae of hornblende (Jaffe et al. 1968). Some Ca-poor metabasites contain a large amount of cummingtonite.

8B-6   MUSCOVITE AND BIOTITE

In high-pressure metamorphism, muscovite occurs in metabasites of the glaucophane-schist, greenschist and epidote–amphibolite facies. In medium- and low-pressure metamorphism, muscovite occurs in metabasites of the greenschist facies, but disappears in those of the epidote–amphibolite and higher facies by reactions with amphibole or other associated minerals as exemplified by the reaction:

muscovite + calcic amphibole → biotite + anorthite + quartz + H_2O. (8B-4)

Muscovite of metabasites is phengitic in low-temperature zones and becomes higher in Al_2O_3, Na_2O and TiO_2 contents with increasing temperature (Ernst, 1972b).

Biotite, on the other hand, occurs in metabasites of a wider range of temperature extending from the low greenschist to the low granulite facies. Biotites in metabasites of the greenschist or epidote–amphibolite facies are either green or brown, whereas those of higher facies are brown or reddish brown (Phillips, 1930; Wiseman, 1934; Katada, 1965). In the Scottish Highlands, biotite occurs in some chlorite zone metabasites. In other words, biotite begins to occur at a lower temperature in metabasites than in metapelites (Phillips, 1930; Wiseman, 1934).

8B-7 ALMANDINE GARNET

In medium- and high-pressure metamorphism, epidote amphibolites and amphibolites commonly contain almandine garnet with 6–10 per cent CaO. Most almandines in metapelites have a much lower CaO content. This difference is apparently due to a chemical difference of their host rocks. In the Dalradian metamorphics of Scotland, almandine begins to occur in metabasites roughly at the same temperature as in metapelites, that is, from the beginning of the almandine zone for the metapelites (Wiseman, 1934).

On the other hand, almandine is very rare or absent in ordinary metabasites of the low-pressure series in the Orijärvi area (Eskola, 1914), central Abukuma Plateau (Miyashiro, 1958), Ryoke belt (Katada, 1965), and northern Michigan (James, 1955). Higher rock-pressure would favour the formation of almandine garnet in metabasites.

The formation of almandine is, however, very sensitive to factors other than rock-pressure. Almandine does not occur in medium-pressure metabasites of the Sulitjelma area (Vogt, 1927), Woodsville quadrangle (Billings and White, 1950), and Wilmington complex near Philadelphia (Ward, 1959), conceivably owing to their unfavourable composition.

It is common for some metabasites to contain almandine and others not to do so even in the same outcrop. Wiseman (1934) and Binns (1965b) have shown that almandine-bearing metabasites have a higher Fe^{2+}/Mg ratio than the associated almandine-free ones. A higher content of MnO and a lower content of CaO would also favour the formation of almandine in metabasites.

Hsu (1968) has shown that reducing conditions are essential for the formation of almandine. This may be the main factor controlling almandine formation in metabasites. Since metabasites do not contain organic matter or graphite, their oxidation state is highly variable (§7B-10).

8B-8   CALCIC CLINOPYROXENE

In calcareous metasediments, diopside begins to occur roughly at the threshold of the amphibolite facies. The diopside usually forms by the reaction of tremolite, calcite and quartz. Similarly, if a metabasite contains calcite and quartz together with a calcic amphibole, calcic clinopyroxene (diopside or salite) would begin to form near the threshold of the amphibolite facies. This is indeed the case in some metamorphic terranes.

In many other terranes, however, calcic clinopyroxene begins to occur in metabasites in the middle or a higher part of the amphibolite facies. The temperature where calcic clinopyroxene begins to occur depends largely on the bulk chemical composition of the rocks. In the granulite facies, where hornblende decomposes, calcic clinopyroxene is a major component of metabasites (e.g. Binns, 1965b).

When calcic clinopyroxene is associated with hornblende or orthopyroxene, the CaO content of the former tends to decrease with rising temperature, conceivably owing to the decreasing width of the miscibility gap between the associated minerals with rising temperature. In metabasites of the amphibolite facies, the clinopyroxenes are usually salites, whereas in those of the granulite facies, they are usually augites (Shido, 1958, p. 186–8; Binns, 1965b; Bard, 1969).

Binns (1965b) has found that calcic clinopyroxenes in granulite facies metabasites of the Broken Hill area contain two kinds of exsolution lamellae: one composed of orthopyroxene parallel to (100) of the host and the other composed of clinohypersthene (very poor in CaO) parallel to (001).

8B-9   ORTHOPYROXENE

In ordinary metabasites, orthopyroxene begins to form at a higher temperature than calcic clinopyroxene. The first occurrence of orthopyroxene in association with clinopyroxene in metabasites is usually considered to mark the beginning of the granulite facies. The pyroxenes form by gradual decomposition of hornblende.

Orthopyroxene, however, may form by other reactions also. The reaction temperature should vary with the type of reaction and the Fe^{2+}/Mg ratio of the rock, as well as with P_s, P_{H_2O} and P_{O_2}. Figure 8B-5 gives experimentally determined high temperature stability limits for four amphiboles in the presence of an aqueous fluid. The Mg-end members of calcic amphiboles (tremolite and pargasite) are stable up to a much higher temperature than the Mg-end member of anthopyllite. Tremolite is stable up to a much higher temperature than ferrotremolite. Thus, if an amphibolite contains much anthopyllite or cummingtonite, these minerals would decompose to form orthopyroxene at a much lower temperature than hornblende of ordinary amphibolite. If Fe-rich amphibolites

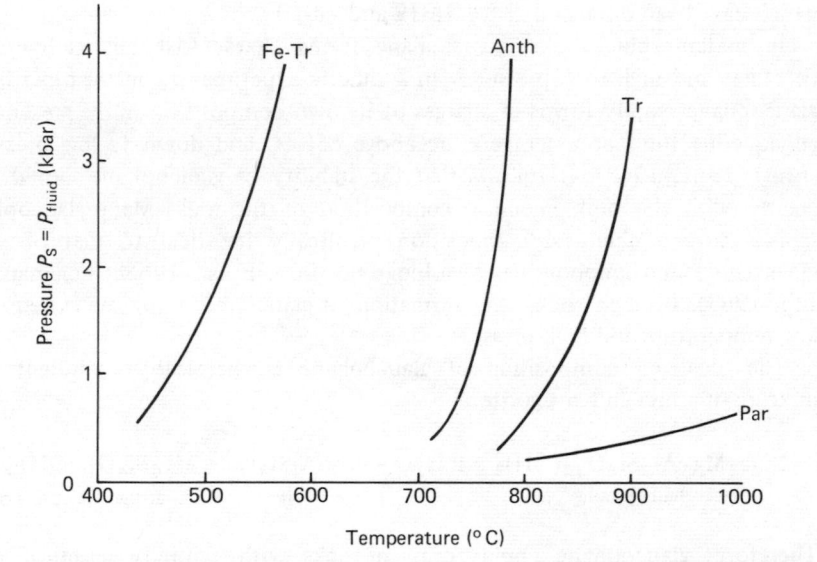

Fig. 8B-5 High temperature stability limit of anthophyllite (Anth), tremolite (Tr), ferrotremolite (Fe-Tr), and pargasite (Par) in the presence of an aqueous fluid. The ferrotremolite curve is for P_{O_2} controlled by the fayalite-magnetite-quartz buffer. (Greenwood, 1963; Boyd, 1959; Ernst, 1966.)

are mixed with Mg-rich ones, the former may produce orthopyroxene at a lower temperature than the latter.

Some orthopyroxenes in rocks of the granulite facies are very high in Al_2O_3 content (up to about 10 per cent) and show strong pleochroism (Howie, 1965; Burns, 1966). Exsolution lamellae composed of augite were found in some granulite facies orthopyroxenes (Binns, 1965b).

When orthopyroxene coexists with calcic clinopyroxene in metamorphic rocks, the partition of Mg and Fe^{2+} between the two minerals indicates an approach to chemical equilibrium. This problem was discussed by a large number of authors (e.g. Mueller, 1960; Kretz, 1961; Howie, 1965).

8B-10 GLAUCOPHANE

Glaucophane and stilpnomelane, both common in metabasites in high-pressure terranes, will be treated in this and the next section. Lawsonite, aragonite and jadeite occur in metabasites of some high-pressure terranes, but will not be discussed here, because they are controlled by similar factors in metagraywackes, which have been discussed in §§ 7B-19 and 7B-20.

In metamorphic rocks, glaucophane forms characteristically at low temperatures and high rock-pressures. In synthetic experiments, on the other hand, glaucophane readily forms in charges of its own composition in the presence of an aqueous fluid at a temperature above 850 °C and down to low pressures (Ernst, 1961, 1968). It follows that the stability of glaucophane should vary greatly with the bulk chemical composition of the rock. Many glaucophane schists have a chemical composition practically identical to that of some greenschists and amphibolites (Washington, 1901; Ernst, 1963a; Coleman and Lee, 1963). In such rocks, the formation of glaucophane appears to require a low temperature and high pressure.

The idealized composition of glaucophane is chemically equivalent to a mixture of albite and antigorite:

$$2 \, Na_2Mg_3Al_2Si_8O_{22}(OH)_2 + 2 \, H_2O = 4 \, NaAlSi_3O_8 + Mg_6Si_4O_{10}(OH)_8.$$
$$\text{glaucophane} \qquad\qquad\qquad \text{albite} \qquad\quad \text{antigorite} \qquad (8B-5)$$

Therefore, glaucophane could form in rocks with ordinary chemical compositions, in which otherwise most or all the Na_2O would have entered plagioclase.

Natural glaucophanes, however, are usually solid solutions between idealized glaucophane and riebeckite $Na_2Fe_3^{2+}Fe_2^{3+}Si_8O_{22}(OH)_2$ (e.g. Miyashiro, 1957b). In a transitional condition between the glaucophane-schist and greenschist facies, some metabasites have glaucophane-bearing mineral assemblages and others greenschist assemblages; the former tend to have higher Fe^{3+}/Fe^{2+} ratios (Iwasaki, 1963, p. 32). The P–T field for the stability of glaucophane in metabasites may become wider by the presence of the riebeckite component.

Synthetic glaucophane has two polymorphs: I and II. Glaucophane I is stable on the high-temperature and low-pressure side of the field of glaucophane II. Natural glaucophane belongs to polymorph II (Ernst, 1963b, 1968).

Glaucophane is sometimes stably associated with actinolite or barroisite, or rarely with hornblende. The existence of wide miscibility gaps between glaucophane and the latter amphiboles has been demonstrated by Klein (1969) and Himmelberg and Papike (1969).

8B-11   STILPNOMELANE

Stilpnomelane is widespread in, but not confined to, high-pressure metamorphic rocks. It is rare in medium-pressure and is extremely rare in low-pressure metamorphic rocks. So far as I am aware, the sole discovery of stilpnomelane in low-pressure metamorphic terranes has been made in slightly metamorphosed basaltic pyroclastics in the lowest temperature zone (zone Ia in fig. 3-4) of the Shiojiri area in the Ryoke belt, Japan (Katada and Sumi, 1966). In medium-pressure metamorphic terranes, stilpnomelane was found, for example, from an epidiorite (Wiseman, 1934, p. 376) and metagraywacke (Mather and Atherton, 1965; Mather, 1970), both in a low-temperature part of the Scottish Highlands, and from slates and phyllites of the northern Appalachians (Zen, 1960).

In high-pressure metamorphic terranes, stilpnomelane is a common mineral in metabasites, metapelites, metapsammites and metacherts of the glaucophane-schist, greenschist and epidote–amphibolite facies. Stilpnomelane is more common in the metapsammites than in the other rocks. In high-pressure facies series, stilpnomelane appears to form most readily in the glaucophane-schist and greenschist facies. When the series extend downwards to prehnite–pumpellyite facies or upwards to the epidote-amphibolite facies, the mineral is less common or absent in the extended parts of the series (Iwasaki, 1963; Hashimoto, 1968; Ernst and Seki, 1967; Niggli, 1960, 1970).

The stability of stilpnomelane, however, varies greatly with the chemical composition of the rock. It is known to occur even in granophyre patches associated with the marginal olivine gabbro of the shallow gabbroic intrusion of Skaergaard, Greenland (Wager and Deer, 1939). It has long been known to occur in iron ores, and especially commonly in weakly metamorphosed iron formations.

Chemically, an Fe^{2+}-rich, dark green variety called ferrostilpnomelane is distinguished from an Fe^{3+}-rich, reddish brown variety, called ferristilpnomelane. The two may occur in different parts of a single hand specimen of schist (e.g. Yui, 1962). Hutton (1938) assumed that ferristilpnomelane forms by the oxidation of ferrostilpnomelane accompanied by loss of H, that is, by the substitution $Fe^{3+}O^{2-} \rightleftharpoons Fe^{2+}(OH)^-$, as in oxyhornblende. However, Hashimoto (1969) has emphasized that there is no systematic difference in H_2O^+ content between the two varieties and there appears to exist a discontinuous change in the oxidation state. Brown (1971) has given strong evidence suggesting that the oxidation of ferrostilpnomelane takes place after metamorphism.

8B-12 OXIDES AND SULPHIDES OF IRON

The frequency of occurrence of iron oxides and sulphides in metabasites of high-pressure and low-pressure metamorphic terranes in Japan is summarized in table 7B-1.

If we compare these minerals in glaucophane-schist facies metabasites with those in amphibolite facies metabasites, haematite is common in the former but is absent in the latter, whereas magnetite and ilmenite are rare or absent in the

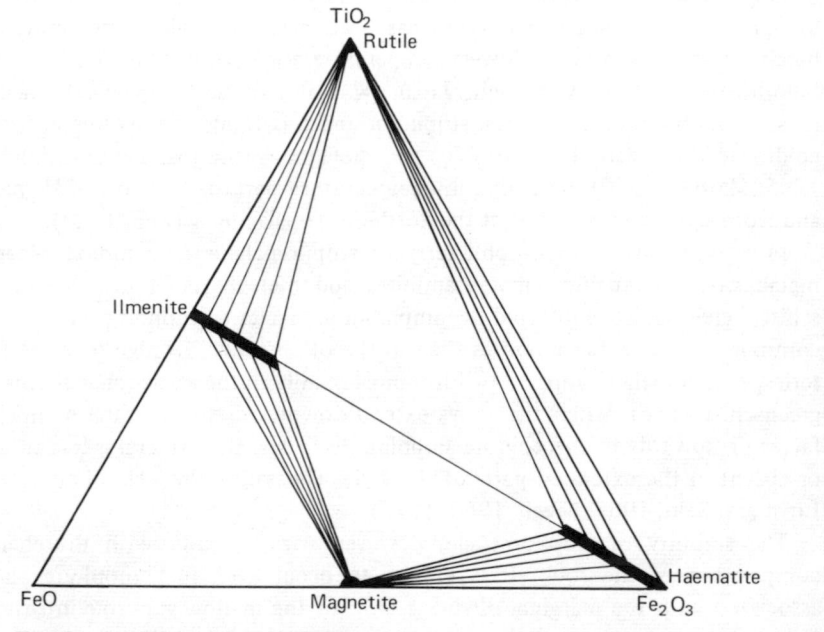

Fig. 8B-6   Paragenetic relations of oxides of iron and titanium in some metamorphic rocks. (Chinner, 1960; Banno and Kanehira, 1961.)

former but are common in the latter. This difference may be regarded as a progressive reduction with rising temperature.

In the low-temperature part of high-pressure metamorphic terranes, metabasites commonly contain haematite, whereas associated metapelites contain no haematite or magnetite. Thus, metapelites are evidently in a more strongly reducing condition than the associated metabasites. This is probably due to the presence of organic matter in the former.

Haematite, magnetite and ilmenite belong to the $FeO-Fe_2O_3-TiO_2$ system, as shown in fig. 8B-6. There is a continuous series of solid solution between ilmenite and haematite at high temperatures above 1000 °C, but a miscibility gap exists under metamorphic conditions. If we neglect the effect of this solid solution, the magnetite–haematite assemblage has a definite value of P_{O_2} under a definite P_s and T, as was discussed in § 2-7. If we consider the effect of solid solution, however, the state of the system is fixed only when the three phases magnetite, haematite and ilmenite coexist (Chinner, 1960).

The TiO_2 content of magnetite in some igneous rocks may be as high as 20 per cent, but that of magnetite in metamorphic rocks is usually less than 4 per cent (Buddington, Fahey and Vlisidis, 1955; Heier, 1956).

For a detailed account of the phase relations in the Fe_2O_3–FeO–TiO_2 system, refer to Nagata (1961).

Chapter 9

Progressive Regional Metamorphism of Limestones

9-1 COMPOSITIONS OF LIMESTONES AND FLUIDS

Compositions of Limestones

Most limestones are mainly composed of three components: $CaCO_3$, $MgCO_3$ and SiO_2. Since the number of components is so small, the relationship between the chemical and mineral compositions in metamorphosed limestones is relatively simple. This greatly facilitates theoretical treatment.

If a limestone is composed of pure $CaCO_3$, metamorphism would produce crystalline limestones (marbles) which tend to become coarser grained with increasing temperature. In most cases, there will be no other changes at all. The temperature of metamorphism is not high enough to dissociate calcite into CaO and CO_2. (In nature, the mineral lime CaO has only been reported as occurring in limestone inclusions in a lava of the volcano Vesuvius.) In some high-pressure metamorphism, however, calcite in limestone may be transformed into aragonite.

If an original limestone contains calcite and dolomite but no quartz, metamorphism may decompose only the dolomite as follows: dolomite → calcite + periclase + CO_2. This decomposition could take place at temperatures several hundred degrees lower than that of $CaCO_3$ under otherwise similar conditions, and the decomposition product periclase is not rare in dolomitic limestones subjected to contact metamorphism. Periclase, usually in small octahedral grains, almost always suffers hydration to form brucite $Mg(OH)_2$ at some later time. Brucite-calcite rocks produced in this way are called predazzite.

If an original limestone contains calcite and quartz, contact or regional metamorphism at relatively high temperatures would produce wollastonite by the reaction: calcite + quartz = wollastonite + CO_2. Wollastonite is a common mineral in limestones in amphibolite and pyroxene-hornfels facies areas. In some rare cases of contact (and not regional) metamorphism, siliceous limestones may contain tilleyite, spurrite, rankinite and larnite.

As a more general case, an original limestone is assumed to be composed of calcite, dolomite and quartz. The metamorphism of such a siliceous dolomitic limestone produces a variety of silicate minerals. In the metamorphism of siliceous dolomitic limestones of the Precambrian terrane in Finland, Eskola (1922) found three steps characterized by the first appearance of the following minerals in order of rising temperature: tremolite → diopside → wollastonite.

Some metamorphosed dolomitic limestones contain forsterite and lack quartz. In some cases of contact metamorphism, monticellite, akermanite, tilleyite, spurrite, rankinite, merwinite, and larnite may form, depending on the temperature, P_{CO_2} and other factors.

If an original limestone contains some Al_2O_3 besides CaO, MgO, and SiO_2 it may produce epidote, hornblende, grandite (Ca-garnet), vesuvianite, and calcic plagioclase. Epidote is usually stable from the greenschist or even lower facies. Grandite begins to appear in the epidote-amphibolite facies in some areas and in the amphibolite facies in others.

Some metamorphic rocks derived from impure calcareous sediments are mainly composed of plagioclase and hornblende. This indicates the possibility that some ordinary amphibolites may also be metamorphic derivatives of calcareous sediments. When metamorphism is accompanied by a degree of metasomatism or metamorphic differentiation, the chemical composition of the resultant amphibolites may be similar to that of amphibolites of basic igneous origin. This problem was discussed in detail by Orville (1969) and Vidale (1969).

If K_2O is present, muscovite, phlogopite and/or K-feldspar would be produced by metamorphism.

Distribution of P_{CO_2}

In progressive metamorphism, limestones are subjected to a series of decarbonation reactions. At the same time, the associated pelitic and psammitic rocks undergo dehydration reactions. Thus, CO_2 and H_2O are liberated together.

The temperatures of decarbonation reactions vary with P_{CO_2} and the composition of the fluid phase if any. The composition of the fluid varies considerably within a short distance in some metamorphic areas, as was clearly demonstrated by Melson (1966) and Carmichael (1970). In such areas, the isograds for decarbonation reactions may intersect with isograds for dehydration reactions.

Our knowledge of the frequency and extent of uneven distribution of CO_2 in metamorphic areas is still very meagre. However, uneven distribution is conceivably very common (§ 4-4). Therefore, decarbonation reactions cannot be used as a reliable geothermometer.

9-2   LIMESTONES IN LOW-PRESSURE REGIONAL METAMORPHISM

Ryoke Belt

Metamorphosed limestones in the Shiojiri area of the Ryoke belt, Japan, were described by Katada (1965). The lowest temperature limestones contain calcite, dolomite, quartz, and plagioclase with subordinate white mica and tourmaline. With rising temperature, tremolite and diopside begin to occur at nearly the same temperature. This is at variance with observations in many other areas of the world where tremolite begins to form at a lower temperature than diopside. Grandite begins to form at a higher temperature than diopside, and wollastonite and scapolite at a still higher temperature (fig. 3-4).

The grandite + quartz assemblage persists to the highest temperature zone, though it is accompanied by wollastonite in some rocks. Wollastonite is accompanied by calcite and quartz.

Central Abukuma Plateau.

Most limestones in this area are rather pure $CaCO_3$ rocks aand show only a progressive increase in grain-size with rising temperature. In low-temperature zones, most grains of calcite measure $0·05-0·1$ mm in diameter, whereas in higher temperature zones they measure $1-10$ mm.

Small parts of the limestones, however, are impure. Further, the metabasites in this area frequently contain calcic bands (Miyashiro, 1953*a*, 1958). These rocks show the following order of the first appearance of minerals: tremolite → diopside → wollastonite (fig. 3-5).

In the high-temperature part of the area, the assemblage garnet + quartz + wollastonite occurs in some rocks, and the assemblage quartz + calcite + calcic plagioclase + wollastonite occurs in others. This may mean that the garnet–quartz reaction is beginning to take place as follows: grossular + quartz = 2 wollastonite + anorthite. This reaction will be discussed near the end of this chapter.

In some quartz-free limestones of the same zone, forsterite occurs in association with calcite, dolomite, tremolite, diopside and phlogopite.

Pyrenees

In some low-pressure regional metamorphic areas in the Pyrenees and Iberian Peninsula, more or less similar sequences of progressive mineral formation have been observed. For example, Bard (1969) has found in the Aracena area in Spain the following sequence with rising temperature: tremolite → diopside → forsterite → grandite → wollastonite (fig. 7A-4).

9-3 LIMESTONES IN MEDIUM-PRESSURE REGIONAL METAMORPHISM

Scottish Highlands

Harker (1932, p. 252–9) gives a cursory description of limestones in the Scottish Highlands, though information from other areas is indistinguishably mixed. In the biotite zone, limestones contain tremolite, zoisite, muscovite, biotite, quartz and often albite. In the almandine zone, limestones contain grandite, vesuvianite, diopside, hornblende, forsterite and microcline. Skarns are formed at the boundary between a limestone and kyanite schist at Glen Urquhart. They contain epidote, zoisite and calcic amphiboles, and occasionally prehnite and pectolite (Francis, 1958).

In the sillimanite zone, limestones contain diopside, grossular, plagioclase and orthoclase, but no wollastonite. Harker considered that wollastonite does not normally form in regional metamorphism. Indeed, in another medium-pressure metamorphic region (northern Appalachians), wollastonite does not occur even in the low granulite facies (Thompson and Norton, 1968).

However, the absence of wollastonite is not a general feature of regional metamorphism, or a general feature of medium-pressure regional metamorphism. Examples of low-pressure metamorphism, producing wollastonite, have already been given. An example of wollastonite-forming medium-pressure regional metamorphism will be described below.

Nanga Parbat

In Nanga Parbat, northwest Himalayas, calcareous metasediments occur in a terrane mainly composed of metapelites which contain kyanite and sillimanite. They show unusually advanced decarbonation for medium-pressure metamorphism (Misch, 1964). In an amphibolite facies area, three progressive metamorphic zones in terms of calc-silicate assemblages have been distinguished:

1. Calcite–quartz–grossular zone, which is characterized by the presence of grossular and of the calcite + quartz assemblage instead of wollastonite,
2. Wollastonite–grossular–quartz zone,
3. Wollastonite–anorthite zone, in which the reaction of grossular with quartz (i.e. grossular + quartz = 2 wollastonite + anorthite) has been completed. Wollastonite occurs there in contact with anorthite or bytownite.

9-4 LIMESTONES IN HIGH-PRESSURE REGIONAL METAMORPHISM

The Alps

Trommsdorff (1966) has described metamorphosed limestones in the Lepontine and Bergell Alps near the border between Switzerland and Italy. The temperature of metamorphism was lowest in the northwestern part of the areas. The limestones exposed there contain talc and tremolite in such assemblages as talc + calcite + dolomite, talc + calcite + tremolite, tremolite + quartz + calcite, and tremolite + calcite + dolomite. Talc appears to begin to form at a little lower temperature than tremolite.

The temperature rises toward the southeast and soon diopside begins to form, and then forsterite. However, the distribution of forsterite is capricious. Tremolite is still stable in some limestones. In the highest temperature part of the area where the Bergell granite is intruded, wollastonite forms in siliceous limestones. Probably this is a result of contact effects and not of regional metamorphism. Scapolite occurs over the whole area from the tremolite zone to the highest temperature one (see § 7A-8).

Sanbagawa Belt

In the Bessi–Ino area, the lowest-temperature zone, belonging to the glaucophane-schist facies, contains limestone with chlorite and quartz besides calcite. In a higher temperature zone of the lower epidote–amphibolite facies, limestones contain tremolite, muscovite, albite and quartz besides calcite. Limestones of the higher epidote–amphibolite facies contain epidote, grossular, diopside and hornblende (Banno, 1964, p. 290).

Franciscan Formation

In well-recrystallized Franciscan metamorphics of the typical glaucophane-schist facies (type III in Coleman and Lee's nomenclature) in the Cazadero area, California, metamorphosed limestones occur as small beds and lenses composed of aragonite. Here, the pressure was high enough to cause transformation of calcite to aragonite. The rock contains, besides carbonate, pyralspite garnet, green amphibole, stilpnomelane, clinozoisite and quartz (Coleman and Lee, 1962, 1963).

9-5 PROGRESSIVE METAMORPHIC REACTIONS IN SILICEOUS DOLOMITIC LIMESTONES

Introductory

Bowen (1940) made an elegant theoretical analysis of the progressive decarbonation of siliceous dolomitic limestones. A number of thermochemical and synthetic studies have been made on individual decarbonation reactions which take place in the presence of a pure CO_2 fluid (e.g. Weeks, 1956; Danielsson, 1950; Harker and Tuttle, 1956). In recent years, theoretical and experimental works on systems involving a fluid containing both CO_2 and H_2O have been made by Greenwood (1962), Winkler (1967), Metz and Winkler (1964), and others. These works have made important contributions to our understanding of natural decarbonation processes.

Winkler (1967) and Metz and Trommsdorff (1968) give elaborate systematic accounts on this problem. The following discussion is based largely on theirs. The presence of a fluid composed of CO_2 and H_2O is assumed below.

Limestones composed of calcite, dolomite and quartz and their metamorphic derivatives may be regarded as belonging to the five-component system $CaO-MgO-SiO_2-CO_2-H_2O$. Disregarding CO_2 and H_2O, the relevant rocks can be represented in a triangular diagram $CaO-MgO-SiO_2$. The compositions of the siliceous dolomitic limestones which we will now consider fall on the left half of the diagrams in Fig. 9-1.

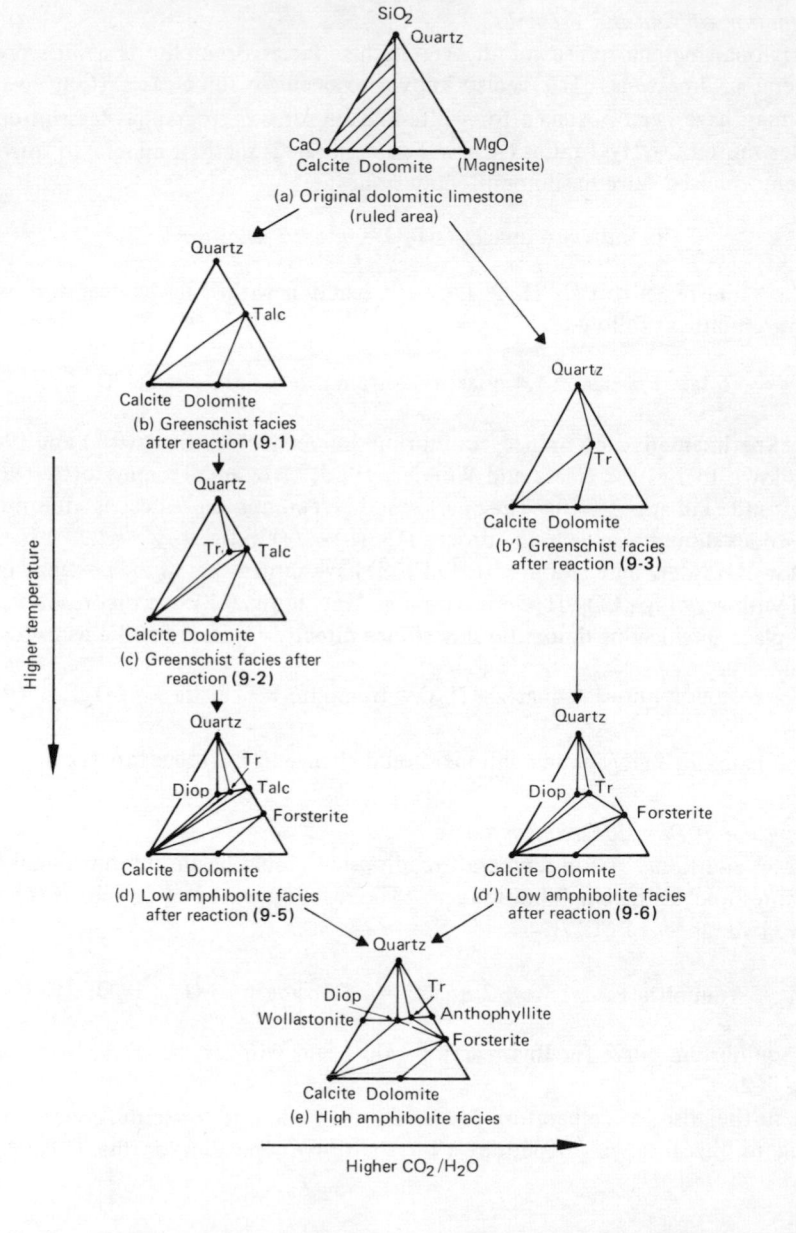

Fig. 9-1 Progressive changes of paragenetic relations in siliceous dolomitic limestones in the presence of a mixture of CO_2 and H_2O. *Abbreviations:* Diop = diopside, Tr = tremolite. Only the rocks plotted on the left half of the triangles are considered. If the fluid is of virtually pure CO_2 composition, hydrous minerals such as talc and tremolite would not form. Such an extreme case is not considered here.

Formation of Talc and Tremolite

In regional metamorphism of the greenschist facies, tremolite is a widespread mineral in limestones. Talc is also known to occur in some cases. (Conceivably talc may have been mistaken for white mica in some petrographic descriptions.) Under most CO_2/H_2O ratios ($< M$ in fig. 9-2), talc is the first mineral to form in metamorphosed siliceous dolomitic limestones:

$$3 \text{ dolomite} + 4 \text{ quartz} + H_2O = \text{talc} + 3 \text{ calcite} + 3 CO_2. \qquad (9\text{-}1)$$

In the same range of CO_2/H_2O, the next reaction with rising temperature will form tremolite as follows:

$$5 \text{ talc} + 6 \text{ calcite} + 4 \text{ quartz} = 3 \text{ tremolite} + 6 CO_2 + 2 H_2O. \qquad (9\text{-}2)$$

The experimentally determined equilibrium curves for reactions (9-1) and (9-2) are shown in fig. 9-2 (Metz and Winkler, 1963; Metz and Trommsdorff, 1968; Metz and Puhan, 1970). The paragenetic relations in siliceous dolomitic limestones should change in the order (a) → (b) → (c) in fig. 9-1.

However, Metz and Trommsdorff (1968) have shown that in the presence of a fluid with very high CO_2/H_2O ratios (range M–N in fig. 9-2), the first reaction to take place in siliceous dolomitic limestones directly forms tremolite as follows:

$$5 \text{ dolomite} + 8 \text{ quartz} + H_2O = \text{tremolite} + 3 \text{ calcite} + 7 CO_2. \qquad (9\text{-}3)$$

In this case, the paragenetic relations should change in the order (a) → (b').

Formation of Diopside and Forsterite

With an additional rise in temperature, diopside should begin to form roughly at the threshold to the amphibolite facies. The reaction would be as follows for the above two ranges of CO_2/H_2O:

$$\text{tremolite} + 3 \text{ calcite} + 2 \text{ quartz} = 5 \text{ diopside} + 3 CO_2 + H_2O. \qquad (9\text{-}4)$$

The equilibrium curve for this reaction (Metz and Winkler, 1964) is also shown in fig. 9-2.

A further rise in temperature causes the formation of forsterite, conceivably owing to the following reactions (9-5) or (9-6) depending on the CO_2/H_2O ratio:

$$\text{talc} + 5 \text{ dolomite} = 4 \text{ forsterite} + 5 \text{ calcite} + 5 CO_2 + H_2O. \qquad (9\text{-}5)$$

This reaction takes place for most values of CO_2/H_2O in the fluid, and the paragenetic relations after this reaction are shown in fig. 9-1(d).

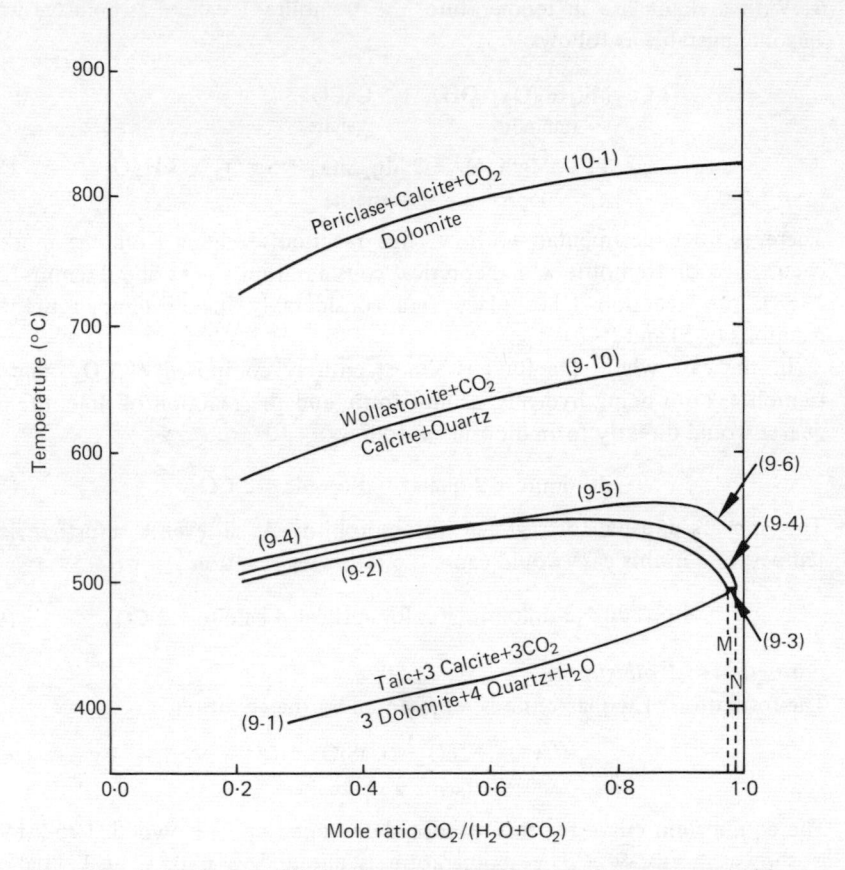

Fig. 9-2   Isobaric equilibrium curves for progressive metamorphic reactions in limestones in the presence of a fluid consisting of CO_2 and H_2O at a total pressure of 1 kbar. Figures in parentheses represent the numbers of equations in the text. (Based on Winkler, 1967; Metz and Winkler 1963, 1964; Metz and Trommsdorff, 1968, and Greenwood, 1962.)

$$\text{tremolite} + 11\ \text{dolomite} = 8\ \text{forsterite} + 13\ \text{calcite} + 9\ CO_2 + H_2O. \quad (9\text{-}6)$$

This reaction takes place only for very high CO_2/H_2O ratios, and the paragenetic relations after this reaction are shown in fig. 9-1(d′).

The equilibrium temperatures for (9-2), (9-4) and (9-5) as shown in fig. 9-2 are very close to each other, especially at low CO_2/H_2O ratios. In other words, tremolite, diopside and forsterite begin to form nearly at the same temperature. (This is at variance with our ordinary petrographic experience on regional metamorphic limestones.)

With a slight rise in temperature, the tremolite + calcite assemblage would become unstable as follows:

$$3 \, Ca_2Mg_5Si_8O_{22}(OH)_2 + 5 \, CaCO_3$$
$$\text{tremolite} \qquad\qquad\qquad \text{calcite}$$

$$= 11 \, CaMgSi_2O_6 + 2 \, Mg_2SiO_4 + 5 \, CO_2 + 3 \, H_2O. \qquad (9\text{-}7)$$
$$\text{diopside} \qquad \text{forsterite}$$

There is no experimental work on this reaction. Judging from the mode of occurrence of tremolite and theoretical consideration (Metz and Trommsdorff, 1968), this reaction takes place at a considerably higher temperature than reactions (9-5) and (9-6).

In the case where the fluid is almost entirely composed of CO_2, talc and tremolite, both being hydrous, cannot form, and the reaction of dolomite with quartz would directly form diopside as follows:

$$\text{dolomite} + 2 \, \text{quartz} = \text{diopside} + 2 \, CO_2. \qquad (9\text{-}8)$$

This appears unusual for regional metamorphism. At all events, a further rise in temperature in this case would cause the following reaction:

$$\text{diopside} + 3 \, \text{dolomite} = 2 \, \text{forsterite} + 4 \, \text{calcite} + 2 \, CO_2. \qquad (9\text{-}9)$$

Formation of Wollastonite

The formation of wollastonite is represented by the equation:

$$CaCO_3 + SiO_2 = CaSiO_3 + CO_2. \qquad (9\text{-}10)$$
$$\text{calcite} \quad \text{quartz} \; \text{wollastonite}$$

The equilibrium curve for this reaction determined by Greenwood (1962, 1967) is shown in fig. 9-2. The temperature is about 580–680 °C at 1 kbar and 610–750 °C at 2 kbar for a wide range of CO_2/H_2O in the coexisting fluid.

The paragenetic relations in such high amphibolite facies conditions are shown in fig. 9-1(e). Anthophyllite is shown here instead of talc (Greenwood, 1963; Winkler, 1967, p. 39).

9-6 METAMORPHIC REACTIONS IN ALUMINA-CONTAINING SILICEOUS LIMESTONES

The high-temperature stability limits for grossular and the grossular + quartz assemblage are marked by the following reactions:

$$2 \, Ca_3Al_2Si_3O_{12} = 3 \, CaSiO_3 + CaAl_2Si_2O_8 + Ca_2Al_2SiO_7 \qquad (9\text{-}11)$$
$$\text{grossular} \qquad\quad \text{wollastonite} \quad\; \text{anorthite} \qquad \text{gehlenite}$$

$$Ca_3Al_2Si_3O_{12} + SiO_2 = 2 \, CaSiO_3 + CaAl_2Si_2O_8. \qquad (9\text{-}12)$$
$$\text{grossular} \qquad\quad \text{quartz} \;\; \text{wollastonite} \quad\; \text{anorthite}$$

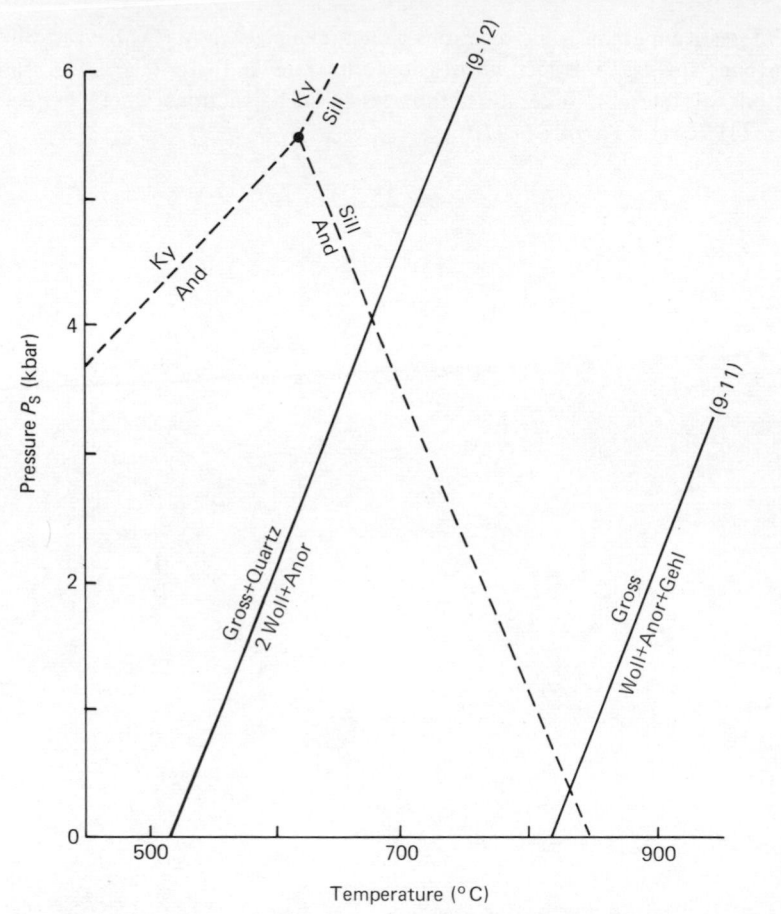

Fig. 9-3 High-temperature stability limits of grossular and the grossular + quartz assemblage under low P_{CO_2}. (Based on Newton, 1966*a*; Storre, 1970, and thermochemical calculations from Robie and Waldbaum's, 1968, constants.) The broken lines represent the stability fields of Al_2SiO_5 minerals.

As shown in fig. 9-3, the equilibrium curve for (9-11) lies at temperatures about 300 °C higher than that for (9-12). Many grossular-bearing limestones contain no quartz. The reaction of grossular with quartz with rising temperature was most clearly demonstrated in the Nanga Parbat area (§ 9-3).

The grossular + quartz assemblage is stable only under a relatively low P_{CO_2}. Increase in P_{CO_2} or in the CO_2/H_2O ratio of the intergranular fluid would change the assemblage into calcite and less calcic minerals such as zoisite or anorthite.

Progressive metamorphic reactions which take place in Al_2O_3-bearing siliceous limestones involve zoisite, clinozoisite, anorthite and grossular. The stability relations of relevant mineral assemblages in such reactions were discussed by Storre (1970) and Kerrick (1970).

Chapter 10

Progressive Contact Metamorphism and Pyrometamorphism

10-1 CONTACT METAMORPHISM OF PELITES TO PRODUCE ANDALUSITE AS THE ONLY Al_2SiO_5 MINERAL

The physical conditions of contact metamorphism are diverse (Miyashiro, 1961b, p. 306–7; Reverdatto, Sharapov and Melamed, 1970). A classification of contact metamorphism in terms of the prevailing rock-pressure was given in §4-5. The first three sections of this chapter deal with descriptions of pelitic contact metamorphic rocks on the basis of this classification.

The early days of metamorphic petrology had two monumental works on contact metamorphism by Rosenbusch (1877) and Goldschmidt (1911). They both treated the most common type of contact metamorphism, i.e. that characterized by andalusite and cordierite (without sillimanite, kyanite and staurolite).

Barr-Andlau Area, France

Rosenbusch (1877) investigated the contact aureole around a Hercynian granite mass of the Barr-Andlau area in Vosges, Alsace. This area was recently re-studied by Bosma (1964). In the southern part of the aureole, about 0·5–1·5 km wide, the early Palaeozoic Steige slates show three zones of advancing recrystallization inwards as follows:

1. Spotted slates or phyllites
2. Spotted schists
3. Hornfelses.

The Steige slates are highly aluminous (Al_2O_3 = 20–27 per cent). The first sign of metamorphism is the generation of black spots, resulting in spotted slates or phyllites. The rocks are mainly composed of quartz and white mica, with subordinate plagioclase, chlorite, tourmaline, haematite and magnetite.

At first, the spots are very small, but soon become bigger up to about 3 mm in size. Rosenbusch considered the dark colour of spots as being due to local concentration of opaque minerals. Some students have considered the possibility that originally the spots had been porphyroblasts of cordierite which later altered into other minerals. X-ray investigation by Bosma has demonstrated the presence in the spots of quartz, muscovite and haematite, occasionally with feldspar and minerals showing 7·1 Å and 14·5 Å reflections. No cordierite has been found. Nonetheless, he considers that the spots were originally cordierite porphyroblasts containing inclusions of quartz, feldspar and iron ore, and that later alteration of the cordierite produced muscovite and the 7·1 Å and 14·5 Å minerals.

The spotted schists are transitional from spotted slates (or phyllites) to hornfelses. Though they have spots, the schists are much less fissile than spotted slates. Biotite and andalusite begin to occur. The latter mineral always forms outside the spots.

The hornfelses have neither fissility nor spots. They all contain andalusite porphyroblasts and a few contain cordierite porphyroblasts also. Muscovite, biotite, quartz, feldspars, tourmaline and opaque minerals occur. Bosma considers that the hornfelses of the inner aureole probably did not pass through the stage of spot formation owing to their rapid heating by the intrusive granite which afforded no time for recrystallization at low temperature.

Oslo Area, Norway
Permian volcanic and plutonic rocks are exposed in the Oslo graben cutting through Precambrian terranes in Norway. The plutonic masses, including alkali syenite, nepheline syenite and alkali granite, were intruded in Palaeozoic sedimentary rocks within the graben, giving rise to contact metamorphic effects (Barth, 1945; Oftedahl, 1959).

Goldschmidt (1911) studied well-recrystallized hornfelses in the inner aureoles of the Oslo (Christiania) area. Though the extent of the aureoles depends largely on the size of the relevant intrusions, the metamorphism in the inner aureoles were found to be in the pyroxene-hornfels facies regardless of the size and rock types of the intrusions. The outer aureoles are poorly recrystallized and have not been well studied, but could belong to the amphibolite facies.

It has been shown that the mineral compositions of hornfelses in the inner aureole vary regularly with their chemical compositions. As shown in fig. 11-1(c), Goldschmidt classified hornfelses ranging from pelitic to calcareous compositions into ten classes:

Class 1: Andalusite–cordierite hornfels.
Class 2: Plagioclase–andalusite–cordierite hornfels
Class 3: Plagioclase–cordierite hornfels
Class 4: Plagioclase–orthopyroxene–cordierite hornfels
Class 5: Plagioclase–orthopyroxene hornfels
Class 6: Plagioclase–diopside–orthopyroxene hornfels

Class 7: Plagioclase–diopside hornfels
Class 8: Grossular–plagioclase–diopside hornfels
Class 9: Grossular–diopside hornfels
Class 10: Grossular–wollastonite–diopside hornfels.

He considered that quartz and orthoclase could occur in any of these types. However, it is likely that grossular is unstable in the presence of quartz, judging from the recent experimental data on the reaction: grossular + quartz = 2 wollastonite + anorthite (§9-6; fig. 9-3). Biotite occurs in classes 1–7. Muscovite does not occur in any class. Hornfelses of class 10 may contain vesuvianite. Pelitic hornfelses belong to classes 1–5, whereas classes 7–10 represent calcareous rocks.

10-2 CONTACT METAMORPHISM OF PELITES TO PRODUCE SILLIMANITE WITH OR WITHOUT ANDALUSITE

Sillimanite is not as common in contact aureoles as in low-pressure regional terranes. This is probably because many instances of contact metamorphism represent a lower rock-pressure than low-pressure regional metamorphism, and the transformation from andalusite to sillimanite takes place at higher temperature with decreasing rock-pressure (fig. 2-1). If sillimanite occurs it does so in the inner aureole, whereas andalusite occurs in the outer aureole.

Four areas will be reviewed in order of increasing widespread occurrence of sillimanite.

Comrie Area, Scottish Highlands

The Carn Chois diorite was intruded into greywackes and slates of the Dalradian Series in the Comrie area near the southern boundary of the Scottish Highlands. The mass of diorite, about 3 km across, is mainly composed of biotite, hornblende and andesine. The country rocks had been metamorphosed in the chlorite-zone grade before the intrusion. The slate outside the contact aureole is mainly composed of muscovite, chlorite, quartz and iron ores.

Three zones of progressive recrystallization have been distinguished (Tilley, 1924).

(*a*) Zone of spotted slates. The first sign of contact metamorphism in slates is the development of minute spots and the hardening of the rock. It begins at a distance of about 400 m from the contact. The spots are less than 1 mm across and are dark grey or black. Both spots and matrix are composed of the same minerals as the original slate. The darker colour of the spots is due to higher proportions of chlorite and muscovite.

(*b*) Zone of biotite development. The first new mineral produced by the contact effect is biotite, which occurs as little brown flakes within spots as well as in the matrix. Chlorite, muscovite and biotite show a tendency to parallelism with the original cleavage. On a closer approach to the contact, the preferred orientation becomes weaker, and the fissility is lost. The spots disappear and the grain size increases.

(c) Zone of cordierite hornfelses. Cordierite begins to occur at a distance of 140 m from the contact, and soon becomes very widespread. In this zone, recrystallization is complete. Orthoclase and biotite occur in most hornfelses. Muscovite and chlorite do not occur. Some hornfelses contain quartz, cordierite and orthopyroxene, whereas others contain no quartz but have cordierite, spinel and corundum.

Completely recrystallized pelitic and psammitic hornfelses were divided into two groups:

1. Quartz-bearing hornfelses, which show nearly the same mineral assemblages as the Oslo Hornfelses.
2. Quartz-free hornfelses, in which the decrease of silica changes andalusite into corundum first, and then cordierite into spinel. Formation of olivine by desilication of orthopyroxene has not taken place.

In this aureole, andalusite is the common Al_2SiO_5 mineral. Although sillimanite was found in some hornfelses near the contact, no distinct sillimanite zone was noticed.

In the associated grits, the first sign of contact metamorphism is hardening of the rocks. Spots do not form. As the contact is approached, little biotite flakes form, and then andalusite and cordierite appear. Some hornfelses contain spinel or corundum.

The Dalradian Series in the area contains small masses of metabasite (epidiorite), which show the actinolite + chlorite + sodic plagioclase + epidote assemblage outside the contact aureole. Within the aureole, the rocks become biotite amphibolites, and then hornblende–orthopyroxene–clinopyroxene–biotite–plagioclase hornfelses. In some rocks, hornblende almost disappears. Thus, the outer and inner aureoles are in the amphibolite and the pyroxene–hornfels facies respectively.

Arisu Area, Northeast Japan

The Cretaceous Tono granodiorite was intruded into Late Palaeozoic sedimentary rocks in the Kitakami Mountains, northern Honshu. The granodiorite mass, containing hornblende and biotite as the main mafic minerals, measures about 40 km across. The contact aureole is 3–5 km wide, being situated about 110–130 km NNE of the city of Sendai.

Seki (1957, 1961b) described the progressive metamorphism of the pelites, limestones and basic volcanics of the Arisu area within the aureole. The progressive metamorphic changes are summarized in fig. 10-1. He divided the aureole into four zones in terms of mineral changes of the metapelites:

(*a*) Chlorite-muscovite zone. Chloritoid occurs in some rocks. This zone is present in the aureole on the western side of the granodiorite, but is absent on the eastern side. The rocks are phyllitic.

(*b*) Biotite–andalusite zone. Biotite begins to occur.

(*c*) Cordierite–almandine zone. Andalusite and gedrite are common. The gedrites have unusually high Al_2O_3 and FeO content (Seki and Yamasaki, 1957). Cummingtonite occurs rarely. Spinel and corundum may occur in the same thin section as quartz, but not in direct contact with it.

(*d*) Sillimanite zone. Muscovite, biotite, cordierite, garnet and microcline are widespread. The garnet contains 82–83 mol per cent of Alm and 9–11 per cent of Pyr. The host rocks have Fe^{2+}/Mg ratios as high as 1·9–2·3 (Onuki, 1968).

The metabasites and limestones in this aureole will be described in § § 10-6 and 10-7.

Steinach Area, Northeast Bavaria, Germany
The contact aureole surrounding the Leuchtenberg granite massif in the Steinach area was investigated by Okrusch (1969, 1971). The rocks outside the aureole are Hercynian banded gneisses and mica schists, which formed by regional metamorphism in the amphibolite facies. These rocks occasionally contain garnet, sillimanite and staurolite. The Leuchtenberg granite (Carboniferous) was intruded into the regional metamorphic rocks with resultant formation of a contact aureole, about 1 km wide.

The contact metamorphism has decomposed garnet and sillimanite of regional metamorphic origin, and has produced cordierite and andalusite instead in the outer aureole. Thus, the characteristic mineral assemblage of the outer aureole is muscovite + biotite + andalusite + cordierite + quartz + plagioclase.

Andalusite is the only polymorph of Al_2SiO_5 in the outermost part of the outer aureole, while both andalusite and sillimanite occur in the rest of the aureole. The sillimanite/andalusite ratio generally increases with closer approach to the granite border. The maximum abundance of andalusite has been observed at a distance of 200–400 m from the granite border. This means that the sillimanite isograd passed across the andalusite zone during the contact metamorphism. Cordierite and garnet of contact metamorphic origin tend to increase as the granite border is approached. In the transitional zone from the outer to the inner aureole, muscovite breaks down by reaction with quartz and biotite to produce sillimanite, cordierite, garnet and K-feldspar. In the inner aureole, K-feldspar can coexist with sillimanite, cordierite and garnet. Thus, the characteristic mineral assemblage of hornfelses in the inner aureole is: sillimanite + cordierite ± almandine ± K-feldspar + biotite + quartz + plagioclase.

The ratio Fe/(Mg + Fe) of cordierite is 0·48–0·49 in the transitional grade from the outer to the inner aureole, and 0·50–0·54 in the inner aureole. The contact metamorphic garnet at the outermost part of the aureole is a fine-grained almandine-spessartine nearly free from zoning, containing 14–18 per cent MnO and 23–25 per cent FeO. In the transitional zone from the outer to the inner aureole, some grains of garnet are zoned with increasing MnO content to the rim and others are zoned with decreasing MnO content to the rim. In the inner aureole, the zoning of garnets is very weak. They are usually typical almandine containing 2–6 per cent MnO and 33–37 per cent FeO.

Glen Clova, Scottish Highlands

The Dalradian sediments in the Glen Clova area were subjected to regional metamorphism of the sillimantie grade of the Barrovian series. Subsequently, adamellite, granodiorite, diorite and picrite of the Newer Granite suite were intruded into them, producing polymetamorphic contact aureoles. In the inner part of these aureoles, the amphibolite formed by regional metamorphism has been transformed into two pyroxene-plagioclase hornfelses, and the metapelites have been changed into orthoclase-cordierite-biotite-hornfelses often with hypersthene, sillimanite, almandine, spinel or corundum. Thus, the inner aureoles belong to the pyroxene-hornfels facies (Chinner, 1962), if not to the granulite facies. These aureoles are different from the Oslo hornfels zones in the common occurrence of sillimanite and almandine.

Almandine occurs commonly in the original regional metamorphic rocks. When subjected to the contact metamorphism, the original almandine in some rocks was entirely replaced by aggregates of cordierite, orthoclase, biotite and spinel, whereas that in other rocks with higher Fe^{2+}/Mg ratios partly persisted in the contact aureoles.

Almandine by itself is stable under the physical conditions of the pyroxene–hornfels facies (fig. 7B-9). The composition field of the rocks which form almandine, however, is more restricted under such conditions than in regional metamorphism of the medium-pressure type. Hence, partial or complete decomposition of regional metamorphic almandine takes place in contact aureoles, as shown in fig. 7B-8.

10-3 CONTACT METAMORPHISM OF PELITES TO PRODUCE STAUROLITE AND KYANITE

Contact metamorphism that produces staurolite and kyanite has been described in some areas of the world. Since staurolite is stable over a wide range of rock-pressure, staurolite-bearing aureoles do not necessarily represent particularly high pressures. When kyanite occurs in contact aureoles, it is usually associated with andalusite and sillimanite. Therefore, P–T conditions near the triple point of the Al_2SiO_5 system probably prevailed in a part of such aureoles.

The possibility of metastable crystallization of Al_2SiO_5 minerals cannot be precluded. Even if this is the case, however, the prevailing rock-pressure was probably relatively high in kyanite-bearing aureoles, as will be discussed below.

Kwoiek Area, British Columbia, Canada

This area, about 130 km northeast of Vancouver, has a contact aureole formed on the eastern flank of the Coast Range batholith. A roof pendant or screen of metamorphic rocks, about 10 km wide, is bordered by hornblende-biotite or biotite quartz diorite. The major rock type within the pendant is a metagraywacke. Metabasites containing pillow structure are also present.

The contact aureole along the major batholith border is about 6 km wide, and the metamorphism ranges from the chlorite to the sillimanite grade (Hollister, 1969a b). Garnet, staurolite, kyanite and sillimanite are common and andalusite has been found in one locality. The following progressive reactions took place in aluminous metagraywackes:

muscovite + chlorite + quartz = garnet + biotite + H_2O

muscovite + chlorite + ilmenite

$\qquad$ = staurolite + garnet + biotite + quartz + H_2O

staurolite + quartz = garnet + sillimanite + H_2O.

Andalusite, kyanite and sillimanite can occur singly or in combinations of any two. The most common pair is kyanite + sillimanite. The textural relations suggest the existence of two sequences of polymorphic transformations:

(a) Andalusite $\rightarrow$ kyanite $\rightarrow$ sillimanite

(b) Andalusite $\rightarrow$ fibrolite (fine-grained mats of sillimanite needles).

Hollister (1969b) has considered that both sequences formed with rising temperature at a similar and constant rock-pressure which is higher than the pressure of the triple point of the Al_2SiO_5 system (fig. 2-1). In sequence (a), andalusite crystallized as a metastable form within the stability field of kyanite, and subsequently stable kyanite formed, followed by stable sillimanite. In sequence (b), andalusite crystallized first as a metastable form, and subsequently fibrolite crystallized as another metastable form, both within the stability field of kyanite.

Garnet grains are usually strongly zoned. The MnO content is as high as 6-18 per cent in the grain centre, and decreases outwards to a value as low as 1-3 per cent, but may increase slightly in the outermost few microns, whereas the FeO content increases outwards. The early formed garnet contains a large part of the MnO present in the rock. Diffusion of material in garnet is negligible under the conditions of this aureole, and so the garnet, once crystallized, is removed from the reacting system, with resultant depletion of MnO from the reacting part of the rock. The new garnet layer to form in equilibrium with the depleted surroundings should be very poor in MnO. Thus, strong zoning could form even without any appreciable change in the partition relation of elements during the growth (Hollister, 1966, 1969a).

Staurolite shows a more complicated pattern of zoning. Sectors of a crystal formed by growth in different directions have different compositions, and within each sector the composition changes from centre to margin (Hollister, 1970; Hollister and Bence, 1967).

Biotite and chlorite are more homogeneous, and would have continuously internally re-equilibrated during their growth with changing external conditions.

A series of Hollister's papers on rocks and minerals of this area represents a successful use of the electron probe X-ray microanalyser for better under-standing of the mechanism of metamorphic reactions.

Donegal Granites, Northwestern Ireland

Caledonian granitic masses were intruded into already metamorphosed Pre-cambrian pelitic rocks in Donegal on the northwest coast of Ireland. The regionally metamorphosed pelites outside the contact aureoles are mostly chlorite–muscovite schists, occasionally containing biotite. Some aureoles con-tain staurolite and kyanite besides andalusite and sillimanite (Pitcher and Read, 1960, 1963; Naggar and Atherton, 1970).

The Thorr pluton and another granodioritic mass have produced strong metamorphic effects on the country rocks. The contact aureole of the Thorr pluton is about 2 or 3 km wide. The plutonic masses are heterogeneous. In the aureole produced by the adamellites, the first mineralogical effect is the appearance of chlorite porphyroblasts. Then, biotite begins to occur, being followed by andalusite. In the advanced stage, cordierite appears, and the rocks becomes andalusite-cordierite hornfelses. Sillimanite occurs only in the inner-most zone. In the aureole produced by dioritic rocks, on the other hand, muscovite flakes and andalusite porphyroblasts are accompanied by cordierite and some sillimanite in the outer aureole. In the inner aureole, sillimanite is abundant, and the hornfelses are extensively mobilized.

The Ardara diapiric granodiorite is about 8 km across, consisting of a core of granodiorite surrounded by a rim of foliated biotite-hornblende tonalite. It was emplaced by a forceful shouldering aside of the country rocks. The contact aureole is about 2 km wide. In the outermost part of the aureole, biotite porphyroblasts appear. On a nearer approach to the intrusion, andalusite and then sillimanite begin to occur. The narrow innermost zone is rich in cordierite. The andalusite is often accompanied by staurolite, and very rarely by a little kyanite. The sillimanite is accompanied by garnet.

The Main Donegal granite proper is the largest intrusion in this area, and is exposed over an area of about 11 by 60 km. It was forcefully intruded by lateral wedging combined with horizontal stretching. The mass is composed of grano-diorite showing banding parallel to the steep country-rock walls. The banding is interpreted as being due to flow of consolidating magma. The aureole is 1·5-2·5 km wide. The intruding magma exerted a marked outward push and horizontal drag on the aureole rocks, producing folds, strain-slip cleavages and schistosity in the aureole.

The metapelites in the outer aureole are characterized by porphyroblasts of chlorite, muscovite and biotite. Choritoid was found at one locality. On a closer approach to the contact, garnet, staurolite, kyanite, and andalusite begin to occur. In the innermost part, sillimanite occurs. In this aureole, staurolite schists are widespread. Kyanite schists are also common. Some schists contain both kyanite and andalusite, or both andalusite and sillimanite. Cordierite is virtually absent.

It may appear that kyanite forms in aureoles subjected to higher mechanical effects due to intrusion. However, Naggar and Atherton (1970) consider that its formation depends not on the mechanical effects but mainly on the lower temperature and the lower FeO/MgO ratios of the metamorphosed rocks. In rocks with higher ratios, garnet and staurolite takes the place of kyanite.

In addition to the two areas reviewed above, the occurrence of kyanite and staurolite in contact aureoles has been reported from the Burke area in the northern Appalachians (Woodland, 1963) and from Ghana (Lobjoit, 1964). Staurolite, andalusite and sillimanite occur in the contact aureole of the Santa Rosa Range, Nevada, which has been described in detail by Compton (1960).

10-4   POSSIBLE EXAMPLE OF PARTIAL MELTING IN A CONTACT AUREOLE

The Caledonian intrusion of the Cashel–Lough Wheelaun area is composed of basic and ultrabasic rocks and was emplaced into pelitic rocks in Connemara near the west coast of Ireland. The rocks outside the contact aureole are Caledonian regional metamorphic rocks of the sillimanite grade. The aureole is about 1·5 km wide, being composed of hornfelses and partly mobilized pelites.

The regional metamorphic rocks outside the aureole are biotite–andesine–quartz–sillimanite–garnet schists. Within the aureole and as inclusions in the intrusion occur cordierite–sillimanite–labradorite–magnetite–biotite hornfels and spinel–magnetite–cordierite hornfels with occasional corundum or ortho-pyroxene.

The compositional changes accompanying the formation of the hornfelses were studied by Leake and Skirrow (1969) and by Evans (1964). Unhornfelsed pelites and hornfelses have the following composition ranges:

	Unhornfelsed pelites per cent	Hornfelses per cent
SiO_2	42·6-51·6	32-49
Na_2O	0·5-2·5	0·2-2·6
K_2O	3·7-6·3	1·2-6·0.

It appears that Si, Na, K, Rb, Cs, Li, Ba and La were removed from hornfelses. This change was considered due to partial melting of the pelites followed by removal of the resultant acidic liquid leaving the solid residue behind. The materials removed in an early stage of partial melting resemble K-rich granite in composition, whereas those in higher temperatures resemble granodiorite.

10·5 PYROMETAMORPHISM OF PELITIC AND PSAMMITIC ROCKS

Pyrometamorphism takes place commonly in xenolithic inclusions in volcanic rocks or minor intrusions, and rarely in the country rocks around minor intrusions of basic or intermediate compositions. Intrusions giving such marked heat effects to the country rocks could represent a conduit for continuously rising magma. A continuous flow may have maintained a high temperature for a considerable period of time.

A pyrometamorphic zone in country rocks is usually less than 20 m wide, that is, much narrower than ordinary contact aureoles. Though the metamorphic rocks show intense mineralogical reconstitution, they are commonly far from chemical equilibrium. Minerals characteristic of the sanidinite facies such as sanidine, anorthoclase, high-temperature plagioclase, tridymite, cristobalite and mullite form. Partial melting takes place. Two examples of the progressive pyrometamorphism of the pelitic and psammitic rocks will be reviewed below.

Sithean Sluaigh, Scotland

In Sithean Sluaigh in Argyllshire, Scotland, a gabbroic plug penetrates through Dalradian phyllites. The plug, which is thought to be the remnant of a volcanic pipe, is made up of a core of olivine-augite gabbros surrounded by a ring of dolerite. The latter has been interpreted as having been intruded between the gabbro and phyllite. The plug is about 100 m across.

The thermal aureole is about 10 m wide (Smith, 1969). The rocks outside the aureole are phyllites of pelitic and psammitic compositions, sporadically containing almandine and biotite. At 10-9 m from the contact on the northeastern side of the plug, the muscovite of the phyllite has broken down to a mat of very fine needles of sillimanite and mullite set in an amorphous matrix.

At 4-3 m from the contact, all the mica and chlorite have broken down to give poorly crystallized cordierite and spinel. Sanidine rims form around the decomposed mica. Almandine begins to break down to form a very fine-grained aggregate probably of orthopyroxene, cordierite and spinellides. So far, individual grains tend to decompose as more or less isolated chemical systems.

At 1·6 m from the contact, the decomposed minerals tend to lose their original outlines. Rims of sanidine and orthopyroxene surround quartz. Breakdown products of muscovite are aggregates of spinel, corundum and cordierite.

At 1–0·5 m from the contact, intergrowths of quartz and feldspar and prismatic orthopyroxene are widespread. Diopsidic pyroxene begins to appear. The grain size becomes coarser and granophyre veins occur.

At 15 cm from the contact, plagioclase grains possibly derived from igneous rocks are rimmed by sanidine. Patches of brown devitrified glass occur interstitially.

In other parts of the aureole close to the contact, mullite–spinel–pseudobrookite rocks and spinel–corundum rocks occur. Granophyre veins are common, and some of them contain quartz paramorphs after tridymite.

The granophyre veins are not thought to be a magmatic differentiate but a product of partial melting of phyllites. Some phyllites have a compositional layering, and only the quartzo–feldspathic layers appear to have been melted to form a granophyric texture. When the siliceous melts were removed, the solid residues were highly desilicated aluminous and ferromagnesian rocks rich in corundum, spinel and mullite

Karroo Dolerite, South Africa

A sill of the Karroo Dolerite near Heilbron, South Africa, has fused overlying arkose (Ackermann and Walker, 1960). Such fusion is relatively rare for the dolerite. The arkose is flat-lying and the dolerite sill is about 15 m thick. The arkose is mainly composed of quartz, orthoclase and subordinate oligoclase. Some almandine and ilmenite were found.

The rocks harden towards the contact. Vitrification of arkose is noticed within a distance of 6 m from the contact. The first glass is formed where grains of K-feldspar and quartz come into contact with one another. It is light in colour. With advancing melting, the colour deepens and the refractive index rises. It contains cordierite microlites. The fusion is accompanied by various mineralogical changes including the transformation of orthoclase to a nearly uniaxial sanidine and the formation of cordierite, tridymite and pyroxene.

Near the contact, pyroxene appears in the glass, and the original K-feldspar is completely resorbed, while some of the quartz and plagioclase still remain. The resultant rock is a green or black buchite containing up to 70 per cent of glass. Mixing of partly fused sediments with fluid dolerite took place locally in a zone less than 0·6 m wide.

10-6 CONTACT METAMORPHISM OF BASIC ROCKS

The existing data on metabasites in contact aureoles are not so abundant as those on metapelites. Three aureoles will be reviewed here, two of which have a metabasite zone with the assemblage actinolite + calcic plagioclase in a low-temperature part.

Iritono Area, Northeast Japan

The Iritono area is in the central Abukuma plateau, about 180 km NNE of Tokyo. The intrusion is heterogeneous and mainly composed of granodiorite and diorite. The rocks outside the contact aureole are regionally metamorphosed basic schists in the greenschist facies in the eastern half of the area and in the amphibolite facies in the western. The contact aureole, less than 1 km wide, has been divided into three progressive zones A′, C′ and D′ (Shido, 1958, pp. 199–201).

Zone A′. On entering the contact aureole, the metabasites are soon recrystallized nearly completely and lose their original fissility. The outermost zone is characterized by metabasites with the actinolite-calcic plagioclase assemblage with or without chlorite, quartz and biotite. No epidote is present. The calcic plagioclase is labradorite or andesine. In the higher temperature part of this zone, some metabasites contain a little blue–green hornblende in association with actinolite, and others contain diopsidic pyroxene. The associated metapsammites are biotite hornfelses with occasional muscovite or garnet.

Zone C′. This zone is characterized by metabasites with the hornblende + plagioclase assemblage, occasionally containing biotite, cummingtonite and diopsidic pyroxene. The hornblende is brownish green or brown. (Note that in the regional metamorphism of the Abukuma Plateau a zone with blue–green hornblende usually occurs between that with actinolite and that with brownish green hornblende, whereas in this contact aureole a zone characterized by the predominance of blue–green hornblende does not occur.)

The associated metapsammites in zone C′ are biotite hornfelses commonly with garnet and/or cummingtonite.

Zone D′. Where the dioritic facies of the intrusion is in contact with the country rocks, orthopyroxene is formed in metabasites up to a distance of several metres from the contact. This part is designated as zone D′. The rocks show the orthopyroxene + clinopyroxene + hornblende + plagioclase assemblage commonly with cummingtonite and biotite. Orthopyroxene does not form in the metabasites in contact with granodiorites.

Arisu Area, Northeast Japan

As shown in fig. 10-1, three progressive metamorphic zones have been distinguished in terms of mineral changes in metabasites (Seki, 1957; especially 1961*b*, pp. 74–75).

1. *Greenschist facies zone.* Actinolite, chlorite, sodic plagioclase, muscovite, biotite and epidote occur.
2. *Actinolite-calcic plagioclase zone.* This zone resembles zone A′ of the Iritono aureole. The plagioclase is andesine or labradorite. Epidote is rare and chlorite is present.
3. *Amphibolite facies zone.* The metabasites are mainly composed of hornblende and plagioclase.

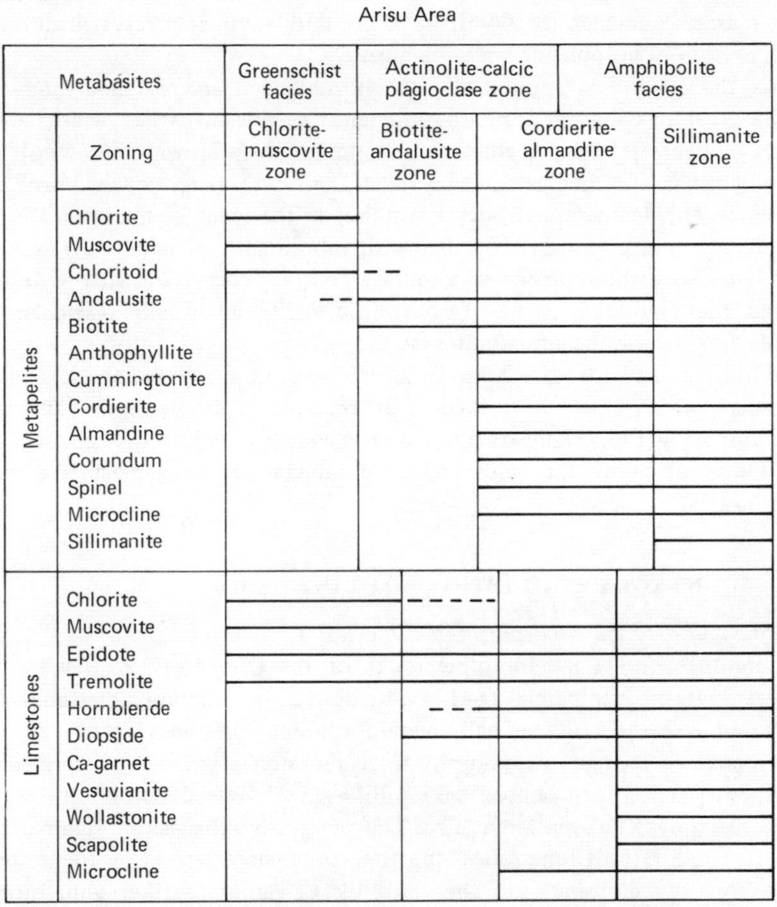

Arisu Area

Metabásites	Greenschist facies	Actinolite-calcic plagioclase zone		Amphibolite facies
Zoning	Chlorite-muscovite zone	Biotite-andalusite zone	Cordierite-almandine zone	Sillimanite zone

Metapelites

Chlorite				
Muscovite				
Chloritoid				
Andalusite				
Biotite				
Anthophyllite				
Cummingtonite				
Cordierite				
Almandine				
Corundum				
Spinel				
Microcline				
Sillimanite				

Limestones

Chlorite				
Muscovite				
Epidote				
Tremolite				
Hornblende				
Diopside				
Ca-garnet				
Vesuvianite				
Wollastonite				
Scapolite				
Microcline				

Fig. 10-1 Progressive mineral changes in the Arisu contact aureole, Kitakami Mountains, Japan. (Seki, 1957, 1961*b*.)

Tilbuster Area, Australia

A Late Palaeozoic monzonite mass has been intruded into Palaeozoic formations in the Tilbuster Area, New South Wales, Australia (Spry, 1955; Binns, 1965a). The monzonite is mainly composed of plagioclase and orthoclase with subordinate biotite, hornblende and quartz. The contact aureole is only about 0·5 km wide. The pelitic rocks adjacent to the contact contain andalusite, cordierite and orthoclase.

Basic lavas occur abundantly in the aureole. Some of them are basaltic and others spilitic in composition. The first sign of metamorphism in the ordinary basic lavas of the outer aureole is the formation of a very pale bluish green hornblende with ragged or fibrous 'actinolite-like' habit by reaction between the plagioclase and clinopyroxene of original basalts. A chemical analysis has shown that it is a hornblende (Si = 6·50 on 23 O). With advancing recrystallization, the rock becomes a hornblende-andesine hornfels.

As the contact is approached, the hornblende becomes moderately deep bluish green in colour and then dark green, its habit less ragged, and its amount tends to decrease. Finally close to the contact, deep brown or brownish green hornblende was formed as stout crystals. The Na_2O content of the hornblendes increases with temperature. Such hornfelses in the inner aureole are made up of hornblende and plagioclase together with subordinate calcic clinopyroxene and iron ore. No orthopyroxene was found. The plagioclase is usually andesine or labradorite. Hornfelses derived from spilitic rocks contain sodic plagioclase. The whole aureole is in the amphibolite facies.

Within a distance of 200 m from the contact many metabasites contain grossular, which would have formed by reaction of calcite-bearing amygdales. The aureole and especially its outer part is widely scapolitized. Scapolite occurs as fillings of joints and veins and as a mineral replacing plagioclase in the metabasites.

10-7 CONTACT METAMORPHISM OF LIMESTONES

Ordinary Contact Metamorphism of Limestones

Among the contact metamorphic rocks of the Oslo area, Norway (§ 10-1), classes 7–10 of Goldschmidt (1911) were derived from impure limestones. Many features are shared with regionally metamorphosed limestones.

Progressive contact metamorphism of limestones has been described in a number of areas. For example, Seki (1961b) has described a progressive series in the Arisu area, Kitakami Mountains. The progressive changes are summarized in fig. 10-1. As regards limestones, the first zone characterized by tremolite, the next zone by diopside, and the third by Ca-garnet together with diopside. Microcline also begins to occur in the third zone. In the innermost fourth zone, wollastonite and vesuvianite occur. Scapolite is also confined to this zone.

It is interesting that within the second zone, there is a small area where wollastonite occurs. No intrusion has been found in this area to account for the appearance of wollastonite. It has been ascribed to a local decrease of P_{CO_2}.

The progressive sequence of calc-silicate minerals in the Arisu area is tremolite → diopside → Ca-garnet → wollastonite. This is very similar to that observed in regional metamorphism, especially of the low-pressure type.

Advanced Decarbonation in Contact Metamorphism

There are a number of contact aureoles where decarbonation reactions have proceeded to greater degrees than in the above case. In such aureoles, periclase, monticellite, akermanite, tilleyite, spurrite, rankinite, merwinite, and larnite may be produced (Bowen, 1940; Tilley, 1951).

Periclase marble or its hydration product (i.e. predazzite) appears to begin to form in a high-temperature part of the pyroxene-hornfels facies in contact metamorphism. W. Johannes determined the equilibrium temperature of the following reaction for a wide range of the CO_2/H_2O ratio for the coexisting fluid at 1 kbar (Winkler, 1967, p. 42):

$$dolomite = calcite + periclase + CO_2. \qquad (10\text{-}1)$$

The reaction takes place generally in the range of 700–850°C as shown in fig. 9-2. For more information about the hydration reaction of periclase, see, in particular, Fyfe (1958).

Decarbonation reactions leading to the formation of such rare minerals as monticellite, akermanite, spurrite and merwinite, have been experimentally investigated in the presence of a nearly pure CO_2 phase by Walter (1963, 1965) and by Tuttle and Harker (1957). The equilibrium temperatures are considerably higher than those for the reactions to form wollastonite and periclase.

Monticellite, akermanite, tilleyite, spurrite, rankinite, merwinite, and larnite are products of extremely advanced decarbonation reactions. If these minerals form in the presence of a fluid composed almost entirely of CO_2, the temperatures of formation would be high enough to class them in the characteristic minerals of the sanidinite facies (Reverdatto, 1970). However, they occur also in some contact aureoles that do not appear to have reached so high a temperature. In some contact aureoles, CO_2 could be actively removed possibly in solution in flowing water. Such a situation would result in an unusually low P_{CO_2}, leading to the generation of above-mentioned minerals.

In contact metamorphic limestones, P_{CO_2} and P_{H_2O} may be so highly variable as to produce a great mineralogical diversity (Agrell, 1965).

Crestmore, California

The Crestmore limestone is a well-documented example showing advanced decarbonation. It is situated about 90 km east of Los Angeles. Monticellite, gehlenite, tilleyite, spurrite, and merwinite have been found (Burnham, 1959; Walter, 1965; Carpenter, 1967).

Masses of dolomitic limestones were engulfed and metamorphosed by a quartz–diorite pluton which is part of the Cretaceous Southern California batholith. A layer, less than 1 m thick, of contact skarns mainly composed of diopside, grossular and wollastonite formed between the limestone and quartz diorite. Later, a small pipe-like body of quartz–monzonite porphyry was intruded and a thicker skarn zone, up to about 20 m thick, formed between the porphyry and the already metamorphosed limestone. The quartz-monzonite porphyry is made up of andesine phenocrysts set in a microcrystalline ground-mass of andesine, orthoclase and quartz with minor green clinopyroxene, hornblende and biotite.

The limestones are now composed of alternating layers of coarsely crystalline calcite marble and brucite marble. The brucite is a hydration product of periclase.

The thicker skarn zone was formed by the metasomatic introduction of great amounts of SiO_2 and Al_2O_3 from the quartz-monzonite porphyry into the surrounding limestones. It may be divided into the following three subzones listed in order from the contact outward:

1. Garnet subzone, consisting of uniform garnet–wollastonite–diopside rock.
2. Vesuvianite subzone, consisting of vesuvianite, garnet, wollastonite, diopside, monticellite, wilkerite, crestmorite, phlogopite, and xanthophyllite.
3. Monticellite subzone, consisting of monticellite, gehlenite, tilleyite, spurrite, merwinite, forsterite, scawtite and dark green spinel.

These subzones have gradational contacts between one another. The outer boundary of the monticellite subzone against the crystalline marble is very sharp and represents the limit of Si and Al metasomatism.

Conceivably, a large amount of water escaping from the crystallizing quartz–monzonite porphyry carried the SiO_2 and Al_2O_3 to cause the metasomatism. The subzones closer to the igneous body have higher SiO_2 and Al_2O_3 contents. The differences in mineral composition between subzones probably resulted from the chemical differences due to metasomatism together with some differences in P_{CO_2} and temperature.

Chapter 11

Metamorphic Facies and Facies Series

11-1 DEVELOPMENT OF THE CONCEPT OF METAMORPHIC FACIES

Eskola's Concept

Goldschmidt's (1911) study of the contact metamorphism of the Oslo area and Eskola's (1914) study of the regional metamorphism of the Orijärvi region revealed that well-recrystallized metamorphic rocks approached chemical equilibrium at a stage of their history and that the mineral compositions they acquired then have usually been preserved over long subsequent periods of time. In a group of metamorphic rocks that have reached chemical equilibrium under the same definite P-T conditions, the mineral compositions of the rocks depend only on their bulk chemical compositions, regardless of their previous history. There exists a regular relationship between the mineral and chemical compositions of such metamorphic rocks. Eskola (1915) attempted to visualize the relationship by his ACF and $A'KF$ diagrams.

Since the metamorphism at Oslo differs in P-T conditions from that at Orijärvi, the relationship between the mineral and chemical compositions of the rocks in the former area differs from that in the latter area. When rocks having the same chemical compositions are subjected to metamorphism under different P-T conditions, the mineral compositions of the resultant rocks can differ from one another.

Eskola (1915) defined a *metamorphic facies* as including all the rocks recrystallized under the same P-T conditions (range). Therefore, each metamorphic facies corresponds to a definite P-T range and is characterized by a regular relationship between the chemical and mineral compositions of the rocks. This has led to the establishment of a sound empirical classification of metamorphic rocks in terms of P-T conditions. Eskola (1915, p. 115) wrote as follows: 'A metamorphic facies includes rocks which . . . may be supposed to have been metamorphosed under identical conditions. As belonging to a certain facies we regard rocks which, if having an identical chemical composition, are composed of the same minerals. It may expressly be pointed out, that this conception does not postulate any supposition whatsoever concerning the genetic, pre-metamorphic relations of the rocks, and that every facies may include all possible chemical and genetical varieties.'

Eskola (1920) proposed five metamorphic facies named greenschist, amphibolite, hornfels, sanidinite, and eclogite facies. In his final systematic account, Eskola (1939) added the epidote–amphibolite, glaucophane-schist and granulite facies, and changed the name of the hornfels facies into the pyroxene–hornfels facies.

Eskola (1920, 1939) thought that there were igneous rocks having virtually the same mineral assemblages as some metamorphic rocks. For example, hornblende and pyroxene gabbros have virtually the same mineral assemblages as amphibolites and two pyroxene-bearing basic hornfelses, respectively. Therefore, he formulated a group of igneous facies, corresponding to the metamorphic ones, such as the hornblende gabbro facies and the gabbro facies. The term *mineral facies* was proposed so as to comprise both metamorphic and igneous facies (Eskola, 1920, 1939).

Though the main merit of the facies classification comes from the inference that each facies represents a restricted range of temperature and pressure, Eskola in his later years did not like to base the definition of metamorphic facies on such an inference. Thus, Eskola (1939) attempted to give an excessively positivistic definition which was based only on directly observable chemical and mineral compositions. This attempt stemmed probably from the positivistic spirit he had aquired in his youth under the influence of F. W. Ostwald, Ernst Mach and others (cf. Eskola, 1954).

Metamorphic Facies and Progressive Metamorphism

Eskola's (1920) metamorphic facies was a static scheme of classification of metamorphic rocks. Each facies was defined on the basis of paragenetic relations observed in different parts of the world. For example, the amphibolite facies was mainly based on the Orijärvi region, the pyroxene–hornfels facies on the Oslo area, and the eclogite facies on some areas in southwestern Norway. Therefore, he did not need to consider the problems related to the transitional relations and boundaries between facies.

Vogt (1927) and Barth (1936) studied progressive metamorphism in relation to metamorphic facies. Since the mineralogical changes are generally gradual, the facies boundaries must be defined more or less arbitrarily. Vogt defined a number of boundaries and contributed in particular toward the establishment of the epidote–amphibolite facies between the greenschist and amphibolite facies.

In areas of intermediate or transitional states between two metamorphic facies, rocks having mineral assemblages apparently characteristic of one and of the other facies usually occur side by side. These different assemblages usually result from slightly different chemical compositions. In areas transitional between the amphibolite and the typical granulite facies, for example, amphibolites and two-pyroxene-bearing metabasites usually occur intermingled, as described by de Waard (1965) and Himmelberg and Phinney (1967).

There are wide regions exhibiting intermediate states between the greenschist and the typical glaucophane-schist facies. Greenschists and glaucophane schists, both of broadly basic composition, occur intermingled. Iwasaki (1963, pp. 32–3) studied such an area in the Sanbagawa belt, and showed that glaucophane schists had higher Fe^{3+}/Fe^{2+} ratios than the associated greenschists.

The greenschist and the amphibolite facies are very widespread. The transitional states from these to other facies (e.g. the glaucophane-schist or the granulite facies) are treated as belonging to the latter facies in this book, since such a definition reduces the sizes of the greenschist and amphibolite facies.

The gradual mineralogical changes are largely due to the effects of solid-solution minerals in metamorphic rocks. Therefore, the precise definitions of metamorphic facies boundaries must be based on an understanding of solid-solution effects. This was emphasized in particular by Ramberg (1952, p. 136), who defined a metamorphic facies as follows: Rocks formed or recrystallized within a certain P–T field, limited by the stability of certain critical minerals of defined composition, belong to the same metamorphic facies.

Mobile Components and Metamorphic Facies

Eskola (1915, 1920, 1939) did not clearly discuss the behaviour of H_2O in metamorphism. Presumably he considered that an aqueous fluid was present and its pressure was equal to that acting on the solids. He had no clear idea of mobile components.

H_2O, CO_2 and possibly some other components appear to be more or less mobile in rock complexes undergoing metamorphic recrystallization. When their mobilities are high, their chemical potentials become externally controlled variables, just as are temperature and rock-pressure (Thompson, 1955). The contents of such mobile components in a rock vary in response to external conditions, whereas those of fixed components are intrinsic to the rock (§ 5-1). Thus, rocks having the same chemical composition in terms of the fixed components are recrystallized to different mineral compositions under different external conditions. Under definite external conditions, a definite relationship exists between the chemical composition of a rock in terms of the fixed components and its mineral composition, which can be represented by composition-paragenesis diagrams (Korzhinskii, 1959).

Thus, we come to the following definition, essentially given by Korzhinskii (1959, p. 64): Rocks which formed under such similar external conditions that the relation between their chemical composition in terms of the fixed components and their mineral composition is marked by the same regularities, can be considered as belonging to the same metamorphic facies.

This definition has real significance only when combined with the knowledge of what components are highly mobile during metamorphism. If only H_2O is highly mobile, a metamorphic facies should represent a volume in the $T-P_s-P_{H_2O}$ space. If H_2O and CO_2 are highly mobile, a facies should represent a 'volume' in the $T-P_s-P_{H_2O}-P_{CO_2}$ 'space'. (Whether CO_2 may be regarded as being mobile, depends partly on the areal extent of a metamorphic zone which should be regarded as belonging to a facies, as will be commented upon below.) Our knowledge on the mobility of CO_2 is very meagre. In this book, a metamorphic facies is regarded tentatively as representing a volume in the $T-P_s-P_{H_2O}$ space. Figure 3-12 shows a rough projection of such volumes onto the P_s-T plane.

A strictly defined relationship between the chemical and mineral compositions of rocks holds only under a set of definite values of external conditions. A metamorphic facies, as we now use it, does not represent such a set of definite values, but represents a set of definite *ranges* of external conditions. Usually the ranges are chosen so that a progressive metamorphic zone representing a metamorphic facies could be shown on a map. Hence, the composition-paragenesis diagrams corresponding to a metamorphic facies represent only approximate or average relationships under a set of definite ranges of external conditions.

Facies Names

Eskola's (1939) eight metamorphic facies are greenschist, epidote–amphibolite, amphibolite, pyroxene hornfels, granulite, sanidinite, glaucophane-schist and eclogite facies. Coombs *et al.* (1959) and Coombs (1960, 1961) proposed two additional facies named zeolite and prehnite–pumpellyite facies. (The latter facies was originally called the prehnite–pumpellyite–metagraywacke facies. We delete 'metagraywacke' after recent usage.) Thus, we have ten facies names, and we use only these facies names in this book.

That zone in contact aureoles which is characterized by the actinolite + calcic plagioclase assemblage in metabasites (§ § 4-5 and 10-6) does not belong to any of the above ten facies. The available data on it, however, are not sufficient to establish a new facies. Therefore, such a zone is simply referred to as the actinolite-calcic plagioclase zone in this book.

Many other facies names were proposed in the past. In certain progressive metamorphic terranes, several mineral zones have been set up and a set of metamorphic facies corresponding to the individual zones has been proposed. However, the mineralogical characteristics of progressive metamorphism are very diverse. If we continue to propose a set of facies names for each area, the total number of names will grow to a terribly large figure. Since we are not very successful in determining the $P-T$ relations between zones of different areas, the growing number of facies names will simply mean an increasing burden to our memory without any essential progress in our knowledge. I will refrain from any more advanced subdivision of metamorphic facies.

It is very important to note that metamorphic facies have been defined on the basis of the relationships between the chemical and mineral compositions of a group of rocks, and not on their geologic settings. Certain zones of regional metamorphism show virtually the same relationship between the chemical and mineral compositions as certain zones of contact metamorphism. All such zones should belong to the same metamorphic facies and should have the same facies name regardless of the difference in their geologic settings.

If one wishes to distinguish a greenschist facies, for instance, of regional metamorphism from that of contact metamorphism under two different facies names, one must find a difference in paragenetic relations between the two greenschist facies. Proposal of such different names is justifiable only when the new names are defined on the basis of distinct paragenetic relations, and not on geologic settings.

Classifications of Facies Series

Metamorphic facies series as observed in progressive metamorphic terranes all over the world have been broadly classified into three baric types (§ 3-3). Facies series belonging to the same baric type, however, are still diversified, and may well be further classified into subtypes. The sound basis of such subdivisions will be ultimately found in petrogenetic grids (§ 3-3). However, it is probably too early to discuss such subdivisions from the viewpoint of a general classification of metamorphism.

A very tentative fivefold classification mainly based on the field association of facies is shown in table 11-1, where each of the low-pressure and high-pressure types is divided into two subtypes. However, the basis of this classification is not strong enough. Therefore, we will use the threefold classification based on the baric types in this chapter.

The zeolite, prehnite–pumpellyite, greenschist, epidote-amphibolite and amphibolite facies represent $P-T$ fields so wide as to extend over two or three baric types. That part of such large facies which belongs to one baric type will be treated as a subfacies, and will be designated by such names as either the low-pressure subfacies of the greenschist facies or the low-pressure greenschist facies. Zwart et al. (1967) and den Tex (1971) adopted the term facies group to represent what is here regarded as a metamorphic facies extending over two or three baric types. This terminology may prove to be useful in future.

In the following classification and descriptions of metamorphic facies, it is assumed beforehand that one progressive sequence of mineral zones belongs to one baric type. For example, if andalusite occurs in a part of a progressive sequence of zones, the whole sequence is assumed to belong to the low-pressure type. A greenschist facies zone within the sequence is regarded as belonging to

Table 11-1 *Tentative fivefold classification of metamorphic facies series*

Name	Occurrence	Examples	Facies series
Low-pressure I	(a) Some zeolite-facies terranes	(a) Akaishi Mountains (Matsuda and Kuriyagawa, 1965)	Zeolite → prehnite-pumpellyite → greenschist → actinolite calcic plagioclase zone → amphibolite → pyroxene hornfels → sanidinite
	(b) Contact metamorphism at lower P_s	(b) Oslo (Goldschmidt, 1911), Iritono (Shido, 1958), Arisu (Seki, 1961b)	
	(c) Pyrometamorphism	(c) Sithean Sluaigh (Smith, 1969)	
Low-pressure II	(a) Some zeolite facies terranes	(a), (b) Tanzawa Mountains (Seki et al. 1969a)	Zeolite → prehnite-pumpellyite → greenschist → amphibolite → granulite
	(b) Contact metamorphism at higher P_s	(b) Nakoso (Shido, 1958)	
		(c) Shiojiri-Takato (Ono, 1969a, b; Katada, 1965), Aracena (Bard, 1969), Uusimaa (Parras, 1958)	
Medium-pressure	(a) Some zeolite facies terranes	(a) Taringatura (Coombs et al., 1959)	Zeolite → prehnite-pumpellyite → greenschist → epidote-amphibolite → amphibolite → granulite
	(b) Regions of Barrovian zones	(b) Scottish Highlands (Wiseman, 1934; Harker, 1932) Northern Appalachians (Thompson and Norton, 1968)	
High-pressure I	Glaucophanitic terranes without a jadeite–quartz zone	Bessi-Ino (Banno, 1964), Iimori (Kanehira, 1967), Katsuyama (Hashimoto, 1968) Lepontine Alps (Niggli, 1970)	Prehnite-pumpellyite → glaucophane-schist → greenschist → epidote-amphibolite → amphibolite; possibly eclogite in some parts
High pressure II	Glaucophanitic terranes with a jadeite–quartz zone	Kanto Mountains (Seki, 1958) Panoche Pass (Ernst, 1965)	(Prehnite-pumpellyite →) glaucophane-schist → greenschist; possibly eclogite in some parts

the low-pressure subfacies of the greenschist facies, even though a conclusive paragenetic criterion may not be available for it.

Since this assumption has not been proved, the characterization of many of the subfacies given below is incomplete from the logical viewpoint. Moreover, subfacies cannot be regarded as having been established, so long as strict definitions are not given on the basis of paragenetic relations. Accordingly, the classification and description given below are entirely tentative. A great deal of work is needed in future to prove or disprove the basic assumptions.

11.2   METAMORPHIC FACIES OF THE LOW-PRESSURE TYPE

Zeolite Facies
This is the low-pressure subfacies of the zeolite facies. The zeolite facies rocks of the Akaishi and Tanzawa Mountains and the Mogami area, all in Japan, probably belong to this subfacies (§ 6A-2).

We have discussed in § 6B-1 that the progressive metamorphism of Ca-zeolite takes place in four steps, which are characterized by the following zeolites:

  (a) Stilbite, chabazite, gismondine
  (b) Heulandite, chabazite, thomsonite, (scolecite)
  (c) Laumontite, scolecite, thomsonite
  (d) Wairakite, thomsonite

In each step, only the first-mentioned zeolite could ceoxist with quartz in the presence of a fluid of a virtually pure H_2O composition.

The associated phyllosilicates are smectite, mixed-layer minerals (smectite–vermiculite–chlorite) and celadonite. Discrete crystals of chlorite can occur only in the high temperature part.

Our knowledge of the paragenetic relations of minerals in this subfacies is meagre.

Prehnite-Pumpellyite Facies
The low-pressure subfacies of the prehnite–pumpellyite facies occurs probably in the Akaishi and Tanzawa Mountains (§ 6A-2). Prehnite, pumpellyite, epidote, chlorite, muscovite and albite are stable.

Greenschist Facies
In burial metamorphic terranes of the Akaishi and Tanawa Mountains, a zone of the greenschist facies occurs (§ 6A-2). In contact aureoles, a zone of the low-pressure greenschist facies has only been observed rarely (§ 4-5). In low-pressure regional metamorphism, on the other hand, a zone of the greenschist facies is well developed in many areas, as exemplified by the Shiojiri area (fig. 3-4) of the Ryoke metamorphic belt.

Albite, chlorite, actinolite, muscovite and epidote are widespread. Muscovite may be phengitic. Biotite occurs throughout the zone of this subfacies in many areas (e.g. Katada, 1965). In other areas, however, the zone of this subfacies can be subdivided into a chlorite zone (without biotite) and a biotite zone (e.g. Tobschall, 1969). Garnets of the almandine–spessartine series may occur in some manganiferous rocks.

Actinolite–Calcic Plagioclase Zone and Epidote–Amphibolite Facies

In the greenschist facies metabasites, the plagioclase is albite and the calcic amphibolite is actinolite, whereas in the amphibolites the plagioclase is intermediate or calcic and the calcic amphibole is hornblende.

In many contact aureoles, the change of plagioclase takes place at a lower temperature than that of calcic amphibole, thus resulting in the formation of a zone of the actinolite-calcic plagioclase assemblage in the area intermediate between those of the two facies (§ 4-5). In low-pressure regional metamorphism, on the other hand, the two mineralogical changes usually take place nearly at the same temperature, thus resulting in a direct transition from the greenschist to the amphibolite facies (§ 3-3), whereas in medium-pressure regional metamorphism, the change of calcic amphibole usually takes place at a lower temperature than that of plagioclase, thus resulting in the formation of epidote-amphibolite in a zone intermediate between those of the two facies.

However, the two changes are controlled by a number of factors, and hence an epidote-amphibolite facies zone could form in some cases of low-pressure metamorphism.

Amphibolite Facies

The low-pressure subfacies of the amphibolite facies is very widespread in contact as well as in low-pressure regional metamorphism. Hornblende, cummingtonite and clinopyroxene occur in metabasites. The colour of hornblende changes with increasing temperature from blue–green through green to brownish green and brown. Cummingtonite and clinopyroxene (usually salite) are more common in the high-temperature part. The plagioclase is usually oligoclase, andesine or labradorite.

In metapelites, the stable Al_2SiO_5 mineral is andalusite in the low-temperature part and sillimanite in the high. Muscovite is stable in the lower part of this subfacies, and biotite is present throughout. Cordierite and almandine are common only in metapelites. The stable K-feldspar is microcline in the low-temperature part, but appears to be orthoclase in the high.

Calcareous rocks may contain wollastonite, grossularite, diopside and sometimes vesuvianite.

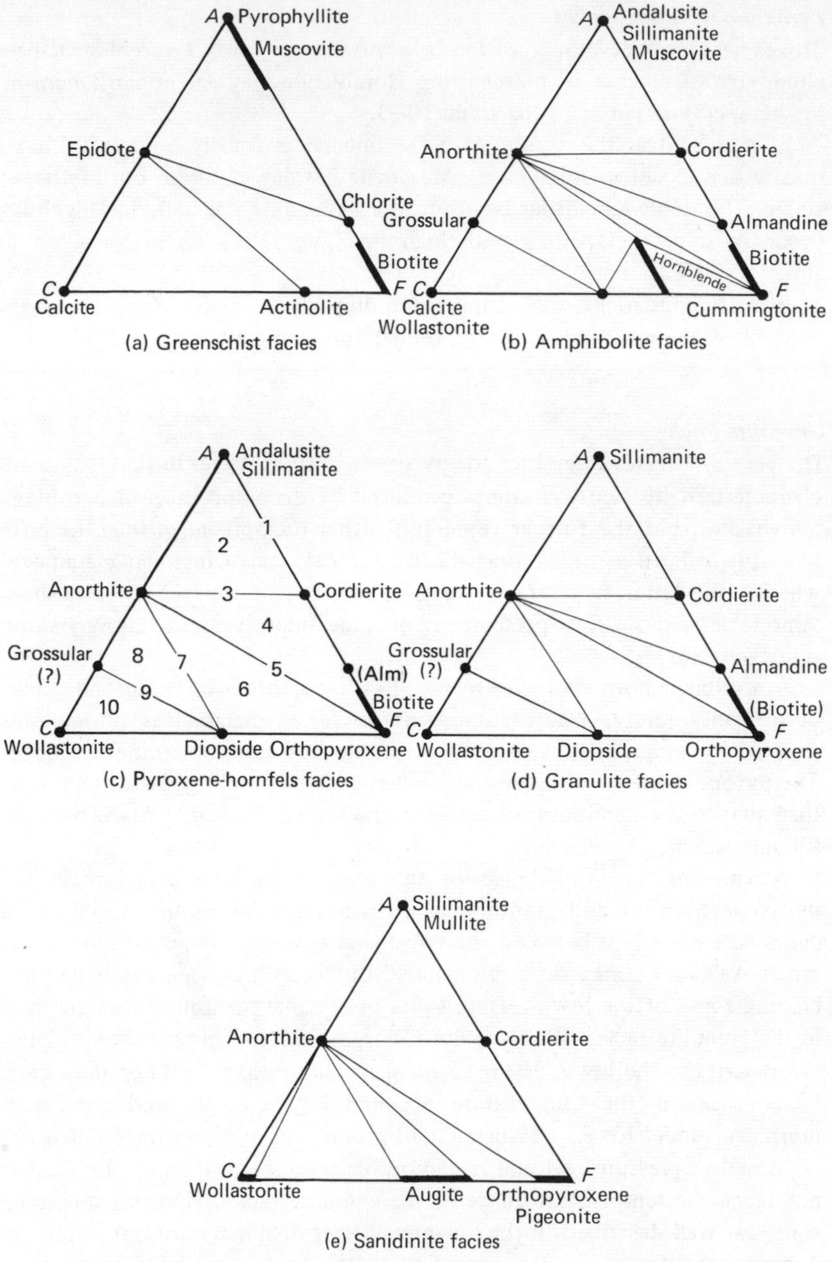

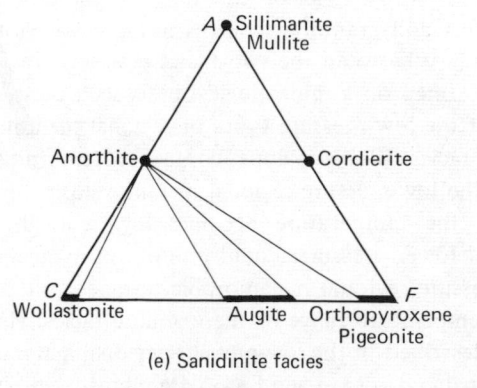

Fig. 11-1 *ACF* diagrams for the low-pressure facies series.

Pyroxene-Hornfels Facies

This facies is well developed in the Oslo and Comrie contact aureoles. Ortho- and clinopyroxenes occur in metabasites. Hornblende may accompany them in the low-temperature part (§§ 10-1 and 10-2).

In metapelites, the stable Al_2SiO_5 mineral is usually andalusite, but sillimanite may occur in some cases. Muscovite has decomposed, but biotite is still stable. The stable K-feldspar is orthoclase. Almandine is usually lacking, but may occur in some metapelites with high Fe^{2+}/Mg ratios, as in the Glen Clova aureole (§10-2).

In calcareous rocks, wollastonite and diopside are stable. Ca-garnet may not be stable in the presence of quartz, though it is stable by itself (§ 10-1).

Granulite Facies

The granulite facies resembles the pyroxene-hornfels facies in that they are both characterized by two pyroxenes produced by decomposition of hornblende in metabasites, but the former represents higher rock-pressures than the latter. It has usually been assumed that the former belongs to regional metamorphism whereas the latter to contact. However, since some contact aureoles show the same facies series as low-pressure regional metamorphism, the above assumption is not tenable.

In pyroxene-hornfels facies metapelites, cordierite is common and almandine occurs only rarely, whereas in granulite facies metapelites of low-pressure regional metamorphism, not only cordierite but also almandine is widespread. The pyrope content of almandine, however, is below 30 per cent, that is, lower than that in the medium-pressure granulite facies. The stable Al_2SiO_5 mineral is sillimanite.

At present our knowledge of the mineral parageneses of rocks of the pyroxene–hornfels and granulite facies is not precise enough to permit us to define the boundary between the two facies precisely. In future, the boundary might well be defined on a mineralogical basis such that all the two-pyroxene-bearing zones of the low-pressure types of regional metamorphism are included in the granulite facies. If this is done, contact metamorphism showing the same facies series as the low-pressure regional metamorphism could produce granulite facies rocks if the temperature becomes high enough, and contact metamorphism under lower pressures could produce pyroxene-hornfels facies rocks.

Most low-pressure regional metamorphic terranes such as the Ryoke belt do not reach the temperature range of the granulite facies. However, the granulite zone was well described in the low-pressure regional metamorphic terrane of the Aracena area, Spain (§§ 7A-4 and 8A-3). Here biotite and hornblende are stable up to the highest temperature part. If the temperature of metamorphism had become still higher, these minerals would have disappeared. The well-documented granulite-facies rocks of Lapland (Eskola, 1952) and Uusimaa (Parras, 1958), both in Finland, probably belong to the low-pressure type.

The mutual relation between the subfacies of the granulite facies as well as their relation to the eclogite facies will be discussed in the next chapter (§ 12-2).

Sanidinite Facies

The sanidinite facies has been considered to represent a *P–T* range of pyrometamorphism, which is nearly the same as that of crystallization of volcanic rocks (§ 10-5). The range is probably on the high-temperature side of the pyroxene–hornfels facies. The sanidinite facies has minerals characteristic of volcanic conditions, such as tridymite, cristobalite, sanidine, anorthoclase and high-temperature plagioclase. The rocks of this facies occur in inclusions within volcanic rocks and some minor intrusions as well as in some country rocks adjacent to the latter (e.g. Thomas, 1922; Agrell and Langley, 1958).

In this facies, the alkali feldspars form a continuous series of solid solution from the K to the Na end-member. Orthopyroxene and/or pigeonite may occur. Sillimanite and/or mullite may form. Wollastonite forms a solid solution (Ca, $Fe^{2+})SiO_3$. Glass may occur.

The sanidinite facies has usually been considered to represent a very low pressure. Certainly tridymite and cristobalite are characteristic of low pressures. However, it is not clear whether all the rocks of this facies were formed at such low pressures. Pyrometamorphic rocks are mostly out of equilibrium, and hence the determination of paragenetic relations in this facies is very difficult.

The definition of the low-temperature limit of the sanidinite facies has not been clearly discussed. Sanidine and the continuous series of alkali–feldspar solid solutions appear to be stable above about 600–650 °C (Tuttle and Bowen, 1958). The temperature of 600–650 °C, however, is probably in the amphibolite facies (§ 3-5). Accordingly, sanidine and alkali–feldspar solid solution would be stable in the pyroxene–hornfels and perhaps also in the granulite facies. Some K-feldspars in rocks of the pyroxene-hornfels facies in the Comrie area (Tilley, 1924) and of the granulite facies in the Wilmington complex (Ward, 1959) have been observed to show sanidine-like optics. It is conceivable that K-feldspars in pyroxene–hornfels and granulite facies rocks were sanidine at the peak of metamorphic temperature, but changed into orthoclase or even into microcline at a declining stage of metamorphism. If so, the presence of sanidine simply indicates that the relevant rocks were more rapidly cooled than some orthoclase-bearing ones, and not necessarily that they formed at higher temperatures.

For the same reason, high-temperature plagioclase also would not be a safe indicator of metamorphic facies. The stability relations of pigeonite and mullite are in dispute. Thus, the precise definition of the sandinite facies is yet to be found.

11-3   METAMORPHIC FACIES OF THE MEDIUM-PRESSURE TYPE

Zeolite Facies

The classical zeolite-facies terrane of Taringatura (§ 6A-4) would belong to the medium-pressure subfacies, as also would the terrane in the central Kii Peninsula (§ 6A-4).

It has been shown that the four dehydration steps in Ca-zeolite assemblages are realized in some low-pressure zeolite facies areas. On the other hand, only two steps characterized by heulandite and by laumontite were found in Taringatura, and only one step characterized by laumontite was observed in Central Kii. The analcime + quartz assemblage is stable in the heulandite step, while albite occurs instead in the laumontite step.

Smectite and celadonite are associated with zeolites. Chlorite and prehnite were found in the zeolite facies zone of central Kii, but not of Taringatura.

Prehnite-Pumpellyite Facies

In the prehnite–pumpellyite facies zone of the Taringatura area, prehnite, pumpellyite, chlorite and albite are stable, whereas in that of central Kii, epidote is stable in addition. With rising temperature, prehnite disappears and actinolite appears instead. A further rise in temperature causes the disappearance of pumpellyite, thus resulting in a greenschist facies zone.

Greenschist Facies

This is the medium-pressure subfacies of the greenschist facies, as observed in the central Kii Peninsula and in the chlorite and the biotite zone of the Barrovian series.

Chlorite, actinolite, epidote, albite and calcite are common in metabasites. Muscovite and biotite occur in some metabasites and more commonly in metapelites. The occurrence of biotite in metapelites is confined to the higher temperature part. Thus, a chlorite-zone and a biotite-zone subfacies may be distinguished. The muscovite is usually phengitic in composition. Stilpnomelane, chloritoid, and garnets of the almandine–spessartine series are common.

In Eskola's (1939) definition, actinolite was erroneously removed from the list of greenschists-facies minerals. Really, it is common in metabasites of this facies. Under a high P_{CO_2}, actinolite is decomposed to form calcite and chlorite. With a further increase in P_{CO_2}, calcite and chlorite would react to form dolomite or ankerite, as has been described for the Woodsville quadrangle (§ 8A-6). Therefore, actinolite greenschist represents a relatively low P_{CO_2} if all the other factors are the same. The name of the actinolite-greenschist facies has been used in some cases in order to emphasize the presence of actinolite in relevant greenschist facies rocks.

Whether rocks recrystallized at the same T, P_s and P_{H_2O}, but at different P_{CO_2}, belong to the same metamorphic facies depends mainly on the mobility of CO_2. If CO_2 is highly mobile, not only T, P_s and P_{H_2O} but also P_{CO_2} should be externally controlled. According to Korzhinskii's definition (§ 11-1), variations in these conditions could lead to different metamorphic facies. If CO_2 is not so mobile, it may be regarded as a fixed component. Though this problem has not been well investigated, it seems that rocks recrystallized at different P_{CO_2} occur intermingled within a mineral zone of an ordinary size in many cases. If it is the case, they might well be classed in the same metamorphic facies.

Epidote–Amphibolite Facies

The medium-pressure subfacies of the epidote–amphibolite facies roughly corresponds to the almandine zone of the ordinary Barrovian series. It corresponds, however, to the higher chlorite and the biotite zone in the Woodsville quadrangle (§ 8A-6).

This facies is characterized by the albite + epidote + hornblende assemblage in metabasites. The hornblende is usually blue–green. Since the problem of the peristerite solvus (§ 7B-8) has not been settled yet, it is not possible to make an unequivocal statement concerning the composition of the 'albite'. Almandine may occur in metabasites with high Fe^{2+}/Mg ratios.

Since actinolite occurs in greenschist facies metabasites and coexists with hornblende in the transitional zone to an epidote–amphibolite facies zone, the boundary between the greenschist and the epidote–amphibolite facies lies at the temperature where the dominant calcic amphibole changes from actinolite to hornblende (§ 8B-4).

In Al_2O_3-poor rocks such as limestones, actinolite or tremolite may occur even in this and the next facies.

Metapelites contain quartz, plagioclase, biotite, muscovite, paragonite, chlorite, and sometimes chloritoid.

Amphibolite Facies

The medium-pressure subfacies of the amphibolite facies roughly corresponds to the kyanite and the sillimanite zone in the Barrovian series.

The metabasites of this facies are mainly composed of plagioclase and hornblende. The plagioclase is usually andesine or labradorite, and the hornblende is green or brown. Epidote is absent and salite may occur. Almandine occurs in amphibolites of this subfacies of some areas. However, the occurrence of almandine depends on the Fe^{2+}/Mg ratio, P_{O_2} and other factors, and the mineral is absent in amphibolites of this subfacies of other areas (§ 8B-7).

Metapelites in the low-temperature part of the amphibolite facies contain muscovite, biotite, almandine, kyanite and staurolite. In the high-temperature part, muscovite is decomposed by reaction with quartz, sillimanite takes the

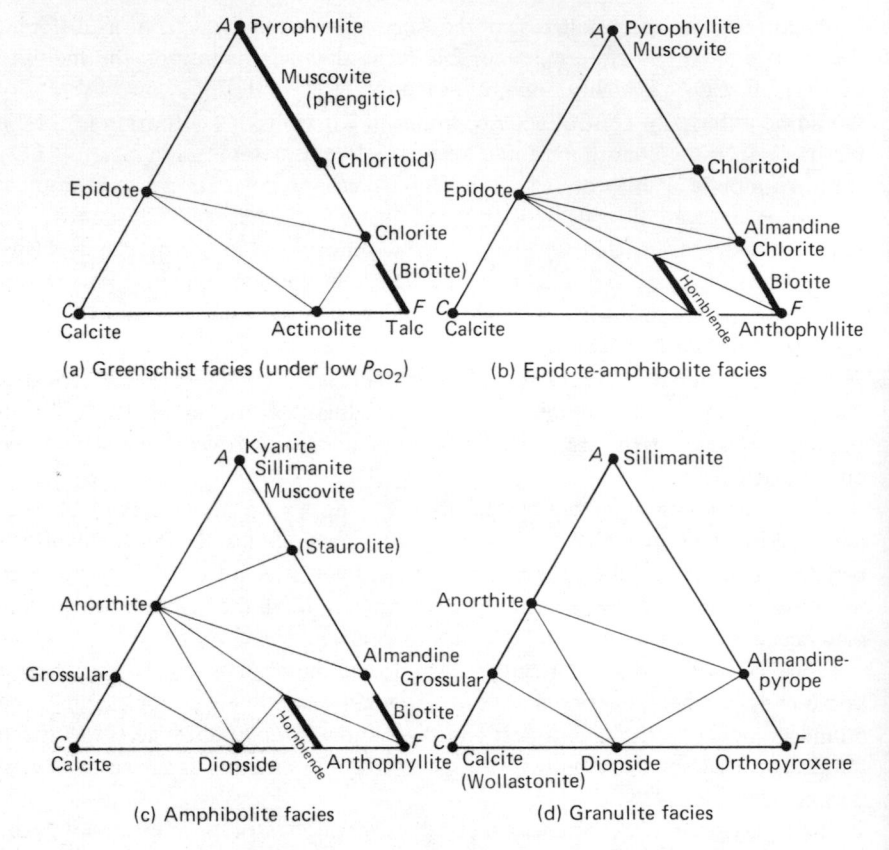

Fig. 11-2 *ACF* diagrams for the medium-pressure facies series.

place of kyanite, and staurolite disappears. Biotite and almandine are stable as far as the highest part.

In most of this subfacies, K-feldspar cannot coexist with kyanite, sillimanite and almandine, whereas it can in the highest temperature part where the muscovite + quartz assemblage is no longer stable.

In calcareous rocks, diopside and Ca-garnet occur commonly, and wollastonite occurs occasionally (§ 9-3).

Granulite Facies

The medium-pressure subfacies of the granulite facies is characterized by metabasites with ortho- and clinopyroxenes formed by the decomposition of hornblende. Biotite also tends to decompose to form orthopyroxene. The decomposition reactions are gradual, and hence usually there is a wide transitional zone where hornblende, biotite, orthopyroxene and clinopyroxene coexist. Typical granulite facies areas with no amphibole and mica are rare. It has therefore been proposed that the granulite facies be divided into two subfacies: a low-temperature subfacies with hornblende and biotite, and a high-temperature subfacies without them (Turner and Verhoogen, 1960; de Waard, 1965). Hornblende and biotite are stable to a higher temperature in quartz-free rocks than in quartz-bearing ones (fig. 12-2).

In metabasites of the medium-pressure-granulite facies, ortho- and clino-pyroxenes are strongly pleochroic and are sometimes high in Al_2O_3 content. Plagioclase is common, but tends to decrease in amount in garnetiferous varieties, since garnet forms partly at the expense of plagioclase.

Almandine–pyrope garnet is common and sometimes abundant in metabasites as well as in metapelites. Metapelites are mainly composed of quartz, plagioclase, K-feldspar, garnet, and usually sillimanite and biotite. Probably orthoclase or sanidine is the stable form of K-feldspar, but microcline is also common, conceivably owing to retrogressive change.

In the low-pressure subfacies of the granulite facies, cordierite is common and the Fe^{2+}/Mg ratio of almandine is high, whereas in the medium-pressure subfacies, cordierite is absent, and the Fe^{2+}/Mg ratio of almandine can be lower. The pyrope content of almandine may be as high as 55 per cent (Eskola, 1957). The high pressure stability limit of cordierite is close to that of sillimanite (fig. 7B-7). Accordingly, the usual absence of cordierite in the medium-pressure granulite facies would not be due to the instability of cordierite by itself but would be due mostly to the dwindling of the rock-composition field over which cordierite is stable with rising pressure.

The mineral assemblages of calcareous metamorphic rocks are not well investigated. The calcite + quartz assemblage is stable in some cases, but wollastonite in others. The grossular + quartz assemblage may be unstable, but Ca-garnet by itself is stable (fig. 9-3).

Since partial melting in the presence of an aqueous fluid begins to occur in the amphibolite facies, metamorphism at higher temperatures should have caused a considerable extent of melting provided that an aqueous fluid was present in the rocks. Therefore, some granulite-facies rocks should represent solid residues of partial melting, and others should have recrystallized without melting under the conditions of $P_{H_2O} \ll P_s$.

Granulite facies metamorphic rocks occur commonly in Precambrian terranes in association with anorthosite massifs, which are sometimes of batholithic dimensions. The genetic significance of the association is not clear, though some anorthosites may represent solid residues of partial melting.

Sometimes areas of the granulite facies merge into amphibolite facies regions. More commonly, such areas are in fault contact with other rocks and granulite facies rocks formed at greater depths could have been brought up into juxtaposition with overlying rocks. Though most granulite facies rocks occur in shield regions, some are present in Phanerozoic orogenic belts. Many of the latter, however, appear to be uplifted basement rocks.

Oliver (1969) has given a review of the modes of occurrence of granulite-facies rocks in the world (fig. 4-3).

11-4 METAMORPHIC FACIES OF THE HIGH-PRESSURE TYPE

Prehnite-Pumpellyite Facies

The high-pressure subfacies of the prehnite–pumpellyite facies has been observed, for example, in the Katsuyama area (§ 6A-5). Prehnite, pumpellyite, epidote, albite, chlorite, actinolite and quartz occur in metabasites, whereas muscovite, chlorite, prehnite, epidote, albite and quartz occur in metapelites. This facies resembles the greenschist facies except for the presence of prehnite and pumpellyite. A transitional state between this subfacies and the glaucophane-schist facies is commonly observed in the lowest temperature part of typical high-pressure metamorphism in California (e.g. fig. 3-8).

Glaucophane-Schist Facies

Lawsonite and the jadite + quartz assemblage occur in the typical parts of the glaucophane-schist facies. Glaucophane (including crossite) forms not only in such typical parts but also in wide transitional states toward the prehnite–pumpellyite, greenschist and epidote–amphibolite facies. In this book, all such transitional states are included in the glaucophane-schist facies. Moreover, the stability of lawsonite is so strongly influenced by P_{H_2O} and P_{CO_2} that the mineral cannot be a reliable indicator of rock-pressure (§ 4-4).

The temperature conditions for the facies are especially variable, ranging probably from 200–550 °C, thus covering the temperature ranges of the prehnite–pumpellyite, greenschist and epidote–amphibolite facies, but being on the high-pressure side. It appears that the formation of lawsonite and the jadeite + quartz assemblage is confined to temperatures below 400 °C (Taylor and Coleman, 1968).

In addition, aragonite, chlorite, actinolite, muscovite (phengitic), paragonite, stilpnomelane, albite, pumpellyite, epidote, chloritoid, spessartine-almandine, piemontite, sphene, and rutile occur in this facies. Some metamorphic rocks and especially metacherts contain aegirine, riebeckite, piemontite, or deerite. Many so-called glaucophane schists are nearly devoid of schistosity.

Nearly all glaucophane schists occur in Palaeozoic and younger orogenic belts. They are usually associated with serpentinite belts. Van der Plas (1959) has given a comprehensive review of the localities of glaucophane and lawsonite in the world (fig. 4-2).

Greenschist Facies

A zone of the greenschist facies occurs on the high-temperature side of a zone of the glaucophane-schist facies in many areas. Chlorite, actinolite, calcite, epidote, albite and quartz occur in metabasites, whereas chlorite, muscovite, albite and quartz occur in metapelites. It is noted that the common occurrence of stilpnomelane and the absence of biotite characterize the high-pressure subfacies of the greenschist facies as distinct from its medium- and low-pressure subfacies.

Epidote-Amphibolite Facies

The high-pressure subfacies of the epidote–amphibolite facies is known, for example, in the Bessi–Ino area (§ 8A-8) and the Iimori area (Kanehira, 1967), both of the Sanbagawa belt.

The assemblage albite + epidote + hornblende occurs in metabasites. These may be associated with chlorite, diopside, garnet, muscovite and biotite. Metapelites contain chlorite, biotite, muscovite, almandine, epidote, albite, calcite and quartz.

This subfacies resembles the medium-pressure subfacies of the same facies. However, epidote amphibolites of the high-pressure series commonly contain muscovite and sometimes barroisite, whereas those of the medium-pressure series do not.

Amphibolite Facies

A wide zone of the amphibolite facies exists in the Pennine nappes of the Central Alps (§ 7A-8; also Wenk and Keller, 1969). The metapelites there contain not only biotite and almandine, but also staurolite, kyanite and even sillimanite. It is conceivable that this zone belongs to the medium-pressure type, though glaucophane occurs in other parts of the Central Alps. No amphibolite facies zone has been reported from typical high-pressure terranes.

Chapter 12

Eclogites and the Eclogite Facies

12-1 ECLOGITES AND ESKOLA'S CONCEPT OF THE ECLOGITE FACIES

The rock name eclogite was coined by Haüy, a founder of crystallography, in 1822. Since the original definition of the name was vague, its meaning was gradually settled through its usage in later years. According to the generally established usage in the first half of the twentieth century, it means a rock mainly composed of omphacite and almandine–pyrope–grossular garnet. Omphacite is a clinopyroxene containing diopside, hedenbergite and jadeite components, and quite frequently some acmite. However, no rigid limitations were given for the composition ranges of these minerals. Thus, some authors called diopside–almandine rocks eclogite. Small amounts of quartz, kyanite, olivine, orthopyroxene, amphibole and rutile may be present, but not plagioclase.

Eskola (1920) proposed the name *eclogite facies* on the basis of this wide definition of eclogite. Since eclogites are basaltic in chemical composition, their unique mineral composition was ascribed to the unique $P-T$ conditions of their formation. Not only the eclogites but also other rocks formed under the same $P-T$ conditions as eclogites were claimed to belong to the eclogite facies. As examples of the latter rocks, he referred to dunite, enstatitite, pyrope–olivine rock, chloromelanitite and jadeitite. Even taking all these rocks into consideration, the composition range is very limited. Rocks of other compositions were unknown in this facies.

Clinopyroxenes high in diopside and hedenbergite components as well as garnets of the almandine–pyrope series are stable in the granulite facies, however. The assemblage of these two minerals, which may be called eclogite in the wide definition, certainly occurs in some granulite-facies areas (e.g. Davidson, 1943). Eskola's idea of the eclogite facies was mainly based on his study of eclogites in southwestern Norway (Eskola, 1921). Some of these eclogites also appear to belong to the granulite facies. Therefore, if the eclogite facies is characterized by traditionally defined eclogite, a part of it overlaps a part of the granulite facies. Since this is very undesirable, we have to change the definition.

Clinopyroxene–almandine rocks occurring in high-pressure terranes have been widely referred to as eclogites. Many of them, if not all, appear to have formed under the same P_s–T conditions as the associated glaucophane schists. However, the eclogite facies differs from all other metamorphic facies in that it has not been found as characterizing a zone in a progressive metamorphic sequence in any area of the world. Thus, the relations of the eclogite facies to the granulite, glaucophane-schist and other facies are not easy to solve.

Judging from the diversity of the modes of occurrence of eclogite facies rocks, the P–T field corresponding to this facies is probably very wide as compared with other facies. It may be divided into some subfacies.

12-2   RELATION BETWEEN THE GRANULITE AND THE ECLOGITE FACIES

Green and Ringwood's Experimental Study

Green and Ringwood (1967) made an experimental study on the transformation of gabbros into eclogites under virtually anhydrous conditions. Plagioclase, pyroxene and possibly olivine are the main constituents of gabbros at low pressures. With an increase in pressure, garnet begins to occur and then increases in amount, whereas plagioclase and olivine decrease and disappear. Thus, a pyroxene–garnet rock, i.e. eclogite, forms.

The P–T relations of such a transformation for quartz–tholeiitic compositions are shown in fig. 12-1. For a medium value of the $Fe^{2+}/(Mg + Fe^{2+})$ ratio, the P–T field below curve B in the figure represents a garnet-free granulite or gabbro assemblage. In this pressure range, the An component of plagioclase is increasingly incorporated into pyroxenes with increase in pressure. This results in the formation of highly aluminous pyroxenes containing Tschermak's molecule: (Ca, Mg) Al_2SiO_6. In the pressure range between curves B and 'Eclog', the reaction between orthopyroxene and plagioclase produces an increasing amount of garnet having an increasing pyrope/almandine ratio with rise in pressure. At higher pressures within this range, the Ab component of plagioclase begins to change into the jadeite component of pyroxene, leading to the disappearance of plagioclase. The Tschermak's molecule of pyroxene decomposes with resultant formation of additional garnet. Thus, typical eclogite with magnesian garnet and jadeite-rich clinopyroxene forms in the P–T field above the curve 'Eclog'.

For undersaturated alkali olivine basalt compositions, the appearance of garnet and the disappearance of plagioclase take place at pressures a few kilobars lower than those for quartz–tholeiite compositions. The higher the Ab content of plagioclase, the higher the pressure necessary for the disappearance of plagioclase (i.e. the formation of eclogite). The higher the $Fe^{2+}/(Mg + Fe^{2+})$ ratio of the rock, the lower the pressure necessary for the formation of garnet.

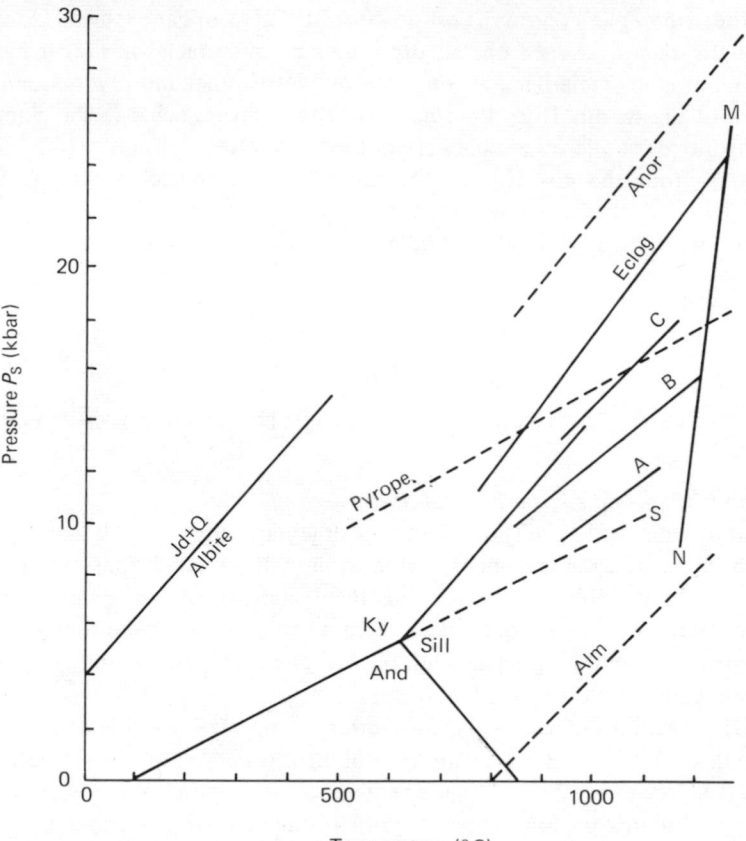

Fig. 12-1   Granulite to eclogite transformation for quartz–tholeiitic compositions under a virtually anhydrous condition (Green and Ringwood, 1967). A, B and C: These curves indicate the first appearance of garnet with rising pressure for quartz-tholeiitic compositions with $Fe^{2+}/(Mg + Fe^{2+}) = 0.9$, 0.4 and 0.1, respectively. Eclog: This curve represents the disappearance of plagioclase with rising pressure. M-N: approximate solidus under dry condition. The low-pressure stability limits of pyrope and almandine end-members are shown by the curves designated as 'pyrope' and 'alm'. The high-pressure stability limit of anorthite is shown by the curve designated 'anor'. The stability fields of kyanite, sillimanite and andalusite and the low-pressure stability limit of the jadeite + quartz assemblage are shown for comparison.

Granulite and Eclogite Facies in Relation to Baric Types

In the temperature range 900–1200 °C, the following stability fields are distinguished in terms of mineral changes in rocks of basaltic compositions with medium $Fe^{2+}/(Mg + Fe^{2+})$ ratios.

1. Just below the curve S in fig. 12-1. This field represents the low-pressure subfacies of the granulite facies. The assemblages calcic plagioclase + orthopyroxene and calcic plagioclase + olivine are stable. Garnet does not occur in metabasites oversaturated or undersaturated with SiO_2. Almandine and cordierite may occur in metapelites. The granulites of Lapland (Eskola, 1952) and Uusimaa (Parras, 1958) belong to this field. The *ACF* diagram for oversaturated rocks and an analogous one for undersaturated rocks are shown in the upper part of fig. 12-2.

2. Between the curves S and B in fig. 12-1. This field represents a low-pressure part of the medium-pressure subfacies of the granulite facies. The calcic plagioclase-orthopyroxene assemblage is stable, but the calcic plagioclase–olivine assemblage is unstable. Garnet may occur in undersaturated metabasites, but does not occur in oversaturated metabasites. Pyroxenes tend to be highly aluminous. The granulites of Scourie, Scotland (O'Hara, 1961) belong to this field. The parageneses are shown in the middle part of fig. 12-2.

3. Between the curves B and 'Eclog' in fig. 12-1. This field represents the high-pressure part of the medium-pressure subfacies of the granulite facies. Almandine-pyrope occurs in both oversaturated and undersaturated metabasites, being formed by reaction between calcic plagioclase and orthopyroxene (fig. 12-2, lower part). The granulites and 'eclogites' of South Harris, Outer Hebrides (Davidson, 1943; Dearnley, 1963) belong to this field.

4. Above the curve 'Eclog' in fig. 12-1. This field represents a part of the eclogite facies distinct from the granulite facies. Calcic plagioclase no longer occurs in metabasites and metapelites. In the pressure field immediately above the curve, anorthite by itself is stable as well as albite by itself. However, they become unstable at higher pressures. The paragenetic relations are schematically shown in fig. 12-3.

It must be emphasized that the pressure necessary for reactions to form eclogites varies with the Fe^{2+}/Mg ratio and the plagioclase composition; hence the transformation should be essentially gradual with a wide transitional *P–T* region.

Petrographic Criteria for the Eclogite Facies

The eclogite facies assemblages as defined above are characterized by the possible presence of quartz or kyanite, as may be schematically shown by the following equations:

Eclogite Facies			Granulite Facies	

$$Mg_3Al_2Si_3O_{12} + CaMgSi_2O_6 + SiO_2 = 4\ MgSiO_3 + CaAl_2Si_2O_8$$

pyrope   diopside   quartz   enstatite   anorthite

$$3\ CaMgSi_2O_6 + 4\ Al_2SiO_5 \qquad = Mg_3Al_2Si_3O_{12} + 3\ CaAl_2Si_2O_8 + SiO_2.$$

diopside   kyanite   pyrope   anorthite   quartz

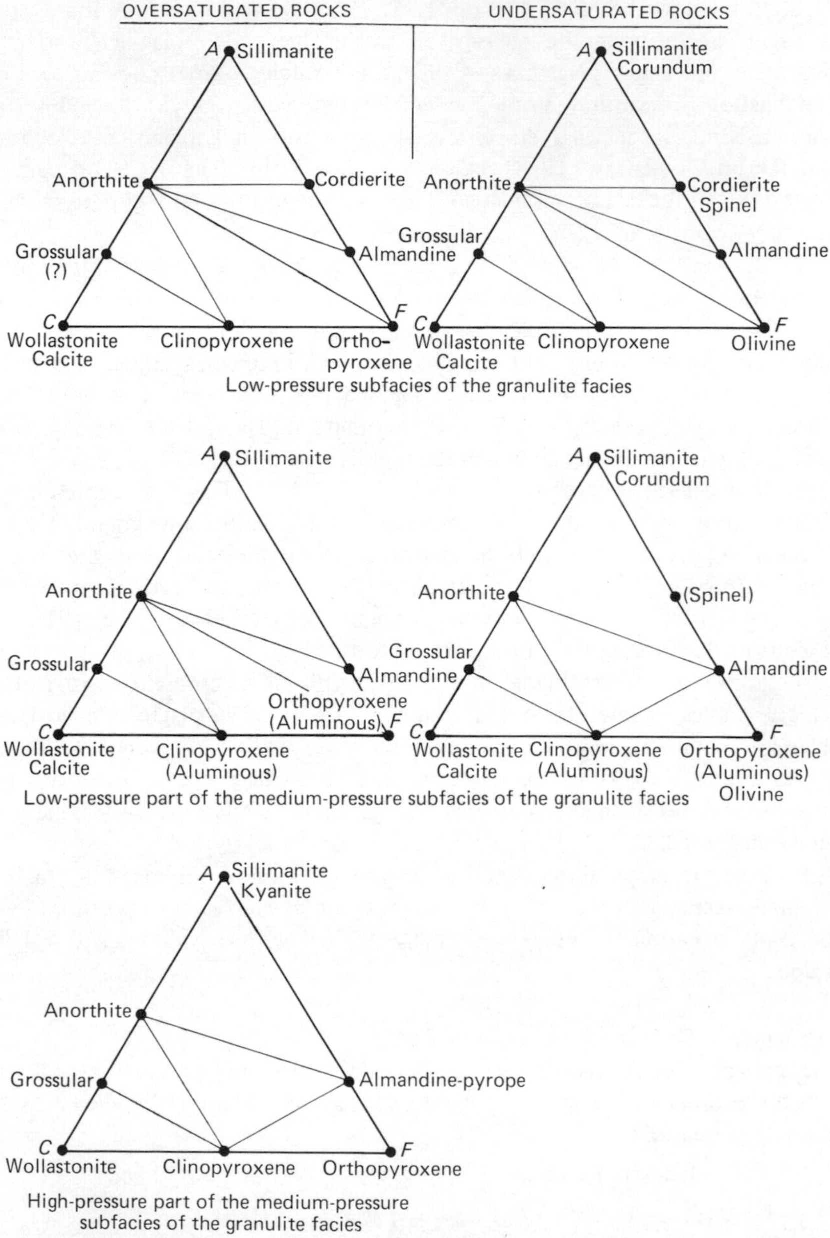

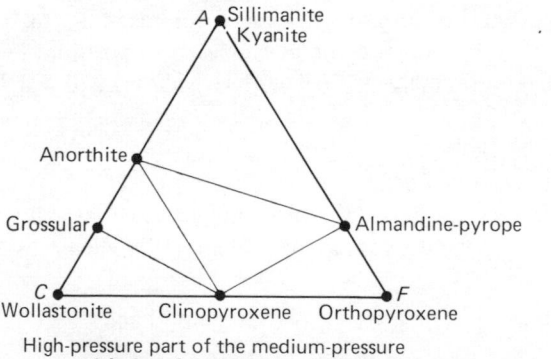

Fig. 12-2 ACF diagrams for rocks oversaturated with SiO_2 (left column) and analogous ones for rocks undersaturated with SiO_2 (right column) in the granulite facies.

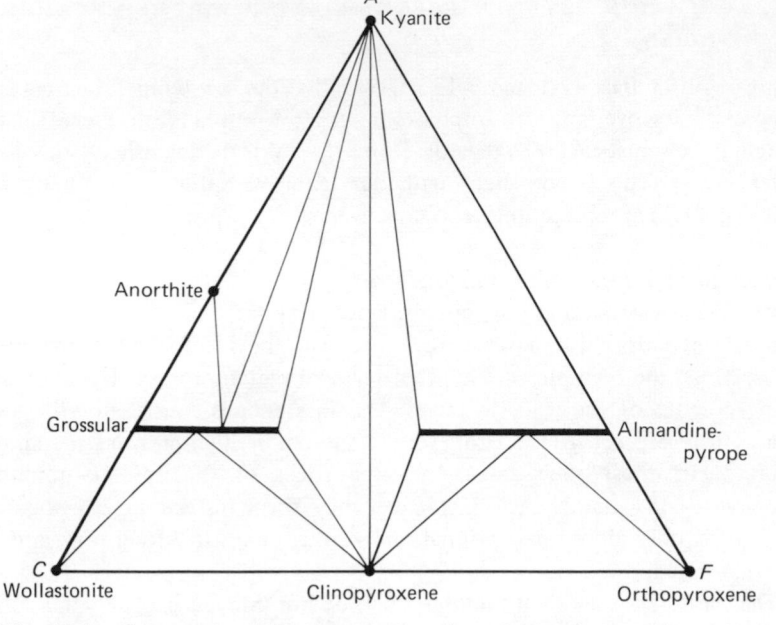

Fig. 12-3 Schematic *ACF* diagram for the eclogite facies. Since the P–T range of the eclogite facies is very wide, various modifications of paragenetic relations are conceivable.

Thus, the occurrence of quartz- or kyanite–eclogite may be used as a petrographic definition of the eclogite facies, if appropriate limits are fixed for the $Fe^{2+}/(Mg + Fe^{2+})$ and the Na/Ca ratios of the rocks.

The Al_2SiO_5 mineral of the eclogite facies is kyanite, whereas that of the granulite facies is commonly sillimanite but rarely kyanite. This observation is consistent with the experimentally determined phase relations of fig. 12-1.

The eclogite facies is characterized only by a group of eclogites with basaltic compositions in which plagioclase is lacking, and clinopyroxene has high ratios of jadeite/Tschermak's component, and occasionally quartz or kyanite occurs.

Eclogites in the broad sense cited at the beginning of this chapter could form under any part of the *P–T* region above curve S in fig. 12-1. However, Green and Ringwood (1967) have proposed that the use of the rock name *eclogite* be confined to the eclogites of the eclogite facies. If this is accepted, the so-called eclogites of the granulite facies should be called *garnet pyroxenites*. Perhaps, this nomenclature will become common in future.

12-3 RELATION BETWEEN ECLOGITE AND HYDROUS METAMORPHIC FACIES

Comparison of figs. 12-1 and 3-12 suggests that the low-temperature part of the eclogite facies overlaps the amphibolite, epidote-amphibolite, greenschist, and glaucophane-schist facies as regards T and P_s, but is distinguished from them by lower P_{H_2O}. This is consistent with our tentative definition of metamorphic facies (§ 11-1) as representing a volume in the $T-P_s-P_{H_2O}$.

Formation of Eclogite in Continental Crust

Figure 3-3 suggests that the condition of 10 kbar and 700 °C is realized in regional metamorphism intermediate between the medium- and high-pressure types. Thus, the high-pressure type of regional metamorphism would be able to form eclogites of the eclogite facies. This inference is consistent with geologic data. However, eclogite is not very common in high-pressure metamorphic terranes. In most such cases, hydrated rocks of the epidote–amphibolite, greenschist, or glaucophane-schist facies may form instead of eclogites, which may form only under exceptional, anhydrous conditions (e.g. Fry and Fyfe, 1969).

The base of a stable continental crust, 35 km thick, is probably near 10 kbar and at 300–700 °C. Accordingly, most of the stable crust will be on the high-pressure side of the curve 'Eclog', that is, within the P_s-T range of the eclogite facies. Virtually no dehydration reactions would take place in non-orogenic crust, and hence P_{H_2O} would be much lower than P_s. If this is true, it may be expected that eclogite is widespread within continental crust. However, the occurrence of eclogite on the surface of continents is not very common. In stable continental crust, there are no penetrative movements, and the rate of recrystallization may be too slow to form eclogite. Alternatively, eclogites once formed in stable lower crust may not be uplifted so as to be exposed on the surface.

Banno (1966) and Green and Ringwood (1967) have given critical reviews of the modes of occurrence of eclogites and related rocks in areas of the granulite, amphibolite, epidote–amphibolite, and glaucophane–schist facies.

The compositions of clinopyroxenes and garnets in eclogites of various modes of occurrence are shown in figs. 12-4 and 12-5.

Eclogites in Gneiss Areas Seemingly of the Amphibolite Facies

Eclogites of the eclogite facies have been reported to occur in some gneiss areas seemingly of the amphibolite facies. Well-documented examples are the eclogites in the Nordfjord area, Norway (Eskola, 1921; Lappin, 1960; O'Hara and Mercy, 1963; Bryhni et al. 1970) and those near Glenelg in the Scottish Precambrian (Alderman, 1936).

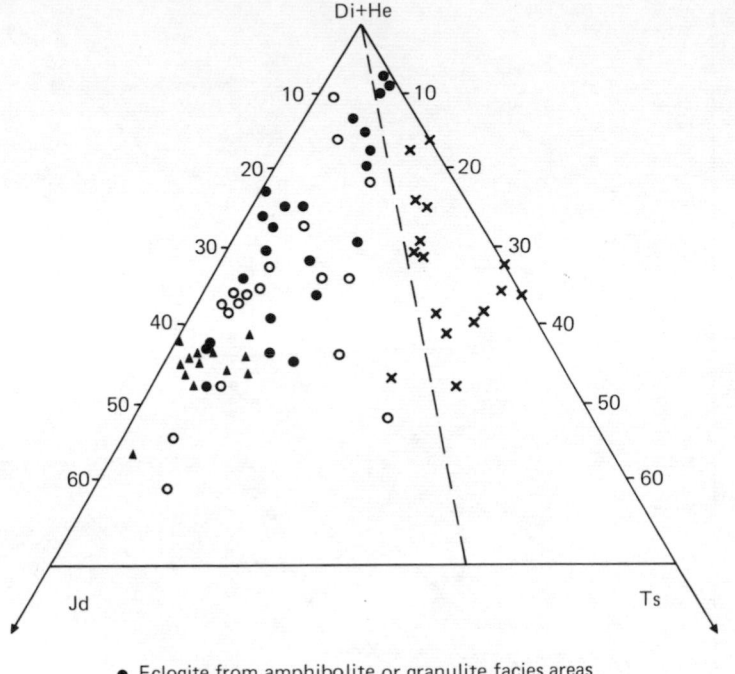

● Eclogite from amphibolite or granulite facies areas

▲ Eclogites from high-pressure metamorphic terranes

○ Eclogite inclusion in kimberlite

✗ 'Eclogite' or granulite inclusion in basaltic rock

Fig. 12-4   Composition of clinopyroxenes from eclogites of various environments, shown in terms of diopside + hedenbergite, jadeite, and Tschermak's component $(Ca, Mg)Al_2 SiO_6$. Jadeite is calculated as the atomic proportion of $Na-Fe^{3+}$, and Tschermak's component is taken as the atomic proportion of Al in four-fold co-ordination. The dotted line represents the 1:2 jadeite to Tschermak's component ratio that may be used to distinguish eclogite- from granulite-facies clino-pyroxenes. (White, 1964; Lovering and White, 1969.)

In such cases it has usually been assumed that the eclogites represent tectonically transported masses and that there is no direct relation between the eclogites and the surrounding metamorphic rocks. This is possible, but has not been proved. It appears largely to have stemmed from the traditional idea that eclogites should have formed at greater depth than ordinary metamorphic rocks.

Some geologists have thought that some such eclogites were intruded and crystallized as igneous masses independent of the surrounding rocks. During the crystallization, conditions of the eclogite facies would have prevailed, but the surrounding rocks would not have been recrystallized possibly owing to the short duration of the conditions.

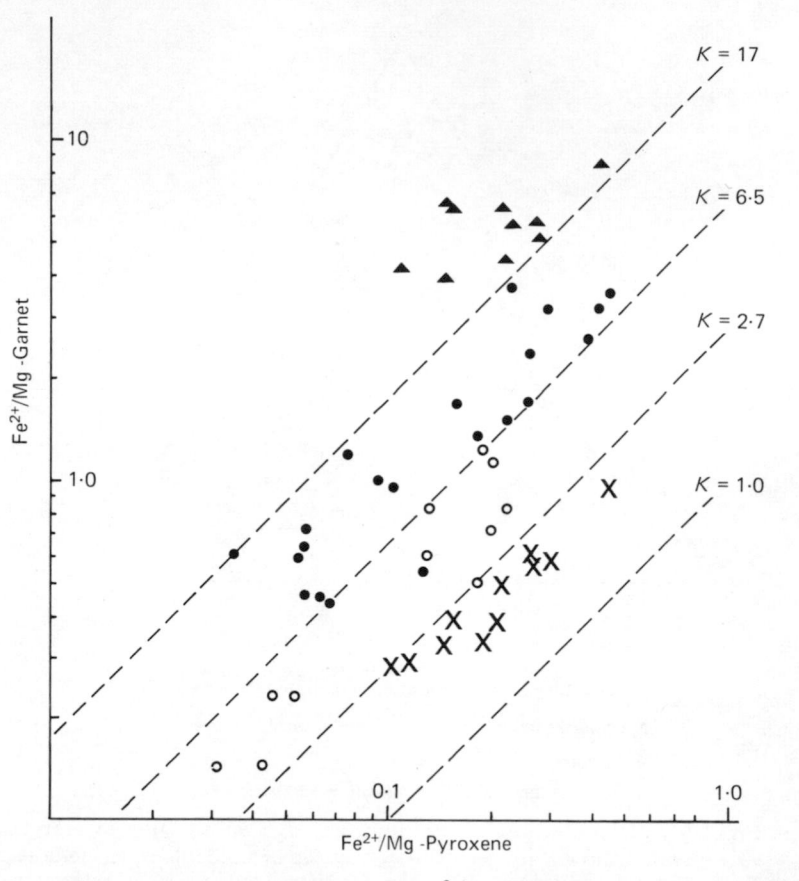

Fig. 12-5   Log-scale distribution of Fe^{2+}/Mg between co-existing garnet and clinopyroxene in eclogites from various environments (Banno and Matsui, 1965; Lovering and White, 1969). Here, Fe^{2+}/Mg is in atomic ratio, and the symbols are the same as in Fig. 12-4. K represents the distribution coefficient for the reaction: Fe-garnet + Mg-pyroxene = Mg-garnet + Fe-pyroxene (§ 2-5).

A third possibility is that not only such eclogites but also the surrounding metamorphic rocks underwent metamorphism in the eclogite facies. The mineralogical differences in quartzo-feldspathic gneisses between the eclogite and the amphibolite facies are slight, and might well be obscured by the superposition of later metamorphism in the amphibolite facies (e.g. Bryhni *et al.* 1970).

A fourth possibility is that P_{H_2O} may have been lower in the eclogites than in the surrounding rocks. This would result in the formation of eclogites within true amphibolite facies terranes, keeping all the rocks in practically the same

P_s-T conditions. Eclogite masses frequently show a transformation into amphibolites towards their margins presumably owing to permeating H_2O. However, if the P_{H_2O} is so uneven, P_{H_2O} cannot be an externally controlled variable. It becomes desirable to treat H_2O as a fixed component in the definition of metamorphic facies, and to regard such eclogites as belonging to the same facies as the surrounding rocks (e.g. amphibolite facies).

At present, we have no decisive evidence for choosing among the above possibilities.

Eclogites in High-Pressure Metamorphic Terranes

Eclogite occurs as isolated blocks (Coleman and Lee's type IV) in metamorphosed areas of the Franciscan formation of California (e.g. § 7A-10). They are usually composed of clinopyroxene, garnet, barroisite, clinozoisite, rutile and sphene. The areas are in the glaucophane-schist facies, and the eclogites are commonly interlayered with glaucophane schists. Lawsonite occurs in some of them as a late-stage mineral. Usually quartz is not present (Coleman and Lee, 1963; Coleman *et al.* 1965). The eclogites appear to have formed under the same P_s-T conditions as the interlayered glaucophane schists.

In the Bessi area of the Sanbagawa belt, Banno (1964, 1966) described the occurrence of eclogite within an epidote–amphibolite mass in the highest temperature part of thè area. The rock contains omphacite, almandine, hornblende (or barroisite), quartz and rutile.

The compositional differences of pyroxene and garnet from eclogites of high-pressure terranes and of gneiss areas are shown in figs. 12-4 and 12-5. Some eclogites in high-pressure terranes contain quartz, and so have the omphacite + garnet + quartz assemblage, characteristic of the eclogite facies, although they have relatively high Fe^{2+}/Mg ratios.

In metamorphosed pillow lavas of the Pennine nappes of the Alps, the cores of some pillows have been converted into eclogites, whereas the matrix between the pillows is of glaucophane schist (Bearth, 1959). It might be thought that during metamorphism P_{H_2O} was lower in the core than in the matrix, resulting in two different mineral compositions forming under the same P_s-T condition. However, the chemical composition of the pillow core could differ from that of the matrix in terms of components other than H_2O. The chemical difference may be mainly responsible for the mineralogical difference.

In the Franciscan and Bessi eclogites, the chemical similarity of the eclogites with the associated metamorphic rocks has not been established. For example, Coleman *et al.* (1965) have noted that the Franciscan eclogites have normative nepheline, whereas the unmetamorphosed Franciscan basalts are not so undersaturated, and that the Franciscan eclogites have higher $(FeO + Fe_2O_3)/MgO$ ratios than most basalts. If the mineralogical differences are due to such chemical differences, the eclogites should belong to the glaucophane-schist or the epidote–amphibolite facies, but not to the eclogite facies. In so far as the relationship between the chemical and mineral compositions in the glauco-phane-schist facies has not been well established, the possibility of the formation of eclogite in this facies cannot be precluded.

12-4 ECLOGITE INCLUSIONS

Eclogites occur as inclusions in basaltic rocks, peridotites and kimberlites. They may be deep-seated crystallization products of their host magmas, or may represent xenoliths derived from the lower crust or the upper mantle. Whatever their origin may be, some of them appear to be in the granulite facies, and others in the eclogite facies.

For a better understanding of the mineral assemblages of such inclusions, the high-pressure phase relations for the composition of 1 anorthite + 2 forsterite are shown in fig. 12-6 (Kushiro and Yoder, 1966). The garnet + forsterite assemblage is stable in the $P-T$ field of the eclogite and high-pressure granulite facies. At lower pressures (part of the granulite facies), spinel appears in association with ortho- and clinopyroxenes.

Eclogite Inclusions in Basaltic Rocks

Well-documented examples of eclogite inclusions in basaltic rocks occur at Salt Lake Crater in Oahu, Hawaii, and at Delegate, in southern New South Wales, Australia. They appear to be eclogites in the granulite facies.

Ejected blocks of eclogite, pyroxenite, peridotite and dunite occur in the olivine–nephelinite tuff at Salt Lake Crater, Hawaii. The pyroxenite, eclogite and olivine eclogite show mutually gradational relations. The eclogite is com-posed of orthopyroxene with exsolution lamellae of clinopyroxene, clinopy-roxene with lamellae of orthopyroxene, opaque spinel, pyrope and pargasitic amphibole. The pyroxenes are high in Tschermak's component, suggesting that the rocks belong to the granulite facies (White, 1964; Lovering and White, 1969; White, 1966; Kuno, 1969).

Green (1966) has accounted for the origin of the eclogite by assuming that nearly monomineralic masses of aluminous clinopyroxene with or without minor quantities of spinel and olivine were formed by accumulation of primary crystals from alkali olivine basalt or basanite magma at a pressure of approximately 13–18 kbar at 1350–1400 °C, and that they were cooled to about 1000 °C at nearly constant pressure to cause exsolution of garnet from the original pyroxene and reaction of spinel and pyroxenes to yield garnet and olivine. All these processes took place on the low-pressure side of curve 'Eclog' but on both sides of curve SY in fig. 12-6, that is, in the high-pressure part of the granulite facies conditions.

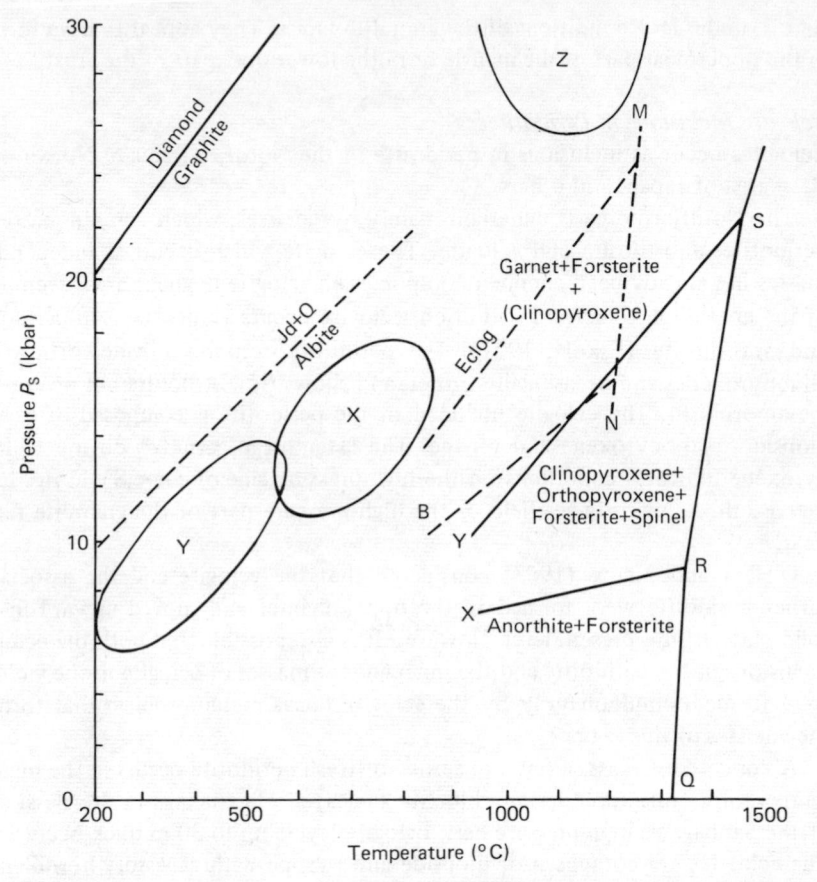

Fig. 12-6   Phase relations for the composition 1 anorthite + 2 forsterite. The line Q-R-S represents the solidus (Kushiro and Yoder, 1966). For comparison, the curves M-N, B and Eclog of Fig. 12-1 are reproduced by broken lines. Possible *P-T* fields of eclogites of gneiss areas (X), of high-pressure metamorphic terranes (Y), and of kimberlites (Z) are shown. (*cf*. Green, 1966.)

Two pipes filled by basanite, olivine nephelinite and their breccias at Delegate, Australia, contain a great variety of inclusions comprising two pyroxene-plagioclase rock, garnet–clinopyroxene–plagioclase rock, eclogite (fassaite eclogite), spinel–garnet–two pyroxene–olivine rock, spinel–garnet–two pyroxene–scapolite–plagioclase rock, and peridotite. The eclogite is mainly composed of clinopyroxene (fassaite) high in Tschermak's component but low in jadeite, pyrope-rich garnet, and brown pargasite (Lovering and White, 1969).

The mineral assemblages suggest that the eclogite and most other inclusions were crystallized or recrystallized probably at 10–15 kbar and 1000–1200 °C, that is, under $P-T$ conditions of the granulite facies. They appear to have formed in the uppermost part of the mantle or in the lowermost part of the crust.

Eclogite Inclusions in Peridotites

Eclogites occur as inclusions in peridotite in the Nordfjord area of Norway, the Bessi area of Japan, and others.

The Nordfjord area is underlain mainly by gneisses, which contain masses of peridotite, anorthosite and eclogite. The eclogite which occurs as independent masses has already been commented upon as an eclogite in gneiss areas seemingly of the amphibolite facies. In addition, eclogite occurs as inclusions in peridotite and anorthosite (Eskola, 1921). The peridotite contains olivine, ortho- and clinopyroxenes and occasionally garnet, and shows flow structure but no contact metamorphism. The eclogite included in the peridotite is composed of garnet, diopside, orthopyroxene and olivine. The assemblage garnet + olivine + clinopyroxene indicates conditions on the high-pressure side of curve SY in fig. 12-6, that is, the eclogite facies field or the high-pressure part of the granulite facies field.

O'Hara and Mercy (1963) considered that the eclogite and the associated garnet peridotite were formed in the upper mantel and moved upward in the solid state to the present level. However, it is also possible that both the eclogite inclusions in the peridotite and the independent masses of eclogite in the vicinity were formed simultaneously by the eclogite facies metamorphism that formed the gneisses of this region.

A concordant mass, a few km across, of fresh peridotite occurs in the highest temperature zone (epidote–amphibolite facies) at Higasiakaisi in the Bessi area of the Sanbagawa high-pressure belt. Eclogite layers up to 30 m thick occur in it. The eclogites are composed of diopside and pyrope with accessory hornblende, olivine, rutile and ilmenite. Some of the eclogites, however, contain omphacite and almandine with a considerable proportion of the pyrope component (Shido, 1959; Banno, 1966; Banno and Yoshino, 1965). Neither quartz nor kyanite occurs. The olivines of the surrounding peridotite show strong preferred orientation. These eclogites could have originally been formed in the upper mantle and subsequently been caught up by the solid intrusion of peridotite and then subjected to slight modification by metamorphism at the present level.

Eclogite Inclusions in Kimberlites

Kimberlite is well known as a source of diamonds. It is a rare ultrabasic rock and occurs as small intrusive bodies (pipes and dikes) in continental regions. Petrographically it is serpentinized and carbonated mica-peridotite of porphyritic texture. The phenocrysts are olivine (about 90 per cent Fo), garnet (almandine-pyrope), pyroxenes (enstatite and diopside), mica (phlogopite-biotite), amphiboles, and picro-ilmenite. Diamond, when present, is sparsely distributed in the groundmass or rarely contained in inclusions of garnet peridotite and eclogite.

The presence of diamond in the groundmass of kimberlite and in inclusions of garnet peridotite and eclogite, strongly suggests that all these rocks came from great depths in the mantle. If we assume that the diamond crystallized in its stability field, their depths of origin must be more than 100 km below the surface. It is likely that they represent the highest pressure part of the eclogite facies conditions.

Ringwood and Lovering (1970) have given experimental evidence suggesting that inclusions of pyroxene-ilmenite intergrowths in African kimberlites were originally homogeneous garnet solid solutions formed at pressures in excess of 100 kbar. If this is accepted, kimberlite magma and some of the xenoliths contained in it could have come up from depths of more than 300 km.

Kimberlite contains inclusions of gneisses, granulites, garnet peridotites, eclogites and so on (Dawson, 1967). The inclusions of gneisses and granulites were probably derived from the walls of the intrusions within the crust. The origin of the garnet peridotite and eclogite inclusions, however, has been a matter of dispute. Many geologists consider that they are xenoliths derived from the mantle (e.g. O'Hara and Mercy, 1963). Others regard them as segregations (cognate inclusions) crystallized out of kimberlite magma at great depths (Williams, 1932).

Kushiro and Aoki (1968) have discussed the possibility that the upper mantle is largely composed of garnet peridotite, which might undergo partial melting followed by crystallization to produce pockets of eclogite. These might be subsequently broken up and brought to the surface by ascending kimberlite magma. The compositional variation in the eclogite inclusions would largely depend on the degree of partial melting and/or crystallization differentiation at high pressures.

Though kimberlite occurs in all the continents, the diamond-bearing type is abundant only in the southern half of Africa and the Yakutia area of Siberia. Extensive mineralogical studies by Russian geologists have greatly contributed to our knowledge.

In the Siberian kimberlites, a peculiar rock called grospydite has been rarely found as inclusions. It is mainly composed of garnet, omphacite and kyanite together with corundum. This rock resembles kyanite eclogite, but differs in the composition of the garnets, which contain 50–80 per cent of the grossular component. The garnets of grospydite and the associated kyanite eclogite form a nearly continuous solid solution series between pyrope and grossular. (The name grospydite has come from an abbreviation of grossular–pyroxene–disthene rock.) There is a regular relationship between the grossular content of the garnet and the jadeite content of the associated clinopyroxene (Sobolev, Kuznetsova and Zyuzin, 1968; Banno, 1967).

12-5 GRANITES AND METAPELITES IN THE ECLOGITE FACIES

What are the mineral assemblages of granitic, psammitic and pelitic rocks which have recrystallized in the eclogite facies? It is possible that such rocks resemble those in the granulite, amphibolite, and epidote–amphibolite facies, and are not easily distinguishable from them.

Green and Lambert (1965) have carried out a high pressure experiment at 950 and 1100 °C on samples of anhydrous adamellite composition. At 950 °C and in the pressure range of the granulite facies (13–15 kbar) and the low-pressure part of the eclogite facies (17–22 kbar), the recrystallized samples are composed of plagioclase, alkali feldspar of the high albite-sanidine series, quartz, pyroxene and garnet. The amount of plagioclase decreases, whereas that of garnet increases with pressure. The pyroxene is entirely or mainly ortho-pyroxene in the pressure range of the granulite facies, but becomes clino-pyroxene at higher pressures. With a further increase of pressure (30–35 kbar), plagioclase disappears, the K/Na ratio of the alkali feldspar increases, and the pyroxene becomes more abundant and higher in jadeite content. Coesite appears at pressures above 32 kbar. At 1100 °C, plagioclase is abundant in the pressure range of both the granulite and the eclogite facies. In other respects, the products are similar to those at 950 °C. The alkali feldspar becomes nearly pure K-end member at 40 kbar. The Na present is contained in pyroxene mainly as the jadeite component.

Thus, the anhydrous assemblages of granitic rocks in the granulite facies resemble those in the low-pressure part of the eclogite facies. They are both similar to assemblages at lower pressures except for the appearance of garnet by the reaction between orthopyroxene and An component.

When P_{H_2O} is high, biotite and/or hornblende may form instead of garnet and pyroxene. Granitic rocks recrystallized under such conditions could not be distinguished from ordinary granitic rocks, unless a detailed study of solid-solution minerals is made. A relatively high jadeite content of pyroxene and a low Fe^{2+}/Mg ratio of garnet, if any, would be diagnostic of the eclogite facies.

More or less similar relations should hold in psammitic and pelitic rocks in the eclogite facies. It is likely, therefore, that some quartzose gneissic rocks associated with eclogite but seemingly in the granulite, amphibolite and epidote-amphibolite facies are actually in the eclogite facies.

III Metamorphism and Crustal Evolution

Outline

Regional and ocean-floor metamorphism occur respectively in orogenic belts and mid-oceanic ridges. Since the P-T distribution during the metamorphism is recorded in the mineral assemblages of metamorphic rocks, petrologic investigation could give a clue to a better understanding of the tectonic processes taking place there. This aspect of the study of metamorphism will be emphasized here in Part III.

This part begins with chapters 13 and 14 where the role of regional metamorphism in the structure of continents will be discussed with reference to North America and Europe respectively, since relatively abundant data are available on these continents. In the following two chapters (chapters 15 and 16), the role of regional metamorphism in the structure of island arcs will be discussed with reference to Japan and other arcs in the Western Pacific Ocean. A long chapter is devoted to the Japanese Islands, partly because paired metamorphic belts, characteristic of the circum-Pacific region, are typically developed there, partly because very detailed petrologic, geologic and geophysical data are available, and partly because I am familar with the Islands.

Then, a summary of main features of metamorphic belts will be given and their tectonic significance will be discussed in chapter 17, with special emphasis on the problem of paired metamorphic belts.

In chapters 13–17, the lateral relations of areas of continental crust will be discussed, whereas the vertical zonation is treated in chapter 18.

In the last two chapters (chapters 19 and 20), our present knowledge of ocean-floor and transform-fault metamorphism will be outlined.

Chapter 13

Metamorphic Belts of North America

13-1 CONSTITUTION OF NORTH AMERICA

There is a wide Precambrian craton in the northern and central parts of North America. Precambrian granitic and metamorphic rocks are exposed in the northern part, which is called the *Canadian Shield,* whereas these rocks are covered by relatively thin, undisturbed, unmetamorphosed Phanerozoic sediments in the central part, which is usually called the *Interior Lowlands* (fig. 13-1).

The craton is bordered on the east by the Appalachian orogenic belt, where Palaeozoic orogenies caused deformation, regional metamorphism and plutonic intrusions.

The Cordilleran orogenic belt lies on the west side of the Shield and Interior Lowlands. There, orogenies took place mainly in Mesozoic and Cenozoic time. Late Mesozoic granitic intrusions took place on a gigantic scale. In both Appalachian and Cordilleran orogenic belts, geosynclinal volcanism involves not only basaltic but also no less abundant andesitic and rhyolitic activity (Butler and Ragland, 1969; Gilluly, 1965; Dickinson, 1962).

The Canadian Shield is bordered on the north by the Palaeozoic Innuitian orogenic belt.

Thus, the arrangement of orogenic belts suggests continental growth by accretion of younger orogenic belts. The oldest part of the Canadian Shield is in its central part. This suggests that continental growth occurred in Precambrian time also.

13-2 METAMORPHISM IN THE CANADIAN SHIELD

Orogenies and Provinces

Four major orogenies have been recognized through the radiometric dating of the Canadian Shield, and are summarized in table 13-1. The Shield may be subdivided into well-defined tectonic and radiometric provinces, each of which underwent a definite major orogeny (fig. 13-1). Such a major orogeny would include a number of events, each of which is comparable to one cycle of

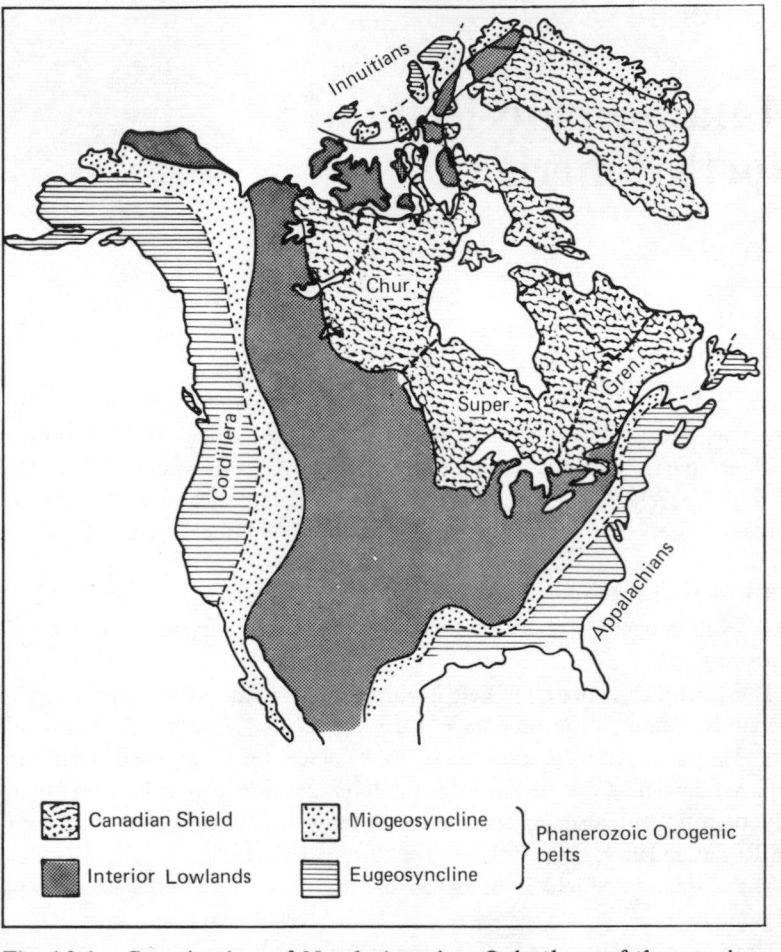

Fig. 13-1   Constitution of North America. Only three of the provinces of the Canadian Shield (Superior, Churchill and Grenville) are shown. (Modified from Kay, 1951, King, 1959, 1969, and others.)

Phanerozoic orogeny. The Grenville province, however, contains not only the metamorphic rocks derived from its own geosynclinal pile but also a great proportion of older crystalline rocks which were more or less modified by the Grenville orogeny. Most of the Appalachian region is underlain by a Precambrian basement consisting of extensions of the Grenville province. Hence, there may have been as great an area of continental crust in Precambrian time as there is at present. Conceivably, the radiometric age pattern may simply show the order of stabilization of provinces from the centre outwards, as claimed by Wynne-Edwards (1969).

Table 13-1 *Orogenies in the Canadian Shield*

Orogeny	Range (m.y.)	Provinces
Grenville orogeny	880–1000	Grenville
Elsonian orogeny	1280–1550	Nain
Hudsonian orogeny (called Penokean in Minnesota)	1640–1820	Churchill, Southern, Bear, Nain
Kenoran orogeny (called Algoman in Minnesota)	2390–2600	Superior, Slave

Note: Based on Stockwell (1964) and King (1969).

The oldest part of the Shield is the Superior province. However, there is some evidence suggesting the former existence of a continental crust that was subjected to orogeny at an earlier time than the Superior province. Graywackes in the Superior province contain quartz, K-feldspar, and other minerals which must have been derived from older granite-gneiss terranes. Moreover, zircons that may have crystallized 3400–3600 m.y. ago have been found in some Precambrian sediments. Hence, continent-forming processes appear to have begun at least 3600 m.y. ago in North America as in Europe and Africa. In North America, such old terranes appear to have been destroyed by later orogenies (Condie, 1967).

The Grenville and Superior Provinces

In spite of the great complexity of the Shield, petrographic studies have been made only in very small parts of it. It is impossible to give a balanced account of the metamorphism of the whole Shield.

The southern half of the Grenville province is relatively well populated and well investigated. It is characterized by the abundance of anorthosites and granulite facies metamorphic rocks, though amphibolite facies rocks are also widely exposed. Almandine is widespread. Areas of the amphibolite and the granulite facies are irregularly intermingled. Wynne-Edwards (1969) has ascribed this variation to metamorphism at similar temperature but varying P_{H_2O}.

The Adirondack Mountains in northern New York State are a southern protrusion of the Grenville exposures. The anorthosites and metamorphic rocks there are well documented (Buddington, 1969; Engel and Engel, 1958, 1960, 1962*a*, *b*; Isachsen, 1969).

It is to be noted that the Grenville province generally shows the highest degree of metamorphism, although it is the youngest of the provinces of the Shield (§ § 8A-7 and 18-2).

The Superior province appears to have been metamorphosed mainly in the greenschist to the low amphibolite facies. The original rocks of this province are mostly poorly sorted graywacke, conglomerate and volcanics of eugeosynclinal character in contrast to the Grenville province where pelite, quartzite and limestones are abundant.

The distribution of baric types in the Shield is not clear, except that no rocks of the glaucophane-schist facies have been found.

13-3 THE METAMORPHIC BELT OF THE APPALACHIANS

Orogenies

The Appalachian geosyncline contains sedimentary sequences ranging from the late Precambrian through the Palaeozoic. Its westernmost zone fringing the eastern margin of the Precambrian craton is miogeosynclinal, and only part of it was metamorphosed during the Palaeozoic orogenies. The region to the east is eugeosynclinal and has been widely metamorphosed, in places to a high degree. The miogeosynclinal zone at least has a Precambrian basement which is a southeastern extension of the Grenville province.

Probably there was a proto-Atlantic ocean between the North American and African continents in late Precambrian time. The two continents approached and then collided with each other to result in the Taconian and Acadian orogenic episodes in Palaeozoic time. The subsequent separation and drifting of the two continents created the present Atlantic Ocean (Wilson, 1966; Dietz and Holden, 1966).

The ages of the three main orogenic episodes in the Appalachians (Rodgers, 1967) are:

Taconian	Middle and Late Ordovician (450-500 m.y.)
Acadian	Middle Devonian (360-400 m.y.)
Alleghanian	Carboniferous and Permian (230-260 m.v.).

The most important phase of regional metamorphism and granite emplacement is the Acadian.

Northern Appalachians

The Appalachians are bisected roughly at the site of New York City into the northern and southern divisions. The structural zones of the northern Appalachians may be summarized as follows (Dietz and Holden, 1966; Zen, 1968; Rodgers, 1968, 1970; Cady, 1968; Bird and Dewey, 1970).

1. *Western miogeosynclinal zone.* This is the westernmost zone mainly composed of deformed and weakly metamorphosed lower Palaeozoic carbonate rocks (dolomite and limestone) accompanied by orthoquartzite. Within it occur

the Taconic and other allochthons consisting of lower Palaeozoic rocks that probably slid as submarine sheets from the east. Near the eastern limit of this zone, the Precambrian basement of Grenville age is exposed as a number of small massifs such as that of the Green Mountains.

The miogeosynclinal carbonate rocks probably represent a great carbonate bank on the Palaeozoic continental shelf somewhat resembling today's Bahama banks.

2. *Eugeosynclinal zone.* This lies to the east of the preceding zone, and is mainly composed of deformed and metamorphosed pelites and graywackes (late Precambrian to Devonian) accompanied by ultrabasic, basic and acidic rocks. It is conceivable that this zone was derived from the sedimentary piles of two continental rises. The western one would have formed in early Palaeozoic time on the east side of the contemporaneous shelf now represented by the western miogeosynclinal zone, whereas the eastern one would have formed in early and middle Palaeozoic time on the west side (oceanic side) of the then stable zone now represented by the very late Precambrian zone (see below).

3. *Very late Precambrian zone.* The easternmost terrane in the Appalachians was formed by deposition followed by orogeny in very late Precambrian time. It is now exposed in areas along the Atlantic coast in Rhode Island, Nova Scotia and Newfoundland. The associated granitic rocks give radiometric ages of about 580 m.y.

Thus, the Appalachian structure was conceivably formed by repeated expansion and contraction of an ocean (Proto-Atlantic Ocean) between the North American and the African plate in a period ranging from very late Precambrian to the end of Palaeozoic time. The very late Precambrian zone, mentioned above, is the oldest part, and was probably formed originally along the continental margin of Africa. In the next stage (i.e. Taconian time), the western miogeosynclinal zone and the western part of the eugeosynclinal zone, mentioned above, appear to have become the main theatre of orogeny and metamorphism. The Acadian orogeny would have affected most of the region, though the metamorphism of a new continental rise deposit in the eastern part of the eugeosynclinal zone is particularly important.

The metamorphic rocks of the northern Appalachians have been well documented (e.g. Barth, 1936, Zen, 1960; Thompson and Norton, 1968; Albee, 1968). As was shown in fig. 7A-6, there are three thermal axes parallel to the elongation of the orogenic belt, and the eastern part of the metamorphic terrane belongs to the low-pressure type, whereas the western part belongs to the medium-pressure type. These two regions may or may not represent extensions of the Caledonian and Hercynian belts in Europe (§ 14-6).

The high-temperature zones are closely associated with masses of the Devonian New Hampshire plutonic series, which may have given strong thermal effects.

Small parts of the metamorphic terrane are younger than Acadian. For example, the regional metamorphism in the Narragansett basin around Providence, Rhode Island, affected Pennsylvanian rocks.

13-4 METAMORPHIC BELTS IN THE CORDILLERAN MOUNTAINS

General Features

The Cordilleran Mountains, extending from the Rocky Mountains to the Pacific Coast, border the craton of the Canadian Shield and Interior Lowlands on the west side. Being as wide as 1500 km, the region includes units of different tectonic and petrographic characteristics. Geosynclinal deposition ranged from late Precambrian to late Mesozoic time with an orogenic culmination in late Mesozoic. The easternmost zone (Rocky Mountains and Colorado Plateau) represents a reactivated Precambrian craton and was not in the pertinent geosyncline. The eastern limit of the geosyncline lies on the western edge of the Colorado Plateau. The geosyncline may be divided into a miogeosynclinal and a eugeosynclinal zone (fig. 13-1).

The Nevadian and Coast Range orogenies, Jurassic to Cretaceous in age, caused deformation, metamorphism and plutonism in the eugeosynclinal zone, whereas the Laramide orogeny, Cretaceous to Tertiary, deformed the miogeosynclinal zone.

At a first glance, the Cordilleran and the Appalachian Mountains are symmetrical with regard to the craton between them. However, the western part of the Cordilleran System has many features that are not present in the Appalachians but are commonly present in circum-Pacific regions. From the viewpoint of metamorphic petrology, the most marked feature is the existence of paired metamorphic belts and of extensive glaucophane-schist terranes in the western Cordillera.

Two belts of metamorphism occur there in a pair. The eastern belt is characterized by gigantic batholiths of granitic rocks in the Sierra Nevada and other places, as shown in fig. 13-2. The western belt is characterized by glaucophane schists of the Franciscan group in the Coast Ranges of California.

In most of the circum-Pacific region, regional metamorphism and intense acidic magmatism took place in late Mesozoic time. The magmatism involved not only the formation of granitic batholiths but also acidic volcanism. The emplacement of gigantic batholiths and associated volcanism in western North America are probably a part of this circum-Pacific event.

It is to be noted that glaucophane schists in western North America are not confined to the Franciscan. In northern Washington State, a belt of Shuksan metamorphic rocks occasionally containing glaucophane extends southeastward from near Bellingham for about 200 km, ultimately disappearing under the Tertiary volcanics of the Cascade Range. The metamorphism appears to be latest

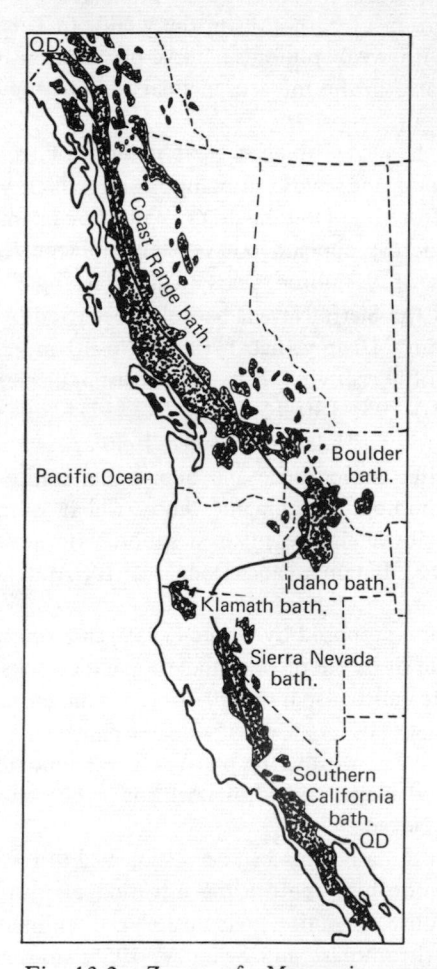

Fig. 13-2 Zone of Mesozoic granitic batholiths in western North America. The names of batholiths are shown. Line QD–QD represents the quartz–diorite line of Moore *et al.* (1961).

Palaeozoic or earliest Mesozoic, i.e. older than the Franciscan. Moreover, this high-pressure belt is in a pair with a metamorphic terrane accompanied by granitic rocks on the east (Misch, 1966). A few more localities of glaucophane schists are known in the Canadian part of the Cordillera (Monger and Hutchinson, 1971).

Mesozoic Granitic Batholiths

In the western United States, andesitic and rhyolitic volcanism was widespread in Palaeozoic and Triassic time, but few granitic plutons were emplaced in this period. Extensive plutonism began in the Jurassic and culminated in the Cretaceous producing the greatest group of batholiths in the world. In Tertiary time, acidic volcanism associated with weak plutonism took place extensively not only in the eugeosynclinal but also in the miogeosynclinal zone to the east (Gilluly, 1965; Dickinson, 1962).

Each of the gigantic batholiths is made up of a great number of smaller plutons, ultrabasic, basic, intermediate and acidic in composition, which were intruded over a considerable length of time (Buddington, 1959). For example, the Boulder batholith, about 70 km across, appears to have been emplaced 78–69 m.y. ago, that is, in a time span of 9 million years (Tilling, Klepper and Obradovich, 1968). The intrusion of the Sierra Nevada batholith occurred over a much longer time span, ranging from 210 m.y. ago (Triassic) to 80 m.y. ago (Cretaceous). Five epochs of granitic activity have been distinguished in California and western Nevada: 210–195, 180–160, 148–132, 121–104, and 90–79 m.y. (Bateman and Eaton, 1967; Evernden and Kistler, 1970).

In these batholiths, quartz diorite, granodiorite and quartz monzonite are most abundant. Granite in the common petrographic sense, which is more potassic than the rocks mentioned above, is present in a subordinate amount (Larsen, 1948; Bateman et al. 1963; Bateman and Dodge, 1970; Anderson, 1952).

In figs. 13-2, the quartz–diorite line proposed by Moore (1959) and Moore et al. (1961) is shown. On the west side of it, the most abundant granitic rocks are quartz diorite [(K-feldspar + perthite)/all feldspars = 0·0–0·1]. On the east side of the line, more potassic granitic rocks (i.e. granodiorite, quartz monzonite and granite proper) are predominant. In other words, the plutonic rocks tend to be more potassic and more acidic toward the interior of the continent. The crust is as thick as 50 km beneath the Sierra Nevada.

Wide regions in the zone of granitic batholiths had been subjected to low- or medium-temperature regional metamorphism before the intrusion of plutons. Stilpnomelane, pumpellyite, staurolite, kyanite, andalusite and sillimanite formed in some schists (e.g. Eric, Stromquist and Swinney, 1955; Compton, 1955; Hietanen, 1961, 1962; Schwarcz, 1969). The contact effects of the plutons were superposed on, and in some areas are indistinguishable from, regional metamorphism.

In the Sierra Nevada, the ages of the deposition of wall rocks range at least from early Palaeozoic to Jurassic. Regional metamorphism is usually in the greenschist facies. Erosion has not advanced enough to expose high temperature regional metamorphic complexes. The presence of great batholiths does not mean the exposure of deep crustal levels. The granitic bodies do not necessarily become bigger with increasing depth. Seismic and gravity data suggest that granitic rocks now exposed on the Sierra Nevada appear to continue only to a depth of about 8 km, below which denser rocks such as gneisses and amphibolites are present (Hamilton and Myers, 1967).

Great zones of serpentinite occur in the Sierra Nevada.

Coast Ranges of California

The Franciscan rocks are eugeosynclinal deposits exposed in the Coast Ranges of California (fig. 13-3). They were probably deposited on the basaltic crust of a deep ocean floor from Late Jurassic to Late Cretaceous time. The total thickness of the accumulations would be 15-20 km. The predominant rock type is graywacke, though basic volcanics, shale, chert and limestone are also included. Abundant serpentinite occurs, but there are no granitic rocks (Bailey *et al.* 1964).

The Coast Ranges are separated from the Sierra Nevada by the Great Valley. A thick sedimentary pile, called the *Great Valley sequence,* is well exposed on the western part of the Valley, though not confined there. It was deposited on the continental shelf and slope at the same time as the Franciscan. In northern California, the Franciscan terrane is thrust partly under the granitic and associated metamorphic terrane of the Klamath Mountains and partly under the Great Valley sequence to the east. The thrust fault, called the *Coast Range thrust,* is commonly at a high angle, dipping eastward (fig. 13-3). Serpentinite masses occur along it (Ernst, 1970; Bailey, Blake and Jones, 1970). The Great Valley sequence is mostly unmetamorphosed but its lower parts show incipient burial metamorphism (Dickinson, Ojakangas and Stewart, 1969).

In southern California, the distribution of rocks has been strongly disturbed by the San Andreas and other strike-slip faults (§ 20-2). Granitic rocks of the zone of batholiths have been cut by the San Andreas and the Nacimiento fault, and the intervening area has been moved northwestward into the Coast Ranges (fig. 13-3).

In the Franciscan, recrystallized and unrecrystallized areas seem to be intermingled. Probably the $P-T$ conditions of metamorphism prevailed on a regional scale, but because of the low temperature, recrystallization took place only in favourable areas. The northwestern zone of the Franciscan is more or less metamorphosed in the zeolite facies with laumontite, whereas the eastern zone is metamorphosed in the glaucophane-schist facies (Bailey *et al.* 1964; Blake, Irwin and Coleman, 1969; Ernst, 1971*a*). The rocks are commonly non-schistose, preserving their original clastic texture.

During the deposition of the Franciscan, the granitic rocks in the Sierra Nevada and Klamath Mountains were continually intruded and some plutons were exposed. K-feldspar and other minerals in the granitic rocks were transported and deposited to enter the sediments of the Great Valley sequences and

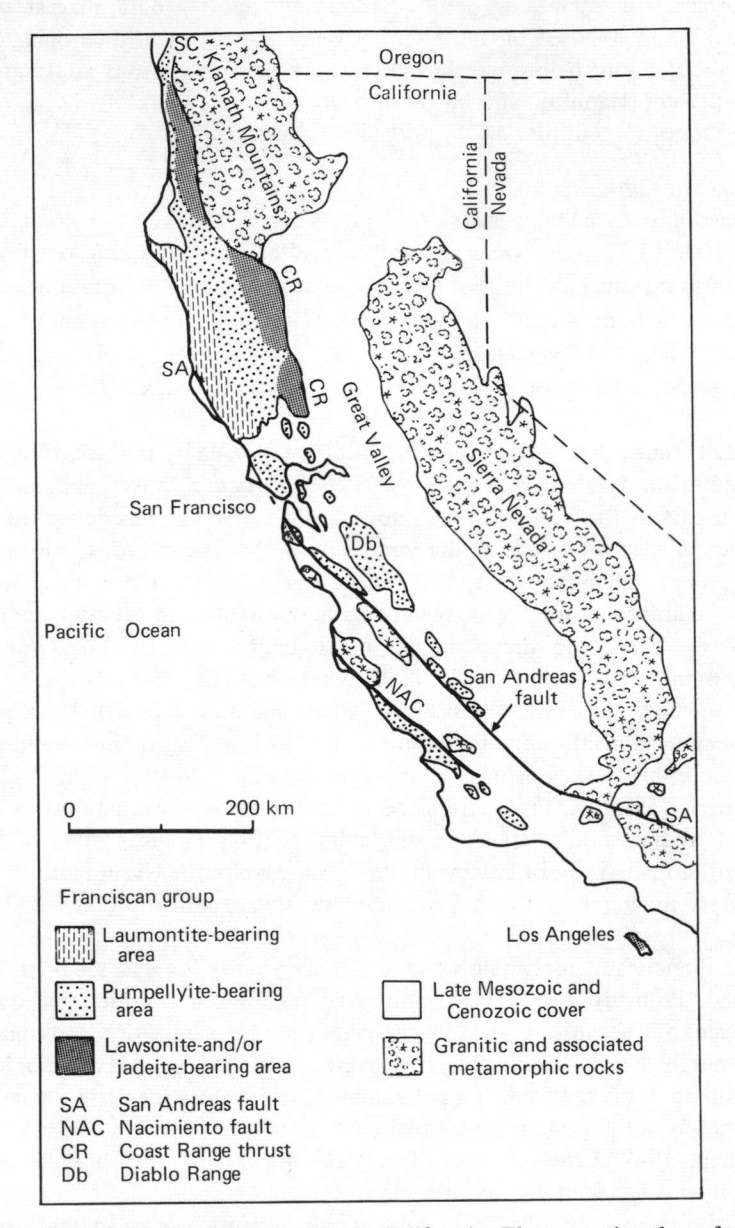

Fig. 13-3   Franciscan rocks in California. The tentative three-fold division shown is mostly based on Bailey *et al.* (1964), Blake *et al.* (1969) and Coleman and Lanphere (1971).

the Franciscan. The western part of the Franciscan is generally younger than the eastern.

Suppe (1969) has shown the existence of two different metamorphic events about 150 m.y. ago (Late Jurassic) and 127–104 m.y. ago (Cretaceous), together with a still younger undated one. This suggests that an early phase of metamorphism took place just after the deposition of older parts of the Franciscan. Then deposition occurred in other parts of the Franciscan and a later phase of metamorphism took place there. Such a sequence of events took place three times or more. It is not clear whether these metamorphic events have a direct genetic connection with the five epochs of granitic activity in California and Nevada mentioned above.

Chapter 14

Metamorphic Belts of Europe

14-1 CONSTITUTION OF EUROPE

There is a wide Precambrian craton, called the *East European Platform* (or Fennosarmatia), extending from the Ural Mountains to central England (Bogdanoff, Mouratov and Schatsky, 1964). Precambrian granitic and associated metamorphic rocks are widely exposed in the north-western part of the Platform, called the *Baltic Shield* (or *Fennoscandia*). Smaller exposures occur in the Ukrainian massif (fig. 14-1). The rest of the Platform is covered by a nearly horizontal layer, commonly 1–3 km thick, of unmetamorphosed sediments ranging from late Precambrian to Cenozoic age. This region is called the *Russian Plate.*

The East European Platform is surrounded on all sides by Phanerozoic orogenic belts. It is bordered on the northwest side by the Caledonian orogenic belt which runs from the British Isles to Norway and west Sweden, and on the south side by the Hercynian orogenic belt and further to the south by the Alpine orogenic belt. The Ural Mountains are on the east side of the Platform.

14-2 METAMORPHISM IN THE BALTIC SHIELD

Age Relations

In the Baltic Shield, the eastern part in the Kola Peninsula and on the west coast of the White Sea (KSB in fig. 14-1) is the oldest terrane, including the Katarchean (3500–3000 m.y.), Saamides (2900–2100 m.y.) and Belomorides (2100–1950 m.y.). On the west side of this region are younger terranes called the *Karelides* and the *Svecofennides* (both 1900–1600 m.y.). The youngest terranes, called the Gothides (1400–1200 m.y.) and the Ripheides (1100–665 m.y.), are further west (fig. 14-1). Thus, the exposed Precambrian becomes younger westwards.

In the East European Platform on the whole, however, the older Precambrian terranes (more than 1600 m.y. old) extend from the eastern three-quarters of the Baltic Shield to the Ukrainian massif. Younger Precambrian terranes are arranged generally surrounding the older (Polkanov and Gerling, 1960), and Phanerozoic orogenic belts surround the whole Platform. Thus, the age pattern is

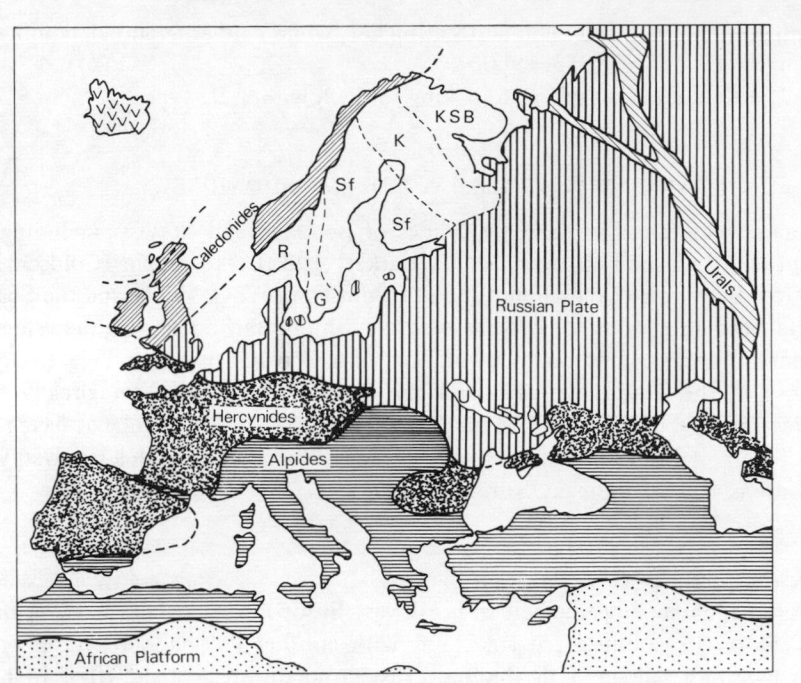

Fig. 14-1   Constitution of Europe. The Baltic Shield and Ukrainian Massif are shown in white, where KSB = Katarchean, Saamides and Belomorides. *Abbreviations:* K = Karelides, Sf = Svecofennides, G = Gothides, R = Ripheides, U = Ukrainian Massif.

more or less concentric, suggesting the possibility of continental growth by accretion.

Svecofennides

The Svecofennides lie in central Sweden and southwestern Finland. The terrane is metamorphosed mostly at relatively high temperatures (amphibolite facies) and is accompanied by an enormous amount of granitic rocks (Geijer, 1963; Eskola, 1963). Because of these characteristics, it was long regarded erroneously as a part of the oldest rocks on earth. Geologic investigations in various areas of the terrane played important roles in the progress of Precambrian geology and geology in general. Above all, Sederholm's (1907, 1923, 1926, 1932) studies of granites and migmatites, and Eskola's (1914, 1915, 1932) studies of the Orijärvi region and granites are monumental in our science (§ 4-1).

Though amphibolite facies rocks are the most widespread, there are some greenschist facies and granulite facies rocks (e.g. Parras, 1958). The metamorphism belongs to the low-pressure facies series. Andalusite occurs in lower temperature parts and sillimanite in higher, while cordierite and almandine are common (Simonen, 1953, 1960b).

The Karelides also appear to belong to the low-pressure type.

14-3 METAMORPHIC BELTS IN THE CALEDONIDES

Metamorphic areas in the Caledonides of Scotland and Norway, including the Scottish Highlands (Barrow, 1912; Harker, 1932), Trondheim (Goldschmidt, 1915), Stavanger (Goldschmidt, 1921), Sulitjelma (Vogt, 1927), and the Sparagmite Formation (Barth, 1938), played very important roles during the establishment of metamorphic petrology before World War II. The classical metamorphic zones in the Trondheim area and the Scottish Highlands have already been shown in figs. 1-2 and 3-1. In the past decade, the Scottish Highlands have again become a centre of a trend of modernized field studies through a renewed wave of intensive work by many British geologists.

Tectonic and Age Relations

The Caledonian orogenic belt in northwest Europe is in contact with the Baltic Shield on its southeastern side, and with another Precambrian craton on its northwestern side along the Moine thrust in northwest Scotland (figs. 14-1 and 7A-10). The latter craton is made up of the Lewisian complex, which before the Mesozoic continental drift was probably in contact with the Precambrian of Greenland, representing a northeastern part of the original Canadian Shield (fig. 14-4). Three orogenic-metamorphic events were distinguished in the Lewisian complex: Scourian (2600–2460 m.y. ago), Inverian (2200–2000 m.y. ago), and Laxfordian (1650–1450 m.y. ago). The rocks exposed are mostly granites and gneisses commonly in the amphibolite and the granulite facies (Sutton and Watson, 1969; Bowes, 1968; O'Hara, 1961).

The northern two-thirds of Great Britain belongs to the Caledonides. The region to the north of the *Highland Boundary fault* (H in fig. 7A-10) is the metamorphosed *Scottish Highlands*, which is bisected by a big strike-slip fault called the Great Glen fault (G in fig. 7A-10). The region between the Highland Boundary and Great Glen faults is called the *Grampian Highlands* where the metamorphosed *Dalradian and Moine Series* are exposed. The latter extends to the north of the Great Glen fault up to the Moine thrust, and is composed of metamorphosed psammites and pelites deposited in late Precambrian time. The Dalradian Series is composed of metamorphosed pelites and psammites, accompanied by limestone and basic volcanics, which were deposited in late Precambrian and Cambrian time.

The Highland Boundary fault was possibly initiated in late Cambrian time and movement continued along it throughout the Caledonian orogeny. The Great

Glen fault is said to involve a strike-slip movement of the order of 100 km, which offsets the metamorphic zones on the opposite sides (Kennedy, 1948).

The Caledonian deformation and regional metamorphism in the Scottish Highlands and in the northernmost part of the Southern Uplands of Scotland (down to the Southern Uplands fault) took place probably between 510 and 490 m.y. ago (mainly Arenigian, i.e. early Ordovician), whereas the southern Caledonides (south of the Southern Uplands fault) were deformed in late Silurian and early Devonian time, though only slightly metamorphosed. Post-climactic slow uplift and erosion took place in the Scottish Highlands throughout late Ordovician, Silurian and early Devonian times. The K–Ar mica ages from the Scottish Highlands range generally from 500 to 400 m.y., and probably represent the time of cooling due to the slow uplift and erosion (Dewey and Pankhurst, 1970).

Metamorphism in the Scottish Highlands
Most of the Caledonian metamorphic belt in Scotland and Norway appears to belong to the medium-pressure facies series. In the Grampian Highlands, however, a low-pressure metamorphic region, called the *Buchan region*, has been clearly distinguished from a medium-pressure region, called the *Barrovian region*, as was shown in fig. 7A-10. There is no marked tectonic line between the two regions, but the difference is simply a result of difference in rock-pressure (Johnson, 1963; Chinner, 1966a). The area investigated by Barrow (§ 3-1) is near the southeastern corner of the Barrovian region.

The distribution of progressive metamorphic zones in the Grampian Highlands is complicated. The sillimanite zone is confined to a relatively small area straddling the Barrovian and Buchan regions. Migmatites are abundant there. Chinner (1966a) has suggested that the generation of sillimanite and migmatite is due to a localized temperature rise which took place at a stage later than the formation of all the other progressive metamorphic zones due to depth-controlled general P-T distribution.

Harte and Johnson (1969) determined the time relation between the phases of deformation and the formation of metamorphic minerals in Barrow's classical area. During the 1st and 2nd phases of deformation, the temperature of metamorphism was low. Then, the temperature increased gradually, and this led to the formation of garnet, staurolite and kyanite between the 2nd and 3rd phases. The highest temperature was reached in a static period between the 3rd and 4th phases of deformation, leading to the generation of sillimanite. Then the temperature decreased to cause some retrogressive changes.

Two groups of granitic rocks, Older and Newer, have been distinguished in the Scottish Caledonides (§ 4-1). The Older Granites were broadly syntectonic and simultaneous with the above regional metamorphism that occurred about 490–500 m.y. ago (Ordovician). They appear to be of metamorphic origin. The

Newer Granites are post-tectonic and 415–390 m.y. old (Late Silurian–Early Devonian).

Southern Uplands of Scotland

Basaltic magmatism of the Caledonian geosyncline was more intense in the Southern Uplands of Scotland (fig. 7A-10) than in the Scottish Highlands (Wiseman, 1934; Read, 1961). Most of the Southern Uplands were only slightly recrystallized, but Ordovician rocks in the Ballantrae area have been found to be metamorphosed in the glaucophane-schist facies. There are associated eclogites and serpentinites (Bloxam and Allen, 1960).

The medium-pressure metamorphic terrane of the Scottish Highlands and the high-pressure terrane of the Southern Uplands could be an example of paired metamorphic belts, though recrystallization in the high-pressure terrane is very limited (Miyashiro, 1967b, p. 427; Dewey and Pankhurst, 1970, p. 362). If this is the case, the Benioff zone in early Ordovician (Arenigian) time would have plunged northwards into the mantle from a trench zone in the position of the Southern Uplands. In later Ordovician time, a new Benioff zone plunging southwards from the trench zone in the same position would have formed, because the middle and late Ordovician volcanism in the Lake district and Wales to the south of the Southern Uplands is mainly of andesitic and rhyolitic compositions and strongly suggests the existence of an island arc (Fitton and Hughes, 1970).

14-4 METAMORPHIC BELTS IN THE WEST EUROPEAN HERCYNIDES

The Hercynides in West Europe were formed between the East European Platform and the African Platform (fig. 14-1). Regional metamorphism took place in wide areas in association with abundant granitic intrusions (table 4-1). As shown in fig. 14-2, they are widely exposed in the so-called Moldanubian zone (including the Bohemian massif, Black Forest, Vosges, and Massif Central), the Pyrenees, and the Ibero–Hesperian zone (western Spain and Portugal).

The regional metamorphism belongs to the low-pressure series in most regions (Zwart, 1967a). Ophiolites are rare and serpentinite is nearly lacking. The radiometric age of metamorphism is about 360–260 m.y. (Late Devonian to Early Permian).

A large number of areas in the Pyrenees and the Ibero–Hesperian zone were described in recent years by Dutch and French geologists (§ 7A-4 and 8A-3).

Most of the Hercynian orogenic belt appears to have older (possibly Precambrian) granitic and metamorphic basements. Hercynian and pre-Hercynian metamorphic rocks are exposed more or less intermingled. The pre-Hercynian basements sometimes contain kyanite, suggesting that they belong to the medium-pressure facies series. In the basements in some areas, eclogites have been found, some of which at least belong to the strictly defined eclogite facies.

L'Ile de Groix off the south coast of Brittany is exceptional in having glaucophane schists belonging to the Hercynides (Cogné, Jeanette and Ruhland, 1966). The significance of this occurrence is not clear.

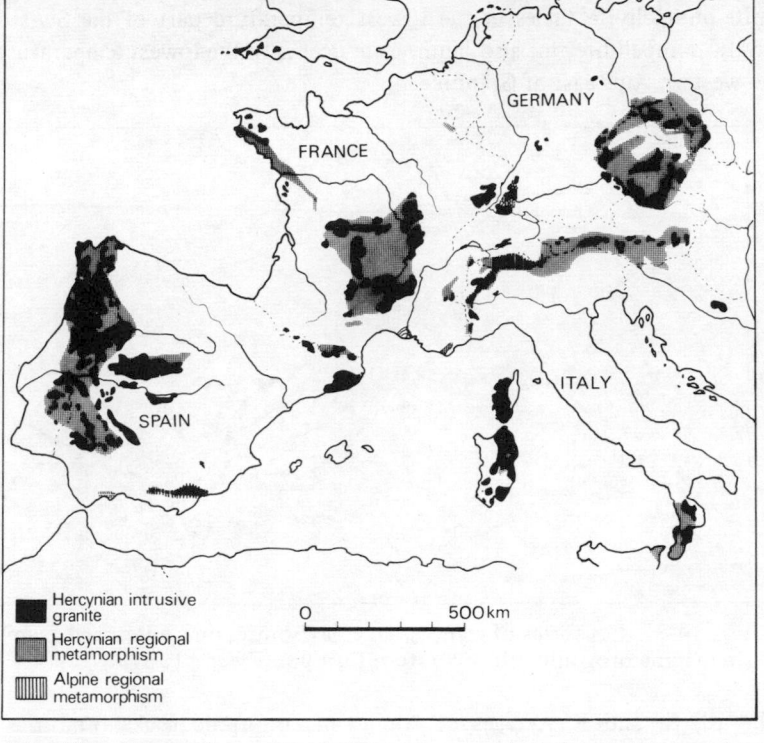

Fig. 14-2   Hercynian and Alpine metamorphic belts in Western Europe. (Zwart, 1967*b*.)

14-5   METAMORPHIC BELTS IN THE WEST EUROPEAN ALPIDES

The Alpine orogenic belt in West Europe was formed generally on the south side of, but largely superposed on, the Hercynides. It shows a sinuous pattern, forming several arcs, for example, in the western Alps. Because of its arcuate pattern, it covers a considerable area, but the individual constituent belts are relatively narrow in contrast to the very wide Hercynides (table 4-1).

Regional metamorphic rocks are widely exposed in the Betic Cordillera in southern Spain (de Roever and Nijhuis, 1964), eastern Corsica (Egeler, 1956), Western and Pennine Alps (Bearth, 1962; Niggli and Niggli, 1965; Niggli, 1970;

Bocquet, 1971), western Balkan Peninsula and islands in the Aegean Sea (van der Plas, 1959). They are of the high-pressure series and mostly in the glaucophane and the greenschist facies (fig. 14-3). In the Swiss Alps, however, the temperature of metamorphism increases southwards, reaching a grade high enough to produce sillimanite (amphibolite facies) in the Lepontine Alps straddling Switzerland and Italy (§ 7A-8 and fig. 7A-11). There is a zone of the prehnite-pumpellyite facies in the lowest temperature part of the Swiss Alps. Not only pumpellyite but also laumontite occurs in the lowest temperature part of the western Alps east of Grenoble.

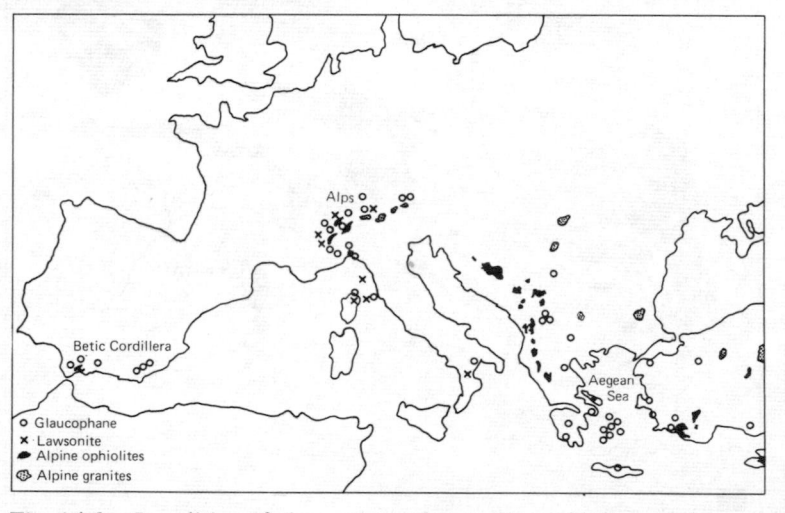

Fig. 14-3   Localities of glaucophane, lawsonite, ophiolites, and granites in the Alpine orogenic belt in Western Europe. (Zwart, 1967*b*.)

The Rb–Sr and K–Ar ages of Alpine metamorphic rocks from the high-temperature part of the Swiss and Italian Alps range from 30 to 11 m.y. depending on their cooling history. The climax of the last phase of the Alpine metamorphism occurred at least 30 m.y. ago, i.e. in Oligocene time (Armstrong, Jaeger and Eberhardt, 1966).

The Alpine belt has older granitic and metamorphic basements. In the Swiss Alps, the Hercynian basement is exposed in wide areas of the Aiguilles-Rouges, Aar, and Gotthard massifs (fig. 7A-11).

The Alpine metamorphic terranes have abundant ophiolites including serpentinite. The Alps are connected eastwards to the Carpathians, though high-pressure metamorphic rocks are not exposed in the latter. A belt of late Tertiary and Quaternary andesite and rhyolite exists in Hungary, southern Slovakia and Transylvania on the south side of the Alpine–Carpathian zone (Pantó, 1968; Karolus, Forgáč and Konečný, 1968). This belt is accompanied by late Tertiary granitic masses, and has unusually high heat-flow values (Boldizsár, 1964; Lee and Uyeda, 1965). The Alpine granites of Bergell and Adamello in southern Switzerland and northern Italy probably belong to this belt, though the former has been intruded into the southernmost part of the Pennine zone. It is likely that this belt of granitic plutons and andesitic volcanism has a hidden low-pressure metamorphic complex in a depth, which is paired with the high-pressure metamorphic complex now exposed in the Pennine zone of the Alps (cf. § 17-2; Miyashiro, 1972b). The descending slab at that time should have been inclined to the south. The recrystallization temperature in the Pennine zone increases southwards, i.e. towards the associated belt of granitic plutons, in the same way as in the Sanbagawa and Franciscan terranes (Ernst, 1971b).

14-6 CORRELATION AND BASEMENTS OF METAMORPHIC BELTS IN THE ATLANTIC REGIONS

Pre-Drift Reconstruction

The orogenic belts in North America and those in Europe show a roughly concentric pattern of radiometric ages, suggesting accretional growth of each continent. Since the present Atlantic Ocean has been formed by continental drift in Mesozoic and Cenozoic time, however, we have to discuss the relationship between the older orogenic-metamorphic belts of North America and Europe.

The pre-drift reconstruction of orogenic belts across the North Atlantic Ocean has been attempted in recent years by Bullard, Everett and Smith (1965) on the basis of submarine topography, by Miller (1965), Fitch (1965) and Wynne-Edwards and Hasan (1970) largely on the basis of radiometric age data, and by Dewey and Kay (1968) and Kay (1969) with special reference to the geologic similarity of Newfoundland with the British Isles. There is general agreement in major points of their interpretations. A tentative reconstruction based on the age relations and metamorphism is shown in fig. 14-4.

The original shapes and mutual relations of orogenic belts older than the Grenville are quite unknown, since they have been too strongly disturbed by later orogenies. The Grenville province in Canada is here correlated tentatively with the Gothides and Ripheides in the Baltic Shield because of the similarity of their ages. The Taconian (Ordovician) and Acadian (Devonian) of the Appalachians are correlated respectively with the Caledonian and Hercynian belts in Europe (§ 13-3).

The Grenville–Riphian, Caledonian–Taconian, and Hercynian–Acadian orogenic belts run across the north Atlantic, but the Caledonian–Taconian belt cuts across the Grenville–Riphian.

It appears that Phanerozoic orogenic belts in the north Atlantic region were formed by mutual movements of the Precambrian continental plates of North

America, Europe and Africa. The Appalachians would have formed by approach and collision of the North American with the African plate. The Caledonian belt of Europe was probably formed by approach and collision of the European with the North American plate (e.g. Dewey, 1969).

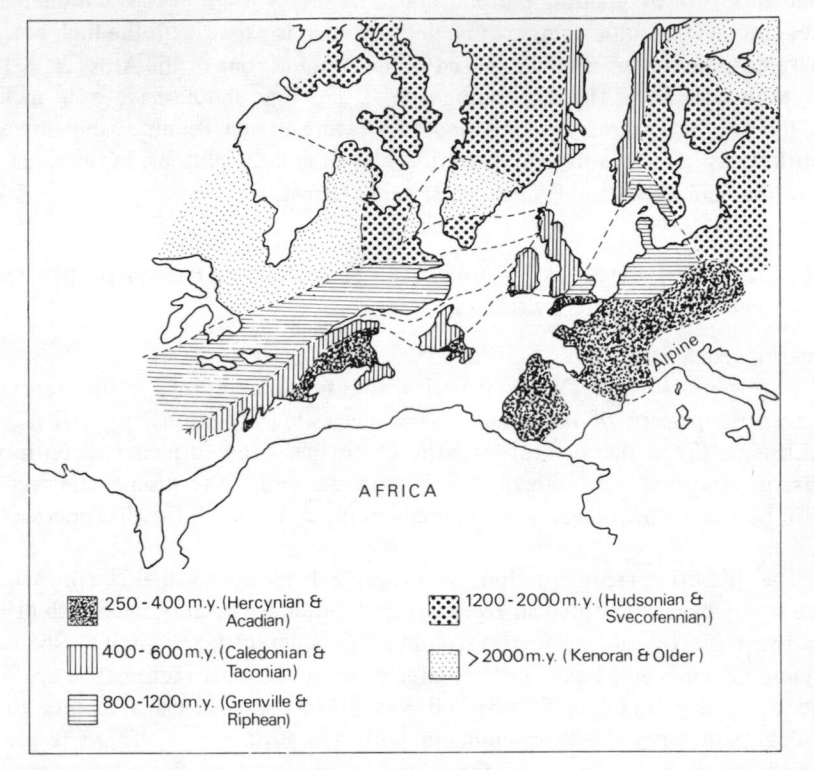

250-400 m.y. (Hercynian & Acadian)

400-600 m.y. (Caledonian & Taconian)

800-1200 m.y. (Grenville & Riphean)

1200-2000 m.y. (Hudsonian & Svecofennian)

>2000 m.y. (Kenoran & Older)

Fig. 14-4 Tentative pre-drift reconstruction of orogenic belts in the North Atlantic region. The fits of continents are based on Bullard *et al.* (1965).

Old Acidic Basements

Large parts of the orogenic belts of the Atlantic regions appear to have formed on older acidic basements, which are exposed, for example, along the west coast of Norway in the Caledonides, in many parts of the Hercynides (§ 14-4), and along the Green Mountains in the Appalachians. Such evidence was regarded by many authors as being against the hypothesis of continental growth (e.g. Wynne-Edwards and Hasan, 1970; Semenenko, 1970). It was claimed that pre-existing acidic materials were eroded and then deposited on other parts of pre-existing basements, and metamorphosed or melted, and that the total area of continents and the total amount of acidic materials on the surface of the earth did not increase by orogenies.

However, such evidence is not necessarily against a secular increase in the amount of acidic material on the surface of the earth. Acidic crusts may have become generally thicker with geologic time. It is likely that parts of orogenic belts in the Atlantic and circum-Pacific regions do not have older acidic basements, and acidic crusts have been enlarged in such places. It appears that the relations observed in the orogenic belts in the Atlantic regions are not valid for all other orogenic belts of the world (Miyashiro, 1967a).

Probably the rate of plate motion in Phanerozoic time was much slower in the Atlantic regions than in the Pacific. This could be the main cause for the rarity of glaucophane schists and paired metamorphic belts in the Atlantic regions (§ 17-2).

Chapter 15

Metamorphic Belts in Japan and its Environs

15-1 PRESENT-DAY ARCS OF THE JAPANESE ISLANDS

Island Arc Structure

A brief account of the present-day tectonics of the Japanese Islands will be given here as a preliminary to detailed discussions of their Mesozoic and older metamorphic belts. The major topography of the Islands is the result of late Tertiary and Quaternary earth movements. It appears, however, that the movements were largely controlled by the thickness of the pre-existing crust which formed by Mesozoic and older orogenies. The Japanese Islands have a thick continental-type crust. The highest area of Japan is the central mountains (called the Japan Alps) of Honshu. Their present height was produced by continual upheaval in the Tertiary and Quaternary. The Quaternary upheaval amounts to 1200 m. On the other hand, this area has the thickest crust in Japan, and the upheaval was controlled probably by the buoyancy of the crust formed by the older orogenies. Therefore, the general shapes of the present-day arcs reflect the features of older orogenic and metamorphic belts.

Japan is composed of four big islands, called, from northeast to southwest, Hokkaido, Honshu (main island), Shikoku and Kyushu, together with a great number of surrounding small islands. Fig. 15-1 shows the present-day island arcs of Japan and its vicinity.

The main arc of Japan may be called the *Honshu Arc,* the trend of which generally represents the direction of Mesozoic orogenic belts. As regards the present-day arc structure, however, the Honshu Arc must be divided into the *Southwest Japan Arc* and the *Northeast Japan Arc*, as shown in the figure. Only the latter is an active island arc at the present day. The Southwest Japan Arc was active during the Cretaceous and older orogenies. Along the axis of this Arc, there is a great fault, called the *Median Tectonic Line* (MTL in fig. 15-1).

The Northeast Japan Arc is in contact on its southern end with another presently active arc, called the *Izu-Bonin Arc*, which protrudes towards the SSE into the Pacific Ocean. There is a great fault between the presently active regions (Northeast Japan and Izu-Bonin Arcs) on the east and the more stabilized

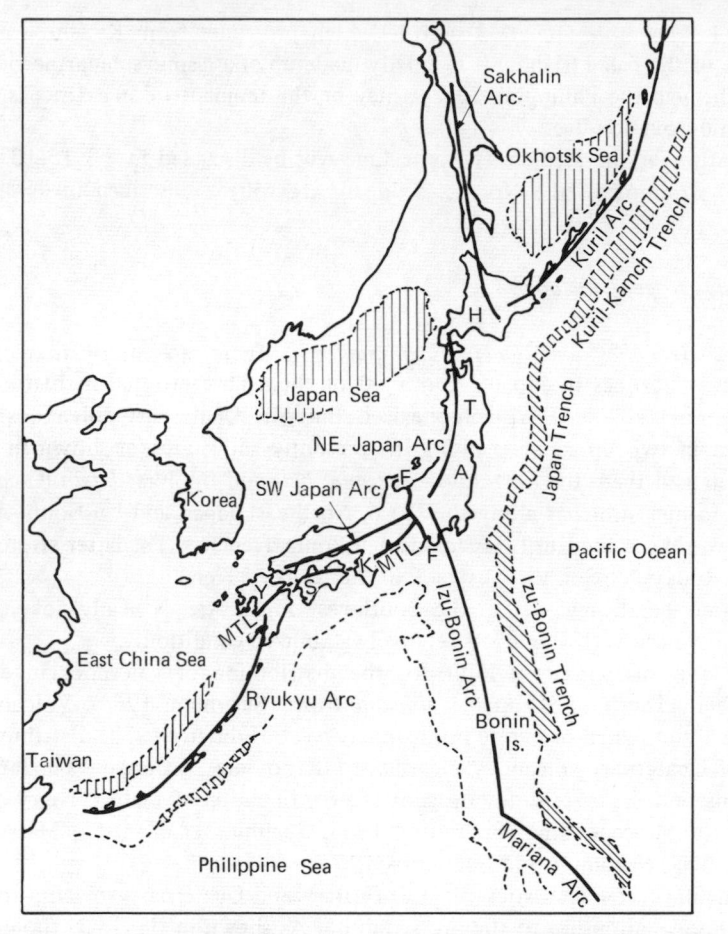

Fig. 15-1 Present-day island arcs in and around the Japanese Islands. Slant ruling indicates trenches, and vertical ruling indicates deep basins in marginal seas. *Abbreviations:* MTL = Median Tectonic Line, F-F = Itoigawa-Shizuoka Line, H = Hokkaido, T = Kitakami Mountains, A = Abukuma Plateau, K = Kii Peninsula, S = Shikoku, Y = Kyushu. The main island containing T, A and K is called Honshu.

Southwest Japan Arc on the west (F-F in fig. 15-1). This fault runs in a NNW direction through the city of Shizuoka across Honshu, and is called the *Itoigawa–Shizuoka Line* after the names of the two cities at both ends. The region immediately east of the Line is called the *Fossa Magna* (fig. 15-8).

The southwestern part of Hokkaido is an extension of the Northeast Japan Arc, whereas the rest of Hokkaido is made up of the *Sakhalin and Kuril Arcs*. The *Ryukyu Arc* connects Kyushu to Taiwan (Formosa).

There is one more arcuate structure which protrudes from Kyushu southeastwards to the Palau Islands. It is mostly made up of aseismic submarine ridges running through the Philippine Sea. It may be the remnant of an extinct island arc or a mid-oceanic ridge.

The nature of the Median Tectonic Line will be discussed in §§ 15-10 and 17-6. The structure of the Northeast Japan Arc will be described in detail in § 17-7.

Present-day Orogeny in Japan

Volcanoes, Trenches and Deep Earthquakes. There are more than 200 Quaternary volcanoes in Japan. About 40 of them have erupted in historical time. Sugimura (1960, 1968) emphasized that the Quaternary volcanoes are distributed in two broad belts along the presently active arcs, as shown in fig. 15-2. He named them the *East Japan volcanic belt* and the *West Japan volcanic belt.* The former stretches along the Kuril, Northeast Japan and Izu-Bonin Arcs on the west side of the Kuril, Japan and Izu-Bonin Trenches. The latter stretches along the Ryukyu Arc on the west side of the Ryukyu Trench.

The intermediate region, i.e. the Southwest Japan Arc, is nearly devoid of Quaternary volcanoes owing to its relatively stabilized condition.

There is a sharp eastern limit for the distribution of volcanoes in each volcanic belt. The limit is called a *volcanic front* (Sugimura, 1968). Volcanoes are especially crowded in a zone immediately west of the front. The distribution pattern of Quaternary volcanoes is similar to that of late Tertiary ones in Japan. This means that the present-day orogeny is a continuation of a late Tertiary one, beginning in Miocene time (Sugimura, 1968; Sugimura *et al.* 1963, Matsuda, 1964; Matsuda, Nakamura and Sugimura, 1967).

The chemical characteristics of late Tertiary and Quaternary volcanic rocks show a close relationship to the arc structure. As shown in fig. 15-3, tholeiitic and calc-alkali rocks occur near the volcanic front (i.e. on the trench side) within the East Japan volcanic belt, whereas tholeiitic and calc-alkali rocks with more alkalic compositions together with some alkali rocks occur on the west side within the belt (Sugimura, 1960, 1968; Kuno, 1959, 1968; Katsui, 1961; Aoki and Oji, 1966; Miyashiro, 1972b). A similar westward increase of alkalic character was observed in the West Japan volcanic belt also. Volcanic rocks of islands in the Japan Sea and Korea are mostly alkalic.

Deep-focus earthquakes occur along a Benioff zone that dips to the west from the trench. The generation of magmas for volcanism and their compositions may be controlled by the plunging lithospheric slab suggested by the Benioff zone (§ 4-2).

It is to be noted that a regular chemical zonation across an island arc occurs in granitic rocks as well. Though most of the granitic rocks of Japan were

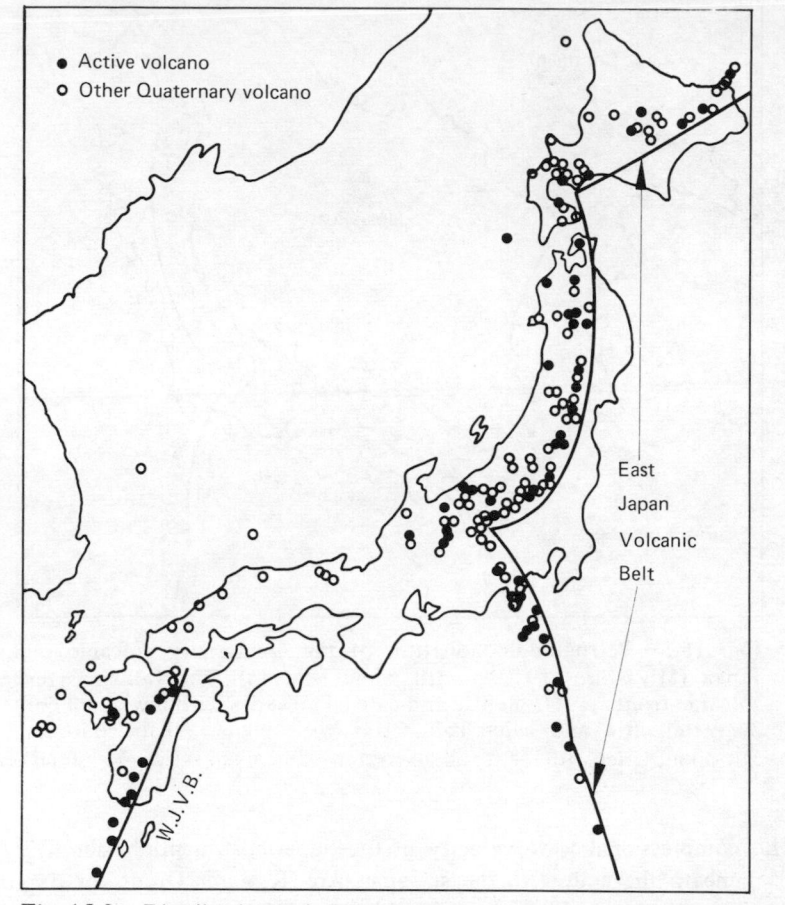

Fig. 15-2   Distribution of Quaternary volcanoes in Japan. The heavy arcuate lines represent the volcanic fronts for the East Japan and the West Japan volcanic belt. (Sugimura, 1960, 1968.)

formed in relation to the Mesozoic and older orogenies, their chemical zonation may be referred to here for comparison. Granitic rocks on the continental side of the Honshu Arc tend to be less calcic than those on the oceanic side, just as the Quaternary volcanic rocks do (Taneda, 1965).

Crust and Upper Mantle. The relatively stable Southwest Japan Arc appears to have a crustal thickness of the order of 30–40 km, whereas the crust of the active Northeast Japan Arc is about 30 km or thinner (e.g. Hashizume *et al.* 1968). Besides near-surface sediments, the crust is generally layered with an acidic upper crust (V_p = c. 6·0 km sec^{-1}) and a more basic lower crust (V_p = c. 6·6 km sec^{-1}).

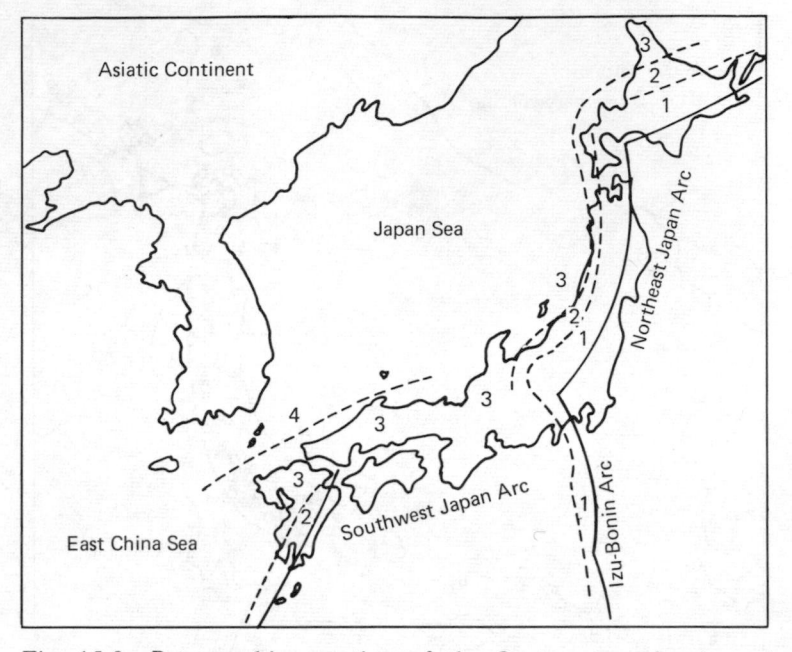

Fig. 15-3 Petrographic zonation of the Quaternary volcanic rocks of Japan (Miyashiro, 1972*b*, with some revision). The full lines represent volcanic fronts. 1 = tholeiitic and calc-alkali series with low alkali contents, 2 = tholeiitic and calc-alkali series with higher alkali contents, 3 = calc-alkali series with higher alkali contents and alkali series, 4 = alkali series.

The compressional wave velocity in the uppermost mantle is about 7·7 km sec^{-1} beneath the active Northeast Japan Arc (Research Group for Explosion Seismology, 1966). Such an anomalously low velocity in the uppermost mantle has been observed in other active regions; e.g. Kamchatka and the Cordilleran orogenic region of the western United States (Fedotov, 1968; Herrin, 1969).

Terrestrial Heat Flow. Fig. 15-4 shows that observed heat-flow values are generally below 40 mW m^{-2} (1 μcal cm^{-2} sec^{-1}) in the Kuril and Japan Trenches and the ocean-side rim of the active Northeast Japan and Kuril Arcs, and that the values increase very rapidly toward the Asiatic continent, reaching the highest values around 100 mW m^{-2} (2·5 μcal cm^{-2} sec^{-1}) in the deep basins of the Japan and the Okhotsk Sea (Uyeda and Vacquier, 1968). Thus, there is a marked contrast between a low and a high heat-flow zone respectively on the oceanic and continental sides of the active arcs.

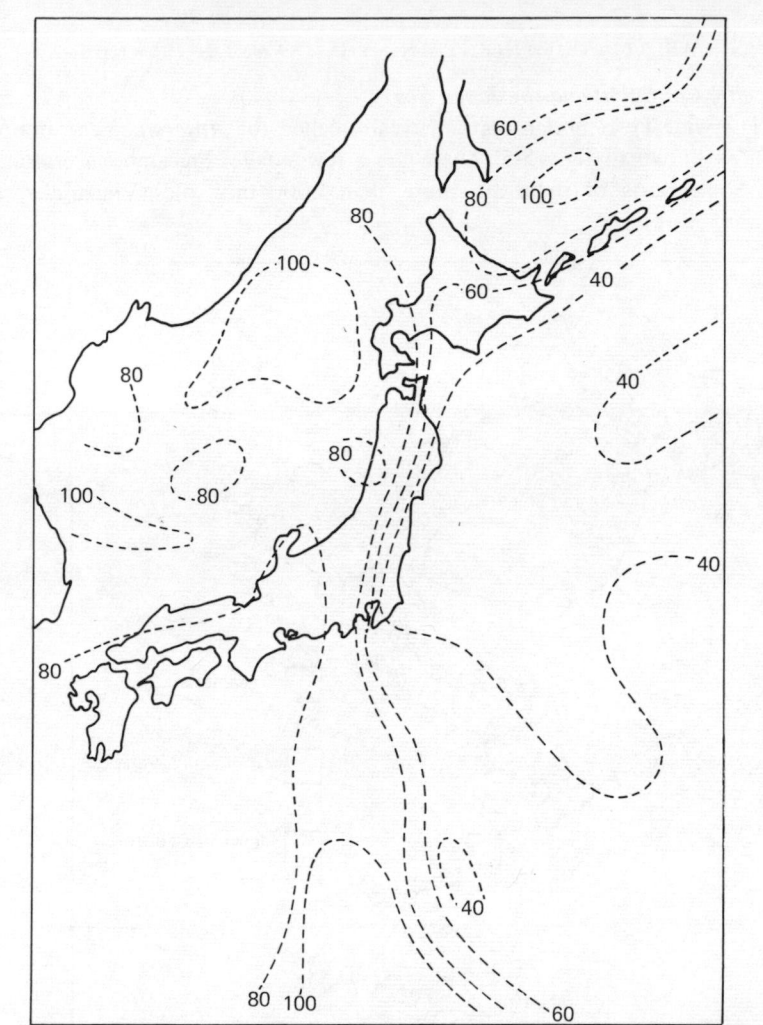

Fig. 15-4   Heat flow distribution in and around the Japanese Islands in mW m^{-2} (Uyeda and Vacquier, 1968). 1 μ cal cm^{-2} sec^{-1} = 42 mW m^{-2}.

A low heat-flow region should generally show lower temperatures than a high heat-flow region at the same depth. Accordingly, a low- and a high-temperature zone run in parallel beneath an active island arc. This contrast would be related to the generation of paired metamorphic belts in the circum-Pacific region, as will be discussed later.

The existence of the contrasted thermal zones has been confirmed also from studies on electrical conductivity anomalies (e.g. Uyeda and Rikitake, 1970).

15-2   THE ASIATIC CONTINENT AND THE JAPANESE ISLANDS

The Asiatic Continent and the Japan Sea

There is a wide Precambrian craton, usually called the *Angaran Shield*, in north
Siberia, as shown in fig. 15-5. There are a few smaller Precambrian cratons in
China. Some parts of them are more than 2000 m.y. old (Vinogradov and
Tugarinov, 1962).

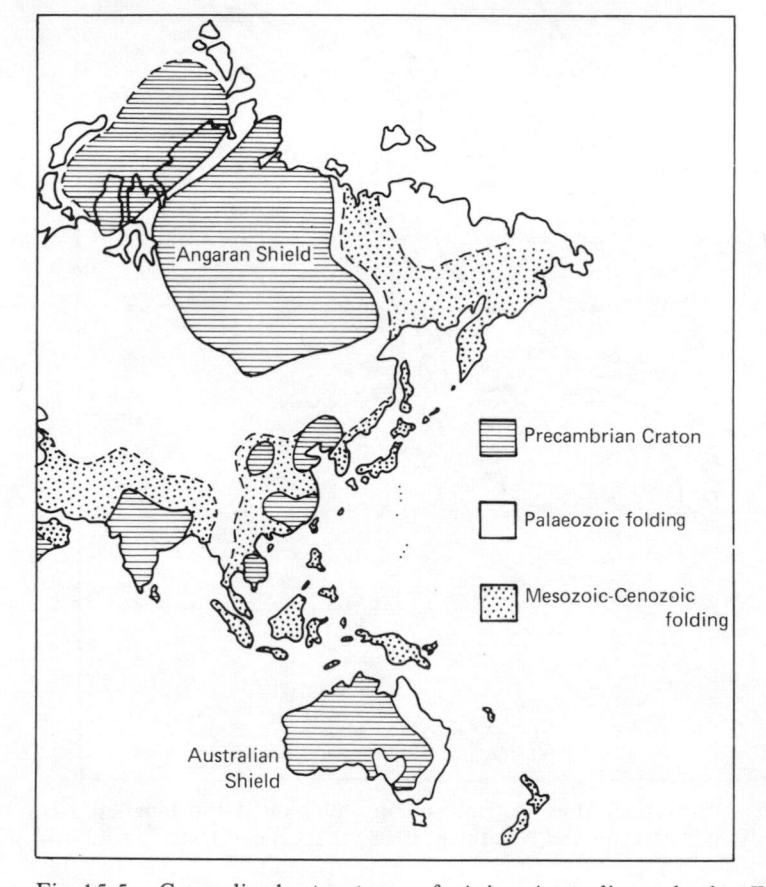

Fig. 15-5   Generalized structure of Asia, Australia and the Western
Pacific.

It is widely believed that the Asiatic Continent has shown accretional growth
around these cratons, and especially around the Angaran Shield. The Himalayan
Mountains were probably formed by collision of the Indian plate against the
Asiatic one, whereas the Japanese Islands represent a front of continental growth
toward the Pacific Ocean.

Small Precambrian terranes are exposed in Korea, and a part of the Japanese Islands certainly has a Precambrian basement. It is likely that these masses were formerly in the coast regions of the then Asiatic Continent, and that the old sialic mass of Japan drifted away from the continent in a relatively young geologic time (e.g. Matsuda and Uyeda, 1971).

A wide northwestern part of the Japan Sea is more than 3000 m deep, where unconsolidated sediments are only a few kilometres thick, and the underlying hard crust ($V_p = 6 \cdot 7$ km sec^{-1}) is of the oceanic type. The origin of the Japan Sea is an important unsolved problem, like that of other marginal seas on the continental side of island arcs (e.g. Menard, 1967).

As regards the Tasman Sea between Australia and New Zealand, van der Linden (1969) has suggested that possible sea-floor spreading from a small mid-oceanic ridge within the Sea might have resulted in the eastern movement of New Zealand. This may be applicable to the Japan Sea, which, however, has no topographic feature suggesting a mid-ocean ridge. Alternatively, the floor of the Japan Sea may have been created not in a definite mid-ocean ridge but at innumerable points therein owing to the materials rising through the mantle beneath the whole area.

There is a remarkable zone of late Mesozoic and early Tertiary acidic-intermediate volcanism and plutonism along the east coast of the Asiatic Continent from the Gulf of Tonkin to the Bering Strait (fig. 15-6). It was called the *East Asiatic volcano-plutonic belt* (Ustiyev, 1965). The late Mesozoic and early Tertiary volcanism and plutonism in Japan show some resemblance to those of the belt (Yamada, 1966; Matsumoto, 1968). It is possible that Japan was on the east coast of the Asiatic continent in early Mesozoic and older time, and that the Japan Sea was formed by the drift of Japan in late Cretaceous and/or Tertiary time.

Precambrian and Palaeozoic Rocks of Japan
Whether a Precambrian basement exists beneath the Japanese Islands has long been a matter of dispute (e.g. Matsumoto *et al.* 1968). Recently, Precambrian Rb-Sr ages (600 ~ 1200 m.y.) have been reported for certain gneissic rocks from the Hida Plateau in central Honshu (Sato, 1968). K-Ar ages of 1600-1640 m.y. have been obtained for biotite and muscovite samples that were separated from metamorphic rock cobbles in Palaeozoic conglomerates in the same region. The cobbles had probably been derived from a metamorphic complex of the Hida Plateau (Shibata, Adachi and Mizutani, 1971). Thus, the existence of Precambrian metamorphic rocks in that region has been established. Radiometric dating has further suggested that a metamorphic event took place in a considerable part of Japan probably around Ordovician time (§ 15-5).

Though the available evidence is still meagre, it appears that phases of geosynclinal sedimentation and/or orogeny took place continually from middle

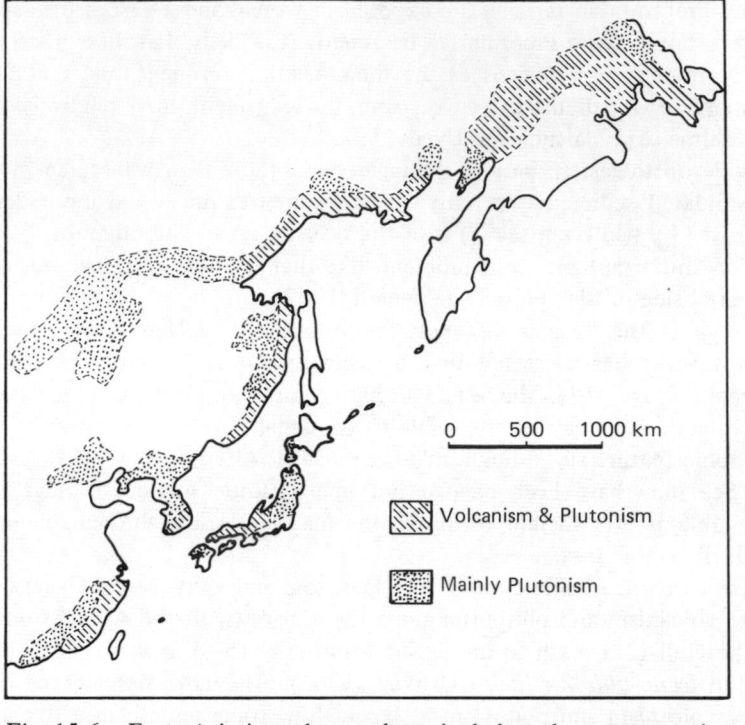

Fig. 15-6   East Asiatic volcano-plutonic belt, where magmatic activity occurred from the Jurassic to the early Tertiary.

Precambrian time to Ordovician on and in the crust that later became the mass of Japan. The width and the thickness of the sialic crust of that time are unknown. The presently available palaeontological record in the Japanese Islands begins with the Silurian.

There was a big geosyncline, called the *Chichibu* or *Honshu* geosyncline, probably along the coast of the Asiatic Continent from Silurian to Permian time. The resultant sedimentary pile is now well exposed in Honshu, Shikoku and Kyushu. The bulk of the terrane exposed at present is Carboniferous and Permian. Regional metamorphism took place repeatedly in this pile.

15-3   PAIRED METAMORPHIC BELTS

The Paired Metamorphic Belts of the Honshu Arc. Detailed petrographic investigation has been made on metamorphic rocks of Japan, as summarized for example in the recently published *Metamorphic Facies Map of Japan* at a scale of 1:2 000 000 (Hashimoto *et al.* 1970).

The metamorphic events in the Jurassic and Cretaceous produced a pair of regional metamorphic belts in the sedimentary pile of the Honshu geosyncline. The pair is composed of the Ryoke (pronounced Ryôké) belt of the low-pressure type on the continental side and the Sanbagawa belt of the high-pressure type on the Pacific Ocean side as has already been shown in fig. 3-10. In this book, the continental side (usually the concave side of an arc) is sometimes called the *inner side*, and the oceanic side (usually the convex side) the *outer side*.

Another pair of metamorphic belts is present on the north side of the Jurassic–Cretaceous pair in Honshu. It is composed of a part of the Hida metamorphic complex on the inner side and the Sangun belt of western Honshu on the outer. A large part of the westward extension of the former lies beneath the Japan Sea (table 15-1). The whole of the Hida complex and its extension are shown as the Hida metamorphic terrane in fig. 3-10. The metamorphism of this pair is probably Permian to Jurassic (table 15-1).

The above-mentioned two pairs were recognized and were called the younger and older pairs by Myashiro (1961a), but recent radiometric dating has revealed that probably there was a still older pair in Honshu. It is made up of another part of the Hida metamorphic complex on the inner side and the circum-Hida metamorphic zone on the outer, both metamorphosed probably in Carboniferous time (table 15-1; fig. 15-8).

Possible Deformations of Arcs. The Jurassic–Cretaceous belts shows a sharp turn of trend at the Itoigawa–Shizuoka Line and Fossa Magna. It is unknown whether this turn is a primary or a secondary feature of the metamorphic belts. If it is primary, Southwest Japan and Northeast Japan should have been two distinct arcs in Jurassic–Cretaceous time. The Izu-Bonin arc may or may not have existed at that time. The oldest known rocks of the Izu–Bonin Arc off Honshu are early Tertiary pyroclastics in the Bonin Islands.

If it is due to secondary bending, the deformation may have taken place by the tectonic movement along the Izu-Bonin Arc during Tertiary time, and/or by that which took place during the drift of the Japanese Islands away from the Continent in late Cretaceous and Tertiary time.

In the Northeast Japan Arc, Cretaceous granitic rocks of the Kitakami Mountains (110–120 m.y.) show remanent magnetization generally in the direction of N 60°-30°W, and those of the Abukuma Plateau (90–100 m.y.) in the direction of N 50°-0°W, as schematically shown in fig. 15-7. On the other hand, Cretaceous granitic and volcanic rocks of the Southwest Japan Arc are generally magnetized in the direction of N 30°–60° E. These data suggest that Northeast Japan made successive counterclockwise rotations with reference to Southwest Japan in Cretaceous and later time (Kawai, Ito and Kume, 1961; N. Kawai, pers. comm. 1970).

Table 15-1 *Tentative chronology of regional metamorphic events in the Honshu Arc of Japan*

Event No.	Geologic age	High-pressure metamorphic area (radiometric age)	Broadly simultaneously metamorphosed area of low-pressure type with granites (radiometric age)
6	Miocene to the present	Not exposed yet (?)	Northeast Japan (3-25 m.y.)
5	Jurassic to Cretaceous	Sanbagawa belt (110 m.y.); Part of Shimanto terrane	Ryoke belt (c. 105 m.y.); Abukuma Plateau (100-120 m.y.)
4	Permian to Jurassic	Sangun belt of western Honshu (160 m.y.)	Hida IV (180-240 m.y.)
3	Carboniferous	Circum-Hida zone including Omi area (320 m.y.)	Hida III (320 m.y.); Maizuru zone (330 m.y.)
2	Ordovician	Kiyama (450 m.y.)	Hida II (500 m.y., Cambro-Ordovician); Kurosegawa zone (430 m.y.)
1	Precambrian	—	Hida I (1,640 m.y.)

Accordingly, the shape of the Northeast Japan Arc during metamorphism would be restored probably by a clockwise rotation of the Abukuma Plateau by about 60°. This rotation gives the Jurassic–Cretaceous pair of metamorphic belts a more natural-looking shape.

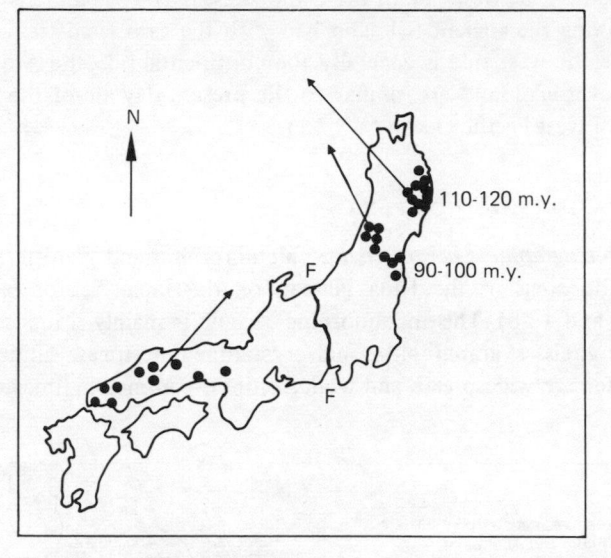

Fig. 15-7 Mean directions of remanent magnetization of Cretaceous granitic and volcanic rocks in three parts of the Honshu Arc. Dots represent the localities of measured specimens. (After N. Kawai and T. Nakajima; Kawai, pers. comm. 1970.)

Thermal Structure and Associated Magmatism. In the two metamorphic belts of the Jurassic–Cretaceous pair, the original thermal structure is fairly well preserved. The two belts are in contact with each other at the Median Tectonic Line. The thermal axis of each belt is near the Line (fig. 15-13).

In the older metamorphic belts, the original thermal structure is not well known because of the later destruction and less advanced investigation.

The relationship between the baric types of regional metamorphism and the characteristics of the associated magmatism as outlined in § 4-1 holds good in each belt of all the pairs. Ophiolites including serpentinites are abundant in high-pressure metamorphic belts and regions on the outer side, whereas granitic bodies are confined to low-pressure metamorphic belts and regions on the inner side. Thus, there are three remarkable belts of ophiolites associated with the three high-pressure metamorphic belts of southwest Japan.

Paired Metamorphic Belts in Hokkaido. There is one more pair of belts in Hokkaido, as shown in fig. 3-10. It is composed of the Hidaka belt of the low-pressure type on the east and the Kamuikotan belt of the high-pressure type on the west. The original sedimentary rocks here are probably Mesozoic and their regional metamorphism is probably Cretaceous and Tertiary.

The main part of Hokkaido, where this pair is exposed, belongs to the Sakhalin Arc situated on the west side of the Okhotsk Sea. The paired belts were formed probably along the ancient Sakhalin Arc with the associated trench on the west side. Since the west side is generally the continental side, the Arc may be regarded as a reversed island arc similar to the present-day arc of the New Hebrides in the southwest Pacific Ocean (§ 17-1).

15-4   HIDA METAMORPHIC COMPLEX

Distribution and Petrographic Characteristics. Metamorphic and granitic rocks are exposed in wide areas of the Hida Plateau on the Japan Sea of central Honshu (figs. 3-10 and 15-8). This metamorphic terrane is mainly composed of quartzo-feldspathic gneisses, amphibolites and crystalline limestones. Sillimanite and pyralspite garnet are widespread, and wollastonite is common in limestones.

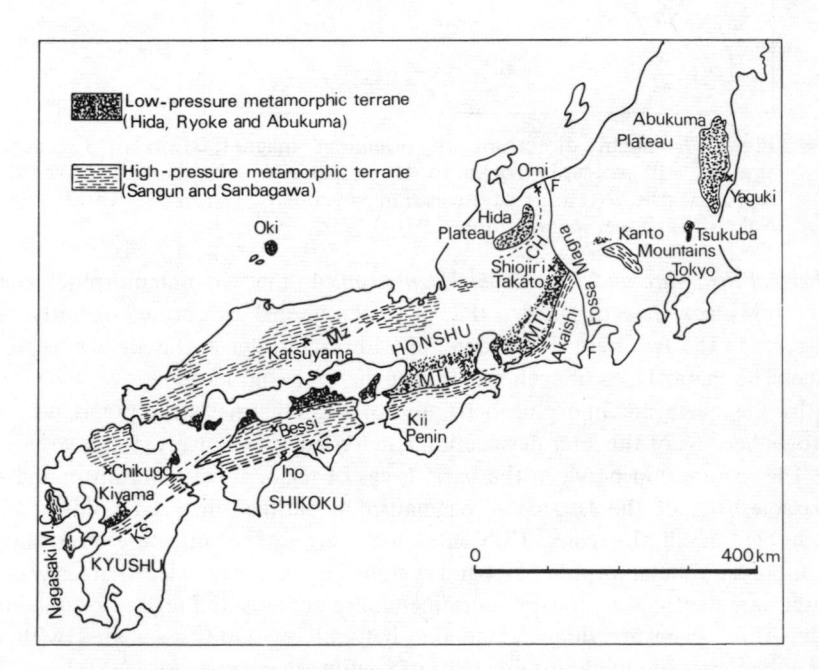

Fig. 15-8   Metamorphic belts and well-documented metamorphic areas in the main part of Japan. *Abbreviations:* CH = circum-Hida metamorphics, F-F = Itoigawa-Shizuoka Line, MTL = Median Tectonic Line, KS = Kurosegawa zone, MZ = Maizuru zone.

Most of the rocks belong to the epidote-amphibolite, amphibolite and granulite facies (Sato, 1968; Suzuki and Kojima, 1970). The distribution of metamorphic temperature in the complex, however, has not been clarified, and is presumably irregular. The metamorphism is of the low-pressure type and possibly partly of the medium-pressure type. Andalusite, kyanite, sillimanite and staurolite occur in the Kurobegawa area near the northeastern end of the complex. No ultramafic rocks occur.

Similar metamorphic and granitic rocks occur in the Oki Islands in the Japan Sea and also within a faulted area, called the Maizuru zone, west of Kyoto (fig. 15-8). Therefore, the Hida metamorphic complex is considered to extend westward in a wide zone, most of which is now covered by the Japan Sea and Palaeozoic and younger sediments.

Age Group. The Hida complex appears to contain metamorphic and plutonic rocks of a number of different ages, ranging from the middle Precambrian to the Jurassic. Radiometric dating made in several recent years has greatly contributed to the elucidation of the long, complicated history, though the age and the nature of individual events have not been established yet.

The Precambrian events are represented by K–Ar ages of about 1640 m.y. (Shibata *et al.* 1971) and a Rb–Sr age of about 1200 m.y. (Sato, 1968). As the details of the events are unknown, they are shown as Hida I in table 15-1. There are four Phanerozoic age clusters around 500 m.y. (Cambro-Ordovician), 320 m.y. (Carboniferous), 240 m.y. (Permian) and 180 m.y. (Jurassic). Tentatively, the complexes corresponding to these clusters are shown as Hida II-IV in table 15-1.

The metamorphic complex of the Hida IV event is here regarded as representing the low-pressure metamorphic belt paired with the Sangun belt. The belt of Hida IV was probably formed on the basement of older metamorphic belts, a large part of which was probably reactivated during the event. The age cluster around 180 m.y. is most marked and widespread all over the Hida Plateau (e.g. Nozawa, 1968). This appears to indicate the age of the consolidation of the Funatsu granite and related regional metamorphism.

15-5   HIGH-PRESSURE METAMORPHIC TERRANES OLDER THAN THE SANGUN BELT

In various parts of Honshu and Kyushu, there are a number of small areas composed of low-temperature metamorphic rocks occasionally containing glaucophane besides the Sangun metamorphic belt to be described later. Recent radiometric dating has revealed that at least some of the small areas were metamorphosed much earlier than the Sangun (table 15-1). Two age groups will be commented upon below.

Kiyama Group (Ordovician)

In central Kyushu, a glaucophane-bearing metamorphic complex is exposed in the Kiyama area (fig. 15-8). It has given an Ordovician age (about 450 m.y.) by the Rb–Sr method (Yamamoto, 1964; Karakida *et al.* 1969).

It is noted that the metamorphic event around 500 m.y. in the Hida complex (i.e. Hida II) is relatively close in age to the Kiyama event. A similar age has been reported on some rocks from the Kurosegawa tectonic zone, which is a line of fault blocks uplifted from the basement within the Paleozoic terrane on the Pacific Ocean side of the Southwest Japan Arc (Matsumoto *et al.* 1968; Hayase and Nohda, 1969).

These data suggest widespread regional metamorphism in and around Ordovician time in Japan. The metamorphism was partly of the high-pressure series and partly associated with granitic activity. However, the shapes and relations of metamorphic belts at that time are not known.

Omi Group (Carboniferous)

There are several small metamorphic areas along an arcuate line on the south and east sides of the Hida metamorphic complex. This group is sometimes called the *circum-Hida metamorphics* (fig. 15-8). They are commonly in the greenschist facies and rarely in the glaucophane-schist and the epidote–amphibolite facies (e.g. Seki, 1959).

The Omi area is at the northeastern end of the arcuate line (Banno, 1958). Metamorphic rocks of this area have given Rb–Sr and K–Ar ages of about 320 m.y., i.e. Carboniferous (Yamaguchi and Yanagi, 1970).

A cluster of isotopic ages in the Hida complex occurs at the same age (i.e. Hida III). Some rocks in the Maizuru zone also give similar ages (Hayase and Ishizaka, 1967). Thus, it is conceivable that widespread regional metamorphism took place in the Carboniferous, and that paired metamorphic belts composed of a Hida and the circum-Hida zone were produced.

The metamorphic rocks of the Chikugo area (fig. 15-8) give a Rb–Sr age of about 270 m.y. (Yamaguchi and Yanagi, 1970). It is not clear whether they belong to the Omi group.

Several separate metamorphic areas, occasionally containing glaucophane, occur on the east side of the Abukuma Plateau and in the Kitakami Mountains; e.g. Yaguki area (fig. 15-8). Their metamorphism was claimed to be pre-Late Devonian (Kanisawa, 1964), but a part of their original rocks was recently found to be Permian. The K–Ar age of a metamorphic rock from one of the areas was 300 m.y. (Carboniferous). Thus, these areas would belong to two or more age groups.

15-6 SANGUN METAMORPHIC BELT

Distribution. A few tens of small areas of schists and phyllites are exposed within a broad zone with an ENE trend in western Honshu (fig. 15-8). The

Palaeozoic terranes lying between these areas also show incipient metamorphism, though they have usually been called 'unmetamorphosed' in the literature. They all are regarded as constituting one and the same metamorphic belt, called the Sangun. Glaucophane occurs occasionally. Serpentinites occur in many places in the zone. The metamorphic rocks are intruded and affected by later granitic rocks, which are genetically related to the Ryoke metamorphism and not to the Sangun.

The age of deposition of the original sediments is at least partly Permian and Carboniferous. On the other hand, it has been claimed that metamorphic rocks in a certain area are unconformably covered by unmetamorphosed Middle Triassic sediments. Thus, the age of Sangun metamorphism appears to be very late Permian or early Triassic, that is, immediately after the deposition of the relevant sediments. A recent Rb-Sr and K-Ar dating of the belt, however, has given a Jurassic age of about 170 m.y. (Shibata and Igi, 1969).

The name Sangun was derived from a mountain in northern Kyushu, where unfortunately the Sangun rocks were mostly subjected to contact metamorphism by Cretaceous granites. Therefore, the type areas of Sangun metamorphic rocks are considered to be in western Honshu in this book without regard to the etymology.

Characteristics. The Sangun metamorphic belt is composed of patchy areas of schistose metamorphic rocks surrounded by poorly recrystallized non-schistose Palaeozoic terranes. The patchy areas, commonly tens of kilometres across, are composed of schists and phyllites in the glaucophane-schist and the greenschist facies (§ 6A-5).Lawsonite is very rare and jadeite does not occur in association with quartz.

The pelitic rocks in the surrounding Paleozoic terranes are usually very poorly recrystallized, but the intercalated basaltic rocks usually show more advanced recrystallization in the prehnite-pumpellyite or the greenschist facies (Nureki, 1969; Hashimoto and Saito, 1970).

The Sangun belt appears to have a number of subparallel thermal axes, though available data are very meagre. The associated serpentinites have a chemical composition similar to ordinary alpine-type masses, and contain relict olivine with 6–10 per cent Fa (Research Group of Peridotite Intrusion, 1967).

15-7 RYOKE METAMORPHIC BELT AND ABUKUMA PLATEAU

Ryoke Metamorphic Belt

The Ryoke metamorphic belt extends from the Shiojiri–Takato area immediately west of the Itoigawa–Shizuoka Line, westward through the Southwest Japan Arc, to central Kyushu (figs. 3-10, 15-8, and 15-9).

A great amount of related granitic rocks occurs in the metamorphic terranes and in a wide region to the north. Most or all of them appear to be late- and post-metamorphic intrusions. The large-scale distribution of metamorphic tem-

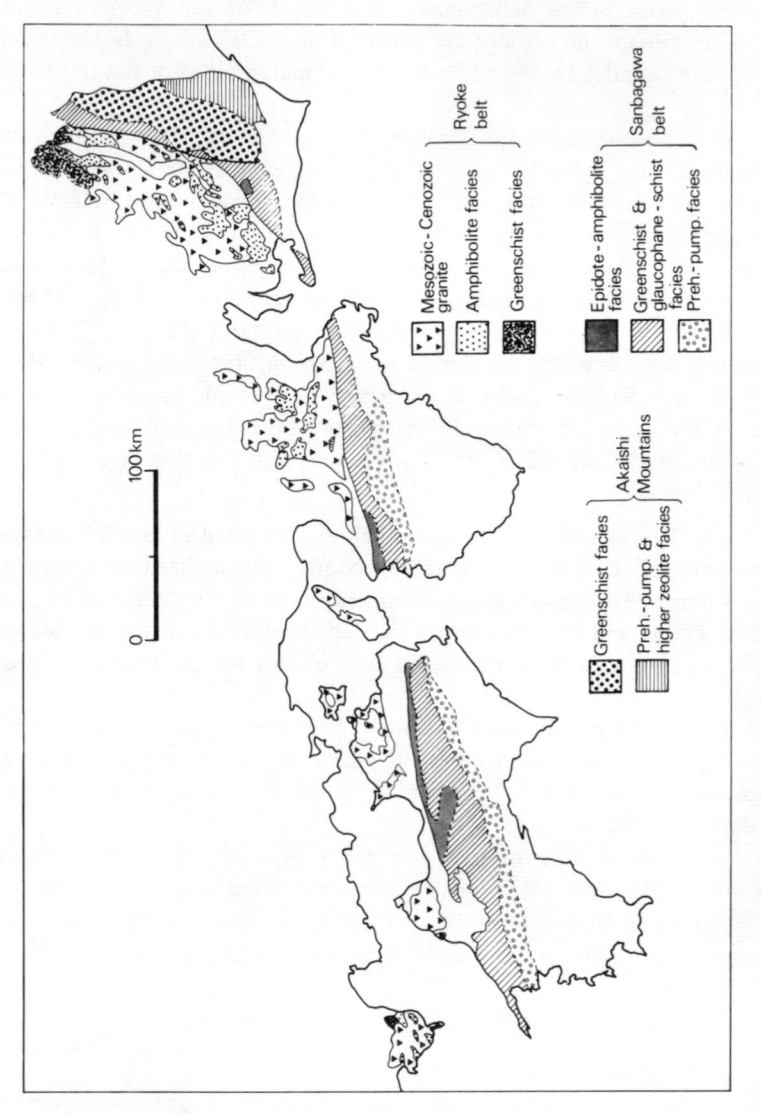

Fig. 15-9 Distribution of metamorphic facies in the Ryoke and San-bagawa belts and the Akaishi Mountains. (Hashimoto *et al.* 1970.)

perature is independent of individual granitic intrusions. Contact metamorphism has been noticed around some intrusions. No peridotite occurs.

The metamorphic rocks are mostly of pelitic and psammitic compositions accompanied by subordinate quartzite and limestone. Metabasites are rare. The metapelites are usually high in K/Na ratio, and in excess Al_2O_3 content, like the virtually unmetamorphosed, late Palaeozoic pelites exposed to the north of the belt.

Most of the well-recrystallized parts of the belt belong to the amphibolite facies, and the rest to the greenschist facies. The belt is of the low-pressure facies series.

In the eastern end of the belt (i.e. Shiojiri–Takato area), the gradual transition from virtually unmetamorphosed slates exposed in the north into schists and gneisses exposed in the south has been well documented, and has already been reviewed in §§ 7A-2 and 8A-2, and fig. 3-4. Andalusite occurs in a medium-temperature zone, and sillimanite in a high-temperature zone. Cordierite is common, but pyralspite garnet is relatively rare.

In the eastern end and many other parts, the temperature of metamorphism increases southward or southeastward across the belt to the Median Tectonic Line, which marks the southern limit of the belt. In the western parts, however, a thermal axis appears to exist within the belt as shown in fig. 15-13 (Miyashiro, 1959, 1961a; Suwa, 1961).

The well-documented Higo metamorphic complex in central Kyushu could represent the western end of the Ryoke belt (Ueta, 1961; Yamamoto, 1962; Tsuji, 1967).

Most of the granitic rocks in southwest Japan were formed in relation to the Ryoke metamorphism (fig. 15-10). The $^{206}Pb/^{238}U$ age of zircons from granites in this belt is about 90 m.y. (Ishizaka, 1969). The Rb–Sr ages of granites are 105–60 m.y., i.e. Cretaceous to early Tertiary (Hayase and Ishizaka, 1967; Yamaguchi and Yanagi, 1970; also Shibata, 1968).

Figure 15-11 shows that the distribution of the K–Ar ages of granitic rocks shows the existence of regional age zones (Kawano and Ueda, 1967a, b). Most granitic rocks in the Ryoke belt belong to a 75–95 m.y. zone (Cretaceous), whereas those in the eastern part and to the north of the belt belong to a 50–65 m.y. zone (early Tertiary). This could mean that granite magmas were formed by events which occurred on a regional scale, and that the theatre of the events moved gradually northwards. Alternatively, these age zones may simply represent the times of their uplift which led to the cooling of granitic bodies.

It is to be noted that all these radiometric data are concerned with granitic rocks only. The true age of regional metamorphism is difficult to establish, because most of the metamorphic rocks may have been more or less affected by post-metamorphic granitic intrusions, and the age of cooling may have been considerably different from that of metamorphic climax.

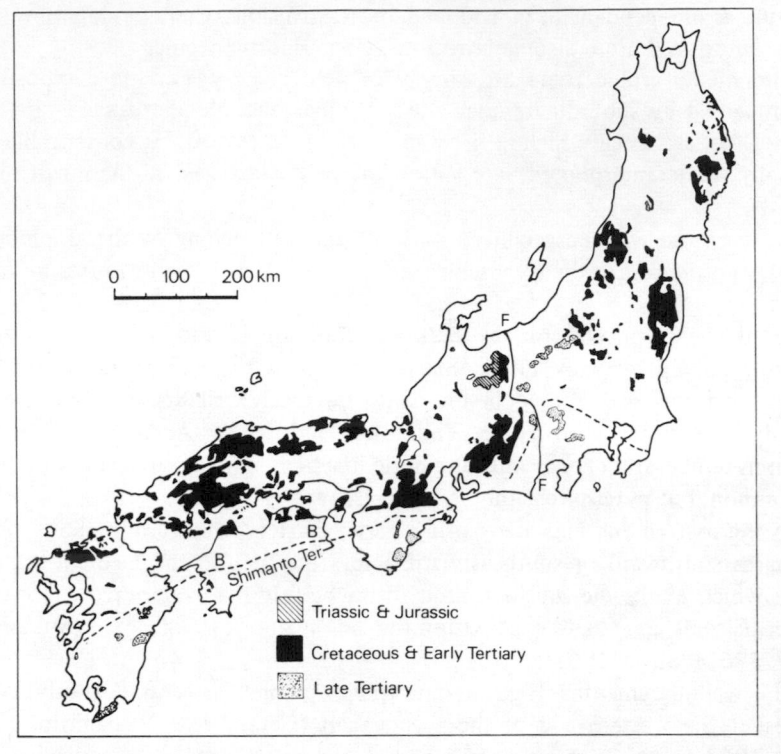

Fig. 15-10   Distribution of granitic rocks in Japan. F–F = Itoigawa–Shizuoka Line. The region to the south of the Butsuzo Line (B–B) is the Shimanto terrane. Triassic and Jurassic granitic rocks are related to a Hida metamorphism (event 4 in Table 15-1), and Cretaceous and early Tertiary ones to the Ryoke. Late Tertiary granitic rocks in the Southwest Japan and Northeast Japan Arcs are related respectively to the Shimanto and Mizuho orogenies.

Thus, the age of the Ryoke metamorphism could be considerably more than 100 m.y. old.

Abukuma Plateau
The eastward extension of the Ryoke belt beyond the Itoigawa–Shizuoka Line and Fossa Magna could be represented by small exposures of granitic rocks to the north of the Kanto Mountains and the Tsukuba metamorphic area, and a great granitic and metamorphic terrane in the Abukuma Plateau (figs. 15-8 and 15-13).

Andalusite and sillimanite occur in some metapelites in these terranes. The temperature of metamorphism increases generally toward the west in Tsukuba and Abukuma (§§ 7A-2 and 8A-2).

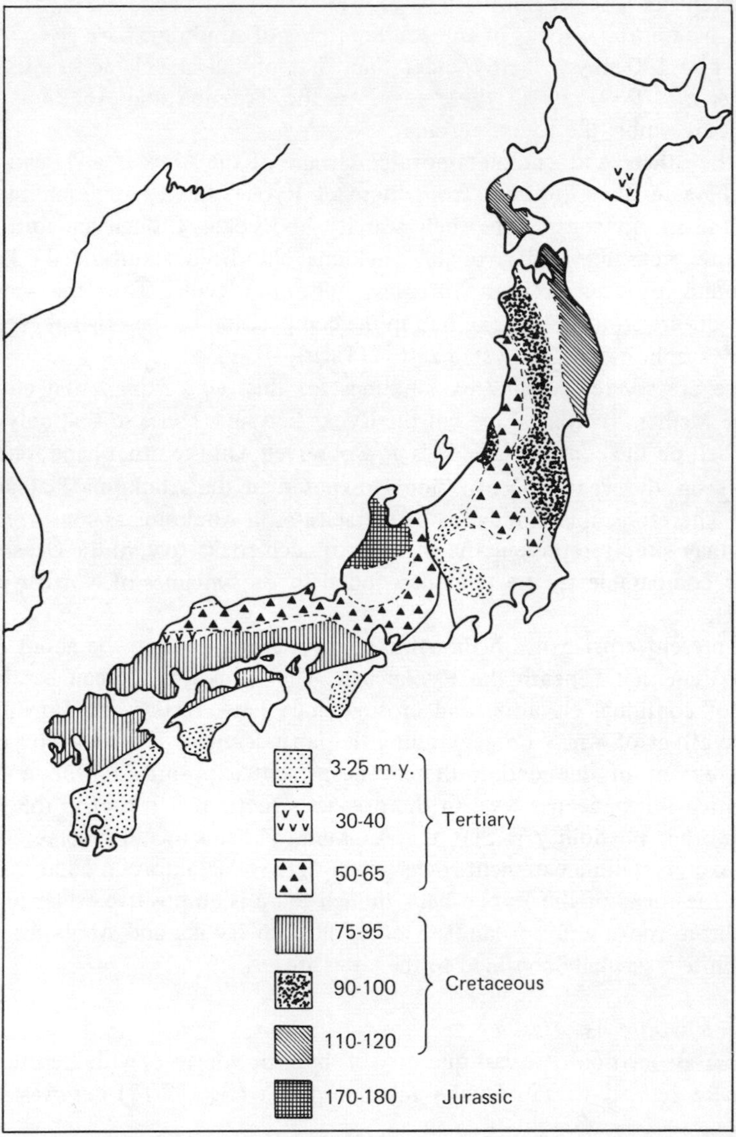

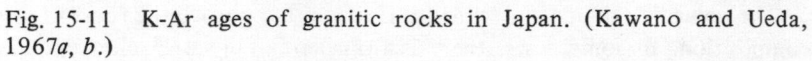

Fig. 15-11   K-Ar ages of granitic rocks in Japan. (Kawano and Ueda, 1967*a, b*.)

The K-Ar ages of granitic rocks from the Abukuma Plateau are 90-100 m.y. (middle Cretaceous), whereas those from a region to the west (i.e. on the continental side) are generally 50-65 m.y. (Kawano and Ueda, 1967a, b).

The metamorphic rocks in the southern part of Abukuma have given a K-Ar age of 120-100 m.y., slightly older than that of the associated granitic ones (Ueda *et al.* 1969). In all these respects, the Tsukuba and Abukuma metamorphics resemble the Ryoke terrane.

On the other hand, the metamorphic terrane of the Abukuma Plateau shows some characteristics different from those of Ryoke. For example, metabasites are abundant, in contrast to their scarcity in Ryoke. Though andalusite and sillimanite occur in both Ryoke and Abukuma, chloritoid, staurolite and kyanite were found to occur rarely in Abukuma, but not in Ryoke. Thus, the Abukuma Plateau differs from the Ryoke belt in the composition of the original rocks and in the *P-T* condition of recrystallization (Tagiri, 1971).

There are several possible explanations for this. One explanation could be that the Median Tectonic Line cut the Ryoke belt lengthwise so that only a half of the belt on the continental side is now observed. On the other hand, only half the belt on the Pacific Ocean side is exposed in the Abukuma Plateau (fig. 15-13). Therefore, the abundance of metabasites in Abukuma as compared with Ryoke may simply represent the increase of such rocks toward the Ocean. This may be comparable to the tendency found in geosynclines of North America (§ 13-4).

The present crust beneath the Abukuma Plateau appears to be about 10 km thinner than that beneath the Ryoke belt. The Abukuma Plateau has been a region of continual elevation and erosion since Late Tertiary time owing to a tectonic effect of a new orogeny along the Japan Trench. This may have caused a great extent of denudation there. The present exposure in Abukuma may represent a much deeper level in the original structure than that in the Ryoke belt. Another possibility is that the Abukuma Plateau includes some areas of pre-Ryoke crystalline basement rocks which were re-metamorphosed at the same time as the rocks of the Ryoke belt, though there is no positive evidence for it.

Gabbroic rocks and cortlandtite occur in both Ryoke and Abukuma, while serpentinite is virtually confined to the latter area.

Granite-Rhyolite Association

The close association of a vast quantity of rhyolitic volcanics with granitic rocks genetically related to the Ryoke metamorphism (fig. 15-12) deserves special attention.

The volcanics are mostly welded tuffs and other pyroclastic rocks of rhyolitic composition. In some cases, there is a smaller amount of dacitic and andesitic rocks. A large mass in central Japan is called the *Nohi rhyolites*, covering an area of about 5000 km^2 with an average thickness presumably of about 2 km.

The extrusion of the rhyolitic rocks appears to have occurred in late Cretaceous time. The rhyolitic rocks are cut and metamorphosed by some granitic intrusions, but unconformably cover other ones (Yamada, 1966; Yamada and Nakai, 1969; Yamada, Kawada and Morohashi, 1971).

The close spatial association, the broad contemporaneity and the chemical similarity of the granitic and rhyolitic rocks suggest the existence of a genetic connection between them (Miyashiro, 1965; Yamada *et al.* 1971). These rocks could represent an ancient volcanic chain (§§4-2 and 15-2).

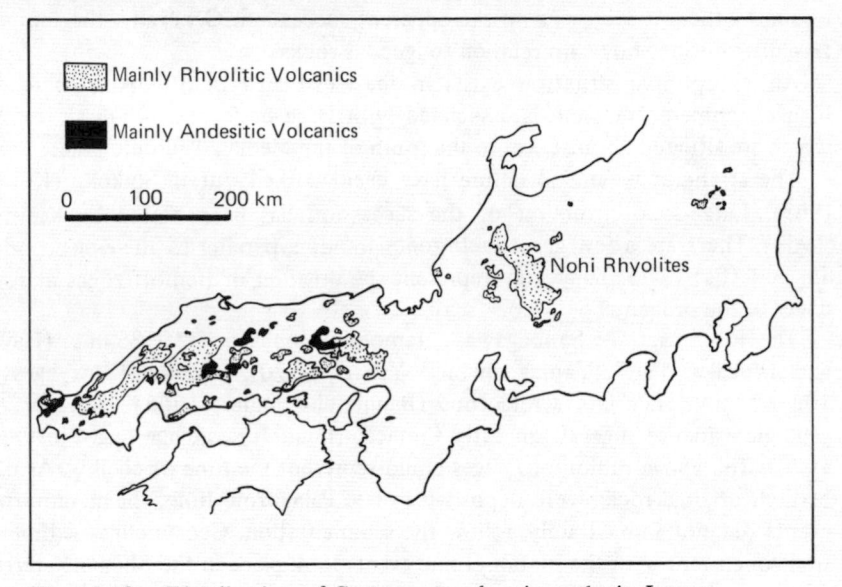

Fig. 15-12   Distribution of Cretaceous volcanic rocks in Japan.

15-8   SANBAGAWA METAMORPHIC BELT

Characteristics. The Sanbagawa (Sambagawa) metamorphic belt runs mostly on the Pacific Ocean side of the Ryoke belt with the Median Tectonic Line between them. The eastward extension of the Sanbagawa belt beyond the Itoigawa-Shizuoka Line is exposed in the Kanto Mountains (figs. 3-10 and 15-8).

Glaucophane occurs occasionally but not abundantly. Jadeite and lawsonite have mostly been found in the eastern half of the belt. Most of the belt belongs to the glaucophane-schist and the greenschist facies. Zones of the epidote-amphibolite facies occur in a few areas in the western half of the belt. Biotite occurs only in rocks of the epidote-amphibolite facies (§§7A-9 and 8A-8).

The metapelites in the Sanbagawa belt are lower in K/Na ratio and in excess Al_2O_3 than those of the Ryoke belt. The absence of paragonite, chloritoid and staurolite in metapelites is probably due to the low Al_2O_3 content and other compositional characteristics of the rocks. Paragonite was found only in some metabasites. Piemontite is a common mineral in metachert. Piemontite schists make good marker horizons for structural analysis of the terrane (e.g. figs. 4 and 10 in Iwasaki, 1963).

The temperature of metamorphism increases generally northward to the Median Tectonic Line. In the opposite direction, the metamorphic terrane usually grades into virtually unmetamorphosed Palaeozoic formations. In many areas, the temperature increases down the apparent stratigraphic succession, but in some others it increases up the apparent succession. Generally, the temperature distribution shows no relation to igneous rocks.

An exceptional situation exists in the Bessi area in Shikoku. There, the highest temperature zone is associated with large basic and ultrabasic masses, which are situated about 6 km to the south of the Median Tectonic Line.

The stratigraphy and structure have been worked out in Shikoku (Kojima, 1963). Large-scale inversion of the succession has been found by Kawachi (1968). There are a few great fault zones trending parallel to the elongation of the belt (fig. 17-1). They may represent the position of Benioff zones at some stages of the orogeny (§ 17-6).

The Rb–Sr ages of Sanbagawa metamorphic rocks are 110–85 m.y. (Hayase and Ishizaka, 1967; Yamaguchi and Yanagi, 1970), and the K–Ar ages are 102–82 m.y., i.e. late Cretaceous (Banno and Miller, 1965). However, the geologic evidence suggests an early Cretaceous and Jurassic age (e.g. Ernst *et al.* 1970). The above radiometric ages could represent the time of cooling. As most of their original rocks were deposited in late Palaeozoic time, the metamorphic events did not immediately follow the sedimentation. Geosynclinal sedimentation concurrent with the metamorphic events took place in the Shimanto terrane on the south side of these metamorphic belts, as will be described later.

Ophiolites. The Sanbagawa belt is accompanied by a large amount of ophiolites in great diversity. Many of them are of the non-sequence type (as defined in §4-1), but some ophiolites along the Mikabu Tectonic Line in the belt show tendencies toward regular sequences. A downward sequence, basaltic pyroclastics → dolerite → gabbro (Suzuki, 1964, 1967), and a symmetrical sequence of basalt → dolerite → gabbro → dolerite → basalt (Suzuki, Sugisaki and Tanaka, 1971) have been observed.

The peridotites and serpentinites along the Mikabu Tectonic Line have higher $(Fe^{2+} + Fe^{3+})/Mg$ ratios than those of the typical non-sequence type ophiolites in other parts of the Sanbagawa belt and the common alpine type ultramafics of the world. Olivine relics of Mikabu serpentinites usually contain 12–20 per cent Fa (Research Group of Peridotite Intrusion, 1967; Hayashi, 1968; Iwasaki, 1969; Miyashiro, 1972*b*).

It is conceivable that the Jurassic–Cretaceous metamorphism in the Sanbagawa terrane may represent only the last major phase of recrystallization in this belt, and have effaced all or most of the mineralogical characteristics in the earlier phases. The abundant extrusion of basaltic rocks of the non-sequence type ophiolites in this belt is simultaneous with the late Palaeozoic sedimentation in the geosyncline. Since ophiolites are characteristic of high-pressure metamorphic terranes, the geosynclinal area of the Sanbagawa belt should have had features characteristic of a high-pressure metamorphic terrane already in late Palaeozoic time. This may mean that underthrusting of a lithospheric slab began in late Palaeozoic time in this zone. If this is the case, the late Palaeozoic geosyncline of Japan should have had two subparallel underthrusting zones corresponding to the Sangun and Sanbagawa belts.

15-9 METAMORPHISM OF THE SHIMANTO TERRANE AND THE MIZUHO OROGENIC BELT

Shimanto Geosyncline and Metamorphism
In early Mesozoic time, the Palaeozoic Honshu geosyncline disappeared, and large parts of the geosynclinal zone were elevated above sea level. Instead, a new big geosyncline was created in middle Mesozoic time on the Pacific Ocean side of the site of the Honshu geosyncline. The thick accumulation in the new geosyncline, called the Shimanto geosyncline, is now exposed in the so-called Shimanto terrane, which includes the southern parts of the Kii Peninsula, Shikoku and Kyushu, in the Southwest Japan Arc (figs. 15-10 and 17-1). This geosyncline persisted through Palaeogene Tertiary time. The eastward extension of the geosyncline is exposed in an area to the south of the Kanto Mountains, where the Mizuho orogeny overprinted it.

Low-temperature regional metamorphism in the prehnite–pumpellyite and greenschist facies took place in large parts of the terrane. Pumpellyite is widespread, and stilpnomelane is rare. However, it is likely that the metamorphism in this terrane had a number of distinct phases with very different geothermal gradients. Different parts probably belong to different baric types.

The early phase of metamorphism is presumably late Jurassic and Cretaceous and simultaneous with the Ryoke and Sanbagawa metamorphism. It is likely that the main phase of the Sanbagawa metamorphism and the early phase of the Shimanto metamorphism are one and the same event. In the central Kii Peninsula, there is a well-documented metamorphic area that is continuous from Palaeozoic rocks in the north into Jurassic ones in the south. A zonal arrangement of greenschist, prehnite–pumpellyite, and zeolite facies rocks is clear generally with lower temperatures of metamorphism in younger rocks (§6A-4). This area appears to represent the early phase of the Shimanto metamorphism, giving a Rb–Sr age of 110 m.y. (Yamaguchi and Yanagi, 1970).

Another metamorphosed area of Shimanto, which is situated in the Akaishi Mountains (§6A-2), may represent a higher geothermal gradient and a younger age. More than ten small granitic masses were intruded into the Shimanto terrane in the Southwest Japan Arc. Their K–Ar ages cluster around two values corresponding to the Miocene: 14 and 21 m.y. (Kawano and Ueda, 1967a, b; Shibata, 1968). It is conceivable that related low-pressure metamorphism took place in Miocene time in some part of Shimanto. It is a matter of definition whether this phase of metamorphism is regarded as belonging to the Shimanto events or to the Mizuho orogeny to be discussed below.

Mizuho Orogeny

A series of orogenic events began in early Miocene time along the Kuril, Northeast Japan, and Izu-Bonin Arcs. Intense volcanism began along these arcs and geosynclinal accumulation with strong folding occurred in northeast Japan. The volcanism has continued through the Pliocene to the present. This new orogeny is called the *Mizuho orogeny* (Sugimura *et al.* 1963; Matsuda *et al.* 1967).

Volcanism, earthquakes, trench formation and other present-day tectonic activity in and around the Northeast Japan Arc as summarized in §15-1 may be regarded as manifestations of this orogeny.

Miocene geosynclinal accumulations, a few kilometres thick, were formed on the basements of Mesozoic and older metamorphic and granitic rocks on the Japan Sea side of northeast Japan. Among the products of the first intense Miocene phase of volcanism, rhyolitic and dacitic rocks were more abundant than basaltic and andesitic rocks. Subsequently, the proportion of andesitic rocks increased remarkably, and the increase has gone on to the present.

To the east of the geosynclinal zone, the Abukuma and Kitakami Mountains continued to rise as blocks from the Miocene to the present. This region is characterized by great positive Bouguer anomalies. The western limit of the region, called the *Morioka–Shirakawa Line*, runs through the cities of Morioka and Shirakawa. Probably the Line represents a big fault, possibly cutting through the whole thickness of the sialic crust (§§17-6 and 17-7).

Considerable parts of the Miocene geosynclinal pile in northeast Japan were metamorphosed in the zeolite, prehnite–pumpellyite, and greenschist facies. Examples of the Mogami area and the Tanzawa Mountains were described in §6A-2. The latter example is exceptional in that metamorphic rocks up to the amphibolite facies were formed. The calculated geothermal gradients in these areas were very high.

Several tens of granitic masses occur over the geosynclinal zone of northeast Japan. Their K–Ar ages are 3–25 m.y. (Miocene and Pliocene). Only the larger masses are shown in fig. 15-10 under the name of late Tertiary granitic rocks. These granites are similar in age to those of the Shimanto terrane of southwest Japan.

15-10   SUMMARY OF THE CHRONOLOGY AND DISTRIBUTION OF METAMORPHIC BELTS IN THE HONSHU ARC

Chronological Summary

A tentative chronology of six major metamorphic events has already been given in table 15-1.

The Hida metamorphic complex has preserved records of the first four events. The corresponding terranes are designated as Hida I-IV in the table. In other words, granite-bearing metamorphic terranes in all these events are exposed within the Hida Plateau. This suggests that there was no great shift of the site of successive metamorphic belts in this period, which would be the main reason why these older metamorphic belts were obscured and destroyed by later events.

The 4th and 5th events showed successive southward shifts of the site of the resultant metamorphic belts. This resulted in relatively good preservation of the metamorphic belts recording these events.

There is no evidence for the formation of glaucophane schists in the Precambrian event. However, the mineral certainly occurs in the Ordovician Kiyama complex of the 2nd event. In the 3rd event, the Hida and circum-Hida terranes were perhaps in a pair. The 4th and 5th events formed two conspicuous pairs of metamorphic belts. The high-pressure belt of the 6th event, if formed, is not exposed yet.

The first five events took place along the Honshu Arc, leading to the formation of a thick continental type crust, whereas the Southwest Japan Arc has remained outside the main theatre of the 6th event. Presumably the Japanese Islands drifted away from the Asiatic Continent in the period between the 5th and 6th events.

Nature of the Median Tectonic Line

The relation between the Median Tectonic Line and the thermal axes of the Ryoke metamorphic belt is shown in fig. 15-13. It shows that in the eastern part of the Southwest Japan Arc, most of the outer half (southeastern half) of the Ryoke belt and the inner half (northwestern half) of the Sanbagawa belt has been lost by fault movement along the Median Tectonic Line. That the Line is a sharp southern limit for distribution of Cretaceous and early Tertiary granitic rocks (fig. 15-10), also suggests that the lost zone is of a considerable width (say, 30 km or more).

Since the fault cuts the metamorphic belts and granites sharply, the main movement must have taken place after the completion of Jurassic and Cretaceous metamorphism and plutonism, though this does not preclude the possibility that the initiation of the Line took place in an older time, say, in the Cretaceous. Cataclasis took place on the fault, and some rocks along the eastern part of the Line have been interpreted by many authors as protoclastic intrusives.

It is widely believed that movement along the Line took place in a number of distinct phases showing different patterns of movement. The initial phase, usually called the Kashio, is widely believed to be thrusting of the Ryoke terrane over the Sanbagawa, though there is no positive evidence for the existence of this phase. It is also conceivable that the Line was initiated as a Benioff-zone fault along which the Sanbagawa terrane was thrust under the Ryoke terrane (e.g. Ernst, 1971*b*). Another possibility is that the Line was initiated simply as the boundary fault between the already metamorphosed Ryoke and Sanbagawa

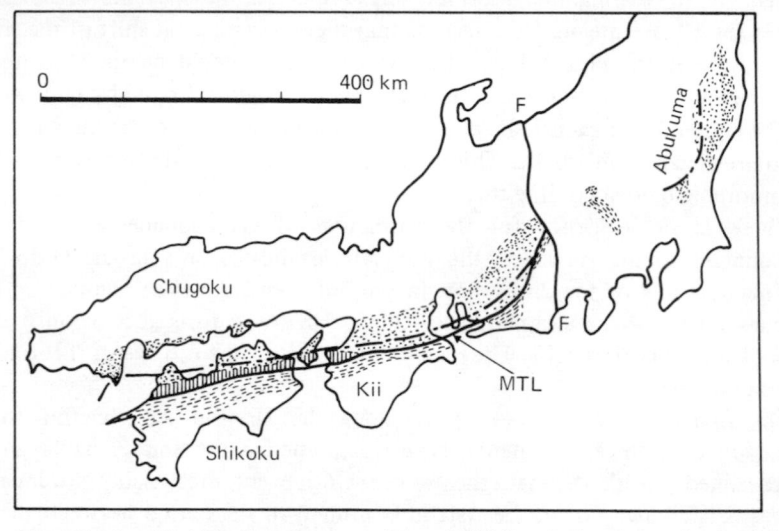

Fig. 15-13   The Median Tectonic Line (MTL) and the thermal axes (chain lines) of the Ryoke and Abukuma metamorphic terranes (dotted). The vertical hatching represents the Upper Cretaceous Izumu Group. F–F = Itoigawa–Shizuoka Line. Dashed area = Sanbagawa belt.

blocks, and that the Sanbagawa was uplifted along the fault so as to be ultimately exposed on the surface (§17-6). If this is the case, the Median Tectonic Line is an ancient analogue of the Morioka–Shirakawa Line of present-day Japan (§15-9 and 17-6).

In the last 100 000 years, a continual right-lateral movement has taken place along the central part of the Line at an average rate of several millimetres per year (Kaneko, 1966; Okada, 1970). However, the eastern end, cut by the Itoigawa–Shizuoka Line, does not show marked off-setting. Hence, the total amount of strike-slip movement since the formation of the Itoigawa–Shizuoka Line in late Tertiary time is very limited. It is possible, however, that the Line was initiated as an early Tertiary strike-slip fault accompanied by some vertical movement. Miyashiro (1972*b*) has shown that the main features of the thermal structures of the Ryoke and Sanbagawa belts can be explained by this hypothesis.

Hidaka Metamorphic Belt

The Hidaka belt is mainly composed of metapelites and metapsammites accompanied by migmatites, granites and gabbros. An axis of maximum metamorphic temperature is exposed within it (fig. 15-14). At the western margin of the belt, a zone of schists is in thrust contact with an unmetamorphosed sedimentary terrane. The belt grades eastwards into an unmetamorphosed terrane (e.g. Hunahashi, 1957).

Granitic rocks occur on the eastern side, whereas gabbroic ones are more abundant on the western side of the axis. A fresh mass of peridotite is present in the westernmost part (Nagasaki, 1966). The asymmetrical distribution of plutonic rocks is similar to that in California (§13-4).

The original sedimentary rocks are probably Mesozoic, and the metamorphism occurred probably in latest Mesozoic or Tertiary time. The K–Ar ages of granitic rocks are about 36–17 m.y., i.e. Oligocene and Miocene (Shibata, 1968; Kawano and Ueda, 1967b).

Some metapelites contain andalusite and sillimanite. Cordierite is common. The metamorphism is of the low-pressure type.

Kamuikotan Metamorphic Belt

The original rocks of the Kamuikotan belt are mainly clastic sediments and basaltic volcanics, Mesozoic in age. They are accompanied by a magnificent serpentinite belt. The age of metamorphism is latest Mesozoic or Tertiary.

Glaucophane, lawsonite and the jadeite + quartz assemblage are widespread, though petrographic study has been made only in small parts of the belt (Banno and Hatano, 1963; Tazaki, 1964). Some rocks of the epidote-amphibolite facies also occur (Shido and Seki, 1959).

The northward extension of the Kamuikotan belt is exposed in Sakhalin (Sobolev et al. 1967).

15-12 METAMORPHIC BELTS OF THE RYUKYU ARC AND TAIWAN

In a few areas near Nagasaki in the westernmost part of Kyushu, a group of low-temperature metamorphic rocks occasionally containing glaucophane occurs in association with serpentinite. It is called the *Nagasaki metamorphic complex* (fig. 15-8). The general structural trend of this complex is N–S. On the west side of these areas, there is a thrust trending N–S. Granitic rocks with cataclastic structure are exposed on islands to the west of the thrust.

The Rb–Sr and K–Ar ages of the Nagasaki metamorphic complex are 94–60 m.y. (middle Cretaceous–early Tertiary). The K–Ar age of a hornfels related to the granitic rock to the west is 81 m.y. (Ueda and Onuki, 1968; Shibata, 1968).

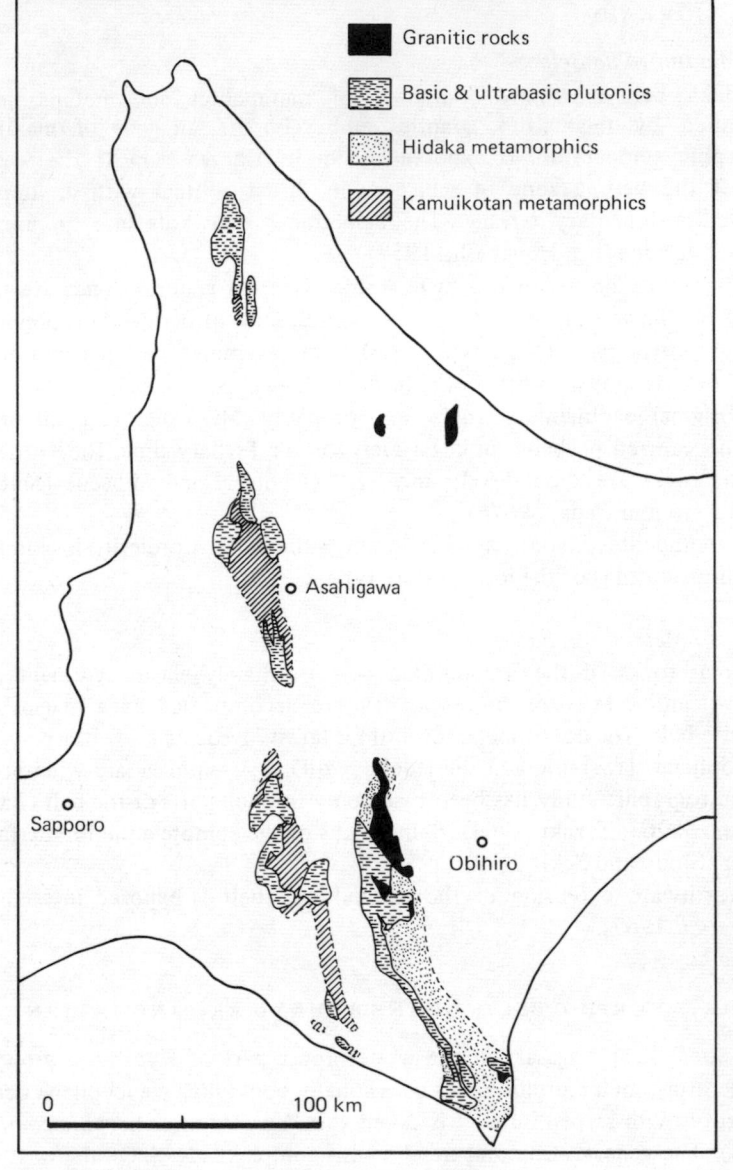

Granitic rocks

Basic & ultrabasic plutonics

Hidaka metamorphics

Kamuikotan metamorphics

Asahigawa

Sapporo

Obihiro

0   100 km

Fig. 15-14   Metamorphic belts in Hokkaido.

It is thus conceivable that the Nagasaki complex and the granitic complex represent paired metamorphic belts along the Ryukyu Arc, which connects Kyushu to Taiwan (Miyashiro, 1965).

Since the islands of the Ryukyu Arc are mostly small, the structure of the Arc is not well exposed. Konishi (1963) distinguished four structural zones along the Arc, correlating them with zones in southwest Japan. A zone that is relatively well exposed in Okinawa Island contains metamorphic rocks of the greenschist and epidote-amphibolite facies. Another zone exposed in Ishigaki and the adjacent islands contains glaucophane schists (Foster, 1965; Kuroda and Miyagi, 1967). There may be some other submerged zones with metamorphic rocks.

Some metamorphic belts in the Ryukyu Arc could be continuous with the rocks of Taiwan beneath the sea floor. On the eastern slope of the Central

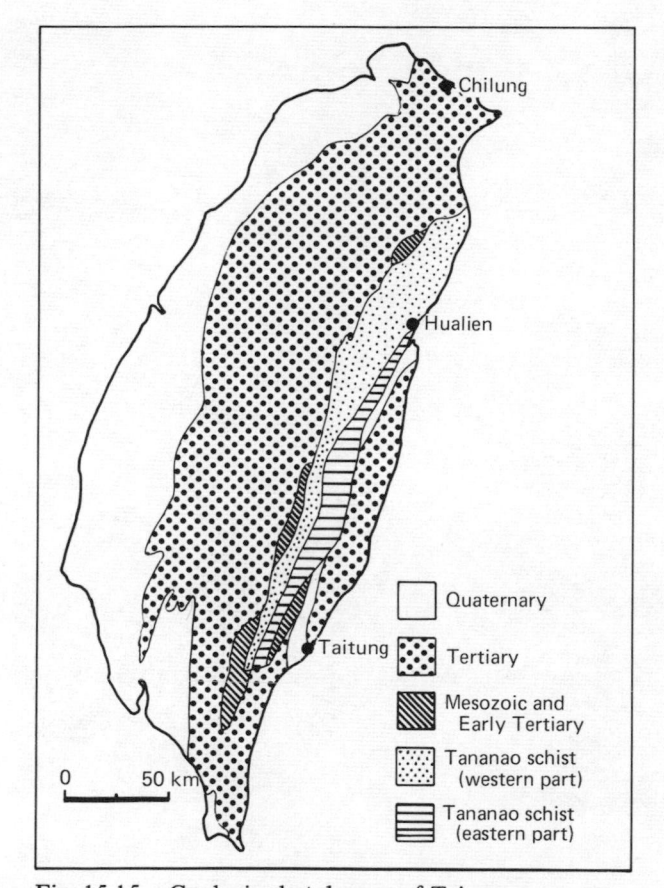

Chilung

Hualien

Taitung

0   50 km

☐ Quaternary

▦ Tertiary

▨ Mesozoic and
    Early Tertiary

▤ Tananao schist
    (western part)

▤ Tananao schist
    (eastern part)

Fig. 15-15   Geologic sketch map of Taiwan.

Mountain Range of Taiwan, a regional metamorphic complex called the *Tananao schists* is exposed with a NNE trend (fig. 15-15). A part of the original sediments is Permian, and K-Ar dating of the associated quartz diorites gave Cretaceous and Tertiary ages. Most of the Tananao schists are in the greenschist facies, but a small part near the northern end is in the amphibolite facies (Yen, 1962, 1963; Yen and Rosenblum, 1964).

Generally speaking, the eastern part of the Tananao schist terrane is mostly in the chlorite zone with abundant serpentinite and occasionally with glaucophane, whereas the western part ranges from the chlorite schists to gneisses of the amphibolite facies. The relation of the western and eastern parts shows some resemblance to paired metamorphic belts (Yen, 1959, 1963).

Chapter 16

Metamorphic Belts in the Southwest Pacific

16-1 CHARACTERISTICS OF THE SOUTHWEST PACIFIC REGIONS

We discussed the metamorphic belts in Japan and the surrounding West Pacific regions in the preceding chapter, and now we are dealing with metamorphic belts in the southwest Pacific region. They show many features in common. Paired metamorphic belts occur in New Zealand and Celebes.

The Australian continent is made up of orogenic belts that tend to become younger toward the east. New Zealand, lying further to the east, tends to be still younger. This relation appears to suggest accretional growth of a continent. Australia is treated here from this viewpoint.

16-2 THE AUSTRALIAN CONTINENT

The Precambrian Shield
There is a wide Precambrian shield with an ordinary continental type crust in the western and central parts of Australia (fig. 16-1). Before the beginning of recent continental drift, it was probably connected with Precambrian regions of Antarctica and India. Seven periods of granitic and metamorphic activity have been provisionally distinguished in the Precambrian from radiometric data. The oldest one is 3050–2900 m.y. ago, and the youngest one 700–650 m.y. ago (Compston and Arriens, 1968).

The Precambrian progressive metamorphism in the Broken Hill area in the western part of New South Wales (§ 8A-7), and that in the East Kimberley area (§ 7A-3) are well documented. In both areas, the highest temperature reached belongs to the granulite facies. Granulite facies rocks occur here and there in the Precambrian terranes (fig. 4-3 and § 18-3). A systematic description of rock types in the Precambrian shield has been given by Joplin (1968), with an introductory statement, 'The areas of Precambrian are vast and the number of geologists relatively few.'

It appears that a large part of the Precambrian metamorphic rocks belongs to the low-pressure series, though rocks of the medium-pressure series also occur.

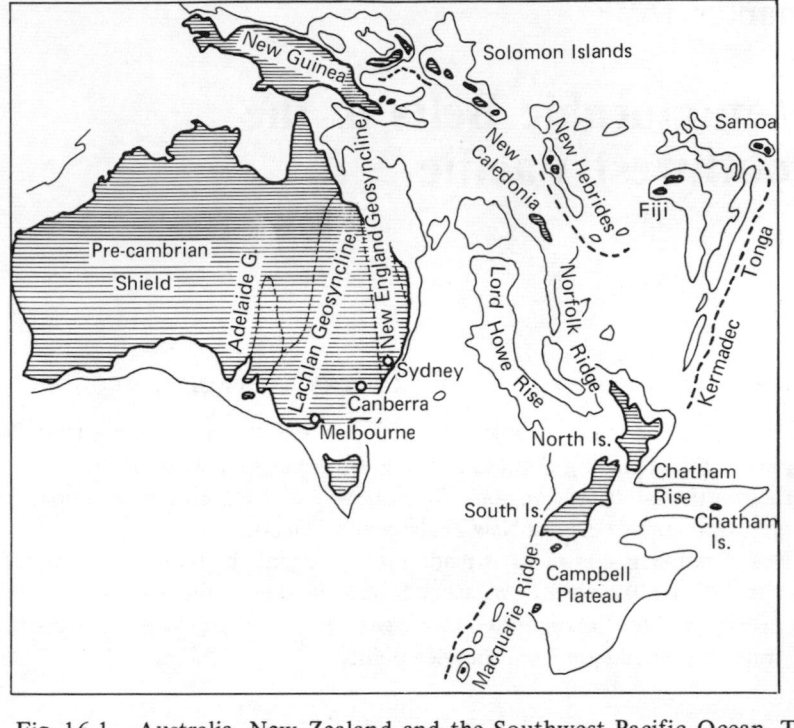

Fig. 16-1   Australia, New Zealand and the Southwest Pacific Ocean. The thick lines are the coast lines and the thin lines 1000 fathom (1830 m) isobaths. Trenches are indicated by thick broken lines. The Tonga and Kermadec Trenches are on the east side of the Tonga and Kermadec Islands, and another trench is on the west side of the Macquarie Ridge. The New Hebrides Trench is on the continental side of the New Hebrides Islands.

Adelaide Geosyncline

The Adelaide Geosyncline was formed in late Precambrian to early Palaeozoic time around the present site of Adelaide in South Australia. Metamorphism and plutonism took place in early Palaeozoic time (600–400 m.y. ago). The general trend of the orogenic belt is N–S, running through the Precambrian craton near its southeastern margin. Its southern end bends toward the west at Kangaroo Island.

Metamorphic rocks in the Mount Lofty Ranges are well documented (§ 7A-3). The metamorphism is transitional between the low- and the medium-pressure type (§ 7A-3). Relatively small bodies of granitic rocks are intruded into schists and gneisses along the axis of the metamorphic belt (White, 1966; White *et al.* 1967).

Tasman Geosyncline

The highlands in the eastern part of the Australian Continent represent the former position of the Tasman geosyncline in Palaeozoic time. The geosyncline contained a large number of subparallel troughs and ridges with an N–S trend. Some of the troughs and ridges had intense volcanism, whereas others did not.

The geosyncline tends to become younger eastwards. The western part of the geosyncline (called the *Lachlan geosyncline*) has sedimentary piles ranging from Cambrian to Middle Devonian with the maximum deformation in Middle Devonian time. The eastern part (called the *New England geosyncline*) has sedimentary piles ranging from Ordovician to Late Permian with the maximum deformation in Late Permian time (Crook, 1969). The easternmost part of the continent near Brisbane has a still younger sedimentary pile extending from the Permian to Early Cretaceous, after which it was deformed. The deposition in this region is roughly simultaneous with that in the New Zealand geosyncline. Volcanism in the Tasman geosyncline is basic and acidic. There are three big serpentinite belts.

The K–Ar ages of granitic rocks show a regular eastward decrease from Silurian to Permian in the Melbourne-Canberra-Sydney region of southeastern Australia (Evernden and Richards, 1962).

The thick sedimentary piles in the Tasman geosyncline were subjected to widespread low-temperature recrystallization in the zeolite, prehnite-pumpellyite and greenschist facies (e.g. Packham and Crook, 1960).

The Hill End trough and the Cowra trough run in a N–S direction to the east and west of Canberra respectively with the Canberra–Molong ridge between them. They are all within the Lachlan geosyncline. Smith (1969) has described recrystallization in these zones in an area about 180 km north of Canberra. The rocks on the ridge are non-schistose and mainly in the prehnite-pumpellyite facies, whereas those in the Hill End trough are schistose and tend to show a higher temperature of recrystallization up to the biotite zone (greenschist facies).

In a region south of Canberra, two high-temperature metamorphic belts run in a N–S direction in the Hill End and Cowra troughs. They are of the low-pressure type, being associated with granitic rocks. The Cooma area in the Hill End trough (Joplin, 1942, 1943, 1968; Pidgeon and Compston, 1965), and the Wantabadgery–Tumbarumba area in the Cowra trough (Vallance, 1953, 1960, 1967) are well documented. The metamorphic rocks in these areas are mostly schists and gneisses of pelitic and psammitic composition, producing andalusite, cordierite and sillimanite (§§ 4-1 and 7A-3).

The occurrence of glaucophane has been reported from a few localities in New South Wales and Queensland in the easternmost part of the Tasman geosyncline. Especially in Port Macquarie, New South Wales, glaucophane is accompanied by lawsonite (Joplin, 1968, p. 100). Their geologic relations and significance are not clear.

16-3 NEW ZEALAND

Island Arc Structure

New Zealand may be a front of eastward growth of the Australian Continent (fig. 16-1). It stands in a line with the island arc of the *Kermadec and Tonga Islands*. The North Island of New Zealand is strongly influenced by the activity of this arc, as is clear, for example, from the presence of many Quaternary volcanoes on the island and also from the Hikurangi Trench to the east. Deep earthquakes take place with a Benioff zone dipping westward. The South Island, on the other hand, has been more stabilized. It is mainly composed of Mesozoic and older metamorphic and plutonic rocks. However, the southern part of South Island has late Tertiary and Quaternary volcanism. This is probably an expression of the weak orogenic activity of the *Macquarie Rise,* which is an island arc extending southwestwards from South Island and is accompanied by a shallow trench on the west side (Summerhayes, 1967; Miyashiro, 1972*b*).

The South Island is cut obliquely by a big strike-slip fault, named the *Alpine fault,* which offsets the Cretaceous metamorphic belts. The total amount of right-lateral shift is estimated to be of the order of 480 km. This fault could be a transform fault connecting the Tonga–Kermadec Trench with the Macquarie Trench (e.g. Morgan, 1968). Cataclastic metamorphism along this fault will be reviewed in § 20-3.

The Tasman Sea, which is on the continental side of New Zealand, has a deep basin with an oceanic type crust, just as the Japan Sea on the continental side of Japan. On the north of the Tasman Sea, there are two great submarine ridges, called the *Lord Howe Rise* and the *Norfolk Ridge,* which continue right through New Zealand and extend further toward the southeast possibly to the Campbell Plateau and the Chatham Rise respectively. These ridges could be surface manifestations of old orogenic belts (Cullen, 1967; van der Linden, 1967, 1969). Landis and Coombs (1967) and van der Linden (1969) considered that the ridge connecting the Lord Howe Rise and the Campbell Plateau represents the Palaeozoic Buller geosyncline, which will be commented upon later, though Aronson (1968) has suggested the possibility of its being of Precambrian age. The Norfolk Ridge–Chatham Rise could represent the late Palaeozoic–Mesozoic New Zealand geosyncline (van der Linden, 1969).

Detrital zircons from a pre-Devonian sedimentary formation in western South Island of New Zealand have shown $^{207}Pb/^{206}Pb$ ages of 1170–1480 m.y. It suggests the presence of some Precambrian crystalline rocks in and/or near New Zealand (Aronson, 1968). Evidently, however, Precambrian rocks have not been found in New Zealand.

Paired metamorphic Belts

Landis and Coombs (1967) have shown the existence of paired metamorphic belts in the South Island. This Island may be divided into western and eastern provinces, which represent respectively a low-pressure metamorphic belt accompanied by abundant granitic rocks and a high-pressure metamorphic belt with some glaucophane-schist facies rocks. A major high-angle fault, called the *Median*

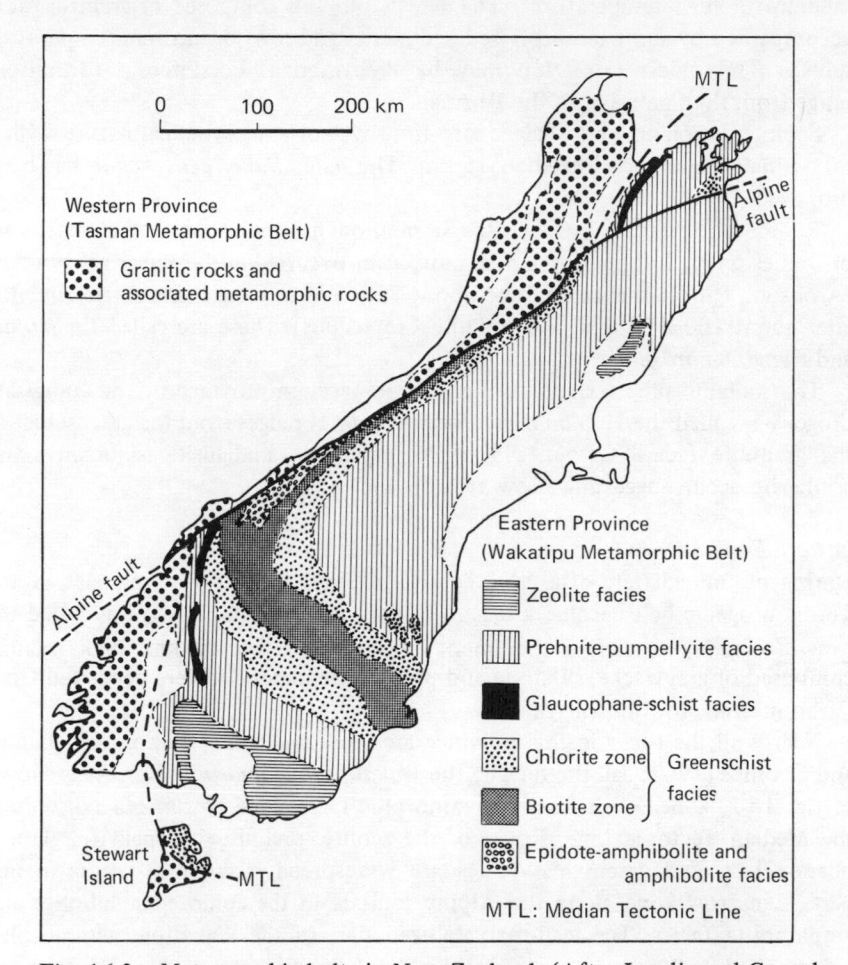

Fig. 16-2   Metamorphic belts in New Zealand. (After Landis and Coombs, 1967.)

Tectonic Line, has been found between the two. It is to be noted that the southernmost part of the fault shows strong bending toward the Campbell Plateau (fig. 16-2).

The geologic and geophysical similarity of the paired belts with those in Japan has been emphasized by Hattori (1968).

Western Province

This province is divided into the South and North blocks by the Alpine fault. The South block is mainly composed of metamorphic and associated granitic rocks. The metamorphic rocks are gneisses and amphibolites recrystallized at medium or high temperatures. The North block is composed of granitic rocks accompanied by unmetamorphosed sediments and low- or medium-temperature schists. The oldest formation may be Precambrian. Fossiliferous formations range from the Cambrian to the Permian.

Sedimentation in early Palaeozoic time was of a geosynclinal nature with a total thickness of not less than 15 km. The name *Buller geosyncline* has been proposed.

Radiometric dating by the Rb–Sr method has indicated that at least two orogenies with plutonism and metamorphism occurred in the Buller geosyncline (Aronson, 1968): the earlier one about 350–370 m.y. ago (Devonian) and the later about 100–120 m.y. ago (middle Cretaceous). These are called the *Tuhua* and *Rangitata orogenies* respectively.

The metamorphic terrane formed in the western province by the Rangitata orogeny is called the *Tasman metamorphic belt*. It ranges from the greenschist to the granulite facies. Garnet, chloritoid, staurolite, andalusite, sillimanite and cordierite occur, suggesting a low-pressure facies series.

Eastern Province

During or immediately after the Tuhua orogeny, the region lying east of the Tuhua orogenic belt became a basin for intense sedimentation. This is called the *New Zealand geosyncline*. Sediment piles, more than 10 km thick, mainly composed of graywacke, siltstone and pyroclastic materials, were deposited from Carboniferous to Jurassic time.

Nearly all the rocks in this province are more or less metamorphosed. Landis and Coombs (1967) call the terrane the *Wakatipu metamorphic belt*. As is shown in fig. 16-2, zones of different metamorphic facies tend to run generally along the Median Tectonic Line. Rocks of the zeolite, prehnite–pumpellyite, glaucophane-schist, and greenschist facies are widespread. Lawsonite occurs in one part. A narrow zone along the Alpine fault is in the epidote-amphibolite and amphibolite facies. The well-recrystallized part of the Wakatipu metamorphic belt is called the *Haast schist group,* which includes the *Otago schists* (Hutton and Turner, 1936; Hutton, 1940; Brown, 1967) in the southern part of the South Island and the *Alpine schists* along the Alpine fault (Reed, 1958; Mason, 1962; Cooper and Lovering, 1970).

The zeolite and prehnite–pumpellyite facies regions in the southern and central parts of South Island have become famous through the studies of Coombs (1954, 1960) and Coombs *et al.* (1959), as reviewed in § 6A-4. Stilpnomelane is common and piedmontite occurs rarely. Progressive increase of metamorphic temperature towards the west is clearly noticed in the Alpine schists, whereas the Otago schists show an axis of maximum metamorphic temperature.

Gabbro and serpentinite of Permian age occur in the western part of this province near the Median Tectonic Line. Small amounts of dioritic and granitic rocks probably of Permian age also occur along the western margin of the province. There are no plutonic bodies in the eastern part of the New Zealand geosyncline.

Landis and Coombs (1967) have emphasized that the New Zealand geosyncline is auto-cannibalistic. Thus, sedimentation is immediately followed by burial metamorphism in the zeolite or prehnite–pumpellyite facies. Some of the rocks formed in this way are exposed to erosion and transported into a newly-formed adjacent depositional basin. Then the new deposits are subjected to another phase of burial metamorphism. Sedimentation, burial metamorphism and erosion proceeded simultaneously.

The K–Ar ages of schists of the prehnite-pumpellyite and greenschist facies are 140–100 m.y., i.e. early to middle Cretaceous (Harper and Landis, 1967). This corresponds to the Rangitata orogeny. This regional metamorphism took place after the Late Jurassic completion of sedimentation and the accompanying burial metamorphism. It is to be noted that metamorphic rocks of the glaucophane-schist facies have given Cretaceous ages though their age of deposition is Late Permian. They would have formed by the Rangitata orogeny rather than by early phases of burial metamorphism.

The metamorphic climax of the Haast schist group is considered to have occurred about 130–140 m.y. ago. This value is observed in the high prehnite-pumpellyite facies rocks. With increasing grade of metamorphism the observed K–Ar and Rb–Sr ages tend to decrease down to 100 m.y., perhaps because the higher grade rocks maintained a higher temperature for a longer period. Metamorphic rocks exposed at a distance less than 30 km from the Alpine fault show anomalously low ages.

The Wakatipu metamorphic belt is obliquely cut by the Alpine fault. Thus, a zone of glaucophane-schist facies rocks and serpentinites, for example, occurs to the east of the fault in the southern part of South Island. Its northward continuation appears to the west of the fault near the northern end of the Island.

The New Zealand geosyncline extends northward to form the basement of North Island. The northward extension of the serpentinite belt in the geosyncline shows a gradual change of direction toward the northwestern projection of North Island. The Median Tectonic Line seems to extend roughly along the west coast of North Island. The extension of the New Zealand geosyncline appears to continue to the Norfolk Ridge.

Chatham Island, 800 km east of South Island, is known to be composed of pumpellyite-bearing metamorphic rocks similar to those of South Island. The New Zealand geosyncline appears to have continued to this Rise.

Metamorphic History of New Zealand and Vicinity

A tentative account on the metamorphic history of New Zealand will be given below.

A Precambrian crystalline basement may have formed along the Lord Howe Rise, the South Island of New Zealand and the Campbell Plateau. Probably the Tuhua orogeny and the accompanying metamorphism took place again roughly along the same zone during the Devonian. Some granitic and metamorphic rocks now exposed in the western province of South Island were formed in this period.

Van der Linden (1969) has shown the possible existence of an extinct or dormant mid-oceanic ridge trending north-south in the Tasman Sea. It is conceivable that the Lord Howe Rise was formed originally along the east coast of Australia, and later, possibly in Tertiary time, moved eastward by the generation of the Tasman Sea.

The New Zealand geosyncline was formed on the northeast side of the old orogenic belt, that is, along the Norfolk Ridge, the North and South Islands of New Zealand and the Chatham Rise. Thic sediment piles were deposited from the Carboniferous to Jurassic. Burial metamorphism took place broadly at the same time as deposition and erosion. Metamorphism accompanying the Rangitata orogeny occurred in both western and eastern provinces probably in the Late Jurassic to Early Cretaceous, leading to the formation of paired metamorphic belts.

The trend of the Median Tectonic Line in North Island and in Stewart Island (south of South Island) suggests a NW–SE trend for the paired metamorphic belts. The northwest projection of North Island represents an effect of the island arc structure formed in this period.

In Late Cretaceous and Tertiary time, the tectonic activity along the arc connecting North Island of New Zealand to the Kermadec and Tonga Islands became intense. Great strike-slip faults were formed on the islands and ocean floors in connection with this activity. The Alpine fault, being one of the faults, strongly changed not only the rock distribution but also the shape of South Island so as to conform with the direction of present-day activity along the North Island–Tonga line.

16-4 CELEBES

The island of Celebes belongs to an orogenic belt of late Mesozoic to Cenozoic age. The island shows a remarkable four-armed morphology, which is due to a connected double arc with its concave side towards the Pacific Ocean (fig. 16-3).

The north arm, the western part of central Celebes, and the south arm together constitute the inner arc, where late Mesozoic and Tertiary granitic rocks are widespread in association with biotite-rich schists (van Bemmelen, 1949).

The granitic masses of the inner arc are cut abruptly on the eastern side by a great fault, called the Median Line, trending north-south. Mylonite has been formed along the Line.

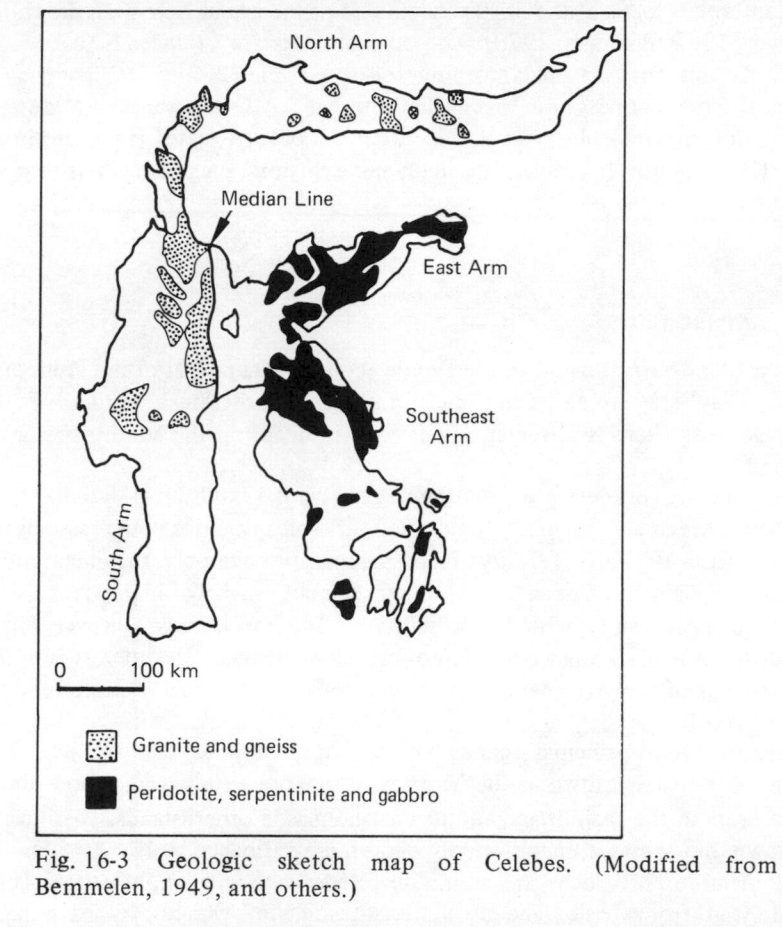

Fig. 16-3 Geologic sketch map of Celebes. (Modified from van Bemmelen, 1949, and others.)

The east arm, the central and eastern parts of central Celebes, and the southeast arm, which all lie to the east of the Median Line, together constitute the outer arc. In this arc, basic and ultrabasic plutonic rocks of Mesozoic and Cenozoic ages are widely exposed, together with glaucophane-schist facies rocks. Basic volcanic rocks are abundant in the geosynclinal pile. In some parts jadeite has been found in association with quartz (de Roever, 1955a). Thus, the inner and outer arcs constitute paired metamorphic belts.

16-5 NEW CALEDONIA

New Caledonia is elongated in a NW–SE direction, measuring about 400 km in length (fig. 16-1). It is noted for extensive development of glaucophane schists and serpentinite. Peridotite-serpentinite masses occur in a line on the southwest side of the island. Their tectonic setting is in dispute (Avias, 1967).

Metamorphic rocks at the northwestern end of the island have been described (Coleman, 1967; Brothers, 1970). The original rocks are Cretaceous to Eocene. The K–Ar ages of the metamorphic rocks are 21–38 m.y. (Oligocene to Miocene). Three zones have been distinguished. All the zones are due to high-pressure metamorphism as is clear from the occurrence of lawsonite in all zones. The metamorphic rocks are usually non-schistose, preserving their original structure.

16-6 YAP ISLANDS

The Yap Islands are situated in the Pacific about 1300 km east of the Philippine Islands. They are on an island arc forming a southeastern border of the Philippine Sea. There is a trench, about 8500 m deep, on the southeast side of the island.

The island arcs bordering the Philippine Sea (i.e. the Izu-Bonin, Mariana, Yap, and Palau Arcs) are generally composed of volcanic rocks and associated sediments from the early Tertiary to the present. For example, the oldest rocks exposed on Guam and Saipan are Eocene (or possibly slightly older) pyroclastics (basaltic, andesitic and dacitic in composition). The Yap Islands, however, differ from all the rest in having a basement of crystalline schists. Our interest is in the characteristics of regional metamorphic rocks produced in such a place very far from any continents.

There are four principal islands in the Yaps, close to one another. The metamorphic rocks, known as the Yap formation, are exposed in an area about 10 km across in the main island and in smaller areas in other islands. Their age is not known, but they are unconformably covered by Miocene beds.

The metamorphic rocks are actinolite greenschists and amphibolites. They were derived from basic igneous materials, some of which may have been abyssal tholeiites. The amphibolites contain blue–green hornblende. Schistosity is well developed and generally strikes northeastward, roughly parallel to the elongation of the arc. Relict minerals are rare. The temperature of metamorphism increases southeastward. The metamorphic rocks are accompanied by small masses of serpentinite (Tayama, 1935; Johnson, Alvis and Hetzler, 1960; Shiraki, 1971).

The Miocene breccias overlying the schists are reported to contain fragments not only of schist, amphibolite and serpentinite, but also of amphibole granite, amphibole syenite, and gabbro. The sources of the latter rocks are not known. The breccias are covered by Miocene volcanics made up of basaltic and andesitic pyroclastics and lavas. Since then, there has been no volcanism in the Yaps in spite of the presence of a deep trench to the east.

Chapter 17

Tectonics of Regional Metamorphic Belts

17-1 TECTONIC CLASSIFICATION OF METAMORPHIC BELTS

Three Main Categories

From the viewpoint of plate tectonics, most or all of the regional metamorphic belts of Phanerozoic time and possibly also of the Precambrian appear to have formed in the following three tectonic settings (Miyashiro, 1972b).

1. *Metamorphic belts formed in continental margins.* This type of belt could be forming now at the Pacific margin of South America, where the oceanic plate of the Pacific is being underthrust. The Mesozoic paired belts of the Franciscan and Sierra Nevada (Miyashiro, 1961a; Hamilton, 1969a, b) and the late Palaeozoic paired belts in Chile (González–Bonorino, 1971) could have formed in a similar way on the western margin of North and South America, respectively.

2. *Metamorphic belts formed beneath ordinary island arcs.* This type of belt could be forming now beneath the Northeast Japan and Kurile Arcs, where the oceanic plate of the Pacific is being underthrust. The Mesozoic paired belts in New Zealand (chapter 16) and the late Palaeozoic and Mesozoic paired belts in southwest Japan (chapter 15) may have formed in a similar way beneath ordinary island arcs, though it is also conceivable that they may have formed on the margin of the nearby continents of that time and drifted afterwards to form island arcs.

3. *Metamorphic belts formed beneath reversed island arcs.* This type of belt would be forming now beneath the New Hebrides Islands, where the plate of the adjacent marginal sea is being underthrust (cf. Isacks and Molnar, 1971). The paired metamorphic belts in Hokkaido (fig. 3-10) could have been formed by the eastward underthrusting of the marginal sea plate beneath the Sakhalin Arc.

Paired metamorphic belts can form in any of the three categories. All the metamorphic belts in these categories may be expected to show a tendency to be paired only if the velocity of plate descent is rapid (§17-2).

Orogenic Belts due to Continental Collision

Some orogenic belts such as the Alps, Urals and Himalayas are believed to have formed by the collision of two continental plates. In the initial stage of continental collision, the two plates are separated by an oceanic plate. The two continental blocks approach each other by the consumption of the intervening oceanic plate due to underthrusting along one trench zone or more. In this stage, the tectonic settings and resultant metamorphic belts, some of which may be paired, should belong to the above-discussed three categories. The alps and Urals have glaucophane-schist belts, which would have formed in this stage.

In the final stage, the intervening ocean is completely lost resulting in a direct collision of the two continental blocks. The buoyancy of the continental blocks would counteract and halt the descent of one plate beneath another. Hence, the characteristics of orogeny in this stage should differ from those of any of the above-discussed three categories. At present we cannot tell whether metamorphism on a large scale takes place at this stage or not. This stage may now be realized in the Himalayas, where there has been no sign of igneous activity (Gilluly, 1971). Metamorphic rocks are exposed in the Himalayas, but most of them appear to be of Precambrian age and formerly belonged probably to the Precambrian metamorphic basement of the Indian subcontinent (Petrushevsky, 1971). There may be some younger metamorphic rocks.

17-2   VARIATION IN THE RATE OF PLATE CONVERGENCE AND THE RELATION BETWEEN PAIRED AND UNPAIRED METAMORPHIC BELTS

Secular Variation in the Baric Type of Regional Metamorphism in Relation to Plate Motion

Low-pressure regional metamorphic rocks occur in any geological time whereas high-pressure metamorphic belts and in particular glaucophane schists formed mostly in the Phanerozoic (§4-3). Most of the typical glaucophane-schist facies rocks were produced by Mesozoic and Cenozoic orogenies.

As was discussed in §4-3, this relation can be explained by assuming that in younger geologic time oceanic plates have become thicker, and inclinations of descending slabs have become steeper, and/or the velocity of plate underthrusting has become more rapid. Such changes should cause lower geothermal gradients in subduction zones with resultant formation of metamorphic rocks of the high-pressure type (Miyashiro, 1972b; Ernst, 1972a).

Contrast in the Rate of Plate Motion between the Pacific and Atlantic

Though the pattern of spreading in the Pacific should have undergone a number of changes since early Mesozoic time, the rapidity of the plate motion has apparently continued to the present. The recent rate of spreading in the East

Pacific Rise is as high as 2–6 cm/year in contrast to the rate of 1–2 cm/year in the Mid-Atlantic Ridge, and it has been demonstrated that the contrast has existed at least since late Mesozoic time (Heirtzler *et al*. 1968).

Presumably, the plate motion in the Pacific was fairly rapid in the Palaeozoic so that atypical high-pressure metamorphic belts were formed in many parts of the circum-Pacific region, and it became very rapid in the Mesozoic so as to produce typical high-pressure belts. It may be considered that arc-trench systems of group I in table 4-4 can produce typical high-pressure belts and paired belts.

It appears that the Atlantic Ocean has closed and opened repeatedly, and this caused the orogenies of the Appalachian, Caledonian and even older belts (Wilson, 1966; Dietz and Holden, 1966; Dewey, 1969). We may consider that usually the velocity of plate descent in the Atlantic regions was not high enough to produce well-developed belts of glaucophane schists. Metamorphic belts of the medium-pressure type could have formed instead.

However, the medium- and low-pressure Caledonian metamorphic belt of the Scottish Highlands may be regarded as being paired with the poorly developed glaucophane-schist area of the Southern Uplands of Scotland. This pair was formed probably in the Early Ordovician. Presumably the Benioff zone at that time was inclined to the north, but a new Benioff zone inclined to the south may have been formed in Late Ordovician time so as to produce a volcanic arc to the south (Fitton and Hughes, 1970).

Possible Effects of the Descent of Mid-Oceanic Ridges beneath Continental Plates. Recent studies in marine geophysics have shown that mid-oceanic ridges descended beneath island arcs and continental margins in the Pacific regions in Mesozoic and Cenozoic times. This has posed a new problem to metamorphic geology, because the lithospheric plate in a ridge must be still hot, and the underthrusting of a hot plate cannot produce a very low geothermal gradient even if its rate is rapid, but may cause particularly intense or widespread magmatism in the arcs or continental margins.

Moreover, the rate of underthrusting of oceanic plates may be unusually rapid over a period preceding the descent of some mid-oceanic ridges beneath island arcs or continental margins, because the rate should be increased by the rate of creation of new lithosphere on the ridges. It may cause high-pressure metamorphism (Miyashiro, 1973; S. Uyeda and A. Miyashiro in preparation).

Atwater (1970) has shown that the mid-oceanic ridge between the Farallon and Pacific plates descended beneath the North American plate in middle Tertiary time. The Franciscan high-pressure metamorphism took place tens of million years prior to the ridge descent, possibly owing to the rapid underthrusting in this period. R. L. Larson and C. G. Chase (in press) have shown that the mid-oceanic ridge between the Kula and Pacific plates descended beneath Japan and the East Asiatic Continent in late Cretaceous times. The Sanbagawa high-pressure metamorphism took place tens of million years prior to the ridge descent.

In both cases, high-pressure metamorphism was halted about 30 or 40 million years prior to the ridge descent. Conceivably, the descending oceanic lithospheres in and near the ridges were still hot enough to increase the geothermal gradients in the subduction zones, and thereby to put an end to high-pressure metamorphism. For a period around the ridge descent, magmatism took place over an unusually wide zone in the Western United States and in East Asia (Miyashiro, 1973; S. Uyeda and A. Miyashiro, in preparation).

Nature of Apparently Unpaired Metamorphic Belts

Most of the metamorphic belts in the Atlantic regions and Europe are apparently unpaired. We may expect, however, that if a high-pressure metamorphic belt occurs, it could be accompanied by a belt of low-pressure metamorphism, since the descending slab of the former belt should tend to create the latter. A most remarkable example of a young, apparently unpaired high-pressure belt is in the Pennine zone of the Alps. It is of interest whether this belt fulfils the above expectation.

The Alpine Ranges continue eastwards to the Carpathians. Andesitic and rhyolitic rocks of late Tertiary and Quaternary ages are abundant in Hungary, southern Slovakia and Transylvania, which are on the south side of the Alpine–Carpathian Ranges. Tertiary granite rocks also occur in the andesitic-rhyolitic belt, though at the western end of the belt the Bergell granite has been intruded into the Pennine zone. This belt of granite, andesite and rhyolite shows usually high heat-flow values, and may be a surface manifestation of a hidden low-pressure metamorphic complex which is paired with the high-pressure belt exposed in the Pennine zone of the Alps (§14-5).

The Hercynian metamorphic belts in Western Europe are mostly of the low-pressure type (Zwart, 1967a, 1969). They might be regarded as a typical example of unpaired metamorphic belts. However, a part of the Hercynides may have paired belts. Hercynian glaucophane schists occur in the Ile de Groix to the south of Brittany. It is possible that an extension of the Ile de Groix metamorphic complex forms a Hercynian high-pressure belt which is now covered mostly by the sea, and is paired with a Hercynian low-pressure belt.

In this connection, it is noteworthy that in some pairs the high-pressure metamorphic belt is much narrower than the total area of the associated low-pressure metamorphic and granitic rocks, and hence it may readily be lost sight of owing to younger sediment covers, submergence beneath the sea, or later recrystallization at medium- or low-pressure conditions. A good example is the Jurassic–Cretaceous pair of southwest Japan. The Sanbagawa high-pressure metamorphic belt of the pair is only 50 km wide. The contemporaneous granitic activity took place not only in the Ryoke low-pressure belt but also in a wide region to the north. In Jurassic–Cretaceous time, this region of granitic activity was probably a part of the East Asiatic volcano-plutonic belt (fig. 15-6) on the east coast of the Asiatic Continent (cf. Lipman *et al.* 1971).

If the rate of plate underthrusting is too slow to cause high-pressure metamorphism, the contrasting characteristics of the paired belts will be obscured with resultant formation of an apparently unpaired belt. If the rise of magma and aqueous fluid is stopped at a very deep level, a low-pressure metamorphic belt may not form (Miyashiro, 1973).

17-3 SUMMARY OF GEOLOGIC FEATURES OF METAMORPHIC BELTS

High-Pressure Metamorphic Belts

1. High-pressure metamorphic belts are characterized by geothermal gradients as low as about $10 \,^{\circ}C \, km^{-1}$ or even lower (§3-4).

The width of a metamorphic belt measured from the thermal axis to the outer limit of the lowest temperature zone is usually few tens or several tens of kilometres.

2. Most of the rocks belong to the laumontite grade of the zeolite facies, the prehnite–pumpellyite, glaucophane-schist and greenschist facies. The character of metamorphic facies series usually varies from area to area even in a single metamorphic belt. There is the possibility that the lowest temperature part (i.e. the zones of the zeolite facies and a part of the prehnite–pumpellyite facies) of apparently progressive metamorphic terranes may actually have been recrystallized at a distinctly older or younger time than the associated higher temperature part.

3. In the Kanto Mountains of the Sanbagawa belt, the jadeite + quartz assemblage occurs only in the middle grade part of the terrane, whereas albite is stable in both lower and higher grade parts (§§3-4 and 7A-9; fig. 3-7). This is probably due to the upward convex shape of the geothermal curve in fig. 3-3 (§ 3-4). The facies series: prehnite–pumpellyite → glaucophane schist → greenschist (or epidote-amphibolite) in the Katsuyama area (§6A-5) and in some western parts of the Sanbagawa belt also suggest an upward convex geothermal curve. In the Franciscan terrane, albite is present in the poorly recrystallized part, and jadeite occurs in the well-recrystallized parts, which probably represent higher temperature zones (§7A-10). The Franciscan, therefore, may be regarded as corresponding to the lower half of the Kanto Mountains series (zones I–IV in fig. 3-7).

4. The temperature of metamorphic recrystallization commonly tends to increase toward the major fault which is the boundary between the high-pressure and the associated low-pressure metamorphic belt; e.g. the Median Tectonic Line in Southwest Japan, the Alpine fault in New Zealand, and the Coast Range thrust in California. However, the temperature decreases from a thermal axis to

both sides, for example, in the Katsuyama area (§6A-5; Hashimoto, 1968). Therefore, the temperature increase toward the major fault probably means that the original high-pressure metamorphic complexes were cut by the major faults into two units of roughly comparable sizes, and the unit on the low-pressure terrane side was lost. Hence, the nature of the major faults is of vital importance in the history of high-pressure metamorphic terranes.

A high-pressure metamorphic terrane is commonly composed of a number of smaller areas which were metamorphosed at somewhat different times, as will be summarized below. It is an unsolved problem why the whole terrane shows a relatively simple thermal structure in spite of the complex age structure. Our petrographic investigations may not have been thorough and precise enough to reveal a really complex thermal structure.

5. The time relation between sedimentation and metamorphism differs in different regions. Auto-cannibalism is common in the geosynclines pertinent to high-pressure metamorphism.

In the Franciscan and New Zealand geosynclines, it has been shown that the apparently single metamorphic terrane is actually composed of a number of sedimentary piles in different basins, that the sedimentation of individual piles was immediately followed by their metamorphism, and that the deposition and metamorphism in one pile were broadly simultaneous with upheaval and erosion in another. It is thus possible that rocks recrystallized in an earlier phase of metamorphism are exposed and eroded, and the resultant materials are transported and deposited in another basin to undergo a later phase of metamorphism (Suppe, 1969; Landis and Combs, 1967).

Such burial metamorphism in New Zealand may be distinct from the later regional metamorphism which produced greenschist and amphibolite facies rocks during the Rangitata orogeny, as there appears to be a temperature fall between the two events. However, the latest phase of burial metamorphism may have merged with the latter event.

In Japan, the Sangun metamorphism in western Honshu may have immediately followed sedimentation of a Palaeozoic geosynclinal pile, whereas the Sanbagawa metamorphism (or the main phase of it) would have taken place in Palaeozoic rocks at a distinctly later time (Juzassic–Cretaceous).

6. Graywackes are abundant, but typical aluminous pelitic rocks are usually scarce in high-pressure terranes.

7. Usually, high-pressure metamorphic belts are not accompanied by granitic rocks (e.g. Franciscan and Sanbagawa). The Alpine metamorphic belt in Western Europe has granitic rocks, which, however, are in a very small amount, and may actually belong to the associated andesitic belt to the south (§17-1).

8. Basaltic volcanics are abundant in geosynclinal piles. Gabbros are also common. More or less serpentinized peridotite occurs in association with these basic rocks. Some of the basaltic volcanics occur as thin beds conformable with

surrounding graywackes, and could be the result of volcanism simultaneous with sedimentation, possibly in a geosyncline. On the other hand, there are big ophiolite complexes showing a regular sequence of basalts, gabbros and serpentinites commonly in a strongly disturbed part of the metamorphic terrane. These may represent fragments of the oceanic crust conceivably transported from ocean floors (§ 4-1).

9. The Franciscan group was probably deposited on an oceanic crust, having no acidic crystalline basement. Atypical high-pressure terranes such as those of the European Alps and the Sangun belt have an acidic crystalline basement, as now exposed, for example, in the Gotthard massif and the Maizuru zone respectively. Such basements may have originally been thinner than the acidic basements of low-pressure metamorphic terranes.

Low-Pressure Metamorphic Belts

1. Low-pressure metamorphic belts are formed under geothermal gradients probably higher than $25\,^{\circ}C\ km^{-1}$ (§3-4). As a consequence, the width of a metamorphic belt measured from the thermal axis to the outer limit of the lowest temperature zone is usually relatively narrow (commonly up to 15 km). Where the whole metamorphic terrane is much wider, it may include two or more thermal axes, or alternatively it may have an unusually wide axial zone representing approximately constant temperatures.

2. Zonal mapping is usually easy, and rocks of the amphibolite facies usually form the widest zone. It appears that partial melting occurs commonly in the high amphibolite facies and higher zones.

3. The time relation between sedimentation and metamorphism is diverse. Regional metamorphism immediately followed geosynclinal sedimentation in eastern Australia, but in the Ryoke region metamorphism took place about 100 m.y. later than the end of sedimentation in the pertinent terrane. In the latter case, however, a new basin for sedimentation was created on the ocean side of the inactive geosyncline, and sedimentation continued there during the Ryoke metamorphism. The sediments of the old geosyncline were involved in the new tectonic movement, partly as a non-crystalline basement to the new geosyncline.

4. Low-pressure metamorphic belts are always accompanied by abundant granitic rocks. The emplacement of granitic rocks occurred mostly in a declining phase or after the end of regional metamorphism. In the Ryoke belt, the age difference between the regional metamorphism and the youngest granitic rocks related to it is probably of the order of 50–70 m.y.

Post-metamorphic granites commonly form hornfelsic aureoles. Late-metamorphic granites are usually surrounded by schistose metamorphic rocks apparently similar to the regional metamorphic rocks of the area. However, a slight increase of recrystallization temperature is noticed toward the contact in some areas, and especially when the intrusion is more basic (e.g. quartz dioritic) in composition.

5. Rhyolitic and/or andesitic volcanic rocks may occur abundantly in association with granitic rocks related to low-pressure regional metamorphism. This is particularly common in the circum-Pacific regions; e.g. the Ryoke belt of Japan, western North America and eastern Australia (§§4-2 and 15-7). Island arcs with andesitic volcanoes could be the surface manifestation of the formation of a low-pressure metamorphic complex in a crust.

6. Basaltic magmatism of the geosynclinal stage is usually scarce, and peridotite and serpentinite are very rare.

7. Pelitic rocks showing a high maturity are usually abundant in the terranes.

Paired Metamorphic Belts

1. Paired metamorphic belts are composed of two parallel belts of contrasting characters. Many paired belts are composed of a low-pressure belt on the continental (i.e. inner) side and a high-pressure belt on the oceanic (i.e. outer) side. In other pairs, the inner belt may be composed of low- and medium-pressure metamorphic terranes, and the outer belt may be a complex of high- and medium-pressure metamorphic terranes. Rarely, a high-pressure belt occurs on the continental side (§17-1).

2. The inner belt of a pair in New Zealand and in California appears to have been metamorphosed roughly in the same period as the associated outer belt. In the Alpine pair, the high-pressure metamorphism of the Pennine zone begun tens of million years earlier than the andesitic–granitic magmatism of the Hungarian zone. Precise age relations are not known in Palaeozoic pairs.

The age of cooling as revealed in isotopic dating may differ considerably from that of the preceding thermal climax, and the thermal climaxes of the two belts in a pair may be nearly simultaneous in many cases.

3. The age of deposition of the original rocks is generally younger in the high-pressure belt than in the associated low-pressure belt in New Zealand and California, whereas it appears to be similar in the Jurassic–Cretaceous pair in southwest Japan.

4. Polyphase recrystallization and auto-cannibalism have been observed in many high-pressure regional metamorphic belts and in some small-scale low-pressure terranes (e.g. §6A-2), but not in typical low-pressure regional metamorphic terranes. However, the K–Ar ages of granitic rocks in the Sierra Nevada and Mesozoic–Early Tertiary Japan show several clusters, which suggest the possibility that low-pressure metamorphic recrystallization and associated granitic activity also may have distinct phases. It is not clear whether there is any close genetic connection between the phases of paired belts.

5. The original distribution of rocks has been greatly modified by later faults in many regions. The two belts of a pair are now in direct contact with each other on a major fault in the Jurassic–Cretaceous of southwest Japan and in New Zealand, but have an intervening zone of virtually unmetamorphosed rocks in Hokkaido.

17-4 TECTONIC SIGNIFICANCE OF MEDIUM-PRESSURE REGIONAL METAMORPHISM

Many metamorphic belts of the medium-pressure type are associated with granitic rocks, just like those of the low-pressure type. However, the thermal effect of individual granitic masses is as obscure as that in low-pressure belt, if not more so.

If the high temperature conditions in low-pressure terranes are generally controlled by the rise of granitic magma and aqueous fluid, the maximum temperature reached may be mainly controlled by the temperature of rising magma and aqueous fluid, being nearly independent of the depths of intrusion. If this is the case, low-pressure regional metamorphism should take place in regions where numerous bodies of magmas rise to shallow depths, whereas medium-pressure metamorphism should take place in regions where the main level of intrusive bodies is deeper.

It is thus conceivable that the difference between the low-pressure terranes and such medium-pressure terranes is not intrinsic but comes simply from a difference in the level of abundant granitic intrusion.

This view is strengthened by the common gradation of a low-pressure area into a medium-pressure one as observed in the Scottish Highlands and the northern Appalachians. There is no tectonic demarcation line between such areas. It need not be considered important whether the inner belt of a pair is entirely of the low-pressure series or contains low- and medium-pressure series areas.

This discussion applies only to the medium-pressure terranes accompanied by granitic rocks. There are other medium-pressure terranes which are not accompanied by granitic rocks, and occur as parts of metamorphic belts that contain rocks of the high-pressure series in other parts. As discussed in §17-2, this type of medium-pressure terrane could form where the tectonic sinking was not rapid enough to cause high-pressure metamorphism.

If this is the case, the medium-pressure metamorphic terranes may be classified into those genetically connected with low-pressure terranes and those genetically connected with high-pressure terranes. Thus, in tectonics, only low- and high-pressure regional metamorphism may be of prime importance, and medium-pressure metamorphism may simply be treated as special variants of low- and high-pressure metamorphism.

17-5 TEMPERATURE DISTRIBUTION IN OROGENIC BELTS

Geothermal Gradients and Tectonic Sinking

The unusually low geothermal gradient of high-pressure metamorphism is probably caused by the rapid tectonic sinking of a geosynclinal pile due to the pull of a descending lithospheric slab. When this is almost simultaneous with sedimentation as in the Franciscan, rapid sedimentation could cooperate with the sinking to decrease the geothermal gradient. Where the sinking was not rapid enough to cause high-pressure metamorphism, a medium-pressure complex should be produced.

It is not simple to determine when tectonic sinking began, because recrystallization in the last major phase may efface the evidence for the earlier history of the terrane, as discussed in relation to the Sanbagawa belt (§15-8).

If underthrusting halts for some reason and a high-pressure metamorphic complex is kept at a definite depth for a long time, a stationary temperature distribution will gradually be approached, resulting in an increase of the geothermal gradient. Thus, the prevailing $P-T$ condition would change toward that of medium-pressure metamorphism, ultimately leading to the destruction of the previously-formed high-pressure mineral assemblages. The destruction of glaucophane observed in many parts of the Alpine metamorphic belt (e.g. de Roever and Nijhuis, 1964) may have occurred in this way. Therefore, the occurrence of a high-pressure metamorphic belt means that the terrane was not only rapidly sunk but also uplifted tectonically so that previously-formed high-pressure mineral assemblages are preserved.

The time of beginning and the duration of low-pressure metamorphism may depend mainly on the way of rise of granitic magma and aqueous fluid. If this is the case, high- and low-pressure metamorphism in a pair of belts may not be simultaneous, as exemplified by the Alpine pair (§ 17-3).

Calculation of the Temperature Distribution in a Geosynclinal Pile and in a Descending Lithosphere

Most metamorphic reactions may be regarded as equilibrium processes to a good approximation. On the other hand, heat conduction in a big geologic body is a very slow process, and the time lag of the temperature rise in a sinking geosynclinal pile or in a descending lithospheric slab may actually control the highest temperature attained by and geothermal gradients in the pertinent body.

The most recent method for the estimation of the temperature distribution in the geologic body is to solve the heat conduction equation by a digital computer using some assumptions on the crustal and subcrustal structures, the moving velocity of the body, the thermal conductivity (diffusivity) of the material, and the amount of heat generation in the mantle and crust. Comparing the calculated heat flow with the observed one, the model is successively revised and improved until the calculated thermal state becomes compatible with geologic and geophysical features of the area.

Minear and Toksöz (1970), Oxburgh and Turcotte (1971), and Hasebe, Fujii and Uyeda (1970) made detailed calculations of the temperature distribution in a lithospheric slab descending from a trench zone and its neighbourhood along the lines mentioned above. According to their results, a very low or even a negative temperature gradient is expected in the lithosphere during the first several tens of millions of years of plunging. When the shear-strain heating or frictional heating is intense along the upper surface of the plunging body, this part may be heated up so strongly that a large amount of magma is produced. Since the basic layer of the original oceanic crust would be present along the upper surface, the resultant magma would be andesitic and its rise would cause an increase in the volume of the continental crust above.

Another approach, free from any complexity due to the model adopted, is the use of the thermal relaxation time τ, which is one of the simplest measures of the rate of temperature variation in a body heated from the outside. The thermal relaxation time of heat conduction for a body of size H and of thermal diffusivity k, is represented by H^2/k, which represents the approximate time needed to heat the body to $1/e$ or $1/2.72$ of the possible maximum heating under the given conditions. When we take $k = 10^{-2}$ cm^2 sec^{-1}, which is a well-accepted order of magnitude for the thermal diffusivity of the material in the earth's crust, the thermal relaxation time for a crustal or lithospheric layer 10, 30 or 100 km thick is roughly 3, 30 or 300 m.y. respectively (Shimazu, 1961).

The development of a geosyncline is usually a process of the order of 100 m.y. in duration. A geosynclinal pile, 10–20 km thick, for example, will be heated in accordance with the slow sinking during its development with no marked time lag. On the other hand, when such a pile descends rapidly within a period of a few million years, unusually low geothermal gradients will form in it. This may be the case with individual phases of the Franciscan metamorphism.

If a lithosphere, 100 km thick, plunges in a period of less than 100 m.y., a marked time lag in heating should take place, leading to the occurrence of unusually low geothermal gradients and high-pressure regional metamorphism.

17-6 MAJOR FAULTS, SLAB SURFACES, AND THE UPLIFT OF HIGH-PRESSURE METAMORPHIC COMPLEXES

Three Classes of Major Faults

Metamorphic belts are usually cut by a large number of faults. Our interpretation of the origin and history of metamorphic belts depends largely on our interpretation of the nature of the major faults. The following three classes of major faults are probably of particular interest in this connection.

1. Transform faults (Wilson, 1965). The San Andreas fault of California and the Alpine fault of New Zealand appear to belong to this class. If the angle between the strike of a strike-slip fault and the axis of a metamorphic belt is small, the strike-slip movement will be difficult to determine. It is possible that some large strike-slip faults have been overlooked in metamorphic terranes.

2. Upper surface of a descending lithospheric slab, which would be a large fault cutting through the crust and upper mantle. The upper surface may represent a Benioff zone. However, there is some ambiguity in this correlation, and hence such faults will be referred to by the name *slab-surface faults.* As high-pressure metamorphism has been regarded as being due to underthrusting of a lithospheric slab, slab-surface faults, if exposed, could be discovered in or near high-pressure metamorphic terranes. Ernst (1970, 1971*b*) suggested that the Coast Range thrust of California and the Median Tectonic Line of Japan may be such faults.

3. Faults, along which crustal blocks that contain high-pressure metamorphic complexes are uplifted. A high-pressure complex formed at a considerable depth must be subsequently uplifted to be exposed on the surface.

The western segment of the Median Tectonic Line of southwest Japan (figs. 17-1 and 15-13) is the fault boundary between the Sanbagawa metamorphic terrane to the south and the upper Cretaceous Izumi Sandstone Group to the north. The Izumi Group was deposited on an eroded surface of the Ryoke

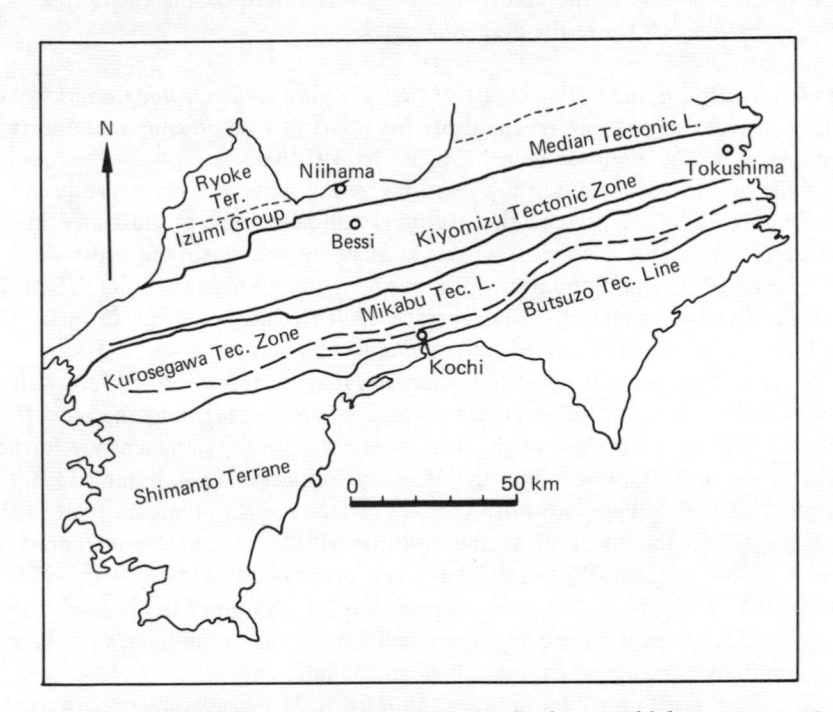

Fig. 17-1 Major faults in and around the Sanbagawa high-pressure metamorphic terrane in Shikoku, Japan. Approximately the region between the Median Tectonic Line and the Kurosegawa Tectonic Zone belongs to the Sanbagawa belt, though the southern limit of recrystallization is obscure.

metamorphic terrane which is paired with the Sanbagawa. Hence, this segment of the Median Tectonic Line was formed distinctly later than the Ryoke and Sanbagawa metamorphism, and the Sanbagawa high-pressure terrane should have been uplifted along the Line, though this does not preclude the possibility that the movement of the Line involved a strike-slip component.

The Northeast Japan Arc (i.e. northeast Honshu) has been active since the Miocene. As is schematically shown in fig. 3-11, it is composed of the outer non-volcanic arc and the inner volcanic arc with a large tectonic line, called the *Morioka–Shirakawa Line* (S in the figure) between them. The Line was detected by a gravity survey by Tsuboi *et al.* (1953-6), though the fault is entirely covered by younger sediments. The outer arc is characterized by high values of positive Bouguer anomalies. It is conceivable that the outer arc has been uplifted since the Miocene along this fault, with resultant approach to the surface of a Miocene high-pressure metamorphic complex in the crust on the oceanic side of this fault. Since the isostatic anomalies in the outer arc are of the order of +100 mgal, the uplift cannot be ascribed to the buoyancy of a thickened crust. The uplift could be due to the continentward push exerted by the Pacific plate on the prismatic block above the descending slab.

Models for the Mechanism of Uplift of High-Pressure Metamorphic Complexes
Three alternative possible mechanisms for uplift of high-pressure metamorphic complexes will be discussed below (Miyashiro, 1972*b*).

1. High-pressure metamorphic recrystallization may take place mainly in the descending oceanic crust and the overlying sediment mass, as illustrated in fig. 3-11. The resultant metamorphic rocks may be added to the crust on the continental side, and be uplifted by movement along a high-angle fault (S in the figure). In this model, the slab-surface fault would occur at or near the oceanic-side limit of the high-pressure metamorphic terrane.

2. A high pressure metamorphic complex may be formed in the foot wall of a slab-surface fault and subsequently it may be uplifted (*a*) along the same fault after the halt of the descent of the lithospheric slab, or (*b*) along a newly formed high-angle fault presumably like the Morioka–Shirakawa Line. In model (*a*), the major fault which represents both the slab surface and the boundary fault of the uplifted block, should occur at the continental-side limit of the high-pressure metamorphic terrane. This model has been proposed by Ernst (1970, 1971*b*). Model (*b*) means that a high-pressure complex which formed in the foot wall of a slab-surface fault is subsequently scraped off, added to the hanging wall, and then uplifted along a newly formed high-angle fault.

3. If the position of the descending slab shifts successively oceanwards, a high-pressure metamorphic complex formed at an older time should become a part of the hanging wall of a subsequently formed slab-surface fault and could be uplifted along a newly formed high-angle fault. In the Franciscan terrane,

metamorphism involves a number of distinct phases of recrystallization (Suppe, 1969), and the rocks in the western part are generally younger than those in the eastern. This may be the result of successive oceanward shifts of the descending slab (Hsü, 1971). A younger high-pressure metamorphic belt tends to form on the oceanic side of an older high-pressure belt, as exemplified in Japan (fig. 3-10) and the Kamchatka Peninsula (Dobretsov and Kuroda, 1969). This means that a new Benioff zone tends to form on the oceanic side of an older one.

In models (1) and (3), a slab-surface fault should appear at or near the oceanic-side limit of the high-pressure metamorphic terrane. In the Sanbagawa high-pressure terrane of Shikoku (fig. 17-1), the Kurosegawa Tectonic Zone (Ichikawa *et al.* 1956) is a group of large subparallel faults exposed near the oceanic-side limit of the metamorphic terrane (Banno, 1964), and might be a slab-surface fault. The Kiyomizu Tectonic Zone, which is exposed about 20 km north of Kurosegawa, might also represent a slab surface. This zone is not a simple fault but a zone, up to 1 km wide, of 'papery' schists probably representing an unusually strong shear movement contemporaneous with the Sanbagawa metamorphism (Kojima and Suzuki, 1958).

Distance between Paired Metamorphic Belts
In the present-day island-arc regions, there exists a non-volcanic zone, usually 100-250 km wide, between a volcanic belt and the associated trench. Dickinson (1971) has proposed the name *arc-trench gap* for such a zone. The arc-trench gaps of some present-day arcs are occupied by uplifted mountains, whereas those of others are occupied by troughs where active sedimentation is taking place. Since high-pressure metamorphism takes place in and near the trench zone, the resultant metamorphic complex should have originally been situated about 50-200 km from the low-pressure metamorphic complex beneath the associated volcanic belt.

Among the above-discussed models for the mechanism of uplift of high-pressure metamorphic complexes, model (3) may apparently reduce the distance between paired belts. In this case, the descending slab shifts oceanwards, and as a consequence the volcanic zone also should shift oceanwards with the resultant approach of the underlying low-pressure complex to an earlier-formed high-pressure complex.

The two metamorphic belts of the pairs in Hokkaido and California are separated from each other by a zone of virtually unmetamorphosed rocks, which may correspond to the arc-trench gaps. On the other hand, in the Jurassic–Cretaceous pair of southwest Japan, the Ryoke low-pressure belt is in direct contact with the Sanbagawa high-pressure belt along the Median Tectonic Line. During the metamorphism, the trench was situated probably to the south of the Sanbagawa belt, that is, in the Shimanto terrane (fig. 17-1), and hence the distance between Ryoke and Shimanto may represent the arc-trench gap

(Dickinson, 1971). The Sanbagawa metamorphism is the recrystallization not of the contemporaneous geosynclinal sediments but of their basement. However, it is still an important question how the high-pressure complex has come into direct contact with the low-pressure complex. The direct contact may have been caused by an oceanward shift of the descending slab, as in the above model (3). However, it is more likely that the direct contact is due to a large strike-slip movement along the Median Tectonic Line (fig. 7 in Miyashiro, 1972*b*).

17-7 TECTONICS OF ISLAND ARCS AND PAIRED METAMORPHIC BELTS IN THE MAIN PART OF JAPAN

Double Arcs

The Northeast Japan Arc is an example of a presently active double arc. As shown in figs. 3-11 and 17-2, it is composed of the following three parallel structural elements from east to west: (*a*) the Japan Trench on the outermost side, (*b*) the Kitakami Mountains and Abukuma Plateau which represent the non-volcanic outer arc, mainly composed of Palaeozoic sedimentary rocks and Mesozoic granites, and (*c*) the chains of Quaternary volcanoes which constitute the volcanic inner arc. In this case, the double arc structure is slightly obscured as the valley between the volcanic and the non-volcanic arc is above sea level. Figure 17-2 shows 500-meter contours above sea level to reveal the topographic characteristics of the double arc.

The Ryukyu Islands also are a double arc, in which a trench is on the eastern side, the non-volcanic outer arc lies in the median zone, including all the bigger islands of Ryukyu, and the volcanic inner arc, being on the western side, is represented by small volcanic islands and submarine volcanoes.

On the other hand, in the double arc of the Kuril Islands, the volcanic inner arc is better developed than the non-volcanic outer one. The Izu-Bonin Arc also has a double arc structure, in which the Bonin Islands constitute an outer arc. To the north of Bonin, an outer arc does not exist. The Kuril and Izu-Bonin Arcs have deep trenches on the outermost zone.

The volcanic front (§15-1, fig. 15-2) approximately represents the ocean-side limit of the volcanic inner arc. The distance between the trench axis and the volcanic front is about 200–300 km in the above double arcs.

Thus, many of the present-day island arcs are double arcs, and we may consider that a double arc is a typical state of presently active island arcs. At an early stage of its development, the non-volcanic outer arc may not be formed with resultant formation of a single arc.

Northeast Japan Arc

As a well-investigated example of active double arcs, we will review here the Northeast Japan Arc. The Arc has been active from the Miocene to the present

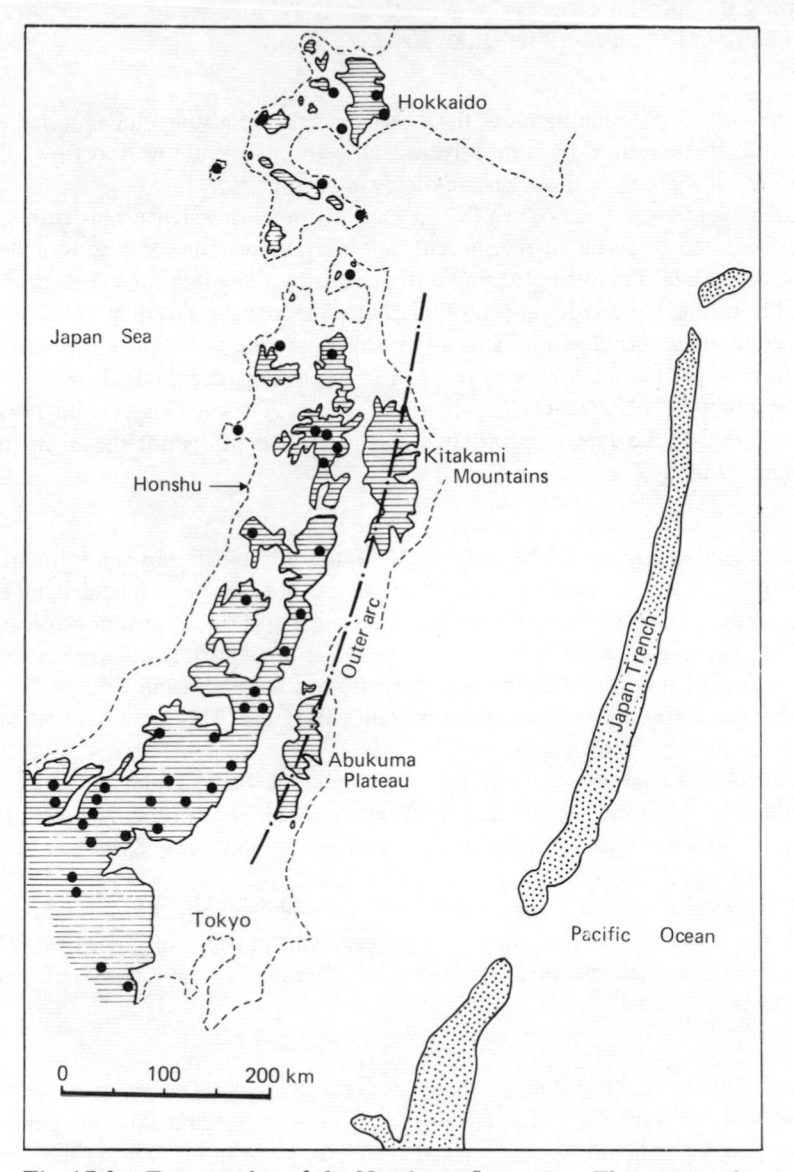

Fig. 17-2 Topography of the Northeast Japan Arc. The present-day coast line is shown by dotted lines. Horizontal ruling indicated areas more than 500 metres above sea level. Quaternary volcanoes are shown by solid circles. The areas more than 7,000 metres deep in the Japan Trench are stippled.

owing to the Mizuho orogeny. A generalized E–W cross-section of the Arc is shown in figs. 3-11 and 17-3 (cf. Rikitake *et al.*, 1968).

Trench Zone. This zone includes the Japan Trench and a zone on its immediate west. It is characterized by a negative isostatic anomaly and low heat flow. The epicenters of shallow earthquakes are highly crowded.

Probably this zone has been subjected to continual subsidence and sedimentation from the Miocene to the present, and may be regarded as a geosyncline. The Pacific plate should be plunging into the depths along this zone. This plunge probably brought the sediment piles deep into the mantle, and they would have undergone metamorphism probably of the high-pressure type. The observed low heat-flow zone should correspond to a low geothermal gradient (fig. 17-3).

The plunging plate reaches a depth of about 700 km beneath the northwestern part of the Japan Sea, though such a great depth is not shown in figs. 3-11 and 17-3.

Non-Volcanic Outer Arc. The outer half of the arc-trench gap is a submarine slope with negative isostatic anomalies and has already been included in the above trench zone, whereas the inner half of the gap is an elevated non-volcanic arc with positive isostatic anomalies. In so far as arcuate ridges above sea level are concerned, this zone may be separated from other zones and be treated as a non-volcanic outer arc in contrast to volcanic inner arc. This zone has low heat flow like the above trench zone.

The non-volcanic arc, consisting of the Abukuma Plateau and the Kitakami Mountains, underwent continual uplift as rigid blocks, probably since the Miocene. The western limit of this region is the Morioka–Shirskawa Line discussed above (§17-6).

Although granitic and metamorphic rocks are widely exposed in this arc, they are all the result of the Late Mesozoic orogeny and possibly of older ones. The metamorphic complexes formed since the Miocene, if any, have not been exposed yet.

Volcanic Inner Arc. The volcanic inner arc is made up of mountain ranges in the central and the Japan Sea side zone of Honshu. Miocene and Pliocene granitic masses also occur in this region. Subsidence took place in many areas and small scale geosynclinal accumulations were formed in the Miocene.

The present heights of volcanoes are the result partly of volcanic accumulations and partly of the tectonic upheaval of their basements. In the geologic past, the altitudes of this zone changed by the upheaval and subsidence of its parts. The present high altitude may be rather fortuitous.

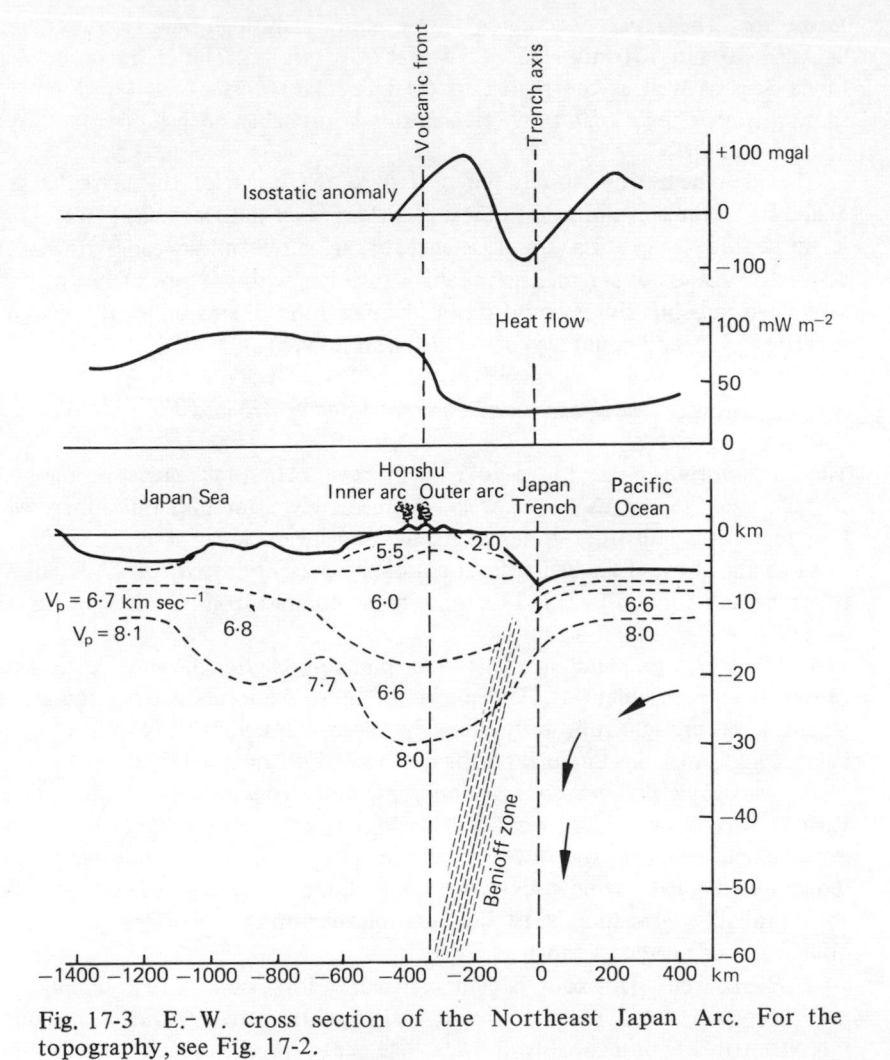

Fig. 17-3 E.-W. cross section of the Northeast Japan Arc. For the topography, see Fig. 17-2.

This zone is characterized by high heat-flow values. Thick Miocene sediments were metamorphosed in the zeolite, prehnite–pumpellyite and greenschist facies. The metamorphism occurred generally under very high geothermal gradients (§ 6A-2). The high values in heat flow and geothermal gradient should be somehow related to volcanism and plutonism in this region. The rise of magmas and the associated aqueous fluid from depth may be partly or entirely responsible for it.

Japan Sea. The Japan Sea region has an abnormally high heat flow, generally between 80 and 130 mW m^{-2} (2–3 μcal cm^{-2} sec^{-1}). The Okhotsk and East China Seas as well as the region to the immediate west of the Izu-Bonin Arc show a similar high heat flow. Hence, this is probably an essential feature of active island arcs.

The high heat flow should be related to the origin of the marginal seas. Shallow earthquakes are rare in these seas (Barazangi and Dorman, 1969). There is no feature suggesting the existence of an active mid-oceanic ridge. The convective rise of upper mantle materials together with the rise of magmas and associated aqueous fluids could create the new lithospheres under the marginal seas (fig. 3-11), and could also cause high heat flow.

Analogy between the Present-day Northeast Japan Arc and the Late Mesozoic Southwest Japan Arc

The metamorphism of the Ryoke–Sanbagawa pair took place probably in Jurassic and Cretaceous time. At and immediately after that time, there were four tectonically distinct zones along the Southwest Japan Arc, which correspond to the above-discussed four zones of the present-day Northeast Japan Arc (Matsuda and Uyeda, 1971). The zones are as follows from south to north (fig. 17-1):

1. *Shimanto geosyncline.* This was probably a trench zone with thick geosynclinal accumulations. The plunge of the oceanic plate from the trench would have brought not only the new Mesozoic sedimentary pile but also Palaeozoic formations under it on the north side into great depths.

2. *Sanbagawa high-pressure metamorphic belt.* This probably represents the Palaeozoic terrane which was under and on the north side of the above geosynclinal pile and was involved in the plunge of the oceanic plate. Subsequently, the Palaeozoic rocks were uplifted and exposed at the surface. This zone probably corresponds to the non-volcanic outer arc of the present-day double arcs of northeast Japan.

3. *Ryoke belt.* This zone is characterized by low-pressure, high-temperature metamorphism and abundant granitic and rhyolitic activity. During the metamorphism this zone probably showed high heat-flow values and volcanism just like the present-day inner arc of northeast Japan. However, the Ryoke metamorphic belt is about 50 km wide in reconstruction, and is narrower than the present-day inner arc, which is about 150 km wide. Therefore, the belt probably corresponds only to an oceanic-side zone of the inner arc of northeast Japan.

4. *Region of granitic activity on the continental side of the Ryoke belt.* The abundance of granitic masses suggests the occurrence of high heat flow at that time. Hence, this zone may be regarded as corresponding to the continental side part of the inner arc of northeast Japan, possibly together with a part of the Japan Sea.

Chapter 18

Metamorphic Structure of Continental Crusts

18-1 CONSTITUTION OF CONTINENTS

The North American and the European Continent have a vast Precambrian craton surrounded by Phanerozoic orogenic belts. On the eastern side of the Asiatic Continent and of the Australian Continent, the outermost zones of Phanerozoic orogenic belts form island arcs which are separated from the continents by marginal seas such as the Japan and the Tasman Sea.

The thickness of the crust, as defined by the depth to the Moho discontinuity, is usually 30–50 km for Precambrian cratons, but is more highly varied and commonly thinner for younger orogenic belts. The rock-pressure is about 10 and 15 kbar at depths of 35 and 50 km respectively.

The observed seismic velocities in the continental crust increase generally downwards:

$$V_p(km\ sec^{-1})$$

	$V_p(km\ sec^{-1})$
Upper crust	5·8–6·3
Lower crust	6·6–7·2

This increase is too great to be accounted for simply by the effect of pressure. Thus, it is believed that the crusts of shields show general changes in chemical composition with depth. Just as the exposed surface of shields are made up of varied rocks, the depths would also be heterogeneous. However, the average composition could become more basic with increasing depth. The thicknesses of the more acidic upper part and the more basic lower part for various regions are given in table 18-1.

There is a substantial difference in heat-flow values between Precambrian cratons (including shields) and Phanerozoic orogenic belts. Lee and Uyeda (1965) have given the following average figures:

	mW m^{-2}	μcal cm^{-2} sec^{-1}
Precambrian shields	38	0·92
Cratons covered by Phanerozoic sediments	64	1·54
Palaeozoic orogenic regions	51	1·23
Mesozoic–Cenozoic orogenic regions	80	1·92

Table 18-1   *Thicknesses of crust*

V_p (km sec^{-1})	Upper crust (5.8-6.3)	Lower crust (6.6-7.2)	Total thickness
	(km)	(km)	(km)
Baltic Shield (Finland)	18	18	36
Russian Plate	18	17	35
Interior Lowlands (N. America)	20	30	50
Appalachians	15	25	40
Alps (West Europe)	20	40	60
Caucasus (Alpine folding)	20	35	55
Sierra Nevada	25	25	50
California Coast Ranges	15	5	20
Northeast Japan (active arc)	15	17	32

Note: After Pakiser and Robinson (1966), Hashizume *et al.* (1968), Closs (1969), and Kosminskaya, Belyaevsky and Volvovsky (1969).

Hence, the temperatures at the same depth are usually higher in Phanerozoic orogenic belts than in Precambrian shields. The present temperature at the Moho beneath the shields would generally be 400–750 °C.

18-2   PRECAMBRIAN SHIELDS

Upper Crust

Precambrian shields are mainly composed of granitic rocks and quartzo-feldspathic gneisses, together with other metasediments and metabasites (§ 1-5).

The average chemical composition of the surface exposures of a shield has been calculated by Sederholm (1925) for the Finnish part of the Baltic Shield and by Shaw *et al.* (1967) and Eade and Fahrig (1971) for the Canadian Shield, as given in table 18-2. The compositions are similar to those of granodiorite, mica gneiss or graywacke in major elements, but differ from granodiorite markedly in some trace elements.

Lower Crust

Since the lower crust of shields is more basic than the upper, many authors have considered that the upper and the lower crusts are composed of granitic and gabbroic (or basaltic) rocks respectively. Poldervaart (1955), Pakiser and Robinson (1966) and Ronov and Yaroshevsky (1969) calculated the average chemical composition of the crust on the basis of such models.

Ringwood and Green (1966*a*, *b*) have, however, criticized such models. If P_{H_2O} is equal or close to P_s in continental crusts, the P_s–T condition of the lower crust as discussed in the preceding section should be usually in the

Table 18-2 *Average chemical composition of surface exposures of Precambrian shields*

	Finland Sederholm (1925)	Canada Shaw *et al.* (1967)	Canada Eade and Fahrig (1971)
SiO_2	67·45	64·93	65·1
TiO_2	0·41	0·52	0·52
Al_2O_3	14·63	14·63	16·0
Fe_2O_3	1·27	1·36	1·4
FeO	3·13	2·75	3·1
MnO	0·04	0·068	0·08
MgO	1·69	2·24	2·2
CaO	3·39	4·12	3·4
Na_2O	3·06	3·46	3·9
K_2O	3·55	3·10	2·89
P_2O_5	0·11	0·15	0·16
H_2O^+ H_2O^-	0·79	$\{ 0·79$ $0·13 \}$	0·8
CO_2	0·12	1·28	0·20
S	–	0·06	–
Cl	–	0·01	–
F	–	0·05	–
C	–	0·02	–
Less	–	0·04	–
Total	99·64	99·708	99·75

Note: Average trace element content for the Canadian Shield is (in p.p.m.): Be 1·3, Ga 14, Cr 99, V 53, Li 22, Ni 23, Co 21, Cu 14, Zr 400, Sr 340, Ba 1070, Rb 118, U 2.5, Th 10.3 (Shaw *et al.* 1967; Shaw, 1968); and Cr 77, Ni 20, Sr 380, Ba 730, U 1.5, Th 10.8 (Eade and Fahrig, 1971).

greenschist, epidote–amphibolite and amphibolite facies, as shown in fig. 3-12. There is some reason to think that the lower crust of Precambrian shields is strongly dehydrated (§ 4-3). Hence, P_{H_2O} may be much lower than P_s. If this is the case, the P_s-T there should usually be in the eclogite and granulite facies (transitional to the eclogite), as shown in fig. 12-6. Accordingly, even if the lower crust has a basic composition, it cannot be gabbroic or basaltic in mineral assemblage, but would be greenschist, epidote–amphibolite, amphibolite, eclogite or garnet granulite.

Table 18-3 shows, however, that such rocks generally have higher seismic velocities than the real lower crust. Therefore, the lower crust cannot be basic in composition. Ringwood and Green have concluded that the lower crust should be either intermediate (e.g. quartz dioritic) in composition or composed of heterogeneous mixtures of acidic and more basic rocks. Though the upper crust also is a mixture of acidic and more basic rocks, the ratio of more basic rocks

Table 18-3 *Compressional wave velocities (V_p) at 2 and 10 kbar in km sec^{-1}*

	At 2 kbar	At 10 kbar
Granite	6·1–6·4	6·3–6·5
Gabbro	7·0–7·1	7·2–7·3
Quartzo–feldspathic gneiss	6·0–6·3	6·3–6·7
Amphibolite	6·8–7·0	7·1–7·3
Epidote amphibolite	7·3–7·4	7·6–7·8
Eclogite	7·8–8·0	8·0–8·2
Dunite	8·0–8·3	8·3–8·4

Note: Compiled from Clark (1966) and Christensen (1965).

would be higher so as to give an intermediate bulk composition for the lower crust.

A possible origin of the crustal layering will be discussed in the next section.

18-3 VERTICAL DIFFERENTIATION OF CONTINENTAL CRUSTS

Partial Melting

The present surface of Precambrian shields is commonly in the amphibolite and granulite facies. During the metamorphism it was probably common for some extent of partial melting to take place. At the same time, on the other hand, the deeper levels within the crust would generally have been at higher temperatures. The lower crust accordingly would have been subjected to a considerable extent of partial melting. It is natural therefore to ascribe the crustal layering partly to partial melting, though there should be many other relevant factors of which we are quite ignorant.

The melts should usually be of granitic composition, and would move upwards through the crust owing to their lower density. They may join to form big granitic bodies, or may remain in a more dispersed state. Upward movement of such melts would increase the acidic character of the upper crust, leaving generally more basic solid residues in the lower crust.

Granulite-Facies Terranes

Ramberg (1951, 1952) has pointed out the possibility that granulite facies metamorphic rocks now exposed on earth might once have been in the lower crust and basified during metamorphism, though he preferred chemical diffusion in the solid state to partial melting as a possible mechanism of basification.

Small-scale basification in a granulite facies zone was demonstrated by Engel and Engel (1958, 1962a) for both metasediments and metabasites of the

Adirondacks (§ 8A-7). On a larger scale, Eade, Fahrig and Maxwell (1966) and Eade and Fahrig (1971) calculated average chemical compositions of amphibolite- and granulite-facies terranes in a part of the Canadian Shield. Granulite facies rocks occur in a great number of separate irregular areas surrounded by amphibolite facies areas. The granulite facies terranes are slightly higher in FeO and CaO, but slightly lower in SiO_2 and K_2O than the amphibolite facies ones.

There are a number of wide terranes of granulite facies rocks in the Australian Shield. They may be classified into the three subfacies representing different rock-pressures that were shown in fig. 12-2. The majority of the granulite facies terranes belong to the low-pressure subfacies and are usually dominated by acidic rocks, though the rest belong to the two higher pressure subfacies and are generally dominated by moderately acidic, intermediate and basic rocks. The latter terranes were probably metamorphosed and partially melted in the lower crust and later tectonically uplifted to their present level. They should show the chemical characteristics of the lower crust. Thus, Lambert and Heier (1968) have made a detailed chemical investigation of such rocks. The rocks of the amphibolite facies are generally similar in chemical composition to those of the lower-pressure subfacies of the granulite facies. On the other hand, the rocks of the two higher pressure subfacies of the granulite facies show an effect of basification: decreases of SiO_2, K_2O, Rb and U, and increases of total iron, MgO and CaO. The K/Rb ratio tends to increase in the process.

The K/Rb ratio of ordinary continental igneous rocks varies in a range of 200–400. Metamorphic rocks in the amphibolite and lower facies show a similar range, whereas those in the granulite facies tend to show higher values. Sighinolfi (1969, 1971) has demonstrated that both K and Rb contents have decreased and the K/Rb ratio has increased in granulite facies basements of the Western Alps and Brazil.

Possible Retrogressive Recrystallization of the Lower Crust

After the formation and modification of a continental crust in orogeny, the region would be stabilized with a consequent decrease of the geothermal gradient. In the upper crust, the decrease of temperature probably results in a general halt of recrystallization, though retrogressive changes usually occur to a limited extent and over a limited area. On the other hand, in the lower crust, the temperature after the stabilization is probably still a few or several hundred degrees Celsius, and hence it is conceivable that retrogressive recrystallization occurs on a grand scale, as was mentioned in § 1-2.

Under such a condition, the granulite facies metamorphic complex of the lower crust would be subjected to eclogite facies conditions, since P_{H_2O} is probably very low. Such a change may be an extremely slow process which could only take place over a long geologic time under non-orogenic conditions.

18-4   PHANEROZOIC OROGENIC BELTS

Since Phanerozoic orogenic belts give much higher heat-flow values than Precambrian shields, there must be a chemical difference in their crust and/or underlying mantle.

The crust of shields could be strongly layered with the lower crust and the underlying upper mantle impoverished in radioactive elements, and a considerable part of the original upper crust containing abundant radioactive elements may have been eroded away. The more radioactive materials removed from the upper part of shields would be mainly deposited in the adjacent Phanerozoic geosynclines. This ultimately contributes to the increase of the radioactive content of the Phanerozoic belts.

It has been widely believed that the Precambrian shields have been eroded to much deeper levels than the Phanerozoic terranes. However, the common occurrence of andalusite in the former casts a serious doubt on this belief. As far as regional metamorphic terranes are concerned, there should be no great difference in the depth of denudation between the Precambrian and the Phanerozoic. Therefore, the common occurrence of granulite facies rocks in the shields but not in the Phanerozoic belts suggests that the metamorphism was generally more intense in the Precambrian terranes than in the Phanerozoic. If this is the case, the crust of the Phanerozoic belts may not have been as strongly layered as that of the Precambrian, and the erosion of the upper layer would not have seriously changed the average compositions of the Phanerozoic belts.

On the other hand, compositional change of continental crusts with geologic age of formation is also possible. Phanerozoic orogenic belts may have differed in composition from early Precambrian ones at the time of their formation. For example, Engel (1963) and Engel and Engel (1964a) have claimed that the first continental crust in the present site of North America was formed on an oceanic crust, and that the early continental crust was similar to the crust in present-day island arcs and their vicinities, the composition of successively added new continental crusts becoming gradually more similar to that of Phanerozoic orogenic belts. In other words, the crust newly added by accretional growth of continents becomes poorer in basic volcanics and richer in acidic volcanics, and the maturity and sorting of sedimentary rocks increases with decreasing age. Graywacke was dominant among the earliest sediments, whereas shale, quartzite and carbonate rocks become important in late Precambrian and younger sediments.

The Franciscan of California, the New Zealand geosyncline and some other younger terranes are rich in basic volcanics and graywacke, and hence are inconsistent with this view.

There are, however, strong opponents to the idea of continental growth and progressive development of acidic crust. Wynne-Edwards and Hansan (1970), for instance, have insisted that new orogenic belts are always formed on sialic basements of older belts, and there is little or no evidence for the significant increase of sial on earth with geologic time, and that the compositional difference between the continental crust and the uppermost mantle was greatest in the earliest geologic time and has decreased since then by upward introduction of basic magma into the crust.

Chapter 19

Ocean-floor Metamorphism and its Significance

19-1 HARD ROCKS OF DEEP OCEAN FLOORS

Deep ocean floors are mostly covered by unconsolidated sediments, commonly a few to several hundred metres thick, which are underlain usually by basaltic rocks.

Basalts are widely exposed on mid-ocean ridges and seamounts. Most of the basalts on the submarine parts of mid-oceanic ridges are olivine tholeiites with K_2O contents below 0.3 per cent and extremely low in Ba, Sr, Pb, Th, U and Zr. Their Na/K and K/Rb ratios are higher than those of most continental basalts. Such basalts have been called oceanic tholeiites or *abyssal tholeiites* (Engel and Engel, 1964*b*; Engel *et al.* 1965).

Abyssal tholeites usually show a relatively small range of compositional variation which is probably partly due to differences in the original magmas and partly due to crystallization differentiation at various depths (Miyashiro, Shido and Ewing, 1969*a*; Shido, Miyashiro and Ewing, 1971). Some tholeiitic magmas form intrusive bodies, in which crystallization differentiation occurs, sometimes leading to the formation of peridotites, ferrogabbros and more acidic rocks (Miyashiro, Shido and Ewing, 1970*b*; Aumento, 1969).

Oceanic islands are commonly composed of basalts and their differentiates of the alkali basalt series. Such basalts occur rarely on the submarine parts of mid-oceanic ridges also.

Peridotites and serpentinites occur on mid-oceanic ridges (Tilley, 1947; Hess, 1962; Melson *et al.* 1967; Miyashiro, Shido and Ewing, 1969*b*). They are especially abundant within transverse fracture zones across the Mid-Atlantic Ridge.

Peridotites and pargasite-bearing varieties of serpentinite of oceanic regions are relatively high in CaO and Al_2O_3 contents and chemically similar to so-called high-temperature peridotite intrusions, which are now widely believed to have risen as solid masses from the upper mantle. On the other hand, pargasite-free serpentinites are lower in CaO and Al_2O_3 than pargasite-bearing ones. Such a compositional variation would be due partly to heterogeneity in the upper mantle, and partly to chemical migration during serpentinization (Miyashiro *et al.* 1969*b*).

Many basaltic rocks and most gabbroic rocks are metamorphosed, however. Ocean-floor metamorphism gives important information on the activity of mid-oceanic ridges. This problem will be treated in the following sections.

19-2 METABASALTS AND METAGABBROS FROM DEEP OCEAN FLOORS

Discovery and Modes of Occurrence

Until recently, it was widely believed that the oceanic hard crust was an essentially igneous structure composed of basalts, gabbros, peridotites and serpentinites. In 1966, however, metamorphosed basalts were reported to have been dredged on the crest of the Mid-Atlantic Ridge near 22°N by Melson *et al.* (1966) and by Melson and van Andel (1966), and on the crest of the Carlsberg Ridge near 5°N in the Indian Ocean by Cann and Vine (1966). In the few succeeding years, metabasalts were found from many other areas on the crest of the Mid-Atlantic Ridge (Aumento and Loncarevic, 1969; Miyashiro *et al.* 1970*a*, 1971).

The metamorphic rocks were all derived from basalts, dolerites and gabbros except for the serpentinites which might well be regarded as metamorphic derivatives of peridotites. The metamorphism must have taken place at some depth, probably beneath the crest of mid-oceanic ridges, since relatively high temperature prevails there. Metamorphic rocks exposed at a great distance from the ridge axis could have been laterally transported by ocean-floor spreading.

Metamorphic rocks occur in the Mid-Atlantic Ridge and Carlsberg Ridge (northern part of the Mid-Indian Ocean Ridge), which both have rough topography, in contrast to the East Pacific Rise with smooth topography which has no exposures of metamorphic rocks. Presumably, metamorphic rocks are present in depth beneath all mid-oceanic ridges, but they are covered by basalts on the East Pacific Rise, whereas intense fault and other tectonic movements on the Mid-Atlantic Ridge resulting in rough topography have led to exposure of metamorphic rocks.

On the Mid-Atlantic Ridge, metamorphic rocks are exposed on the lower walls of the Median Valley and transverse fracture zones. In the latter case, they are accompanied by serpentinites, and appear to be inclusions therein. Probably the serpentinites were intruded in the solid state along the fracture zones. Metamorphic rocks formed at some depth were probably torn apart and brought up to the surface. This interpretation is consistent with the view that the serpentinites were formed by hydration of upper mantle peridotites.

Metabasalts

The majority of metabasalts so far dredged are non-schistose or only weakly schistose, preserving their original igneous textures. Recrystallization is incomplete in most specimens. The original structures of pillow lavas and breccias are commonly noticeable. It suggests that basaltic eruptions produced thick volcanic piles, the lower part of which became sufficiently hot to undergo metamorphic recrystallization. This may be regarded as a kind of burial metamorphism.

Most of the metabasalts belong to the zeolite and greenschist facies. It is not known whether rocks of the prehnite–pumpellyite and epidote–amphibolite facies exist in deep ocean floors. Some metabasalts are in the greenschist facies transitional to the amphibolite facies.

Some of the greenschist facies metabasalts preserve their original chemical compositions as abyssal tholeiites except for an increase of H_2O content, but the rest show a marked decrease of CaO content and a variation of SiO_2 content (Miyashiro et al. 1971). Many metabasalts are spilitic (Melson and van Andel, 1966; Cann, 1969). Virtually all the zeolite facies metabasalts show marked introduction of Na_2O from the outside. Such compositional variations could have been caused by flow of a hot aqueous fluid through the crust. With intense decrease of CaO, the amount of chlorite increases, leading to the formation of chlorite–quartz rocks (Miyashiro et al. 1971).

The detection of chemical migration during metamorphism is relatively easy, because the composition range of abyssal tholeiites is limited.

Metagabbros

Most metagabbros preserve their original texture. They are in the greenschist and amphibolite facies. However, some of them have suffered retrogressive changes in the zeolite facies.

There are unmetamorphosed basalts as well as metabasalts on mid-oceanic ridges, whereas nearly all gabbroic rocks are more or less metamorphosed. The metabasalts are usually in the zeolite or greenschist facies, whereas the metagabbros are usually in the greenschist or amphibolite facies. These two facts suggest the common presence of gabbroic rocks in relatively deep levels beneath the ridges.

In view of the small thickness of the oceanic crust, the temperature of typical amphibolite facies would be too high to be realized within the crust even beneath the crest of mid-oceanic ridges. It follows that some metagabbros were probably consolidated and then recrystallized in the upper mantle.

Most metagabbros preserve their original chemical composition. However, some metagabbros underwent chemical migration similar to that in metabasalts. In other words, introduction of Na_2O occurred in the zeolite facies, and a decrease of CaO and variations of SiO_2 occurred in the greenschist facies (Miyashiro et al. 1971).

It is to be noted that chemical migration is much more intense in the zeolite and greenschist facies than in the amphibolite facies. Amphibolite facies rocks have not undergone the intense metasomatism characteristic of lower temperatures.

19-3   SIGNIFICANCE OF OCEAN-FLOOR METAMORPHISM

Relation to Heat Flow

The crust beneath normal oceanic basins is only about 6 km thick. The temperature at its base would be about 120–180 °C. Thus, fissure filling and incipient metamorphism in the zeolite facies may take place in the lower crust, but marked metamorphic recrystallization cannot. Metamorphism in the greenschist and higher facies would take place at higher temperatures usually in the crust and upper mantle beneath the crest of mid-oceanic ridges.

The observed heat flow on the crest is much higher than in normal oceanic basins. This suggests a higher geothermal gradient and higher temperatures at depth beneath the crest than beneath normal oceanic basins. The temperature in the deeper crust there may reach about 350 °C to correspond to greenschist facies. Amphibolite facies recrystallization could take place in gabbroic pockets in the uppermost mantle.

The heat-flow values measured on the crest are highly variable. Values as high as 300 mW m^{-2} (8 μcal cm^{-2} sec^{-1}) occur mixed with those as low as 40–80 mW m^{-2} (1–2 μcal cm^{-2} sec^{-1}). If all these values are accepted, the high values could be due not only to the general heat conduction but also to some more localized situations. Igneous intrusion is one possible situation of this kind. A flow of a hot aqueous fluid is another. The hot fluid that caused metasomatism should transport heat, causing localized high heat flow. The widespread introduction of Na_2O into zeolite facies metabasites suggests permeation of an hot aqueous fluid. Accordingly, even the apparently general increase of heat flow on the crest may be due not only to heat conduction but also to permeation of a fluid (Miyashiro et al. 1971; Miyashiro, 1972a).

If a rock recrystallized beneath the crest moves laterally at the same depth owing to ocean-floor spreading, the temperature of the rock will gradually decrease, leading possibly to retrogressive changes.

Magnetized layer and Demagnetization

Normal oceanic basins are covered by unconsolidated sediments, whereas the crest of mid-oceanic ridges usually has almost no sediment cover. At any rate, ignoring the sediment cover, the surface layer of the oceanic crust is made up of basalts and dolerites. This layer is responsible for the linear patterns of magnetic anomalies of the ocean floors.

Recent investigations, however, have revealed that the magnetized layer is as thin as about 0·5 km to a few km, and that the underlying layer is virtually demagnetized (Vine, 1966; Heirtzler, 1968). The demagnetization takes place at too shallow a depth to be ascribed to heating above the Curie point. The demagnetized layer of the crust is probably composed of metabasalts, meta-dolerites and metagabbros. During metamorphic recrystallization, the strong thermo-remanent magnetism of igneous rocks should disappear and instead chemical remanent magnetization should appear. The latter, however, is negligible in strength in comparison with the former (Miyashiro *et al.* 1971).

Layers 2 and 3

Seismologically, the oceanic crust beneath normal oceanic basins is usually divided into three layers. Layer 1 represents unconsolidated sediments. Basaltic volcanics containing a considerable amount of pyroclastic or fragmental materials show a seismic velocity corresponding to that of layer 2, and gabbros, metabasalts and metagabbros show a seismic velocity corresponding to layer 3. Accordingly, layers 2 and 3 may be regarded as being composed of such materials in many cases.

However, it is known that layer 3 is absent beneath the crest of Mid-Atlantic Ridge, and instead layer 2 is thicker there than beneath ordinary oceanic crust (Talwani, Le Pichou and Ewing, 1965). This must be accounted for.

The rough topography of the crest of the Mid-Atlantic Ridge is probably a surface manifestation of strong fault movements, which should cause fracturing within the crust and the upper mantle. Intensely fractured metabasites show a marked decrease of seismic velocity in comparison with unfractured rocks. This could lead to the disappearance of layer 3, and the complementary thickening of layer 2.

The crestal crust would move laterally, eventually entering the region where no more fracturing takes place. The openings in the fractured metabasites and in the overlying pyroclastic accumulations would be gradually filled by low-temperature mineral veins. The filling would occur more easily in the lower part of the crust, thus resulting in the generation of a coherent crustal plate, which should be layer 3. Thus, the boundary between layers 2 and 3 in this case is not the same as the boundary between volcanics containing pyroclastic materials and the metamorphosed and unmetamorphosed solid rocks.

In the East Pacific Rise, fault movements are not so intense judging from its smooth topography. There, layer 3 exists beneath the crest and flanks.

Chapter 20

Cataclastic Metamorphism along Transform Faults

20-1 CLASSIFICATION AND FIELD RELATIONS OF CATACLASTIC METAMORPHIC ROCKS

Our knowledge of cataclastic metamorphism in general, and of cataclastic metamorphism along transform faults, is very superficial and scrappy. In view of the tectonic importance of transform faults, a brief review of cataclastic rocks and particularly of those related to transform faults will be given in this chapter.

Since the extent of recrystallization is by definition limited in cataclastic metamorphism, it is usually considered to occur at low temperatures. However, the maximum possible temperature for such metamorphism is unknown. If the rate and amount of deformational movement are great and if the length of time favourable for recrystallization is short, cataclastic rocks may form even at medium temperatures.

Classification

Weak cataclasis produces tectonic breccias, which are mainly made up of angular rock-fragments set in a subordinate matrix of more highly crushed materials. With advanced cataclasis, the proportion of crushed materials increases, and eventually aphanitic rocks form. Large crystals and large rock-fragments which have survived crushing in rocks subjected to advanced cataclasis are called *porphyroclasts*.

Rocks subjected to advanced cataclasis are classified here into five groups (Read, 1964; Higgins, 1971), as follows:

1. *Incoherent cataclastic rocks* are generally believed to form at low confining pressure under near-surface conditions. These include fault breccia, fault gouge and fault pug in order of increased crushing.

2. *Structureless cataclasites* are coherent rocks showing a mortar texture without foliation or preferred orientation. This texture is characterized by the presence of porphyroclastic minerals set in a matrix of finely-crushed material. The name cataclasite was originally coined to represent structureless cataclastic metamorphic rocks by Grubenmann and Niggli (1924), but was used afterwards

to represent cataclastic rocks in general by other authors. The term structureless cataclasite leaves no possibility of confusion.

3. *Mylonites* are coherent rocks showing flow structure (foliation) of cata-clastic origin (Lapworth, 1885; Waters and Campbell, 1935). The flow structure is usually an alternation of materials showing different degrees of cataclasis, or showing different mineral compositions. Mylonites are believed to have been produced by grinding in a definite direction. The names protomylonite, mylonite (in a narrow sense) and ultramylonite have been used to represent the three subclasses of the mylonites in order of increasing degree of cataclasis.

Most rocks of the above three classes show a small extent of recrystallization, which commonly produces greenschist facies minerals such as quartz, chlorite, epidote, and calcite, though minerals of other facies are not rare. If recrystal-lization is more important, the resultant rocks belong to the following class.

4. Cataclastic rocks apparently subjected to a considerable extent of re-crystallization usually show a flow structure, and include mylonite gneiss and blastomylonite in order of increasing mylonitization. The cataclasis and re-crystallization are said to be concurrent, but the available evidence for this is not always sufficient.

5. Rocks may be melted by exceptionally intense frictional heating due to grinding in a mylonite zone. Such a rock has been called pseudotachylite, but most rocks described as pseudotachylites may have been formed in other ways. Truly melted rocks appear to be extremely rare (e.g. Scott and Drever, 1953). Some mylonites subjected to extremely strong grinding, become submicroscopic in grain size, and look dark between crossed polars (nicols), even though they are not true glasses (e.g. Waters and Campbell, 1935).

Most mylonites have been derived from quartzo–feldspathic rocks, though some examples of basic and ultrabasic mylonites are known.

Quartz and micas are easily affected by deformation and cataclasis. Quartz first becomes elongated and strained, and then shattered at the edges of grains or throughout, ultimately becoming fine-grained masses, usually around porphyro-clasts of other minerals. Micas are easily recrystallized but tend to preserve a record of deformation in their shapes. With advancing cataclasis, micas are also ground into finer flakes. Feldspars and especially garnet are resistant to crushing, and as a result commonly form porphyroclasts in cataclastic rocks. A detailed and critical glossary of cataclastic rocks has been given by Higgins (1971).

Field Relations

On and near the surface of the earth, block movement produces a sharp fault plane or a sharply-limited narrow fault zone, which is filled with incoherent cataclastic rocks. At greater depths, a similar block movement appears to produce usually an ill-defined fault zone which is made up of coherent cataclastic rocks.

Higgins (1971) has suggested that within such a fault zone mylonites are formed in a zone where deformational movement is especially intense, and the mylonite zone grades into zones of cataclastic rocks accompanied by a considerable extent of recrystallization (such as blastomylonites and mylonite gneisses), which in turn grade outwards into ordinary schists and gneisses. In such a fault zone, structureless cataclasites usually occur in relatively well-defined zones, which are either parallel or oblique to the zone of mylonites. It is said that such structureless cataclastic rocks are younger than the mylonites and other cataclastic rocks showing flow structure.

In many areas, sharp faults accompanied by incoherent cataclastic rocks cut through the zones of mylonites and structureless cataclasites. These faults appear to represent movements at still later times. The San Andreas fault along which the movement of the San Francisco earthquake occurred, for example, is such a sharp fault cutting through a wide, old zone of mylonites.

A transform fault should extend from the surface to the base of the lithosphere. With increasing depth, the temperature and confining pressure should increase with resultant change in the nature of cataclastic metamorphism. However, we have no clear knowledge of this change.

20-2   THE SAN ANDREAS FAULT ZONE IN CALIFORNIA

The San Andreas fault cuts through California in a SE–NW direction (fig. 13-3). Its total exposed length is about 1000 km. The cumulative right-lateral displacement since the Jurassic is estimated at about 550 km (Hill and Dibblee, 1953).

It is widely believed that this fault is now a transform which connects the Gulf of California segment of the East Pacific Rise with its Northeast Pacific segment, i.e. the Gorda Ridge (Wilson, 1965; Suppe, 1970). The Pacific plate is in contact with the North American plate along this fault. Prior to the middle Tertiary, the whole length of the East Pacific Rise was probably off the west coast of North America. The plate spreading eastwards from the Rise, called the Farallon plate, was being subducted in the trench zone along the west coast of North America, whereas the present Pacific plate was at that time spreading westwards from the Rise. Thus, the Pacific plate was separated from the North American plate by the Farallon plate. The San Andreas fault became the boundary between the Pacific and North American plates in middle Tertiary time when the Farallon plate was consumed (Atwater, 1970). The nature of the San Andreas fault prior to middle Tertiary time is not clear.

In detail, the fault is a complex zone, up to several km wide, of cataclastic rocks cut by a large number of subparallel faults (e.g. Dickinson, 1966). The most recent great movement along a fault within the fault zone took place during the San Francisco earthquake of 1906. Cataclastic rocks observed at the present surface were probably formed at various depths.

The mylonites in the fault zone in an area south of San Francisco were described by Waters and Campbell (1935). The exposed rocks are heterogeneous breccias, which are composed of angular blocks of a great variety of rocks such as metabasalt, serpentinite, quartz diorite, and quartzite, embedded in a dark-coloured aphanitic mylonite matrix. The mylonite matrix grades into the blocks with gradual changes in colour, grain size and the degree of banding. The mylonites show well-developed banding (lamination), which produces a rock cleavage similar to that of slate. The banding usually strikes parallel with the strike of the fault zone, and usually dips nearly vertically, though local variations and contortions appear.

Every stage in the crushing of original coarse-grained rocks to a submicroscopic paste can be observed. In general, crushing begins with the fracturing of the rock into a coherent breccia whose fragments are megascopically visible. Further movement results in the formation of a thin film of crush powder between the mineral grains forming mortar structure. The next stage is the development of a closely spaced series of microscopic shears in the individual mineral grains. Then, films of pulverized material form and grow along the shears, until finally the uncrushed material between adjacent shears is isolated as streaks, lenticles and irregular masses in the powdery matrix. Complete milling of the rock to a submicroscopic powder takes place only rarely. Some bands in extremely finely pulverized mylonites are dark between crossed polars.

In all the mylonites, we find at least some recrystallization in the matrix. The most common newly formed minerals are quartz and chlorite.

20-3   THE ALPINE FAULT OF NEW ZEALAND

The Alpine fault cuts through the South Island of New Zealand in a NE–SW direction (fig. 16-2). The cumulative right-lateral movement since Jurassic time is estimated at about 480 km as mentioned before. The fault could be a transform fault connecting the Tonga–Kermadec trench with another trench on the west side of the Macquarie Rise, though the possibility of other interpretations has not been precluded.

Cataclastic rocks of all the four classes mentioned above, occur in a zone, up to about 2 km wide, along the fault (Reed, 1964). The incoherent rocks have been ascribed to near-surface crushing during relatively recent (probably Quaternary) movement. Structureless cataclasites and mylonites have been ascribed to different types of movement at greater depth in older geologic times. A gradation has been observed from a mylonite zone to the original Haast schists.

Newly formed minerals including chlorite, epidote and carbonate are widespread.

20-4 CATACLASTIC ROCKS ALONG TRANSFORM FAULTS IN OCEAN FLOORS

Cataclastic metamorphic rocks have been found to occur along fracture zones across the Mid-Atlantic Ridge. They could have formed by movement along transform faults (Miyashiro *et al.* 1971). However, the degree of crushing in the examples reported to date is not high.

St Paul's Rocks in the equatorial Atlantic is a group of small islands made up of well-mylonitized peridotites and related rocks (Tilley, 1947; Melson *et al.* 1967). The islands are located near the junction of the St Paul's fracture zone with the crest of the Mid-Atlantic Ridge. The mylonitization may simply be a result of upward intrusion of the ultrabasic mass in the solid state, or alternatively it may be related to the transform fault and/or normal faults in the crest of the Ridge. The banding of the mylonite is vertical and strikes nearly north–south. In other words, the banding is roughly perpendicular to the trend of the fracture zone.

Appendix

History of the Study
of Metamorphism

1. VIEWS ON METAMORPHIC ROCKS IN THE LATE EIGHTEENTH AND
   THE EARLY NINETEENTH CENTURY

Fragments of geological knowledge began in Classical Greece or even at an older time. The science of geology as a field of systematic knowledge was initiated in the middle eighteenth century by J. E. Guettard and Nicolas Desmarest in France. Toward the end of the century, two great theorists appeared: Werner and Hutton. A. G. Werner (1749–1817) was an influential professor of geology and mineralogy (collectively called geognosy by him) at the School of Mines at Freiberg, Saxony. This school was founded in 1765 as one of the oldest institutions for education of professional mining engineers and geologists. Werner and his followers, who were called *Neptunists*, explained all the rocks as being of sedimentary origin (chemical and clastic). They had no ideas of igneous and metamorphic rocks and of tectonic movements.

James Hutton (1726–97) was a leisured gentleman and typical eighteenth-century intellectual living in Edinburgh, who had active interest in all the branches of human knowledge including chemistry, meteorology, agriculture and philosophy. He spend much time in geological studies, and regarded basaltic and granitic rocks as being formed by consolidation of molten materials (magmas). The advocates of this idea were called *Plutonists*. The concept of metamorphism is said to have begun with Hutton's theory, a systematic description of which was given in his book, *Theory of the Earth* (1795). In Hutton's view, some sedimentary rocks were brought to depths of the earth where high temperature and high pressure caused metamorphism of them. The schists and gneisses of the Scottish Highlands, for instance, were regarded as being metamorphic, though he did not introduce such a technical term.

According to Johannsen (1931, p. 185), the term 'metamorphism' was introduced by A. Boué in 1820, while the term 'metamorphic rocks' became popular through the first edition of Charles Lyell's *Principles of Geology* (vol. 2, 1833).

As regards the origin of basalt and granite, the Plutonists had secured a victory by 1820, that is, immediately after the death of Werner. However, the problem as to whether schists and gneisses are metamorphic rocks, primordial igneous rocks or chemical sediments that were deposited in primordial oceans, was not settled until the middle of the nineteenth century. According to the common view at that time, gneisses were formed in the oldest geologic age, and then schists were formed, being followed by phyllites. Fossiliferous sedimentary rocks were considered to have been formed at a still later time.

In the meanwhile, transitional relationships between unmetamorphosed sediments and schists were found at places. Thus, a greater number of geologists gradually came to think that some or all of the phyllites and schists are metamorphic in origin.

2. VIEWS ON REGIONAL METAMORPHISM IN THE SECOND HALF OF THE NINETEENTH CENTURY

The distinction between contact and regional metamorphism is said to have been first noticed in the middle of the nineteenth century by Élie de Beaumont and A. Daubré. The term regional metamorphism was proposed by the latter author.

The geosynclinal theory for mountain building was formulated in the period 1859-1910 by James Hall, J. D. Dana, and E. Haug. The widespread occurrence of regional metamorphic rocks in orogenic belts attracted the attention of all these authors. The high temperature, high pressure, and deformational movement in the depths of geosynclinal piles were considered to be the cause of metamorphism.

Though I am not familiar with the publications of this period, it appears that there were two contrasting views on the cause of regional metamorphism. One school of geologists stressed the importance of high temperature in the depths of the earth and effects of plutonic masses as the agents of regional metamorphism, while the other school emphasized the effects of pressure (hydrostatic or non-hydrostatic) and deformational movement.

The former school included many geologists in Britain and France, some of whom used the name of 'plutonic metamorphism' for regional metamorphism. The effect of water and other materials emanating from the associated plutonic masses was especially emphasized by French authors.

The latter school who accentuated the effects of pressure and deformational movements included many German and Swiss geologists. They used the name of 'dynamic' or 'dislocation metamorphism' for regional metamorphism. The term 'dynamic metamorphism' (or dynamometamorphism) was proposed in 1886 by H. Rosenbusch, who was a great master in descriptive petrography. This term, and the idea attached to it, were propagated by his students and pervaded generations of geologists all over the world.

Furthermore, some later authors began to use the names of dynamic and dislocation metamorphism to denote cataclastic metamorphism (i.e. intense mechanical deformation of rocks). Many other names were introduced to represent the supposedly dominant agents such as thermal and dynamothermal metamorphism. This increased the confusion of nomenclature.

Metamorphic rocks closely associated with granitic rocks are widely exposed in Britain and France, whereas the Alpine metamorphic rocks are rarely accompanied by granitic ones. This difference may have been a part of the factual basis underlying the contrasting views on regional metamorphism.

3. GOLDEN AGE OF MICROSCOPIC PETROGRAPHY

The period of 1870–1900 was the golden age of microscopic petrography. Microscopic observation of thin sections was a powerful new technique in geology at that time, before which rocks had been examined by means of the magnifying glass and chemical analysis. Reliable determination of rock-forming minerals and their textural relations became possible for the first time by the introduction of the microscope. It was then a new finding that plagioclases form a continuous series of solid solution. Thus, the nature and the extent of diversity of the rocks were first clarified. A great number of rock names were coined and systems of rock classifications were proposed.

F. Zirkel (1838–1912) of Leipzig and H. Rosenbusch (1836–1914) of Heidelberg were the two greatest masters in this field who attracted students from all over the world. Since their contributions were so remarkable, microscopic petrography became the major field in the study of rocks. The threefold classification of rocks into igneous, sedimentary and metamorphic was proposed by B. von Cotta in 1862, and was accepted by Zirkel and Rosenbusch. They published voluminous books, among which Rosenbusch's *Elemente der Gesteinslehre* (3rd ed., 1910) is relatively concise and readable.

The last phase of the pre-eminence of microscopic petrography was represented by the great work *Die Kristallinen Schiefer* (1904–6) written by U. Grubenmann (1850–1924) of Zurich. In this book, he classified all the regional metamorphic rocks into twelve groups according to their chemical compositions. The rocks of each group were divided into three categories according to the depth-zones, named epi-, meso- and kata-zones in order of presumably increasing depth of their metamorphism. The epi-zone (i.e. shallow zone) was assumed to be characterized by low temperature, low pressure and strong deformation, the kata-zone by high temperature, high pressure and weak deformations. The meso-zone was assumed to be intermediate. This idea of depth zones had been originally formulated by Becke (1903), but became popular through Grubenmann's book.

All the observed regional metamorphic rocks were assigned to one of the three zones. For example, phyllite, chlorite schist, and glaucophane schist were assigned to the epi-zone, mica schist and amphibolite to the meso-zone, and some of the gneisses, eclogite and jadeite rock to the kata-zone. The assignment was made on the basis of general impressions about the grain sizes and mineral compositions, since the mineral zonation of progressive metamorphic terranes was not known at that time. In Zurich, Grubenmann was succeeded by Paul Niggli, who continued to advocate the basic idea and terminology of his predecessor. Hence, Grubenmann's doctrine dominated a large part of continental Europe almost to the present (e.g. Grubenmann and Niggli, 1924).

In about 1920–30, advocates of new approaches of metamorphic geology had to justify their existence by criticizing the doctrine of Grubenmann and Niggli. For example, Harker criticized Grubenmann's static attitude (as was already mentioned in § 3-1). Eskola also repeatedly criticized Grubenmann's depth-zones in order to show the superiority of his facies classification.

4. DISCOVERY AND SIGNIFICANCE OF PROGRESSIVE METAMORPHIC ZONES

Progressive Metamorphic Zones
Prior to the beginning of study of the progressive mineral changes, the progressive textural changes were investigated by Rosenbusch (1877) in the contact aureole of the Barr-Andlau area in Alsace (§ 10-1). He distinguished two textural zones on a geologic sketch map. Progressive textural zones were later mapped in regional metamorphic terranes of New Zealand and other countries.

Progressive mineral zones were mapped for the first time in 1893 by George Barrow in a part of the regional metamorphic terrane of the Scottish Highlands (fig. 3-1). He was a self-taught mapping geologist (1853–1932) of the British Geological Survey. His work was completed in 1912, and was a most epoch-making contribution to metamorphic petrology. However, it did not attract much attention until the confirmation of his work by Tilley (1925) and Harker (1932).

Without being aware of Barrow's work, V. M. Goldschmidt (1915) investigated the progressive regional metamorphism in the Trondheim area (then called Trondhjem) of Norway (fig. 1-2).

The investigation of mineral zoning did not become popular until the publication of studies by Tilley (1925) and Harker (1932) in the Scottish Highlands, by Vogt (1927) in the Sulitelma area of Norway, and by Barth (1936) in Dutchess County, New York State. Even in the late 1930s, the area and the diversity of the mapped terranes were so small that this method did not effectively work as the connecting link between petrography and geology.

The first successful attempt to demonstrate the geologic significance of progressive metamorphic zones was made by W. Q. Kennedy (1948), who showed the distribution of metamorphic zones in the whole Scottish Highlands and connected it to the structure of the Caledonian orogenic belt.

Alfred Harker (1859–1939)

Harker was not only a renowned preacher of progressive metamorphism, but also a creator of a unique doctrine of metamorphism. He belonged to St John's College and Department of Geology (not that of Mineralogy and Petrology) in the University of Cambridge. In his early years, he played a leading part in the introduction of microscopic petrography into Britain and was a brilliant igneous geologist as the author of the Skye Memoir (1904) and the book *The Natural History of Igneous Rocks* (1909). After the publication of this book, his major interest was directed toward the study of metamorphism. His basic ideas on metamorphism were first formulated in 1918. With an addition of a lot of descriptive material, he wrote the book *Metamorphism* (1932). This beautifully written book exerted a deep influence on those interested in metamorphic rocks all over the world in the 1930s and 1940s.

In the book he gave good descriptions of progressive contact as well as regional metamorphism. Regional metamorphism was described mostly on the basis of his observation on rocks of the Barrovian region (§ 14-3) of the Scottish Highlands, though the metamorphism in the Buchan region was also briefly treated. He believed that most cases of regional metamorphism all over the world would show a great similarity to one another, and that the regional metamorphism observed in the Barrovian region represented the 'normal' kind of regional metamorphism. Probably because of the expected similarity, descriptive data from the Norwegian Caledonides, the Alps, Brittany and other areas outside Britain were mixed into the systematic description of rocks of the Scottish Highlands. His concept of normal regional metamorphism pervaded the thinking of geologists for thirty years.

His doctrine was based largely on the structural and mineralogical difference (or contrast) between regional and contact metamorphism. He ascribed it to the presence of strong non-hydrostatic stress in the former and its absence in the latter. Chlorite, muscovite, almandine, kyanite are more common in or characteristic of regional metamorphic rocks and hence were regarded as stress-minerals, which were defined as being stable only in the presence of non-hydrostatic stress. On the other hand, cordierite and andalusite were considered to be characteristic of contact metamorphic rocks and were classed in anti-stress minerals, which were defined as being stable in the absence of non-hydrostatic stress.

The hypothesis of stress minerals provided geologists with too much of a short-cut explanation, and discouraged the then growing effort under the leadership of Goldschmidt and Eskola to explain the mineral composition of metamorphic rocks on the basis of theoretical chemistry. Harker reviewed Goldschmidt's classification of the Oslo hornfelses, but had reservations about its significance, and almost entirely ignored Eskola's works. Goldschmidt and Eskola belonged to a generation more than twenty years younger than Harker.

Criticism of the hypothesis of stress minerals was inevitable for the establishment of metamorphic petrology along Eskola's lines. For example, my proposal of the inverted-Y model for the phase diagram of Al_2SiO_5 minerals (Miyashiro, 1949) was originally designed to defend Eskola's standpoint against Harker's under the circumstance of those days.

In the 1950s, most of the minerals characteristic of schists were synthesized under conditions which did not include non-hydrostatic stress. Thus, Harker's hypothesis of stress minerals gradually waned and disappeared, though the problem of the thermodynamic effects of non-hydrostatic stress has not been completely solved yet.

5. INVESTIGATION OF EQUILIBRIUM MINERAL ASSEMBLAGES

Theoretical chemistry, including the thermodynamic theory of chemical equilibrium, began to form in the later half of the nineteenth century, and came to be used in igneous petrogenesis from the 1890s, especially through the pioneering works of J. H. L. Vogt. The application of this theory to metamorphic petrology was successfully attempted first by Goldschmidt and then by Eskola, leading to the establishment of a sound basis for the study of the mineral composition of metamorphic rocks.

The two great masters were followed by T. Vogt, T. F. W. Barth, and Hans Ramberg. All these investigators were born or worked in North Europe, and constituted the golden age of North European metamorphic geology in the first half of the twentieth century.

V. M. Goldschmidt (1888–1947)

Victor Moritz Goldschmidt was born in Zurich as a son of a chemistry professor, but was educated in Oslo (then called Christiania) and became a professor of crystallography, mineralogy and petrology there.

In 1911, he published a voluminous work on the pyroxene–hornfels facies contact metamorphism of the Oslo area at the age of 23 (§ 10-1). It contained the first successful application of the phase rule to the study of metamorphic mineral assemblages. Then he calculated from thermochemical data the equilibrium curve for the reaction to form wollastonite under the condition of $P_{CO_2} = P_s$ (Goldschmidt, 1912a). This curve gave the first numerical reference scale for the temperature of metamorphism.

Later he published one of the earliest successful maps of progressive metamorphic zones in the Trondheim area (Goldschmidt, 1915) as mentioned in the preceding section, and a laborious study of metasomatism in the Stavanger area (Goldschmidt, 1921). All these studies were epoch-making in the relevant branches of metamorphic geology. However, after World War I, his interest was turned to the distribution of minor elements in the crust, and to crystal chemistry.

He had a strong influence upon Eskola, Barth and Ramberg. Barth began his scientific career as a co-worker of Goldschmidt in the 1920s. Victor Goldschmidt, the famous morphological crystallographer of Heidelberg, was not blood-related to V. M. Goldschmidt.

Pentti Eskola (1883-1964)

Eskola majored in chemistry as a student in Helsinki, and then became a geologist, hoping to apply theoretical chemistry into geology. From 1908 to 1914, he studied Precambrian metamorphic rocks of the Orijärvi mining area in southwest Finland. There he found the equilibrium mineral assemblages of the amphibolite facies. Comparison of his observation with the mineral assemblages of the Oslo hornfelses led him to the concept of the metamorphic facies (Eskola, 1914, 1915, 1920; § 11-1).

Eskola became a professor in Helsinki. He studied in Goldschmidt's laboratory in Oslo in 1919-20 and in the Geophysical Laboratory, Washington, D.C., in 1921-2. In his later years, he was interested especially in the granite problem. He was an eclectic magmatist (e.g. 1932, 1933). In the first half of the twentieth century, the *Bulletin de la Commission géologique de Finlande* was among the most important journals for metamorphic geologists of the world for the papers published by Sederholm, Eskola and associated geologists.

It appears that the works of Goldschmidt and Eskola had not been fully appreciated in the world prior to World War II. This is observed, for example, in F. Y. Loewinson-Lessing's book *A Historical Survey of Petrology* (1955), in which Harker's doctrine of metamorphism was reviewed, but due attention was not paid to Goldschmidt and Eskola's.

Though Goldschmidt's paper on the Oslo hornfelses was widely circulated throughout the world, his later papers were not. Eskola's (1920) first comprehensive account of the principle of metamorphic facies was published in the *Norsk Geologisk Tidsskrift*, then a young journal with limited circulation. The next systematic description of metamorphic facies (Eskola, 1929) was written in Finnish. Their works, therefore, were not easily accessible to many geologists of the world.

The unpopularity of Goldschmidt and Eskola at that time was partly due to the fact that the concept of chemical equilibrium in metamorphic rocks was entirely foreign to the then prevalent way of thinking among ordinary geologists. The familiarity of these two pioneers with theoretical chemistry was due to their personal circumstances, whereas most other geologists at that time were far from this field of science. Now that the concept of chemical equilibrium has become very popular in petrology, present-day petrologists may not understand how difficult the concept was for many geologists in the past to appreciate.

Eskola made intensive descriptive studies on the rocks of the Orijärvi area and the eclogites of southwest Norway. These rocks did not show progressive relations, and, though he had some personal acquaintance with progressive mineral changes (e.g. Eskola, 1922), the elaboration of the metamorphic facies with reference to progressive metamorphism was made mainly by T. Vogt (1927) and Barth (1936). T. Vogt was a son of J. H. L. Vogt. The petrographic data from the Scottish Highlands were also very useful. Barth's (1936) study on regional metamorphism of Dutchess County, New York, played an important role in the propagation of North European concepts of metamorphic geology to the United States (e.g. see Balk, 1936, p. 690).

At last, Eskola wrote a comprehensive systematic description of his doctrine of metamorphism as a part of the book *Die Entstehung der Gesteine* (Berlin, 1939). This book is the monument of the North European metamorphic geology. However, its circulation outside Germany was hindered by World War II that began shortly after its publication, and the unsold copies were burned during the war.

The doctrine of Eskola became popular around 1950 through the textbooks written by Turner (1948), Turner and Verhoogen (1951), Barth (1952) and Ramberg (1952).

Hans Ramberg (1917–)

Hans Ramberg was born in Norway and was educated in Oslo by Goldschmidt and Barth. He stayed in the University of Chicago in 1948–62. For some years around 1950 he was the most brilliant star in metamorphic petrology. He attempted to break through the old framework of thinking within a storm of applause from young generations. His papers were conspicuous for their novelty. He discussed the thermodynamics of solid solutions, the petrologic effect of bond types, fantastic applications of thermodynamics to large-scale migration within the earth's crust (Ramberg, 1944–5, 1948, 1949, 1951, 1952). He also gave a clear model for the material migration and the behaviour of H_2O in the crust.

The origin of granites was the most loudly discussed problem of petrology in the late 1940s. The time-honoured doctrine of Bowen was no longer attractive to many young men. Ramberg was a most thorough and ingenious advocate of the metamorphic origin of granite.

Glaucophane-Schists and the Diversity of Regional Metamorphism

The problem of glaucophane-schists was of vital importance for the understanding of the diversity of regional metamorphism. Eskola (1929, 1939) displayed keen insight in showing that glaucophane schists and associated rocks belong to a separate metamorphic facies. This insight is especially impressive as glaucophane-schists do not occur in North Europe. This view was accepted by Barth (1952), and Dutch geologists, including Brouwer, Egeler and de Roever. However, Taliaferro (1943), Turner (1948), and Turner and Verhoogen (1951) in California raised objections to this view, ascribing the formation of glaucophane-schists to the effects of peculiar pore solutions emitted from associated basic and ultrabasic rocks. The latter view was more popular than Eskola's in the 1940s and the early 1950s. It was probably a brief discussion written by de Roever (1955b) and the discovery of jadeite in apparently unmetamorphosed gray-wackes of the Franciscan (Bloxam, 1956) that marked a turning point in the general trend of thinking in favour of Eskola's.

Even among the advocates of Eskola's view, the genetic relation of glauco-phane-schist facies metamorphism to other facies was not clear. Many students of glaucophane-schists presumed at that time that the whole terranes had been metamorphosed under virtually the same $P-T$ conditions. The best way for establishing the glaucophane-schist facies was to demonstrate the existence of progressive metamorphism including a zone of the facies. This was made for the first time by Banno (1958; also Miyashiro and Banno, 1958) in the Omi area of Japan. Shortly after this, the progressive nature of the Sanbagawa metamorphism was demonstrated by Seki (1958) in the Kanto Mountains and Banno in the Bessi area (§ 7A-9).

Ernst (1961) demonstrated that glaucophane can be synthesized even at a low pressure and high temperature on its own composition. This finding contributed greatly to the clarification of the point that the formation of glaucophane is controlled by the complicated combination of physical conditions and chemical compositions, whereas jadeite and lawsonite are related more directly to the externally controlled conditions of the glaucophane-schist facies.

The nature and the extent of the diversity of regional metamorphism have been clarified around 1960 (Miyashiro, 1961a; § 3-3). An international project for the cartography of metamorphic belts of the world has been undertaken since 1967 under the leadership of H. J. Zwart of Leiden (e.g. Zwart et al. 1967).

6. PRESSURE AND TEMPERATURE OF METAMORPHISM

Nature of Metamorphic Reactions

Prior to World War II, virtually all the petrologists believed that an aqueous fluid existed in the intergranular space during metamorphism, and that metamorphic reactions and material migration took place through this medium (§ 2-3). Thus, it was agreed that a high pressure on the solid phase was accompanied by the same high fluid pressure.

In the 1940s, many of the enthusiastic advocates of the hypothesis of granitization came to doubt the existence of a fluid phase. A lucid model of metamorphic rocks having no fluid phase but with intergranular H_2O and CO_2 molecules was described by Ramberg (e.g. 1952, p. 174–82). He considered that the rocks in great depths were recrystallized generally to lose open spaces between mineral grains, and so they could not contain a fluid phase. Discrete atoms, ions or molecules can migrate there by diffusion mainly through grain boundaries and mosaic fissures but not by mechanical flow. A truly fluid phase existed not in grain boundaries but as inclusions in minerals and only occasionally in open spaces within rocks. Thus he distinguished between rock-pressure and fluid-pressure.

Danielsson (1950) published a strict thermodynamic calculation of the equilibrium curve of the wollastonite reaction, where P_{CO_2} and P_s were distinguished. This work served well as an example of the treatment of volatile-liberating reactions. Meanwhile, the thermodynamics of such reactions were beautifully treated from a much more general standpoint by J. B. Thompson (1955).

Korzhinskii (1936, 1959) and J. B. Thompson (1955, 1970) formulated the thermodynamics of open systems, which would give a theoretical basis to the chemical treatment of metamorphic processes. The usefulness of thermochemical method was demonstrated in the 1950s, especially in combination with synthetic methods. Detailed work was done, for example, on the stability relation of jadeite.

Synthesis of Metamorphic Minerals

J. H. L. Vogt and A. Harker, the two great pioneers in theoretical igneous petrology, inferred the phase relations of some pyrogenic minerals from metallurgical data and early incomplete experimental studies. The systematic attempt at the accurate determination of such phase relations began with the establishment of the Geophysical Laboratory at the Carnegie Institution of Washington in 1907. The new flow of accurate data from the Geophysical Laboratory was so overwhelming that Harker gave up his effort in this field and turned to metamorphic petrology.

There were some attempts at synthesis of metamorphic minerals before World War II, but abundant data of hydrothermal syntheses began to flow out at about 1950 by the work of H. S. Yoder and others (e.g. Yoder, 1952, 1959; Yoder and Eugster, 1955). The breakdown equilibria of mica and other minerals were determined in the pressure range up to a few kilobars.

From the early days of geology, various phenomena were ascribed to the effects of high pressures within the earth. Goldschmidt and Eskola believed in the importance of the effects of pressure as shown, for example, in their ideas on the origin of eclogite and glaucophane schist. In the pressure range up to a few kilobars, however, the effect of rock pressure on the phase relations between solid minerals is small. For this reason, some of the brilliant experimental workers in the early 1950s were critical of the supposition of the effect of high pressure by geologists.

In 1953, Coes succeeded in making a pressure vessel to be used at high temperatures under a pressure of a few tens of kilobars (Coes, 1953). He synthesized a new form of SiO_2 (later named coesite) and various metamorphic minerals which had not been artificially produced at a few kilobars. This gave great encouragement to contemporary experimenters as regards the effectiveness of high pressure. Shortly after it, Birch and his co-workers quantitatively demonstrated the effects by determination of phase relations involving jadeite and kyanite under a pressure up to a few tens of kilobars (Robertson, Birch and McDonald, 1957; Birch and Le Comte, 1960; Clark, Robertson and Birch, 1957).

Temperature of metamorphism

The relative temperatures of metamorphic facies were easily known from the study of progressive metamorphism and dehydration reactions. The numerical values of temperature were much more difficult to obtain.

Harker (1932, p. 209) stated that the recrystallization of muscovite could take place even in cataclastic metamorphism at 'ordinary temperature' and hence he appears to consider that the low temperature limit of the chlorite zone in the Scottish Highlands probably approaches the 'ordinary temperature' (which probably means the surface temperature).

This view was nearly unanimously supported to the 1950s. Thus, the chlorite zone of regional metamorphism, for instance, was considered to represent a temperature range of about 0–250 °C in textbooks published in the 1950s and early 1960s. The temperatures of other grades were estimated so as to be consistent with this. Thus, the temperature range of the amphibolite facies was considered as about 350–600 °C, too low to cause partial melting of metamorphic rocks. This strengthened the belief of some geologists in the metamorphic origin of granites (see Miyashiro, 1972a).

The first strong impact to revise the estimates of metamorphic temperatures came from the establishment of the zeolite and prehnite-pumpellyite facies by Coombs (1954, 1960, 1961) combined with experimental works on zeolite synthesis. A considerable range of temperature corresponding to these two facies must be interposed between the ordinary surface temperature and the temperature of chlorite-zone metamorphism. Thus, the estimated temperature of the chlorite zone increased by about 300 °C.

The second strong impact came from the synthetically determined temperature of the triple point of Al_2SiO_5. Experiments in the first half of the 1960s gave a temperature of 300 °C or 390 °C for this point. However, R. C. Newton (1966b) gave 520 °C, and then Althaus (1967) and Richardson et al. (1969) gave a temperature around 600 °C for it.

According to this new scale of temperature, partial melting would take place relatively commonly in regional metamorphism. This has shed a new light on the problem of material migration in metamorphism.

From the beginning of the twentieth century, most of the brilliant petrologists have had a romantic vision in the application of physics and chemistry to petrology. Reactions in the earth's crust should be governed by the same laws as for those done in the laboratory. It was expected that petrology should become the experimental physical chemistry applied to the crust. The determination of the temperature and pressure of geologic processes was regarded even as the aim of petrology. This vision appears to have been realized to a certain extent, though much more remains to be done.

In this respect, Winkler's book, *Petrogenesis of Metamorphic Rocks*, (1965, 1967) most clearly represents the present-day atmosphere of petrology. It represents the triumphal song of experimental studies.

7. TECTONIC SIGNIFICANCE OF METAMORPHISM AND THE FUTURE OF METAMORPHIC PETROLOGY IN UNIFIED EARTH SCIENCE

Earth science has begun to change radically in the last decade. Radiometric dating has greatly modified our views on the structure and historical development of orogenic belts and continental crusts. Seismic and gravity studies have clarified the structure of the crust and upper mantle. Particularly important is the recent progress in the study of ocean floors. This has led to the hypotheses of ocean-floor spreading and plate tectonics.

These hypotheses have aroused a new accelerated progress in the study of the solid earth. All branches of geology and geophysics are now beginning to collaborate in the establishment of unified earth science.

It is happy that metamorphic petrology has advanced in the last decade so markedly as to be able to meet the new situation. We can estimate the temperature and pressure of metamorphism with considerable reliability, and such estimated values may be used in the construction of geologic models for orogenic processes and ocean-floor spreading along with geophysical data. Marked advance in the petrologic survey in many regions has clarified the nature and diversity of regional metamorphism and related magmatism. The tectonic significance of metamorphism has been well established. Thus, metamorphic geology is playing an important role in the unified earth science in the framework of plate tectonics (Miyashiro, 1972*a, b*). This book is intended to outline metamorphic geology along this line.

References

References

Ackermann, P. B. and Walker, F. (1960). Vitrification of arkose by Karroo dolerite near Heilbron, Orange Free State. *Q. J. geol. Soc. Lond.* **116**, 239–54.

Agrell, S. O. (1965). Polythermal metamorphism of limestones at Kilchoan, Ardnamurchan. *Min. Mag.* **34**, 1–15.

— and Langley, J. M. (1958). The dolerite plug at Tievebulliagh near Cushendall, Co. Antrim. *Proc. R. Ir. Acad., B* **59**, 93–127.

Albee, A. L. (1962). Relationships between the mineral association, chemical composition and physical properties of the chlorite series. *Am. Mineral.* **47**, 851–70.

— (1965a). A petrogenetic grid for the Fe–Mg silicates of pelitic schists. *Am. J. Sci.* **263**, 512–36.

— (1965b). Distribution of Fe, Mg, and Mn between garnet and biotite in natural mineral assemblages. *J. Geol.* **73**, 155–64.

— (1968). Metamorphic zones in northern Vermont. In *Studies of Appalachian Geology: Northern and Maritime*, ed. E-an Zen *et al.*, 329–341. New York, Interscience.

Alderman, A. R. (1936). Eclogites in the neighbourhood of Glenelg, Inverness-shire. *Q. J. geol. Soc. Lond.* **92**, 488–530.

Althaus, E. (1967). The triple point andalusite-sillimanite-kyanite. *Contr. Mineral. Petrol.* **16**, 29–44.

— (1969). Experimental evidence that the reaction of kyanite to form sillimanite is at least bivariant. *Am. J. Sci.* **267**, 273–7.

Anderson, A. L. (1952). Multiple emplacement of the Idaho batholith. *J. Geol.* **60**, 255–65.

Aoki, K. and Oji, Y. (1966). Calc-alkaline volcanic rock series derived from alkali olivine basalt magma. *J. geophys. Res.* **71**, 6127–35.

Armstrong, R. L., Jaeger, E. and Eberhardt, P. (1966). A comparison of K–Ar and Rb–Sr ages on Alpine biotites. *Earth Planet. Sci. Letters* **1**, 13–19.

Aronson, J. L. (1968). Regional geochronology of New Zealand. *Geochim. Cosmochim. Acta* **32**, 669–97.

Atherton, M. P. (1964). The garnet isograd in pelitic rocks and its relation to metamorphic facies. *Am. Mineral.* **49**, 1331–49.

— (1965). The chemical significance of isograds. In *Controls of Metamorphism*, ed. W. S. Pitcher and G. W. Flinn, 169–202. Edinburgh and London: Oliver and Boyd.

— (1968). The variation in garnet, biotite and chlorite composition in medium grade pelitic rocks from the Dalradian, Scotland, with particular reference to the zonation in garnet. *Contr. Mineral. Petrol.* **18**, 347–71.

Atwater, T, (1970). Implications of plate tectonics for the Cenozoic tectonics of western North America. *Bull. geol. Soc. Am.* **81**, 3513–36.

Aubouin, J. (1965). *Geosynclines.* Amsterdam, Elsevier.

Aumento, F. (1969). Diorites from the Mid-Atlantic Ridge at 45°N. *Science, N.Y.* **165**, 1112–13.

— and Loncarevic, B. D. (1969). The Mid-Atlantic Ridge near 45°N, III. *Can. J. Earth Sci.* **6**, 11–23.

Avias, J. (1967). Overthrust structure of the main ultrabasic New Caledonian massives. *Tectonophysics* **4**, 531–41.

Bailey, E. H., Blake, M. C. and Jones, D. L. (1970). On-land Mesozoic oceanic crust in California Coast Ranges. *U. S. Geol. Surv., Prof. Paper 700-C* 70–81.

— , Irwin, W. P. and Jones, D. L. (1964). Franciscan and related rocks, and their significance in the geology of western California. *Calif. Div. Mines Geol. Bull.* **183**.

Balk, R. (1936). Structural and petrologic studies in Dutchess County, New York, Part I. *Bull. geol. Soc. Am.* **47**, 685–774.

Banno, S. (1958). Glaucophane schists and associated rocks in the Omi district, Niigata Prefecture, Japan. *Jap. J. Geol. Geogr.* **29**, 29–44.

— (1964). Petrologic studies on Sanbagawa crystalline schists in the Bessi-Ino district, central Sikoku, Japan. *Tokyo Univ. Fac. Sci. J. Sec. II*, **15** 203–319.

— (1966). Eclogite and eclogite facies. *Jap. J. Geol. Geogr.* **37**, 105–22.

— (1967). Effect of jadeite component on the paragenesis of eclogitic rocks. *Earth Planet. Sci. Letters* **2**, 249–54.

— and Hatano, M. (1963). Zonation of metamorphic rocks in the Horokanai area of the Kamuikotan metamorphic belt in Hokkaido (in Japanese with English abstract). *J. geol. Soc. Japan* **69**, 388–93.

— and Kanehira, K. (1961). Sulfide and oxide minerals in schists of the Sanbagawa and central Abukuma metamorphic terranes. *Jap. J. Geol. Geogr.* **32**, 331–48.

— and Matsui, Y. (1965). Eclogite types and partition of Mg, Fe and Mn between clinopyroxene and garnet. *Proc. Japan Acad.* **41**, 716–21.

— and Miller, J. A. (1965). Additional data on the age of metamorphism of the Ryoke–Abukuma and Sanbagawa metamorphic belt, Japan. *Jap. J. Geol. Geogr.* **36**, 17–22.

— and Yoshino, G. (1965). Eclogite-bearing peridotite mass at Higasiakaisi-yama in the Bessi area, central Sikoku, Japan. *I. U. G. S. Upper Mantle Symposium*, New Delhi, 1964, 150–60. Copenhagen, 1965.

Barazangi, M. and Dorman, J. (1969). World seismicity maps compiled from ESSA, Coast and Geodetic Survey, epicenter data, 1961–7. *Bull. Seismol. Soc. Am.* **59**, 369–80.

Bard, J. P. (1969). *Le Metamorphisme Regional Progressif des Sierras d'Aracena en Andalousie Occidentale (Espagne)*. University of Montpellier.

— (1970). Composition of hornblendes formed during the Hercynian progressive metamorphism of the Aracena metamorphic belt (SW Spain). *Contr. Mineral. Petrol.* **28**, 117–34.

Barker, F. (1961). Phase relations in cordierite-garnet-bearing Kinsman quartz monzonite and the enclosing schist, Lovewell Mountain quadrangle, New Hampshire. *Am. Mineral.* **46**, 1166–76.

Barrow, G. (1893). On an intrusion of muscovite-biotite gneiss in the south-eastern Highlands of Scotland, and its accompanying metamorphism. *Q. J. Geol. Soc. Lond.* **49**, 330–58.

— (1912). On the geology of lower Dee-side and the southern Highland Border. *Proc. Geol. Ass.* **23**, 274–90.

Barth, T. F. W. (1936). Structural and petrologic studies in Dutchess County, New York. II. *Bull. Geol. Soc. Am.* **47**, 775–850.

—— (1938). Progressive metamorphism of sparagmite rocks of southern Norway. *Norsk Geol. Tidsskr.* **18**, 54–65.

—— (1945). Studies in the igneous rock complex of the Oslo region, II. *Skrifter Utgitt Det Norske Videnskaps–Akademi Oslo. I. Mat.-Naturv. Kl. 1944, No. 9.*

—— (1952, 1962). *Theoretical Petrology.* 1st and 2nd eds. New York: John Wiley.

—— (1961). Abundance of the elements, areal averages and geochemical cycles. *Geochim. Cosmochim. Acta* **23**, 1–8.

Bateman, P. C., Clark, L. D., Huber, N. K., Moore, J. G. and Rinehart, C. D. (1963). The Sierra Nevada batholith: A synthesis of recent work across the central part. *U.S. Geol. Surv., Prof. Paper 414-D.*

—— and Dodge, F. C. W. (1970). Variations of major chemical constituents across the central Sierra Nevada batholith. *Bull. geol. Soc. Am.* **81**, 409–20.

—— and Eaton, J. P. (1967). Sierra Nevada batholith. *Science, N.Y.* **158**, 1407–17.

Bearth, P. (1959). Ueber Eklogit, Glaukophanschiefer und metamorphe Pillow-laven. *Schweiz. Mineral. Petrogr. Mitt.* **39**, 267–86.

—— (1962). Versuch einer Gliederung alpinmetamorpher Serien der Westalpen. *Schweiz. Mineral. Petrogr. Mitt.* **42**, 127–37.

Becke, F. (1903). Ueber Mineralbestand und Struktur der kristallinischen Schiefer. *Compte rendu IX. Session du congrès géologique internat. (Vienne),* Part 2, pp. 553–70.

Billings, M. P. and White, W. S. (1950). Metamorphosed mafic dikes of the Woodsville quadrangle, Vermont and New Hampshire. *Am. Mineral.,* **35**, 629–43.

Binns, R. A. (1964). Zones of progressive regional metamorphism in the Willyama complex, Broken Hill district, New South Wales. *J. geol. Soc. Australia* **11**, 283–330.

—— (1965a). Hornblendes from some basic hornfelses in the New England region, New South Wales. *Min. Mag.* **34**, 52–65.

—— (1965b). The mineralogy of metamorphosed basic rocks from the Willyama complex. Broken Hill district, New South Wales, Parts 1 and 2. *Min. Mag.* **35**, 306–26, 561–87.

—— (1967). Barroisite-bearing eclogite from Naustdal. Sogn og Fjordane, Norway. *J. Petrol.* **8**, 349–71.

—— (1969). Ferromagnesian minerals in high-grade metamorphic rocks. *Geol. Soc. Australia Spec. Paper No. 2* 323–32.

Birch, F. and Le Comte, P. (1960). Temperature-pressure plane for albite composition. *Am. J. Sci.* **258**, 209–17.

Bird, J. M. and Dewey, J. F. (1970). Lithosphere plate-continental margin tectonics and the evolution of the Appalachian orogen. *Bull. geol. Soc. Am.* **81**, 1031–60.

Blake, M. C., Jr, Irwin, W. P. and Coleman, R. G. (1969). Blueschist-facies metamorphism related to regional thrust faulting. *Tectonophys.* **8**, 237–46.

Bloxam, T. W. (1956). Jadeite-bearing metagreywackes in California. *Am. Mineral* **41**, 488–96.

—— and Allen, J. B. (1960). Glaucophane-schist, eclogite and associated rocks from Knockormal in the Girvan-Ballantrae complex, south Ayrshire. *Trans. R. Soc. Edinb.* **64**, 1–27.

Bocquet, J. (1971). Cartes de répartition de quelques minéraux du métamorphisme alpin dans les Alpes franco-italiennes. *Eclogae geol. Helvetiae* **64**, 71–103.

Boettcher, A. L. and Wyllie, P. J. (1967). Revision of the calcite-aragonite transition, with the location of a triple point between calcite I, calcite II and aragonite. *Nature, Lond.* **213**, 792–3.

—— and —— (1968). Melting of granite with excess water to 30 kilobars pressure. *J. Geol.* **76**, 235–44.

Bogdanoff, A. A., Mouratov, M. V. and Schatsky, N. S. (1964). *Tectonics of Europe; Explanatory Note to the International Tectonic Map of Europe,* Scale 1:2 500 000. Moscow: Nauka.

Boldizsár, T. (1964). Terrestrial heat flow in the Carpathians. *J. geophys. Res.* **69**, 5269–275.

Bosma, W. (1964). The spots in the spotted slates of Stiege (Vosges) and Vogtland (Saxony). *Geol. en Mijnbouw* **43**, 476–89.

Bowen, N. L. (1928). *The Evolution of the Igneous Rocks.* Princeton: Princeton University Press.

—— (1940). Progressive metamorphism of siliceous limestone and dolomite. *J. Geol.* **48**, 225–74.

Bowes, D. R. (1968). The absolute time scale and subdivision of Precambrian rocks in Scotland. *Geol. Fören. Stockholm Förh.* **90**, 175–88.

Boyd, F. R. (1959). Hydrothermal investigations of amphiboles. In *Researches in Geochemistry,* ed. P. H. Abelson, 377–96. New York: John Wiley.

—— and England, J. L. (1959). Pyrope. *Geophys. Lab. A. Rep. Director for 1958-1959* 83–7.

Brothers, R. N. (1970). Lawsonite-albite schists from northernmost New Caledonia. *Contr. Mineral. Petrol.* **25**, 185–202.

Brown, E. H. (1967). The greenschist facies in part of eastern Otago, New Zealand. *Contr. Mineral. Petrol.* **14**, 259–92.

—— (1969). Some zoned garnets from the greenschist facies. *Am. Mineral.* **54**, 1662–77.

—— (1971). Phase relations of biotite and stilpnomelane in the greenschist facies. *Contr. Mineral. Petrol.* **31**, 275–99.

Brown, P. E., Miller, J. A. and Grasty, R. L. (1968). Isotope ages of late Caledonian granitic intrusions in the British Isles. *Proc. Yorkshire geol. Soc.* **36**, 251–76.

——, ——, Soper, N. J. and York, D. (1965). Potassium-argon age pattern of the British Caledonides *Proc. Yorkshire geol. Soc.* **35**, 103–38.

Brown, W. L. (1960). The crystallographic and petrologic significance of peristerite unmixing in the acid plagioclases. *Z. Kristallog.* **113**, 330–44.

—— (1962). Peristerite unmixing in the plagioclases and metamorphic facies series. *Norsk Geol. Tidsskr.* **42**, 2. Halvbind (*Feldspar Volume*), 354–82.

Bryhni, I., Green, D. H., Heier, K. S. and Fyfe, W. S. (1970). On the occurrence of eclogite in western Norway. *Contr. Mineral. Petrol.* **26**, 12–19.

Buddington, A. F. (1939). Adirondack igneous rocks and their metamorphism. *Mem. geol. Soc. Am.* 7.

—— (1959). Granite emplacement with special reference to North America. *Bull. geol. Soc. Am.* **70**, 671–747.

—— , Fahey, J. and Vlisidis, A. (1955). Thermometric and petrogenetic significance of titaniferous magnetite. *Am. J. Sci.* **253**, 497–532.

Bullard, E., Everett, J. E. and Smith, A. G. (1965). The fit of the continents around the Atlantic. In *A Symposium on Continental Drift* ed. P. M. S. Blackett *et al. Phil. Trans. R. Soc. Lond. Ser. A* **258**, 41-51.

Burnham, C. W. (1959). Contact metamorphism of magnesian limestones at Crestmore, California. *Bull. geol. Soc. Am.* **70**, 879-920.

Burns, R. G. (1966). Origin of optical pleochroism in orthopyroxenes. *Min. Mag.* **35**, 715-19.

Burst, J. F., Jr (1959). Postdiagenetic clay mineral environmental relationships in the Gulf Coast Eocene. *Proc. Sixth Nat. Conf. on Clays and Clay Minerals,* 327-341. Oxford: Pergamon Press.

Butler, B. C. M. (1967). Chemical study of minerals from the Moine schists of the Ardnamurchan area, Argyllshire, Scotland. *J. Petrol.* **8**, 233-67.

Butler, J. R. and Ragland, P. C. (1969). Petrology and chemistry of meta-igneous rocks in the Albemarle area, North Carolina slate belt. *Am. J. Sci.* **267**, 700-26.

Cady, W. M. (1968). The lateral transition from the miogeosynclinal to the eugeosynclinal zone in northwestern New England and Adjacent Quebec. In: *Studies of Appalachian Geology: Northern and Maritime,* ed E-an Zen *et al.,* 151-161. New York: Interscience.

Campbell, A. S. and Fyfe, W. S. (1965). Analcime–albite equilibria. *Am. J. Sci.* **263**, 807-16.

Cann, J. R. (1969). Spilites from the Carlsberg Ridge, Indian Ocean. *J. Petrol.* **10**, 1-19.

— and Vine, F. J. (1966). An area on the crest of the Carlsberg Ridge: petrography and magnetic survey. *Phil. Trans. R. Soc. Lond. Ser. A* **259**, 198-217.

Capdevila, R. (1968) Les types de métamorphisme "intermédiaires de basse pression" dans le segment hercynien de Galice nord-orientale. *Acad. Sci. C. R.* **266**, 1924-7.

Carmichael, D. M. (1970). Intersecting isograds in the Whetstone Lake area, Ontario, *J. Petrol.* **11**, 147-81.

Carmichael, I. S. E. (1964). The petrology of Thingmuli, a Tertiary volcano in eastern Iceland. *J. Petrol.* **5**, 435-60.

Carpenter, A. B. (1967). Mineralogy and petrology of the system CaO–MgO–CO_2–H_2O at Crestmore, California. *Am. Mineral.* **52**, 1363-41.

Chatterjee, N. D. (1961). The Alpine metamorphism in the Simplon area, Switzerland and Italy. *Geol. Rundschau* **51**, 1-72.

— (1970). Synthesis and upper stability of paragonite. *Contr. Mineral. Petrol.* **27**, 244-57.

Chinner, G. A. (1960). Pelitic gneisses with varying ferrous/ferric ratios from Glen Clova, Angus, Scotland. *J. Petrol.* **1**, 178-217.

— (1961). The origin of sillimanite in Glen Clova, Angus. *J. Petrol.* **2**, 312-23.

— (1962). Almandine in thermal aureoles. *J. Petrol.* **3**, 316-40.

— (1965). The kyanite isograd in Glen Clova, Angus, Scotland. *Min. Mag.* **34** (*Tilley vol.*), 132-43.

— (1966a). The distribution of pressure and temperature during Dalradian metamorphism. *Q. J. geol. Soc. Lond.* **122**, 159-86.

— (1966b). The significance of the aluminum silicates in metamorphism. *Earth Sci. Revs.* **12**, 111-26.

— (1967). Chloritoid, and the isochemical character of Barrow's zones. *J. Petrol.* **8**, 268-82.

Chinner, G. A., Smith, J. V. and Knowles, C. R. (1969). Transition metal contents of $Al_2 SiO_5$ polymorphs. *Am. J. Sci. 267-A (Schairer Vol.)*, 96–113.

Christensen, N. I. (1965). Compressional wave velocities in metamorphic rocks at pressures to ten kilobars. *J. geophys. Res.* **70**, 6147–64.

Clark, S. P., Jr, editor (1966). Handbook of physical constants. Revised ed. *Mem. geol. Soc. America* **97**.

—— , Robertson, E. C. and Birch, F. (1957). Experimental determination of kyanite–sillimanite equilibrium relations at high temperatures and pressures. *Am. J. Sci.* **255**, 628–40.

Closs, H. (1969). Explosion seismic studies in Western Europe. In *The Earth's Crust and Upper Mantle*, ed. P. J. Hart, *Geophys. Monogr.* **13**, 178–88. American Geophys. Union, Washington, D.C.

Coes, L., Jr, (1953). A new dense crystalline silica. *Science, N.Y.* **118**, 131–2.

Cogné, J., Jeanette, D. and Ruhland, M. (1966). L'Ile de Croix. *Serv. Carte géol. Alsace Lorraine Bull.*, **19**, Part 1, 41–95.

Coleman, R. G. (1967). Glaucophane schists from California and New Caledonia. *Tectonophys.* **4**, 479–98.

—— (1971). Plate tectonic emplacement of upper mantle peridotites along continental edges. *J. geophys. Res.* **76**, 1212–22.

—— and Clark, J. R. (1968). Pyroxenes in the blueschist facies of California. *Am. J. Sci.* **266**, 43–59.

—— and Lanphere, M. A. (1971). Distribution and age of high-grade blueschists, associated eclogites, and amphibolites from Oregon and California. *Bull. geol. Soc. Am.* **82**, 2397–412.

—— and Lee, D. E. (1962). Metamorphic aragonite in the glaucophane schists of Cazadero, California. *Am. J. Sci.* **260**, 577–95.

—— and —— (1963). Glaucophane-bearing metamorphic rock types of the Cazadero area, California. *J. Petrol.* **4**, 260–301.

—— , —— , Beatty, L. B. and Brannock, W. W. (1965). Eclogites and eclogites: Their differences and similarities. *Bull. geol. Soc. Am.* **76**, 483–508.

—— and Papike, J. J. (1968). Alkali amphiboles from the blueschists of Cazadero, California. *J. Petrol.* **9**, 105–22.

Compston, W. and Arriens, P. A. (1968). The Precambrian geochronology of Australia. *Can. J. Earth Sci.* **5**, 561–83.

Compton, R. R. (1955). Trondhjemite batholith near Bidwell Bar, California. *Bull. geol. Soc. Am.* **66**, 9–44.

—— (1960). Contact metamorphism in Santa Rosa Range, Nevada. *Bull. geol. Soc. Am.* **71**, 1383–416.

Condie, K. (1967). Composition of the ancient North American crust. *Science, N.Y.* **155**, 1013–15.

Coombs, D. S. (1954). The nature and alteration of some Triassic sediments from Southland, New Zealand. *Trans. R. Soc. N.Z.* **82**, 65–109.

—— (1960). Lower grade mineral facies in New Zealand. *Internat. geol. Congr. 21st Sess. Rep.* Part 13, 339–51. Copenhagen.

—— (1961). Some recent work on the lower grades of metamorphism. *Australian J. Sci.* **24**, 203–15.

—— , Ellis, A. J., Fyfe, W. S. and Tayler, A. M. (1959). The zeolite facies, with comments on the interpretation of hydrothermal synthesis. *Geochim. Cosmochim. Acta* **17**, 53–107.

Coombs, D. S., Horodyski, R. J. and Naylor, R. S. (1970). Occurrence of prehnite-pumpellyite facies metamorphism in northern Maine. *Am. J. Sci.* **268**, 142-56.

— and Whetten, J. T. (1967). Composition of analcime from sedimentary and burial metamorphic rocks. *Bull. geol. Soc. Am.* **78**, 269-82.

Cooper, A. F. and Lovering, J. F. (1970). Greenschist amphiboles from Haast River, New Zealand. *Contr. Mineral. Petrol.* **27**, 11-24.

Crawford, M. L. (1966). Composition of plagioclase and associated minerals in some schists from Vermont, U.S.A., and South Westland, New Zealand, with inferences about the peristerite solvus. *Contr. Mineral. Petrol.* **13**, 269-94.

Crawford, W. A. and Fyfe, W. S. (1964). Calcite-aragonite equilibrium at 100 °C. *Science, N.Y.* **144**, 1569-70.

Crook, K. A. W. (1969). Contrasts between Atlantic and Pacific geosynclines. *Earth Planet. Sci. Letters* **5**, 429-38.

Cullen, D. J. (1967). Island arc development in the southwest Pacific. *Tectonophys.* **4**, 163-72.

Daly, R. A. (1917). Metamorphism and its phases. *Bull. geol. Soc. Am.* **28**, 375-418.

Dalziel, I. W. D. (1966). A structural study of the granitic gneiss of Western Ardgour, Argyll and Inverness-shire. *Scott. J. Geol.* **2**, 125-52.

Danielsson, A. (1950). Das Calcit-Wollastonitgleichgewicht. *Geochim. Cosmochim. Acta* **1**, 55-69.

Davidson, C. F. (1943). The Archaean rocks of the Rodil district, South Harris, Outer Hebrides. *Trans. R. Soc. Edinb.* **61**, 71-112.

Dawson, J. B. (1967). A review of the geology of kimberlite. In *Ultramafic and Related Rocks,* ed. P. J. Wyllie, 241-51. New York: John Wiley.

Dearnley, R. (1963). The Lewisian complex of South Harris with some observations on the metamorphosed basic intrusions of the Outer Hebrides, Scotland. *Q. J. geol. Soc. Lond.* **119**, 243-397.

Dennis, J. G. (1967). *International Tectonic Dictionary—English Terminology.* American Association of Petroleum Geologists, Tulsa, Oklahoma.

den Tex, E. (1971). The facies groups and facies series of metamorphism, and their relation to physical conditions in the earth's crust. *Lithos.* **4**, 23-41.

de Roever, W. P. (1955a). Genesis of jadeite by low-grade metamorphism. *Am. J. Sci.* **253**, 283-98.

— (1955b). Some remarks concerning the origin of glaucophane in the North Berkeley Hills, California. *Am. J. Sci.* **253**, 240-4.

— (1956). Some differences between post-Paleozoic and older regional metamorphism. *Geol. en Mijnbouw* **18**, 123-7.

— (1964). On the cause of the preferential distribution of certain metamorphic minerals in orogenic belts of different age. *Geol. Rundschau* **54**, 933-43.

— and Nijhuis, H. J. (1964). Plurifacial alpine metamorphism in the eastern Betic Cordilleras (SE Spain) with special reference to the genesis of the glaucophane. *Geol. Rundschau* **53**, 324-36.

de Sitter, L. U. (1956). *Structural Geology.* New York: McGraw-Hill.

de Waard, D. (1965). A proposed subdivision of the granulite facies. *Am. J. Sci.* **263**, 455-61.

Dewey, J. F. (1969). Evolution of the Appalachian/Caledonian orogen. *Nature, Lond.* **222**, 124-9.

Dewey, J. F. and Kay, M. (1968). Appalachian and Caledonian evidence for drift in the north Atlantic. In *The History of the Earth's Crust, a Symposium,* ed. R. A. Phinney, 161-7. Princeton: Princeton University Press.

— and Pankhurst, R. J. (1970). The evolution of the Scottish Caledonides in relation to their isotopic age pattern.*Trans. R. Soc. Edinb.* **68**, 361-89.

Dickinson, W. R. (1962). Petrogenetic significance of geosynclinal andesite volcanism along the Pacific margin of North America. *Bull. geol. Soc. Am.* **73**, 1241-56.

— (1966). Structural relationships of San Andreas fault system, Cholame Valley and Castle Mountain Range, California. *Bull. geol. Soc. Am.* **77**, 707-26.

— (1968). Circum-Pacific andesite types. *J. geophys. Res.* **73**, 2261-9.

— (1970). Relations of andesites, granites, and derivative sandstones to arc-trench tectonics. *Rev. Geophys. Space Phys.* **8**, 813-60.

— (1971). Clastic sedimentary sequences deposited in shelf, slope, and trough settings between magmatic arcs and associated trenches. In *Pacific Geology,* ed. M. Minato, vol. 3, 15-30. Tokyo: Tsukiji Shokan.

— and Hatherton, T. (1967). Andesitic volcanism and seismicity around the Pacific. *Science, N.Y.* **157**, 801-3.

—, Ojakangas, R. W. and Stewart, R. J. (1969). Burial metamorphism of the late Mesozoic Great Valley sequence, Cache Creek, California. *Bull. geol. Soc. Am.* **80**, 519-26.

Dietz, R. S. (1963). Alpine serpentines as oceanic rind fragments. *Bull geol. Soc. Am.* **74**, 947-52.

— and Holden, J. C. (1966). Miogeoclines (miogeosynclines) in space and time. *J. Geol.* **74**, 566-83.

Dobretsov, N. L. (1968). Lawsonite-glaucophane and glaucophane schists of the U.S.S.R. and some problems of metamorphism in orogenic belts (in Russian with English abstract). *Internat. geol. Congr. 23rd Sess., Dokl. Soviet Geol., Problem* **3**, 31-9.

— and Kuroda, Y. (1969). Geologic laws characterizing glaucophane metamorphism in northwestern part of the folded frame of Pacific Ocean. *Internat. geol. Rev.* **12**, 1389-407.

Doll, C. G. (1961). *Centennial Geologic Map of Vermont.* Vermont Geol. Survey.

Donnelly, T. W., Rogers, J. J. W., Pushkar, P. and Armstrong, R. L. (1971). Chemical evolution of the igneous rocks of the eastern West Indies: an investigation of thorium, uranium and potassium distributions, and lead and strontium isotopic ratios. *Mem. geol. Soc. America* **130**, 181-224.

Eade, K. E. and Fahrig, W. F. (1971). Geochemical evolutionary trends of continental plates – A preliminary study of the Canadian shield. *Bull. geol. Surv. Can.* **179**.

—, — and Maxwell, J. A. (1966). Composition of crystalline shield rocks and fractionating effects of regional metamorphism. *Nature, Lond.* **211**, 1245-9.

Egeler, C. G. (1956). The Alpine metamorphism in Corsica. *Geol. Mijnbouw* **18**, 115-18.

Engel, A. E. J. (1963). Geologic evolution of North America. *Science, N.Y.* **140**, 143-52.

— and Engel, C. G. (1958, 1960). Progressive metamorphism and granitization of the major paragneiss, northwest Adirondack Mountains, New York. I, II. *Bull. geol. Soc. Am.* **69**, 1369-414; **71**, 1-58.

Engel, A. E. J. and Engel, C. G. (1962a). Progressive metamorphism of amphibolite, Northwest Adirondack Mountains, New York. In *Petrologic Studies (Buddington Vol.)*, ed. A. E. J. Engel *et al.*, 37–82. Geol. Soc. America.

— and — (1962b). Hornblendes formed during progressive metamorphism of amphibolites, northwest Adirondack Mountains, New York. *Bull. geol. Soc. Am.* **73**, 1499–514.

— and — (1964a). Continental accretion and the evolution of North America. In *Advancing Frontiers in Geology and Geophysics*, ed. A. P. Subramaniam and S. Balakrishna, 17–38. Indian Geophysical Union.

— and — (1964b). Composition of basalts from the Mid-Atlantic Ridge. *Science, N.Y.* **144**, 1330–3.

—, — and Havens, R. G. (1964). Mineralogy of amphibolite interlayers in the gneiss complex, northwest Adirondack Mountains, New York. *J. Geol.* **72**, 131–56.

—, — and — (1965). Chemical characteristics of oceanic basalts and the upper mantle. *Bull. geol. Soc. Am.* **76**, 719–34.

Epstein, S. and Taylor, H. P., Jr (1967). Variation of $^{18}O/^{16}O$ in minerals and rocks. In *Researches in Geochemistry*, ed. P. H. Abelson, vol. 2, 29–62. New York: John Wiley.

Eric, J. H., Stromquist, A. A. and Swinney, C. M. (1955). Geology and mineral deposits of the Angels Camp and Sonora quadrangles, Calaveras and Tuolumne Counties, California. *Calif. Div. Mines Spec. Rep* **41**.

Ernst, W. G. (1960). The stability relations of magnesioriebeckite. *Geochim. Cosmochim. Acta* **19**, 1–40.

— (1961). Stability relations of glaucophane. *Am. J. Sci.* **259**, 735–65.

— (1963a). Petrogenesis of glaucophane schists. *J. Petrol.* **4**, 1–30.

— (1963b). Polymorphism in alkali amphiboles. *Am. Mineral.* **48**, 241–60.

— (1963c). Significance of phengitic micas from low-grade schists. *Am. Mineral.* **48**, 1357–73.

— (1964). Petrochemical study of coexisting minerals from low-grade schists, Eastern Sikoku, Japan. *Geochim. Cosmochim. Acta* **28**, 1631–68.

— (1965). Mineral parageneses in Franciscan metamorphic rocks, Panoche Pass, California. *Bull. geol. Soc. Am.* **76**, 879–914.

— (1966). Synthesis and stability relations of ferrotremolite *Am. J. Sci.* **264**, 37–65.

— (1968). *Amphiboles, Crystal Chemistry, Phase Relations and Occurrence*. New York: Springer-Verlag.

— (1970). Tectonic contact between the Franciscan mélange and the Great Valley Sequence – crustal expression of a late Mesozoic Benioff zone. *J. geophys. Res.* **75**, 886–901.

— (1971a). Do mineral parageneses reflect unusually high-pressure conditions of Franciscan metamorphism? *Am. J. Sci.* **270**, 81–108.

— (1971b). Metamorphic zonations on presumably subducted lithospheric plates from Japan, California and the Alps. *Contr. Mineral. Petrol.* **34**, 43–59.

— (1972a). Occurrence and mineralogic evolution of blueschist belts with time. *Am. J. Sci.* **272**, 657–68.

— (1972b). Ca-amphibole paragenesis in the Shirataki district, central Shikoku, Japan. *Mem. Geol. Soc. Am.* **135**, 73–94.

— and Seki, Y. (1967). Petrologic comparison of the Franciscan and Sanbagawa metamorphic terranes. *Tectonophys.* **4**, 463–78.

Ernst, W. G., Seki, Y., Onuki, H. and Gilbert, M. C. (1970). Comparative study of low-grade metamorphism in the California Coast Ranges and the outer metamorphic belt of Japan. *Mem. geol. Soc. Am.* **124**.

Eskola, P. (1914). On the petrology of the Orijärvi region in southwestern Finland. *Bull. Comm. géol. Finlande.* No. 40.

— (1915). On the relations between the chemical and mineralogical composition in the metamorphic rocks of the Orijärvi region. *Bull. Comm. géol. Finlande.* No. 44.

— (1920). The mineral facies of rocks. *Norsk Geol. Tidsskr.* **6**, 143–94.

— (1921). On the eclogites of Norway. *Videnskaps. Skrifter. I. Mat.-Naturv. Kl. 1921,* No. 8.

— (1922). On contact phenomena between gneiss and limestone in western Massachusetts. *J. Geol.* **30**, 265–94.

— (1929). On mineral facies. *Geol. Fören. Stockholm Förh.* **51**, 157–72.

— (1932). On the origin of granitic magmas. *Mineral. Petrogr. Mitt.* **42**, 455–81.

— (1933). On the differential anatexis of rocks. *Bull. Comm. géol. Finlande.* No. 103, 12–25.

— (1939). Die metamorphen Gesteine. In *Die Entstehung der Gesteine,* by Tom F. W. Barth, C. W. Correns and P. Eskola, 263–407. Berlin: Julius Springer. Reprinted in 1960 and 1970.

— (1952). On the granulites of Lapland. *Am. J. Sci. (Bowen Vol.),* 133–71.

— (1954). A proposal for the presentation of rock analyses in ionic percentage. *Acad. Scient. Fennicae Annal. Ser. A,* III. No. 38.

— (1957). On the mineral facies of charnockites. *Madras Univ. J. B, 27 (Centenary No.),* 101–19.

— (1963). The Precambrian of Finland. In *The Precambrian,* ed. K. Rankama, vol. 1, 145–263. New York: Interscience.

Eugster, H. P. (1959). Reduction and oxidation in metamorphism. In *Researches in Geochemistry,* ed. P. H. Abelson, 397–426. New York: John Wiley.

— and Wones, D. R. (1962). Stability relations of the ferruginous biotite, annite. *J. Petrol.* **3**, 82–125.

— and Yoder, H. S. (1954). Paragonite. *Geophys. Lab. A. Rep. Director for 1953-1954,* 111–14.

— and — (1955). The join muscovite-paragonite. *Geophys. Lab. A. Rep. Director for 1954-1955,* 124–6.

Evans, B. W. (1964). Fractionation of elements in the pelitic hornfelses of the Cashel-Lough Wheelaun intrusion, Connemara, Eire. *Geochim. Cosmochim. Acta* **28**, 127–56.

— (1965). Application of a reaction-rate method to the breakdown equilibria of muscovite and muscovite plus quartz. *Am. J. Sci.* **263**, 647–67.

— and Guidotti, C. V. (1966). The sillimanite–potash feldspar isograd in western Maine, U.S.A. *Contr. Mineral. Petrol.* **12**, 25–62.

Evernden, J. F. and Kistler, R. W. (1970). Chronology of emplacement of Mesozoic batholithic complexes in California and western Nevada. *U.S. geol. Surv. Prof. Paper 623.*

— and Richards, J. R. (1962). Potassium–argon ages in eastern Australia. *J. geol. Soc. Aust.* **9**, 1–49.

Fabriès, J. (1963). Les formations cristallines et métamorphiques du Nord-Est de la province de Séville (Espagne). *Sciences de la Terre, Mem. 4.*

Fabriès, J. (1968). Nature des hornblendes et types de métamorphisme. Internat. Mineral Assoc. 5th General Meeting (1966). *Papers and Proc.* 204–11. Mineral. Soc. London.

Fawcett, J. J. and Yoder, H. S., Jr (1966). Phase relationships of chlorites in the system $MgO-Al_2O_3-SiO_2-H_2O$. *Am. Mineral.* **51**, 353–80.

Fedotov, S. A. (1968). On deep structure, properties of the upper mantle, and volcanism of the Kuril-Kamchatka Island Arc according to seismic data. In *The Crust and Upper Mantle of the Pacific Area,* ed. L. Knopoff *et al., Geophys. Mongr.* **12**, 131–9. Washington, D.C.: American Geophys. Union.

Fenner, C. N. (1929). The crystallization of basalts. *Am. J. Sci.* [5] **18**, 225–53.

Fitch, F. J. (1965). The structural unity of the reconstructed North Atlantic continent. In *A Symposium on Continental Drift,* ed. P. M. S. Blackett *et al., Phil. Trans. R. Soc. Lond. Ser. A.* **258**, 191–3.

Fitton, J. G. and Hughes, D. J. (1970). Volcanism and plate tectonics in the British Ordovician. *Earth Planet. Sci. Letters* **8**, 223–8.

Fonteille, M. and Guitard, G. (1968). Le comportement de l'eau dans les roches métamorphiques catazonales des Pyrénées et son influence sur les assocations minérales des paragness d'origine pélitique, en particulier des kinzigites. *Acad. Sci. Paris C. R.* **267**, 1133–5.

Foster, H. L. (1965). Geology of Ishigaki-shima, Ryukyu-retto. *U.S. Geol. Surv. Prof. Paper 399-A.*

Francis, G. H. (1956). Facies boundaries in pelites at the middle grades of regional metamorphism. *Geol. Mag.* **93**, 353–68.

— (1958). Petrological studies in Glen Urquhart, Inverness-shire. *British Museum, Mineral. Bull.* **1**, 123–62.

French, B. M. and Eugster, H. P. (1965). Experimental control of oxygen fugacities by graphite-gas equilibriums. *J. geophys. Res.* **70**, 1529–39.

Frey, M. (1969*a*). Die Metamorphose des Keupers vom Tafeljura bis zum Lukmanier-Gebiet. *Beitr. geol. Karte der Schweiz, neue Folge,* 137.

— (1969*b*). A mixed-layer paragonite/phengite of low-grade metamorphic origin. *Contr. Mineral. Petrol* **24**, 63–5.

— (1970). The step from diagenesis to metamorphism in pelitic rocks during Alpine orogenesis. *Sedimentol.* **15**, 261–79.

Fry, N. and Fyfe, W. S. (1969). Eclogites and water pressure. *Contr. Mineral. Petrol.* **24**, 1–6.

Fyfe, W. S. (1958). A further attempt to determine the vapor pressure of brucite. *Am. J. Sci.* **256**, 729–32.

—, Turner, F. J. and Verhoogen, J. (1958). Metamorphic reactions and metamorphic facies. *Mem. geol. Soc. Am.* **73**.

— and Valpy, G. W. (1959). The analcime-jadeite phase boundary: some indirect deductions. *Am. J. Sci.* **257**, 316–20.

Fyson, W. K. (1971). Fold attitudes in metamorphic rocks. *Am. J. Sci.* **270**, 373–82.

Ganguly, J. (1968). Analysis of the stabilities of chloritoid and staurolite and some equilibria in the system $FeO-Al_2O_3-SiO_2-H_2O-O_2$. *Am. J. Sci.* **266**, 277–98.

— and Newton, R. C. (1968). Thermal stability of chloritoid at high pressure and relatively high oxygen fugacity. *J. Petrol,* **9**, 444–66.

Garlick, G. D. and Epstein, S. (1967). Oxygen isotope ratios in coexisting minerals of regionally metamorphosed rocks. *Geochim. Cosmochim. Acta* **31**, 181–214.

Gass, I. G. (1968). Is the Troodos Massif of Cyprus a fragment of Mesozoic ocean floor? *Nature, Lond.* **220**, 39–42.

Geijer, P. (1963). The Precambrian of Sweden. In *The Precambrian*, ed. K. Rankama, vol. 1, 81–144. New York: Interscience.

Gemuts, I. (1965). Regional metamorphism in the Lamboo complex, East Kimberley area. *W. Aust. geol. Surv. A. Rep. 1964.* 36–41.

Gibbs, G. V. (1966). The polymorphism of cordierite I. *Am. Mineral.* **51**, 1068–87.

Gilluly, J. (1965). Volcanism, tectonism, and plutonism in the western United States. *Geol. Soc. Am. Spec. Paper 80.*

— (1971). Plate tectonics and magmatic evolution. *Bull. geol. Soc. Am.* **82**, 2383–96.

Goldschmidt, V. M. (1911). Die Kontaktmetamorphose im Kristianiagebiet. *Vidensk. Skrifter. I. Mat.-Naturv. K. (1911)*, No. 11.

— (1912a). Die Gesetze der Gesteinsmetamorphose, mit Beispielen aus der Geologie des südlichen Norwegens. *Vidensk. Skrifter. I. Mat.-Naturv. Kl. (1912)*, No. 22.

— (1912b). Ueber die Anwendung der Phasenregel auf die Gesetze der Mineralassoziation. *Centralblatt Mineral. Geol. Palaeont. (1912)*, 574–6.

— (1915). Geologisch-petrographische Studien im Hochgebirge des südlichen Norwegens. III. Die Kalksilikatgneise und Kalksilikatglimmerschiefer im Trondhjem-Gebiete. *Vidensk. Skrifter. I. Mat.-Naturv. Kl. (1915)*, No. 10.

— (1921). Geologisch-petrographische Studien im Hochgebirge des südlichen Norwegens. V. Die Injektionsmetamorphose im Stavanger-Gebiete. *Vidensk. Skrifter. I. Mat.-Naturv. Kl. (1920)*, No. 10.

— (1922). On the metasomatic processes in silicate rocks. *Econ. Geol.* **17**, 105–23.

Goldsmith, J. R. and Newton, R. C. (1969). P-T-X relations in the system $CaCO_3$-$MgCO_3$ at high temperatures and pressures. *Am. J. Sci.* **267-A** (*Schairer vol.*), 160–90.

González-Bonorino, F. (1971). Metamorphism of the crystalline basement of central Chile, *J. Petrol.* **12**, 149–75.

Gorai, M. (1951). Petrological studies on plagioclase twins. *Am. Mineral.* **36**, 884–901.

Grant, J. A. (1968). Partial melting of common rocks as a possible source of cordierite-anthophyllite bearing assemblages. *Am. J. Sci.* **266**, 908–31.

Green, D. H. (1966). The origin of the 'eclogites' from Salt Lake Crater, Hawaii. *Earth Planet. Sci. Letters* **1**, 414–20.

— and Lambert, I. B. (1965). Experimental crystallization of anhydrous granite at high pressures and temperatures. *J. geophys. Res.* **70**, 5259–68.

— and Ringwood, A. E. (1967). An experimental investigation of the gabbro to eclogite transformation and its petrological applications. *Geochim. Cosmochim. Acta* **31**, 767–833.

Green, T. H. and Ringwood, A. E. (1966). Origin of the calc-alkaline igneous rock suite. *Earth Planet. Sci. Letters* **1**, 307–16.

— and — (1968). Genesis of the calc-alkaline igneous rock suite. *Contr. Mineral. Petrol.* **18**, 105–62.

Green, J. C. (1963). High-level metamorphism of pelitic rocks in northern New Hampshire. *Am. Mineral.* **48**, 991–1023.

Greenwood, H. J. (1961). The system $NaAlSi_2O_6$-H_2O-argon: total pressure and water pressure in metamorphism. *J. geophys. Res.* **66**, 3923–46.

Greenwood, H. J. (1962). Metamorphic reactions involving two volatile components. *Geophys. Lab. A. Rep. Director for 1961-1962*, 82-5.

—— (1963). The synthesis and stability of anthophyllite. *J. Petrol.* **4**, 317-51.

—— (1967). Wollastonite: stability in H_2O-CO_2 mixtures and occurrence in a contact-metamorphic aureole near Salmo, British Columbia, Canada. *Am. Mineral.* **52**, 1669-80.

Griffin, J. J., Windom, H. and Goldberg, E. D. (1968). The distribution of clay minerals in the world ocean. *Deep Sea Res.* **15**, 433-59.

Grubenmann, U. (1904-6). *Die Kristallinen Schiefer.* 1st ed. I (1904), II (1906); 2nd ed. (1910). Berlin: Gebrüder Bornträger.

—— and Niggli, P. (1924). *Die Gesteinsmetamorphose. I.* Berlin: Gebrüder Bornträger.

Guidotti, C. V. (1963). Metamorphism of the pelitic schists in the Bryant Pond quadrangle, Maine. *Am. Mineral.* **48**, 772-91.

—— (1968). Prograde muscovite pseudomorphs after staurolite in the Rangeley-Oquossoc areas, Maine. *Am. Mineral.* **53**, 1368-76.

—— (1969). A comment on 'Chemical study of minerals from the Moine schists of the Ardnamurchan area, Argyllshire, Scotland' by B. C. M. Butler, and its implications for the phengite problem. *J. Petrol.* **10**, 164-70.

—— (1970). The mineralogy and petrology of the transition from the lower to upper sillimanite zone in the Oquossoc area, Maine. *J. Petrol.* **11**, 277-336.

Guitard, G. and Raguin, E. (1958). Sur la présence de gneiss à grenat et hypersthène dans le massif de l'Agly (Pyrénées Orientales). *Acad. Sci. (Paris) C. R.* **247**, 2385-8.

Halferdahl, L. B. (1961). Chloritoid: its composition, X-ray and optical properties, stability, and occurrence. *J. Petrol.* **2**, 49-135.

Hallimond, A. F. (1943). On the graphical representation of the calciferous amphiboles. *Am. Mineral.* **28**, 65-89.

Hamilton, W. (1969a). Mesozoic California and the underflow of Pacific mantle. *Bull. geol. Soc. Am.* **80**, 2409-30.

—— (1969b). The volcanic central Andes — A modern model for the Cretaceous batholiths and tectonics of Western North America. In *Proceedings of the Andesite Conference,* ed. A. R, McBirney, *Oregon State Dept. Geol. Mineral Industr. Bull.* **65**, 175-184.

—— and Myers, W. B. (1967). The nature of batholiths. *U.S. Geol. Surv. Prof. Paper 554-C.*

Hara, I., Kanisawa, S., Kano, H., Kuroda, Y., Maruyama, T., Mitsukawa, H., Nureki, T., Umemura, H. and Uruno, K. (1969). Polymetamorphism in the Abukuma Plateau with special regards to the discovery of staurolite and kyanite (in Japanese with English abstract). *Mem. geol. Soc. Japan* **4**, 83-97.

Harder, H. (1956). Untersuchungen an Paragoniten und an Natrium-haltigen Muskoviten. *Heidelberger Beitr. Mineral. Petrogr.* **4**, 227-69.

Harker, A. (1904). Tertiary Igneous Rocks of Skye. *Mem. Geol. Surv.*

—— (1918). The present position and outlook of the study of metamorphism in rock masses. *Q. J. geol. Soc. Lond.* **74**, li-lxxx.

—— (1932). *Metamorphism. A Study of the Transformations of Rock-Masses.* London: Methuen.

Harker, R. I. and Tuttle, O. F. (1956). Experimental data on the $P_{CO_2}-T$ curve for the reaction: calcite + quartz $\rightleftarrows$ wollastonite + carbon dioxide. *Am. J. Sci.* **254**, 239-56.

Harper, C. T. and Landis, C. A. (1967). K-Ar ages from regionally metamorphosed rocks, South Island, New Zealand, and some tectonic implications. *Earth Planet. Sci. Letters* 2, 419-29.

Harry, W. T. (1950). Aluminium replacing silicon in some silicate lattices. *Min. Mag.* 29, 142-9.

―― (1958). A re-examination of Barrow's Older Granites in Glen Clova, Angus. *Trans. R. Soc. Edinb.* 63, 393-412.

Harte, B. and Johnson, M. R. W. (1969). Metamorphic history of Dalradian rocks in Glens Clova, Esk and Lethnot, Angus, Scotland. *Scott. J. Geol.* 5, 54-80.

Hasebe, K., Fujii, N. and Uyeda, S. (1970). Thermal processes under island arcs. *Tectonophys.* 10, 335-55.

Hashimoto, M. (1968). Glaucophanitic metamorphism of the Katsuyama district, Okayama Prefecture, Japan. *Tokyo Univ. Fac. Sci. Jour. Sec. 2* 17, 99-162.

―― (1969). A note on stilpnomelane mineralogy. *Contr. Mineral. Petrol.* 23, 86-8.

――, Igi, S., Seki, Y., Banno, S. and Kojima, G. (1970). *Metamorphic Facies Map of Japan* (scale 1:2 000 000). Geol. Surv. Japan, Kawasaki.

―― and Saito, Y. (1970). Metamorphism of Paleozoic greenstones of the Tamba Plateau, Kyoto Prefecture. *J. geol. Soc. Japan* 76, 1-6.

Hashizume, M., Oike, K., Asano, S., Hamaguchi, H., Okada, A., Murauchi, S., Shima, E. and Nogoshi, M. (1968). Crustal structure in the profile across the northeastern part of Honshu, Japan, as derived from explosion seismic observations, part 2. *Bull. Earthquake Res. Inst.* 46, 607-30.

Hattori, H. (1968). Late Mesozoic to Recent tectogenesis and its bearing on the metamorphism in New Zealand and in Japan. *Geol. Survey Japan Rep.* 229.

Hay, R. L. (1966). Zeolites and zeolitic reactions in sedimentary rocks. *Geol. Soc. Am. Spec. Paper* 85.

Hayase, I. and Ishizaka, K. (1967). Rb-Sr dating on the rocks in Japan, I (in Japanese with English abstract). *J. Jap. Assoc. Mineral. Petrol. econ. Geol.* 58, 201-11.

―― and Nohda, S. (1969). Geochronology on the 'oldest rock' of Japan. *Geochem. J.* 3, 45-52.

Hayashi, M. (1968). Chemical characteristics of the serpentinites in Shikoku (in Japanese with English abstract). *J. Jap. Assoc. Mineral. Petrol. econ. Geol.* 59, 60-72.

Heier, K. S. (1956). Thermometric and petrogenetic significance of titaniferous magnetite – discussion. *Am. J. Sci.* 254, 506-15.

―― (1957). Phase relations of potash feldspar in metamorphism. *J. Geol.* 65, 468-79.

―― (1960). Petrology and geochemistry of high-grade metamorphic and igneous rocks on Langöy, northern Norway. *Norges Geol. Unders. No. 207.*

―― (1961). The amphibolite-granulite facies transition reflected in the mineralogy of potassium feldspars. *Cursillos Conf. Fasc.* 8, 131-7.

Heirtzler, J. R. (1968). Sea-floor spreading. *Scient. Am.* 219, No. 6. 60-70.

――, Dickson, G. O., Herron, E. M. Pitman, W. C. and Le Pichon, X. (1968). Marine magnetic anomalies, geomagnetic field reversals, and motions of the ocean floor and continents. *J. geophys. Res.* 73, 2119-36.

Henley, K. J. (1970). Application of the muscovite-paragonite geothermometer to a staurolite-grade schist from Sulitjelma, north Norway. *Min. Mag.* **37**, 693-704.

Hensen, B. J. (1971). Theoretical phase relations involving cordierite and garnet in the system $MgO-FeO-Al_2O_3-SiO_2$. *Contr. Mineral. Petrol.* **33**, 191-214.

Herrin, E. (1969). Regional variations of P-wave velocity in the upper mantle beneath North America. In *The Earth's Crust and Upper Mantle*, ed. P. J. Hart, *Geophys. Monogr.* **13**, 242-246. Washington, D.C: American Geophys. Union.

Hess, P. C. (1969). The metamorphic paragenesis of cordierite in pelitic rocks. *Contr. Mineral. Petrol.* **24**, 191-207.

Hess, H. H. (1962). History of ocean basins. In *Petrologic Studies (Buddington Vol.)*, ed. A. E. J. Engel *et al.*, 599-620. Geol. Soc. America.

Hey, M. H. (1954). A new review of the chlorites. *Min. Mag.* **30**, 277-92.

Hietanen, A. (1956). Kyanite, andalusite, and sillimanite in the schist in Boehls Butte quadrangle, Idaho. *Am. Mineral.* **41**, 1-27.

— (1961). Metamorphic facies and style of folding in the Belt Series northwest of the Idaho batholith. *Bull. Comm. géol. Finlande No. 196*, 73-103.

— (1962). Staurolite zone near the St. Joe River, Idaho. *U.S. geol. Surv. Prof. Paper 450-C*, 69-72.

Higgins, M. W. (1971). Cataclastic rocks. *U.S. geol. Surv. Prof. Paper 687*.

Hill, M. L. and Dibblee, T. W. (1953). San Andreas, Garlock, and Big Pine faults, California. *Bull. geol. Soc. Am.* **64**, 443-58.

Himmelberg, G. R. and Papike, J. J. (1969). Coexisting amphiboles from blueschist facies metamorphic rocks. *J. Petrol.* **10**, 102-14.

— and Phinney, W. C. (1967). Granulite-facies metamorphism, Granite Falls-Montevideo area, Minnesota. *J. Petrol.* **8**, 325-48.

Holdaway, M. J. (1971). Stability of andalusite and the aluminum silicate phase diagram. *Am. J. Sci.* **271**, 97-131.

Holland, Th. H. (1900). The charnockite series, a group of Archaean hypersthenic rocks in Peninsular India. *India geol. Survey, Mem.* **28**, 119-249.

Hollister, L. S. (1966). Garnet zoning: an interpretation based on the Rayleigh fractionation model. *Science, N.Y.* **154**, 1647-51.

— (1969a). Contact metamorphism in the Kwoiek area of British Columbia: An end member of the metamorphic process. *Bull. geol. Soc. Am.* **80**, 2465-94.

— (1969b). Metastable paragenetic sequence of andalusite, kyanite, and sillimanite, Kwoiek area, British Columbia. *Am. J. Sci.* **267**, 352-70.

— (1970). Origin, mechanism, and consequences of compositional sector-zoning in staurolite. *Am. Mineral.* **55**, 742-66.

— and Bence, A. E. (1967). Staurolite: Sectoral compositional variations. *Science, N.Y.* **158**, 1053-6.

Holmes, A. and Reynolds, D. L. (1947). A front of metasomatic metamorphism in the Dalradian of Co. Donegal. *Bull. Comm. géol. Finlande. No. 140*, 25-65.

Hoschek, G. (1967). Untersuchungen zum Stabilitätsbereich von Chloritoid und Staurolith. *Contr. Mineral. Petrol.* **14**, 123-62.

— (1969). The stability of staurolite and chloritoid and their significance in metamorphism of pelitic rocks. *Contr. Mineral. Petrol.* **22**, 208-32.

Howie, R. A. (1965). The pyroxenes of metamorphic rocks. In *Controls of Metamorphism*, ed. W. S. Pitcher and G. W. Flinn, 319-26. Edinburgh and London: Oliver and Boyd.

Hsü, K. J. (1971). Franciscan mélanges as a model for eugeosynclinal sedimentation and underthrusting tectonics. *J. geophys. Res.* **76**, 1162-70.

Hsu, L. C. (1968). Selected phase relationships in the system Al-Mn-Fe-Si-O; a model for garnet equilibria. *J. Petrol.* **9**, 40-83.

Hunahashi, M. (1957). Alpine orogenic movement in Hokkaido, Japan. *Hokkaido Univ. Fac. Sci. J. Ser. 4,* **9**, 415-69.

Hutton, C. O. (1938). The stilpnomelane group of minerals. *Min. Mag.* **25**, 172-206.

— (1940). Metamorphism in the Lake Wakatipu region, Western Otago, New Zealand. *N. Z. Dep. Sci. Ind. Res. Geol. Mem. No. 5.*

— and Turner, F. J. (1936). Metamorphic zones in northwest Otago. *Trans. R. Soc. N. Z.* **65**, 405-6.

Hutton, J. (1795). *Theory of the Earth with Proofs and Illustrations.* 2 vols. London and Edinburgh. (Reprinted 1959, H. R. Engelmann and Wheldon & Wesley.)

Hyndman, R. D. and Hyndman, D. W. (1968). Water saturation and high electrical conductivity in the lower continental crust. *Earth Planet. Sci. Letters* **4**, 427-32.

Ichikawa, K., Ishii, K., Nakagawa, C., Suyari, K. and Yamashita, N. (1956). The Kurosegawa zone (in Japanese with German abstract). *J. geol. Soc. Japan* **62**, 82-103.

Iiyama, T. (1960). Recherches sur le rôle de l'eau dans la structure et le polymorphisme de la cordiérite. *Bull. Soc. française minéral crystal.* **80**, 155-78.

— (1964). Étude des réactions d'échange d'ions Na-K dans la série muscovite-paragonite. *Bull. Soc. française minéral. crystal.* **87**, 532-41.

Isachsen, Y. W., (ed.) (1969). Origin of anorthosite and related rocks. *New York State Museum and Science Service, Mem. 18.*

Isacks, B. and Molnar, P. (1971). Distribution of stresses in the descending lithosphere from a global survey of focal-mechanism solutions of mantle earthquakes. *Rev. Geophys. space Phys.* **9**, 103-74.

—, Oliver, J. and Sykes, L. R. (1968). Seismology and the new global tectonics. *J. geophys. Res.* **73**, 5855-99.

Ishizaka, K. (1969). U-Th-Pb ages of zircon from the Ryoke metamorphic terrain, Kinki district (in Japanese with English abstract). *J. Jap. Assoc. Mineral. Petrol. econ. Geol.* **62**, 191-7.

Iwasaki, M. (1963). Metamorphic rocks of the Kotu-Bizan area, eastern Sikoku. *Tokyo Univ. Fac. Sci. J., Sec. 2,* **15**, 1-90.

— (1969). The basic metamorphic rocks at the boundary between the Sanbagawa metamorphic belt and the Chichibu unmetamorphosed Paleozoic sediments (in Japanese with English abstract). *Mem. geol. Soc. Japan* **4**, 41-9.

Izawa, E. (1968). Carbonaceous matter in some metamorphic rocks in Japan. *J. geol. Soc. Japan* **74**, 427-32.

Jaffe, H. W., Robinson, P. and Klein, C., Jr (1968). Exsolution lamellae and optic orientation of clinoamphiboles. *Science, N.Y.* **160**, 776-8.

Jakeš, P. and White, A. J. R. (1972). Major and trace element abundances in volcanic rocks of orogenic areas. *Bull. geol. Soc. Am.* **83**, 29–40.

James, H. L. (1955). Zones of regional metamorphism in the Precambrian of northern Michigan. *Bull. geol. Soc. Am.* **66**, 1455–88.

Johannes, W. and Puhan, D. (1971). The calcite–aragonite transition, reinvestigated. *Contr. Mineral. Petrol.* **31**, 28–38.

Johannsen, A. (1931). *A Descriptive Petrography of the Igneous Rocks.* Vol. 1. Chicago: University of Chicago Press.

Johnson, C. G., Alvis, R. J. and Hetzler, R. L. (1960). Military geology of Yap Islands, Caroline Islands. *Intelligence Div., Office of the Engineer, Headquarters U.S. Army Pacific.*

Johnson, M. R. W. (1962). Relations of movement and metamorphism in the Dalradians of Banffshire. *Trans geol. Soc. Edinb.* **19**, 29–64.

—— (1963). Some time relations of movement and metamorphism in the Scottish Highlands. *Geol. en Mijnbouw* **42**, 121–42.

Joplin, G. A. (1942, 1943). Petrological studies in the Ordovician of New South Wales. I, II. *Proc. Linn. Soc. N.S.W.* **67**, 156–96; **68**, 159–83.

—— (1964). *A Petrography of Australian Igneous Rocks.* Sydney: Angus and Robertson.

—— (1968). *A Petrography of Australian Metamorphic Rocks.* Sydney: Angus and Robertson.

Kanehira, K. (1967). Sanbagawa crystalline schists in the Iimori district, Kii Peninsula. *Jap. J. Geol. Geogr.* **38**, 101–15.

——, Banno, S. and Nishida, K. (1964). Sulfide and oxide minerals in some metamorphic terranes in Japan. *Jap. J. Geol. Geogr.* **35**, 175–91.

Kaneko, S. (1966). Transcurrent displacement along the Median Line, Southwestern Japan. *New Zealand Jour. Geol. Geophys.,* **9**, 45–59.

Kanisawa, S. (1964). Metamorphic rocks of the southwestern part of the Kitakami Mountainland, Japan. *Tohoku Univ. Sci. Rept., Ser. 3 9*, 155–198.

Karakida, Y., Yamamoto, H., Miyachi, S., Oshima, T. and Inoue, T. (1969). Characteristics and geologic situations of metamorphic rocks in Kyushu (in Japanese with English abstract). *Mem. geol. Soc. Japan* **4**, 3–21.

Karolus, K., Forgáč, J. and Konečný, V. (1968). Neovolcanics of the West Carpathians. *Internat. geol. Congr. 23rd Session, Guide to Excursion 18AC,* Bratislava.

Katada, M. (1965). Petrography of Ryoke metamorphic rocks in northern Kiso district, central Japan. *J. Jap. Assoc. Mineral. Petrol. econ. Geol.* **53**, 77–90, 155–164, 187–204.

—— and Sumi, K. (1966). Stilpnomelane co-existing with biotite in a Ryoke metamorphic rock. *J. geol. Soc. Japan* **72**, 543–4.

Katsui, Y. (1961). Petrochemistry of the Quaternary volcanic rocks of Hokkaido and surrounding areas. *Hokkaido Univ. Fac. Sci. J. Ser. 4*, **11**, 1–58.

Kawachi, Y. (1968). Large-scale overturned structure in the Sanbagawa metamorphic zone in central Shikoku, Japan. *J. geol. Soc. Japan* **74**, 607–16.

Kawai, N., Ito, H. and Kume, S. (1961). Deformation of the Japanese Islands as inferred from rock magnetism. *Geophys. J.* **6**, 124–30.

Kawano, Y. and Ueda, Y. (1967a). Periods of the igneous activities of the granitic rocks in Japan by K-A dating method. *Tectonophys.* **4**, 523–30.

Kawano, Y. and Ueda, Y. (1967*b*). K-Ar dating on the igneous rocks in Japan (VI) (in Japanese with English abstract). *J. Jap. Assoc. Mineral. Petrol. econ. Geol.* **57**, 177-87.

—— , Yagi, K. and Aoki, K. (1961). Petrography and petrochemistry of the volcanic rocks of Quaternary volcanoes of Northeastern Japan. *Tohoku Univ. Sci. Rep., Ser. 3,* **7**, 1-46.

Kay, M. (1951). North American geosynclines. *Mem. Geol. Soc. Am.* **48**.

—— (1969). Continental drift in North Atlantic Ocean. In *North Atlantic – Geology and Continental Drift, a Symposium,* ed. M. Kay, *American Assoc. Petroleum Geol., Mem.* **12**, 965-73.

Keith, T. E. C., Muffler, P. and Cremer, M. (1968). Hydrothermal epidote formed in the Salton Sea geothermal system, California. *Am. Mineral.* **53**, 1635-44.

Kennedy, W. Q. (1948). On the significance of thermal structure in the Scottish Highlands. *Geol. Mag.* **85**, 229-34.

Kern, R. and Weisbrod, A. (1967). *Thermodynamics for Geologists* (translated by D. McKie). San Francisco: Freeman, Cooper & Co.

Kerrick, D. M. (1968). Experiments on the upper stability limit of pyrophyllite at 1.8 kilobars and 3.9 kilobars water pressure. *Am. J. Sci.* **266**, 204-14.

—— (1970). Contact metamorphism in some areas of the Sierra Nevada, California. *Bull. geol. Soc. Am.* **81**, 2913-38.

King, P. B. (1959). *The Evolution of North America.* Princeton: Princeton University Press.

—— (1969). The tectonics of North America – a discussion to accompany the tectonic map of North America (scale 1:5 000 000). *U.S. Geol. Surv. Prof. Paper* 628.

Kisch, H. J. (1966). Zeolite facies and regional rank of bituminous coals. *Geol. Mag.* **103**, 414-22.

Kitahara, S. and Kennedy, G. C. (1967). The calculated equilibrium curves for some reactions in the system $MgO-SiO_2-H_2O$ at pressures up to 30 kilobars. *Am. J. Sci.* **265**, 211-17.

Klein, C., Jr (1968). Coexisting amphiboles. *J. Petrol.* **9**, 281-330.

—— (1969). Two-amphibole assemblages in the system actinolite-hornblende-glaucophane. *Am. Mineral.* **54**, 212-37.

Köhler, A. (1941). Die Abhängigkeit der Plagioklasoptik vom vorangegangenen Wärmeverhalten. Die Existenz einer Hoch- und Tieftemperaturoptik. *Mineral. Petrogr. Mitt.* **53**, 24-49.

—— (1949). Recent results of investigations on the feldspars. *J. Geol.* **57**, 592-9.

Koizumi, M. and Roy, R. (1960). Zeolite studies. I. Synthesis and stability of the calcium zeolites. *J. Geol.* **68**, 41-53.

Kojima, G. (Kozima, Z.) (1963). On the fundamental structure of the Sambagawa crystalline schist zone (in Japanese with English abstract). *Hiroshima Univ. geol. Rep., No. 12,* 173-82.

—— and Suzuki, T. (1958). Rock structure and quartz fabric in a thrusting shear zone: the Kiyomizu Tectonic Zone in Shikoku, Japan. *Hiroshima Univ. J. Sci., Series C,* **2**, 173-93.

Konishi, K. (1963). Pre-Miocene basement complex of Okinawa, and the tectonic belts of the Ryukyu Islands. *Kanazawa Univ. Sci. Rep.* **8**, 569-602.

Korzhinskii, D. S. (1936). Mobility and inertness of components in meta-somatosis. *Bull. Akad. Nauk. S.S.S.R. Ser, Geol., 1936*, No. 1, 35–60.
—— (1959). *Physicochemical Basis of the Analysis of the Paragenesis of Minerals.* New York: Consultants Bureau.
—— (1965). The theory of systems with perfectly mobile components and processes of mineral formation. *Am. J. Sci.* **263**, 193–205.
Kosminskaya, I. P., Belyaevsky, N. A. and Volvovsky, I. S. (1969). Explosion seismology in the U.S.S.R. In *The Earth's Crust and Upper Mantle*, ed. P. J. Hart, *Geophys. Monogr.* **13**, 195–208. Washington, D.C.: American Geophys. Union.
Kretz, R. (1961). Some applications of thermodynamics to coexisting minerals of variable composition. Examples: orthopyroxene-clinopyroxene and ortho-pyroxene-garnet. *J. Geol.* **69**, 361–87.
Kuno, H. (1959). Origin of Cenozoic petrographic provinces of Japan and surrounding areas. *Bull. volcanol. Ser.* 2, **20**, 37–76.
—— (1960). High-alumina basalt. *J. Petrol.* **1**, 121–45.
—— (1966). Lateral variation of basalt magma type across continental margins and island arcs. *Bull. volcanol.* **29**, 195–222.
—— (1968). Origin of andesite and its bearing on the island arc structure. *Bull volcanol.* **32**, 141–76.
—— (1969). Mafic and ultramafic nodules in basaltic rocks of Hawaii. *Geol. Soc. Am. Mem.* **115**, 189–234.
Kuroda, Y. and Miyagi, H. (1967). Metamorphic rocks of the Ishigakishima, Ryukyu Islands, Japan (in Japanese with English abstract). *Comm. Vol. for Prof. H. Shibata*, 148–52.
Kushiro, I. (1968). Compositions of magmas formed by partial zone melting of the Earth's upper mantle. *J. geophys. Res.* **73**, 619–34.
—— and Aoki, K. (1968). Origin of some eclogite inclusions in kimberlite. *Am. Mineral* **53**, 1347–67.
—— and Yoder, H. S., Jr (1966). Anorthite-forsterite and anorthite-enstatite reactions and their bearing on the basalt-eclogite transformation. *J. Petrol.* **7**, 337–62.
Lambert, I. B. and Heier. K. S. (1968). Geochemical investigations of deep-seated rocks in the Australian Shield. *Lithos* **1**, 30–53.
Lambert, R. St J. (1959). The mineralogy and metamorphism of the Moine schists of the Morar and Knoydart districts of Inverness-shire. *Trans. R. Soc. Edinb.* **63**, 553–88.
Landis, C. A. (1971). Graphitization of dispersed carbonaceous material in metamorphic rocks. *Contr. Mineral. Petrol.* **30**, 34–45.
—— and Coombs, D. S. (1967). Metamorphic belts and orogenesis in southern New Zealand. *Tectonophys.* **4**, 501–18.
Lapadu-Hargues, P. (1945). Sur l'existence et la nature de l'apport chimique dans certaines séries cristallophylliennes. *Bull. Soc. géol. France Ser.* 5, **15**, 255–310.
Lappin, M. A. (1960). On the occurrence of kyanite in the eclogites of the Selje and Åheim district, Nordfjord. *Norsk Geol. Tidsskr.* **40**, 289–96.
Lapworth, C. (1885). The Highland controversy in British geology: its causes, course, and consequence. *Nature, Lond.* **32**, 558–9.
Larsen, E. S., Jr (1948). Batholith and associated rocks of Corona, Elsinore, and San Luis Rey quadrangles, southern California. *Mem. Geol. Soc. America* **29**.

Laves, F. (1954). The coexistence of two plagioclases in the oligoclase composition range. *J. Geol.* **62**, 409-11.

Leake, B. E. (1965). The relationship between composition of calciferous amphibole and grade of metamorphism. In *Controls of Metamorphism*, ed. W. S. Pitcher and G. W. Flinn, 299-318. Edinburgh and London: Oliver & Boyd.

— (1968). A catalog of analyzed calciferous and subcalciferous amphiboles together with their nomenclature and associated minerals. *Geol. Soc. America Spec. Paper* **98**.

— and Skirrow, G. (1960). The pelitic hornfelses of the Cashel-Lough Wheelaun intrusion, County Galway, Eire. *J. Geol.* **68**, 23-40.

Lee, D. E., Coleman, R. G. and Erd, R. C. (1963). Garnet types from the Cazadero area, California. *J. Petrol.* **4**, 460-92.

Lee, W. H. K. and Uyeda, S. (1965). Review of heat flow data. In *Terrestrial Heat Flow*, ed. W. H. K. Lee, *Geophys. Monogr.* **8**, 87-190. Washington, D.C.: American Geophys. Union.

Le Pichon, X. (1968). Sea-floor spreading and continental drift. *J. geophys. Res.* **73**, 3661-97.

Liou, J. G. (1970). Synthesis and stability relations of wairakite, $CaAl_2Si_4O_{12}.2H_2O$. *Contr. Mineral. Petrol.* **27**, 259-82.

— (1971a). Synthesis and stability relations of prehnite, $Ca_2Al_2Si_3O_{10}(OH)_2$. *Am. Mineral.* **56**, 507-31.

— (1971b). Stilbite-laumontite equilibrium. *Contr. Mineral. Petrol.* **31**, 171-7.

Lipman, P. W., Prostka, H. J. and Christiansen, R. L. (1971). Evolving subduction zones in the western United States, as interpreted from igneous rocks. *Science, N.Y.* **174**, 821-5.

Lobjoit, W. M. (1964). Kyanite produced in a granitic aureole. *Min. Mag.* **33**, 804-8.

Loewinson-Lessing, F. Y. (1954). *A Historical Survey of Petrology* (translated by S. I. Tomkeieff). Edinburgh and London: Oliver and Boyd.

Loomis, A. A. (1966). Contact metamorphic reactions and processes in the Mt. Tallac roof remnant, Sierra Nevada, California. *J. Petrol.* **7**, 221-45.

Lovering, J. F. and White, A. J. R. (1969). Granulitic and eclogitic inclusions from basic pipes at Delegate, Australia. *Contr. Mineral. Petrol.* **21**, 9-52.

Lundgren, L. W., Jr (1966). Muscovite reactions and partial melting in southeastern Connecticut. *J. Petrol.* **7**, 421-53.

Luth, W. C., Jahns, R. H. and Tuttle, O. F. (1964). The granite system at pressures of 4 to 10 kilobars. *J. geophys. Res.* **69**, 759-73.

Lyell, C. (1830, 1832, 1833). *Principles of Geology*. 1st ed., 3 vols. London: John Murray.

Lyons, J. B. (1955). Geology of the Hanover quadrangle, New Hampshire-Vermont. *Bull. geol. Soc. Am.* **66**, 105-46.

— and Morse, S.A. (1970). Mg/Fe partitioning in garnet and biotite from some granitic, pelitic, and calcic rocks. *Am. Mineral.* **55**, 231-45.

Macdonald, G. A. (1960). Dissimilarity of continental and oceanic rock types. *J. Petrol.* **1**, 172-7.

MacGregor, M. and Wilson, G. (1939). On granitization and associated processes. *Geol. Mag.* **76**, 193-215.

Mall, A. P. and Singh, C. D. P. (1972). Coexisting plagioclases from basic charnockites of India. *Lithos* **5**, 105-14.

Marmo, V. (1962). On granites. *Bull. Comm. géol. Finlande* **201**, 1–77.

Mason, B. (1962). Metamorphism in the southern Alps of New Zealand. *Bull. Am. Mus. Nat. Hist.* **123**, 213–47.

Mason, R. (1967). The field relations of the Sulitjelma gabbro, Nordland. *Norsk Geol. Tidsskr.* **47**, 237–47.

Mather, J. D. (1970). The biotite isograd and the lower greenschist facies in the Dalradian rocks of Scotland. *J. Petrol.* **11**, 253–75.

— and Atherton, M. P. (1965). Stilpnomelane from the Dalradian. *Nature, Lond.* **207**, 971–2.

Matsuda, T. (1964). Island-arc features and the Japanese Islands (in Japanese with English abstract). *Chigaku-zasshi* **73**, 271–80.

— and Kuriyagawa, S. (1965). Lower grade metamorphism in the eastern Akaishi Mountains, central Japan (in Japanese with English abstract). *Tokyo Univ. Earthquake Res. Inst. Bull.* **43**, 209–35.

—, Nakamura, K. and Sugimura, A. (1967). Late Cenozoic orogeny in Japan. *Tectonophys.* **4**, 349–66.

— and Uyeda, S. (1971). On the Pacific-type orogeny and its model — Extension of the paired belts concept and possible origin of marginal seas. *Tectonophys.* **11**, 5–27.

Matsumoto, Takashi (1968). A hypothesis on the origin of the late Mesozoic volcano-plutonic association in East Asia. In *Pacific Geology*, ed. M. Minato *et al.*, vol. 1, p. 77–83. Tokyo: Tsukiji Shokan.

Matsumoto, Tatsuro, Yamaguchi, M., Yanagi, T., Matsushita, S., Hayase, I., Ishizaka, K., Kawano, Y. and Ueda, Y. (1968). The Precambrian problem in younger orogenic zones: an example from Japan. *Can. J. Earth Sci.* **5**, 643–8.

McKee, B. (1962a). Widespread occurrence of jadeite, lawsonite, and glaucophane in central California. *Am. J. Sci.* **260**, 596–610.

— (1962b). Aragonite in the Franciscan rocks of the Pacheco Pass area, California. *Am. Mineral.* **47**, 379–87.

McNamara, M. J. (1965). The lower greenschist facies in the Scottish Highlands. *Geol. Fören. Stockholm Förh.* **87**, 347–89.

Melson, W. G. (1966). Phase equilibria in calc-silicate hornfels, Lewis and Clark County, Montana. *Am. Mineral.* **51**, 402–21.

—, Bowen, V. T., van Andel, T. H. and Siever, R. (1966). Greenstones from the central valley of the Mid-Atlantic Ridge. *Nature, Lond.* **209**, 604–5.

—, Jarosewich, E., Bowen, V. T. and Thompson, G. (1967). St Peter and St Paul Rocks: a high-temperature, mantle-derived intrusion. *Science, N.Y.* **155**, 1532–5.

— and van Andel, T. H. (1966). Metamorphism in the Mid-Atlantic Ridge, 22 °N latitude. *Marine Geol.* **4**, 165–86.

Menard, H. W. (1967). Transitional types of crust under small ocean basins. *J. geophys. Res.* **72**, 3061–73.

Metz, P. W. and Puhan, D. (1970). Experimentelle Untersuchung der Metamorphose von kieselig dolomitischen Sedimenten. *Contr. Mineral. Petrol.* **26**, 302–14.

— and Trommsdorff, V. (1968). On phase equilibria in metamorphosed siliceous dolomites. *Contr. Mineral. Petrol.* **18**, 305–9.

— and Winkler, H. G. F. (1963). Experimentelle Gesteinsmetamorphose — VII. Die Bildung von Talk aus kieseligem Dolomit. *Geochim. Cosmochim. Acta* **27**, 431–57.

Metz, P. W. and Puhan, D. (1964). Experimentelle Untersuchung der Diopsid-bildung aus Tremolit, Calcit und Quarz. *Naturwissenschaften* **51**, 460.

Miller, J. A. (1965). Geochronology and continental drift — the North Atlantic. In *A Symposium on Continental Drift*, ed. P. M. S. Blackett *et al., Phil. Trans. R. Soc. Lond. Ser. A*, **258**, 180–191.

Minear, J. W. and Toksöz, M. N. (1970). Thermal regime of a downgoing slab and new global tectonics. *J. geophys. Res.* **75**, 1397–419.

Misch, P. H. (1964). Stable association wollastonite-anorthite, and other calc-silicate assemblages in amphibolite-facies crystalline schists of Nanga Parbat, northwest Himalayas. *Beitr. Mineral. Petrogr.* **10**, 315–56.

—— (1966). Tectonic evolution of the northern Cascades of Washington State. In *Tectonic History and Mineral Deposits of the Western Cordillera,* etc. *Can. Inst. Mining Metallurgy Spec. Vol. 8*, 101–48.

Miyagi, K. (1964). Petrology of alkali rocks, sub-alkali rocks and high-alumina basalt from the Green Tuff formations of Northeast Japan (in Japanese with English abstract). *J. geol. Soc. Japan* **70**, 72–87.

Miyashiro, A. (1949). The stability relation of kyanite, sillimanite and andalu-site, and the physical conditions of the metamorphic processes (in Japanese with English abstract). *J. geol. Soc. Japan* **55**, 218–23.

—— (1953*a*). Progressive metamorphism of the calcium-rich rocks of the Gosaisyo-Takanuki district, Abukuma Plateau, Japan. *Jap. J. Geol. Geogr.* **23**, 81–107.

—— (1953*b*). Calcium-poor garnet in relation to metamorphism. *Geochim. Cosmochim. Acta* **4**, 179–208.

—— (1957*a*). Cordierite–indialite relations. *Am. J. Sci.* **255**, 43–62.

—— (1957*b*). The chemistry, optics and genesis of the alkali-amphiboles. *Tokyo Univ. Fac. Sci. J., Sec. 2*, **11**, 57–83.

—— (1958). Regional metamorphism of the Gosaisyo-Takanuki district in the central Abukuma Plateau. *Tokyo Univ. Fac. Sci. J., Sec. 2*, **11**, 219–72.

—— (1959). Abukuma, Ryoke, and Sanbagawa metamorphic belts (in Japanese with English abstract). *J. geol. Soc. Japan* **65**, 624–37.

—— (1960). Thermodynamics of reactions of rock-forming minerals with silica. I–VI. *Jap. J. Geol. Geogr.* **31**, 71–8, 79–84, 107–11, 113–20, 241–6, 247–52.

—— (1961*a*). Evolution of metamorphic belts. *J. Petrol.* **2**, 277–311.

—— (1961*b*). Metamorphism of rocks. In *Constitution of the Earth*, ed. C. Tsuboi (in Japanese), 243–68. Tokyo: Iwanami-Shoten.

—— (1964). Oxidation and reduction in the earth's crust with special reference to the role of graphite. *Geochim. Cosmochim. Acta* **28**, 717–29.

—— (1965). *Metamorphic Rocks and Metamorphic Belts* (in Japanese). Tokyo: Iwanami-Shoten.

—— (1967*a*). Aspects of metamorphism in the circum-Pacific region. *Tectono-phys.* **4**, 519–21.

—— (1967*b*). Orogeny, regional metamorphism, and magmatism in the Japanese Islands. *Medd. fra Dansk Geol. Forening* **17**, 390–446.

—— (1967*c*). Metamorphism of mafic rocks. In *Basalts: Treatise on Rocks of Basaltic Composition,* ed. H. H. Hess and A. Poldervaart, vol. 2, 799–834. New York: Interscience.

—— (1972*a*). Pressure and temperature conditions and tectonic significance of regional and ocean-floor metamorphism. *Tectonophys.* **13**, 141–59.

Miyashiro, A. (1972b). Metamorphism and related magmatism in plate tectonics. *Am. J. Sci.* **272**, 629–56.

—— (1973). Paired and unpaired metamorphic belts. *Tectonophysics* **17**, 241–54.

—— and Banno, S. (1958). Nature of glaucophanitic metamorphism. *Am. J. Sci.* **256**, 97–110.

—— and Haramura, H. (1966). Sedimentation and regional metamorphism in the Paleozoic geosynclinal pile of Japan. *Indian Geophys. Union, 1966, No. 3 (Proc. Symposium on Tectonics)*, 45–55.

—— and Seki, Y. (1958). Enlargement of the composition field of epidote and piemontite with rising temperature. *Am. J. Sci.* **256**, 423–30.

—— and Shido, F. (1970). Progressive metamorphism in zeolite assemblages. *Lithos* **3**, 251–60.

—— and —— (1973). Progressive compositional change of garnet in metapelite. *Lithos* **6**, 13–20.

——, —— and Ewing, M. (1969a). Diversity and origin of abyssal tholeiite from the Mid-Atlantic Ridge near 24° and 30° N latitude. *Contr. Mineral. Petrol.* **23**, 38–52.

——, —— and —— (1969b). Composition and origin of serpentinites from the Mid-Atlantic Ridge near 24° and 30°N latitude. *Contr. Mineral. Petrol.* **23**, 117–27.

——, —— and —— (1970a). Petrologic models for the Mid-Atlantic Ridge. *Deep Sea Res.* **17**, 109–23.

——, —— and —— (1970b). Crystallization and differentiation in abyssal tholeiites and gabbros from mid-oceanic ridges. *Earth Planet. Sci. Letters* **7**, 361–5.

——, —— and —— (1971). Metamorphism in the Mid-Atlantic Ridge near 24° and 30°N. *Phil. Trans. R. Soc. Lond., Ser. A*, **268**, 589–603.

Monger, J. W. H. and Hutchison, W. W. (1971). Metamorphic map of the Canadian Cordillera. *Geol. Surv. Can., Paper 70–33.*

Moore, J. G. (1959). The quartz–diorite boundary line in the western United States. *J. Geol.* **67**, 198–210.

——, Grantz, A. and Blake, M. C., Jr (1961). The quartz diorite line in northwestern North America. *U.S. Geol. Surv. Prof. Paper 424–C*, 87–90; also, 1963, *Prof. Paper 450–E*, 89–93.

Moores, E. M. (1969). Petrology and structure of the Vourinos ophiolitic complex of northern Greece. *Geol. Soc. America, Spec. Paper 118.*

—— and Vine, F. J. (1971). The Troodos massif, Cyprus and other ophiolites as oceanic crust; evolution and implications. *Phil. Trans. R. Soc. London, Ser. A*, **268**, 443–66.

Morgan, W. J. (1968). Rises, trenches, great faults, and crustal blocks. *J. geophys. Res.* **73**, 1959–82.

Müller, G. (1970). Kaledonische Intrusivgesteine des Stavanger-Gebietes. *Contr. Mineral. Petrol.* **27**, 52–65.

—— and Schneider, A. (1971). Chemistry and genesis of garnets in metamorphic rocks. *Contr. Mineral. Petrol.* **31**, 178–200.

Mueller, R. F. (1960). Compositional characteristics and equilbrium relations in mineral assemblages of a metamorphosed iron formation. *Am. J. Sci.* **258**, 449–97.

—— (1967). Mobility of the elements in metamorphism. *J. Geol.* **75**, 565–82.

Muffler, L. J. P. and White, D. E. (1969). Active metamorphism of Upper Cenozoic sediments in the Salton Sea geothermal field and the Salton Trough, southeastern California. *Bull. geol. Soc. Am.* **80**, 157–82.

Nagasaki, H. (1966). A layered ultrabasic complex at Horoman, Hokkaido, Japan. *Tokyo Univ. Fac. Sci. J., Sec. 2,* **16**, 313–46.

Nagata, T. (1961). *Rock Magnetism.* Revised ed., Tokyo: Maruzen.

Naggar, M. H. and Atherton, M. P. (1970). The composition and metamorphic history of some aluminium silicate-bearing rocks from the aureoles of the Donegal granites. *J. Petrol.* **11**, 549–89.

Naylor, R. S. (1968). Origin and regional relationships of the core-rocks of the Oliverian domes. In *Studies of Appalachian Geology: Northern and Maritime,* ed. E-an Zen *et al.* 231–40. New York: Interscience.

Newton, M. S. and Kennedy, G. C. (1968). Jadeite, analcite, nepheline, and albite at high temperatures and pressures. *Am. J. Sci.* **266**, 728–35.

Newton, R. C. (1966*a*). Some calc-silicate equilibrium relations. *Am. J. Sci.* **264**, 204–22.

—— (1966*b*). Kyanite–andalusite equilibrium from 700 to 800°C. *Science, N.Y.* **153**, 170–2.

——, Goldsmith, J. R. and Smith, J. V. (1969). Aragonite crystallization from strained calcite at reduced pressures and its bearing on aragonite in low-grade metamorphism. *Contr. Mineral. Petrol.* **22**, 335–48.

—— and Kennedy, G. C. (1963). Some equilibrium reactions in the join $CaAl_2 Si_2 O_8 - H_2 O$. *J. geophys. Res.* **68**, 2967–83.

Niggli, E. (1960). Mineral-Zonen der Alpinen Metamorphose in der Schweizer Alpen. *Internat. geol. Congr. 21st Sess. (Norden), Rep.* Part 13, 132–8. Copenhagen.

—— (1970). Alpine Metamorphose und alpine Gebirgsbildung. *Fortschr. Miner.* **47**, 16–26.

—— and Niggli, C. R. (1965). Karten der Verbeitung einiger Mineralien der alpidischen Metamorphose in den Schweizer Alpen (Stilpnomelan, Alkali-Amphibol, Chloritoid, Staurolith, Disthen, Sillimanit). *Eclogae Geol. Helvetiae,* **58**, 335–68.

Nockolds, S. R. and Allen, R. (1953, 1954, 1956). The geochemistry of some igneous rock series, I, II and III. *Geochim. Cosmochim. Acta* **4**, 105–42; **5**, 245–85; **9**, 34–77.

Nozawa, T. (1968). Isotopic ages of Hida metamorphic belt; summary and note in 1968 (in Japanese with English abstract). *J. geol. Soc. Japan* **74**, 447–50.

Nureki, T. (1969). Geological relations of the Sangun metamorphic rocks to the 'non-metamorphic' Paleozoic formations in the Chugoku province (in Japanese with English abstract). *Mem. Geol. Soc. Japan* **4**, 23–39.

Offler, R. and Fleming, P. D. (1968). A synthesis of folding and metamorphism in the Mt Lofty Ranges, South Australia. *J. Geol. Soc. Aust.* **15**, 245–66.

Oftedahl, C. (1959). Volcanic sequence and magma formation in the Oslo region. *Geol. Rundschau* **48**, 18–26.

O'Hara, M. J. (1961). Zones ultrabasic and basic gneiss masses in the early Lewisian metamorphic complex at Scourie, Sutherland. *J. Petrol.* **2**, 248–76.

—— and Mercy, E. L. P. (1963). Petrology and petrogenesis of some garnetiferous peridotites. *Trans. R. Soc. Edinb.* **65**, 251–314.

Okada, A. (1970). Fault topography and rate of faulting along the Median Tectonic Line in the drainage basin of the River Yoshino, northeastern Shikoku, Japan (in Japanese with English abstract). *Geogr. Rev. Japan* **43**, 1-21.

Oki, Y. (1961*a*). Metamorphism in the northern Kiso Range, Nagano Prefecture, Japan. *Jap. J. Geol. Geogr.* **32**, 479-96.

—— (1961*b*). Biotite in metamorphic rocks. *Jap. J. Geol. Geogr.* **32**, 497-506.

Okrusch, M. (1969). Die Gneishornfelse um Steinach in der Oberpfalz. Eine phasenpetrologische Analyse. *Contr. Mineral. Petrol.* **22**, 32-72.

—— (1971). Garnet-cordierite-biotite equilibria in the Steinach aureole, Bavaria. *Contr. Mineral. Petrol.* **32**, 1-23.

Oliver, R. L. (1969). Some observations on the distribution and nature of granulite-facies terrains. *Geol. Soc. Aust. Spec. Publ. No. 2*, 259-68.

Ono, A. (1969*a*). Geology of the Ryoke metamorphic belt in the Takato-Sioziri area, Nagano Prefecture (in Japanese with English abstract). *J. geol. Soc. Japan* **75**, 491-8.

—— (1969*b*). Zoning of the metamorphic rocks in the Takato-Sioziri area, Nagano Prefecture (in Japanese with English abstract). *J. geol. Soc. Japan* **75**, 521-36.

—— (1970). Origin of pegmatites (in Japanese with English abstract). *J. geol. Soc. Japan* **76**, 13-21.

Onuki, H. (1968). Almandine hornfelses from the Tono contact aureole, Kitakami Mountainland. *J. Jap. Assoc. Mineral. Petrol. econ. Geol.* **59**, 9-20.

Orville, P. M. (1962). Alkali metasomatism and feldspars. *Norsk Geol. Tidsskr.* **42**, 2. *Halvbind* (Feldspar Vol.), 283-316.

—— (1969). A model for metamorphic differentiation origin of thin-layered amphibolites. *Am. J. Sci.* **267**, 64-86.

Osberg, P. H. (1968). Stratigraphy, structural geology, and metamorphism of the Waterville-Vassalboro area, Maine. *Maine geol. Surv. Bull.* **20**.

—— (1971). An equilibrium model for Buchan-type metamorphic rocks, south-central Maine. *Am. Mineral.* **56**, 570-86.

Osborn, E. F. (1962). Reaction series for subalkaline igneous rocks based on different oxygen pressure conditions. *Am. Mineral.* **47**, 211-26.

Oxburgh, E. R., and Turcotte, D. L. (1971). Origin of paired metamorphic belts and crustal dilation in island arc regions. *J. geophys. Res.* **76**, 1315-27.

Ozawa, A. (1968). *The Green Tuff* (in Japanese). Tokyo: Ratisu.

Packham, G. H. and Crook, K. A. W. (1960). The principle of diagenetic facies and some of its implications. *J. Geol.* **68**, 392-407.

Page, L. R. (1968). Devonian plutonic rocks in New England. In *Studies of Appalachian Geology: Northern and Maritime*, ed. E-an Zen *et al.*, 371-83. New York: Interscience.

Pakiser, L. C. and Robinson, R. (1966). Composition of the continental crust as estimated from seismic observations. In: *The Earth beneath the Continents*, ed. J. S. Steinhart and T. J. Smith, *Geophys. Monogr.* **10**, 620-6. Washington, D. C.: American Geophys. Union.

Palm, Q. A. (1958). Les roches cristallines des Cévennes médianes à hauteur de Largentière, Ardèche, France. *Geol. Ultraiectina* **3**.

Pantó, G. (1968). Cenozoic volcanism in Hungary. *Internat. geol. Congr. 23rd Sess., Guide to Excursion 40C*, Budapest.

Parras, K. (1958). On the charnockites in the light of a highly metamorphic complex in southwestern Finland. *Bull. Comm. géol. Finlande No. 181.*

Peacock, M. A. (1931). Classification of igneous rock series. *J. Geol.* **39**, 54–67.

Petrushevsky, B. A. (1971). On the problem of the horizontal heterogeneity of the earth's crust and uppermost mantle in southern Eurasia. *Tectonophys.* **11**, 29–60.

Phillips, F. C. (1930). Some mineralogical and chemical changes induced by progressive metamorphism in the Green Bed group of the Scottish Dalradian. *Min. Mag.* **22**, 239–56.

Pidgeon, R. T. and Compston, W. (1965). The age and origin of the Cooma granite and its associated metamorphic zones, New South Wales. *J. Petrol.* **6**, 193–222.

Pitcher, W. S. (1965). The aluminium silicate polymorphs. In *Controls of Metamorphism,* ed. W. S. Pitcher and G. W. Flinn, 327–41. Edinburgh and London: Oliver and Boyd.

— and Read, H. H. (1960). The aureole of the Maine Donegal Granite. *Q. J. geol. Soc. Lond.* **116**, 1–36.

— and — (1963). Contact metamorphism in relation to manner of emplacement of the granites of Donegal, Ireland. *J. Geol.* **71**, 261–96.

Poldervaart, A. (1955). Chemistry of the earth's crust. *Geol. Soc. America Spec. Paper 62 (Crust of the Earth),* 119–44.

Polkanov, A. A. and Gerling, E. A. (1960). The pre-Cambrian geochronology of the Baltic shield. *Internatl. Geol. Congr. 21st Sess. (Norden) Rept.,* part 9, 183–91. Copenhagen.

Powers, M. C. (1959). Adjustment of clays to chemical change and the concept of the equivalance level. *Proc. 6th natn. Conf. on Clays and Clay Minerals,* 309–26. Oxford: Pergamon Press.

Pugin, V. A. and Khitarov, N. I. (1968). The system $Al_2O_3-SiO_2$ at high temperatures and pressures (in Russian). *Geokhimiya (1968),* No. 2, 157–65.

Ramberg, H. (1944, 1945). The thermodynamics of the earth's crust. I, II. *Norsk Geol. Tidsskr.* **24**, 98–111; **25**, 307–26.

— (1948). Radial diffusion and chemical stability in the gravitational field. *J. Geol.* **56**, 448–58.

— (1949). The facies classification of rocks: a clue to the origin of quartzo-feldspathic massifs and veins. *J. Geol.* **57**, 18–54.

— (1951). Remarks on the average chemical composition of granulite and amphibolite-to-epidote amphibolite facies gneisses in west Greenland. *Meddr. Dansk Geol. Foren.* **12**, 27–34.

— (1952). *The Origin of Metamorphic and Metasomatic Rocks.* Chicago: University of Chicago Press.

Read, H. H. (1957). *The Granite Controversy.* London: Thomas Murby.

— (1961). Aspects of Caledonian magmatism in Britain. *Liverpool Manchester geol. J.* **2**, 653–83.

Reed, J. J. (1958). Regional metamorphism in southeast Nelson. *N. Z. Geol. Surv. Bull. N. S. 60.*

— (1964). Mylonites, cataclasites, and associated rocks along the Alpine fault, South Island, New Zealand. *N. Z. J. Geol. Geophys.* **7**, 645–84.

Reilly, G. A. and Shaw, D. M. (1967). An estimate of the composition of part of the Canadian shield in northwestern Ontario. *Can. J. Earth Sci.* **4**, 725–39.

Research Group for Explosion Seismology (1966). Explosion seismological research in Japan. In *The Earth beneath the Continents,* ed. J. S. Steinhart and T. J. Smith, *Geophys. Monogr.* **10**, 334-48. Washington, D.C.: American Geophys. Union.

Research Group of Peridotite Intrusion (1967). Ultrabasic rocks in Japan. *J. geol. Soc. Japan* **73**, 543-53.

Reverdatto, V. V. (1970). Pyrometamorphism of limestones and the temperature of basaltic magmas. *Lithos* **3**, 135-43.

—— , Sharapov, V. N. and Melamed, V. G. (1970). The controls and selected peculiarities of the origin of contact metamorphic zonation. *Contr. Mineral. Petrol.* **29**, 310-37.

Richardson, S. W. (1968). Staurolite stability in a part of the system Fe-Al-Si-O-H. *J. Petrol.* **9**, 468-88.

—— , Gilbert, M. C. and Bell, P. M. (1969). Experimental determination of kyanite-andalusite and andalusite-sillimanite equilibria; the aluminum silicate triple point. *Am. J. Sci.* **267**, 259-72.

Rikitake, T. S., Miyamura, S., Tsubokawa, I., Murauchi, S., Uyeda, S., Kuno, H. and Gorai, M. (1968). Geophysical and geological data in and around the Japan arc. *Can. J. Earth Sci.* **5**, 1101-18.

Ringwood, A. E. (1969). Composition and evolution of the upper mantle. In *The Earth's Crust and Upper Mantle,* ed. P. J. Hart, *Geophys. Monogr.* **13**, 1-17. Washington, D. C.: American Geophys. Union.

—— and Green, D. H. (1966*a*). An experimental investigation of the gabbro-eclogite transformation and some geophysical implications. *Tectonophys.* **3**, 383-427.

—— and —— (1966*b*). Petrological nature of the stable continental crust. In *The Earth beneath the Continents,* ed. J. S. Steinhart and T. J. Smith, *Geophys. Monogr.* **10**, 611-19. Washington, D.C.: American Geophys. Union.

—— and Lovering, J. F. (1970). Significance of pyroxene-ilmenite intergrowths among kimberlite xenoliths. *Earth Planet. Sci. Letters* **7**, 371-5.

Robertson, E. C., Birch, F. and MacDonald, G. J. F. (1957). Experimental determination of jadeite stability relations to 25 000 bars. *Am. J. Sci.* **255**, 115-35.

Robie, R. A. and Waldbaum, D. R. (1968). Thermodynamic properties of minerals and related substances at 298.15 K (25.0 °C) and one atmosphere (1·013 bars) pressure and at higher temperature. *U.S. geol. Surv. Bull.* **1259**.

Robinson, P., Ross, M. and Jaffe, H. W. (1971). Composition of the anthophyllite-gedrite series, comparison of gedrite and hornblende, and the anthophyllite-gedrite solvus. *Am. Mineral.* **56**, 1005-41.

Rocci, G. and Juteau, T. (1968). Spilite-keratophyres et ophiolites. *Geol. Mijnbouw* **47**, 330-9.

Rodgers, J. (1967). Chronology of tectonic movements in the Appalachian region of eastern North America. *Am. J. Sci.* **265**, 408-27.

—— (1968). The eastern edge of the North American continent during the Cambrian and early Ordovician. In *Studies of Appalachian Geology: Northern and Maritime,* ed. E-an Zen *et al.,* 141-9. New York: Interscience.

—— (1970). *The Tectonics of the Appalachians.* New York: John Wiley (Interscience).

Ronov, A. B. and Yaroshevsky, A. A. (1969). Chemical composition of the earth's crust. In *The Earth's Crust and Upper Mantle*, ed. P. J. Hart, *Geophys. Monogr.* **13**, 37–57. Washington, D.C.: American Geophys. Union.

Rosenbusch, H. (1877). Die Steiger Schiefer und ihre Contactzone an der Granititen von Barr-Andlau und Hohwald. *Abh. zur geol. Specialkarte von Elsass-Lothringen* **1**, 79–393.

—— (1910). *Elemente der Gesteinslehre.* 3rd ed. Stuttgart: Schweizerbart'sche Verlagsbuchhandlung.

Runcorn, S. K. (1965). Changes in the convection pattern in the earth's mantle and continental drift: evidence for a cold origin of the earth. In *A Symposium on Continental Drift*, ed. P. M. S. Blackett *et al.*, Phil. Trans. R. Soc. Lond. Ser. A **258**, 228–51.

Saha, P. (1961). The system $NaAlSiO_4$ (nepheline)–$NaAlSi_3O_8$ (albite)–H_2O. *Am. Mineral.* **46**, 859–84.

Sander, B. (1930). *Gefügekunde der Gesteine.* Vienna: Springer.

Sato, S. (1968). Precambrian-Variscan polymetamorphism in the Hida massif, basement of the Japanese Islands. *Tokyo Univ. Educ. Sci. Rep. Sec. C*, **10**, 15–130.

Schreyer, W. (1964–6). Synthetische und natürliche Cordierite I, II, III. *Neues Jahrb, Mineral. Abh.*, **102**, (1964), 39–67; **103**, (1965), 35–79; **105**, (1966), 211–44.

—— (1968). A reconnaissance study of the system $MgO-Al_2O_3-SiO_2-H_2O$ at pressures between 10 and 20 kbar. *Geophys. Lab. A. Rep. Director for 1967-1968*, 380–92.

Schwarcz, H. P. (1969). Pre-Cretaceous sedimentation and metamorphism in the Winchester area, northern Peninsular Ranges, California. *Geol. Soc. Am. Spec. Paper 100.*

Scott, J. S. and Drever, H. I. (1953). Frictional fusion along a Himalayan thrust. *Proc. R. Soc. Edinb. Ser. B* **65**, 121–40.

Sederholm, J. J. (1907). Om granit och gneis. *Bull. Comm. géol. Finlande No. 23.*

—— (1923, 1926). On migmatites and associated Pre-Cambrian rocks of southwestern Finland. I, II. *Bull. Comm. géol. Finlande Nos. 58 and 77.*

—— (1925). The average composition of the earth's crust in Finland. *Bull. Comm. géol. Finlande No. 70.*

—— (1932). On the geology of Fennoscandia with special reference to the Pre-Cambrian. *Bull. Comm. géol. Finlande No. 98.*

—— (1967). *Selected Works: Granites and Migmatites.* Edinburgh and London: Oliver & Boyd.

Seifert, F. and Schreyer, W. (1970). Low temperature stability limit of Mg cordierite in the range 1–7 kbar water pressure: a redetermination. *Contr. Mineral. Petrol.* **27**, 225–38.

Seitsaari, J. (1953). A blue–green hornblende and its genesis from the Tampere schist belt. *Bull. Comm. géol. Finlande No. 159*, 83–95.

Seki, Y. (1957). Petrological study of hornfelses in the central part of the Median Zone of Kitakami Mountainland, Iwate Prefecture. *Saitama Univ. Sci. Rep., Ser, B*, **2**, 307–61.

—— (1958). Glaucophanitic regional metamorphism in the Kanto Mountains, central Japan. *Jap. J. Geol. Geogr.* **29**, 233–58.

Seki, Y. (1959). Petrological studies on the circum-Hida crystalline schists. *Saitama Univ. Sci. Rep., Ser. B,* **3**, 209-20.

— (1960). Jadeite in Sanbagawa crystalline schists of central Japan. *Am. J. Sci.* **258**, 705-15.

— (1961*a*). Pumpellyite in low-grade regional metamorphism. *J. Petrol.* **2**, 407-23.

— (1961*b*). Calcareous hornfelses in the Arisu district of the Kitakami mountains, northeastern Japan. *Jap. J. Geol. Geogr.* **32**, 55-78.

— (1965). Behaviour of carbon dioxide in low-grade regional metamorphism (in Japanese with English abstract). *J. Jap. Assoc. Mineral. Petrol. econ. Geol.* **54**, 1-13.

— (1966). Rock alteration in some geothermal areas of Japan (in Japanese with English abstract). *J. Jap. Assoc. Mineral. Petrol. econ. Geol.* **55**, 212-19.

— (1969). Facies series in low-grade metamorphism. *J. geol. Soc. Japan* **75**, 255-66.

—, Oki, Y., Matsuda, T., Mikami, K. and Okumura, K. (1969*a*). Metamorphism in the Tanzawa Mountains, central Japan. *J. Jap. Assoc. Mineral. Petrol. econ. Geol.* **61**, 1-29, 50-75.

—, Onuki, H., Oba, T., and Mori, R. (1971). Sanbagawa metamorphism in the central Kii Peninsula. *Jap. J. Geol. Geogr.* **41**, 65-78.

—, Onuki, H., Okumura, K. and Takashima, I. (1969*b*). Zeolite distribution in the Katayama geothermal area, Onikobe, Japan. *J. Jap. Geol. Geogr.* **40**, 63-79.

— and Yamasaki, M. (1957). Aluminian ferroanthophyllite from the Kitakami Mountainland, north-eastern Japan. *Am. Mineral.* **42**, 506-20.

Semenenko, N. P. (1970). Geochronological aspects of stabilization of continental Precambrian platforms. *Eclogae geol. Helvetiae* **63**, 301-310.

Shaw, D. M. (1965). Geochemistry of pelitic rocks. III. Major elements and general geochemistry. *Bull. geol. Soc. Am.* **67**, 919-34.

— (1968). Radioactive elements in the Canadian Precambrian Shield and the interior of the earth. In *Origin and Distribution of the Elements,* ed. L. H. Ahrens, 855-70. Oxford: Pergamon Press.

—, Reilly, G. A., Muysson, J. R., Pattenden, G. E. and Campbell, F. E. (1967). An estimate of the chemical composition of the Canadian Precambrian shield. *Can. J. Earth Sci.* **4**, 829-53.

Shibata, K. (1968). K-Ar age determinations on granitic and metamorphic rocks in Japan. *Geol. Surv. Jap. Rep. No. 227.*

—, Adachi, M. and Mizutani, S. (1971). Precambrian rocks in Permian conglomerate from central Japan. *J. geol. Soc. Japan* **77**, 507-14.

— and Igi, S. (1969). K-Ar ages of muscovite from the muscovite-quartz schist of the Sangun metamorphic terrain in the Tari district, Tottori Prefecture, Japan. *Geol. Surv. Jap. Bull.* **20**, 707-9.

Shido, F. (1958). Plutonic and metamorphic rocks of the Nakoso and Iritono districts in the central Abukuma Plateau. *Tokyo Univ. Fac. Sci. J., Sec. 2,* **11**, 131-217.

— (1959). Notes on rock-forming minerals (9). Hornblende-bearing eclogite from Gongen-yama of Higashi-Akaisi in the Bessi district, Sikoku. *J. geol. Soc. Japan* **65**, 701-3.

— and Miyashiro, A. (1959). Hornblendes of basic metamorphic rocks. *Tokyo Univ. Fac. Sci. J., Sec. 2,* **12**, 85-102.

Shido, F., Miyashiro, A. and Ewing, M. (1971). Crystallization of abyssal tholeiites. *Contr. Mineral. Petrol.* **32**, 251–66.

—— and Seki, Y. (1959). Notes on rock-forming minerals (11). Jadeite and hornblende from the Kamuikotan metamorphic belt. *J. geol. Soc. Japan* **65**, 673–77.

Shieh, Y. N. and Taylor, H. P., Jr (1969). Oxygen and hydrogen isotope studies of contact metamorphism in the Santa Rosa Range, Nevada and other areas. *Contr. Mineral. Petrol.* **20**, 306–56.

Shimazu, Y. (1961). A geophysical study of regional metamorphism. *Jap. J. Geophys.* **2**, 135–76.

Shiraki, K. (1971). Metamorphic basement rocks of Yap Islands, western Pacific: Possible oceanic crust beneath an island arc. *Earth Planet. Sci, Letters* **13**, 167–74.

Sighinolfi, G. P. (1969). K–Rb ratio in high-grade metamorphism: a confirmation of the hypothesis of a continual crustal evolution. *Contr. Mineral. Petrol.* **21**, 346–56.

—— (1971). Investigations into deep crustal levels: fractionating effects and geochemical trends related to high-grade metamorphism. *Geochim. Cosmochim. Acta* **35**, 1005–21.

Simonen, A. (1953). Stratigraphy and sedimentation of the Svecofennidic, early Archean supracrustal rocks in southwestern Finland. *Bull Comm. géol. Finlande No. 160*, 1–64.

—— (1960a). Plutonic rocks of the Svecofennides in Finland. *Bull. Comm. géol. Finlande No. 189.*

—— (1960b). Pre-Quaternary rocks in Finland. *Bull. Comm. géol. Finlande No. 191.*

Smith, D. G. W. (1969). Pyrometamorphism of phyllites by a dolerite plug *J. Petrol.* **10**, 20–55.

Smith, J. R. (1958). The optical properties of heated plagioclases. *Am. Mineral.* **43**, 1179–94.

—— and Yoder, H. S., Jr (1956). Variations in X-ray powder diffraction patterns of plagioclase feldspars. *Am. Mineral.* **41**, 632–47.

Smith, J. V. (1962). Genetic aspects of twinning in feldspars. *Norsk Geol. Tidsskr.* **42**, 2. Halvbind *(Feldspar Vol.)*, 244–63.

Smith, R. E. (1969). Zones of progressive regional burial metamorphism in part of the Tasman geosyncline, eastern Australia. *J. Petrol.* **10**, 144–63.

Smithson, S. B. and Barth, Tom. F. W. (1967). The Precambrian Holum granite, south Norway. *Norsk Geol. Tidsskr.* **47**, 21–55.

Sobolev, V. S., Dobretsov, N. L., Reverdatto, V. V., Sobolev, N. V., Jr, Ushrakova, E. N. and Khlestov, V. V. (1967). Metamorphic facies and series of facies in the U.S.S.R. *Meddr. Dansk Geol. Forening,* **17**, 458–72.

Sobolev, N. V., Jr, Kuznetsova, I. K. and Zyuzin, N. I. (1968). The petrology of grospydite xenoliths from the Zagadochnaya kimberlite pipe in Yakutia. *J. Petrol.* **9**, 253–80.

Spry, A. (1955). Thermal metamorphism of portions of the Woolomin Group in the Armidale district, part 2. *Proc. R. Soc. N. S. W.* **89**, 157–70.

Steiner, A. (1953). Hydrothermal rock alteration at Wairakei, New Zealand. *Econ. Geol.* **48**, 1–13.

—— (1955). Wairakite, the calcium analogue of analcime, a new zeolite mineral. *Min. Mag.* **30**, 691–8.

Steiner, A. (1963). The rocks penetrated by drillholes in the Waiotapu thermal area, and their hydrothermal alteration. *N. Z. Dept. Sci. Indust. Res. Bull.* **155**, 28–34.

Stewart, F. H. (1946). The gabbroic complex of Belhelvie in Aberdeenshire, *Q. J. geol. Soc. Lond.* **102**, 465–98.

Stockwell, C. H. (1964). Fourth report on structural provinces, orogenies and time-classification of rocks of the Canadian Precambrian Shield. In *Age Determinations and Geological Studies, Part 2. Geol. Surv. Can. Paper 64-17*, 1–21, 26–29.

Storre, B. (1970). Stabilitätsbedingungen Grossular-fürender Paragenesen im System $CaO-Al_2O_3-SiO_2-CO_2-H_2O$. *Contr. Mineral. Petrol.* **29**, 145–62.

Stout. J. H. (1971). Four coexisting amphiboles from Telemark, Norway. *Am. Mineral.* **56**, 212–24.

Sturt, B. A. (1962). The composition of garnets from pelitic schists in relation to the grade of regional metamorphism. *J. Petrol.* **3**, 181–91.

—— and Harris, A. L. (1961). The metamorphic history of the Loch Tummel area, central Perthshire, Scotland. *Liverpool Manchester geol. J.* **2**, 689–711.

Sugi, K. (1930). On the granitic rocks of Tsukuba district and their associated injection-rocks. *Jap. J. Geol. Geogr.* **8**, 29–112.

Sugimura, A. (1960). Zonal arrangement of some geophysical and petrological features in Japan and its environs. *Tokyo Univ. Fac. Sci. J., Sec. 2*, **12**, 133–53.

—— (1968). Spatial relations of basaltic magmas in island arcs. In *Basalt: Treatise on Rocks of Basaltic Composition*, ed. H. H. Hess and A. Poldervaart, Vol. 2, 537–71. New York: Interscience.

—— Matsuda, T., Chinzei, K. and Nakamura, K. (1963). Quantitative distribution of late Cenozoic volcanic materials in Japan. *Bull. volcanol.* **26**, 125–40.

Summerhayes, C. P. (1967). New Zealand region volcanism and structure. *Nature, Lond.* **215**, 610–11.

Suppe, J. (1969). Times of metamorphism in the Franciscan terrain of the Northern Coast Ranges, California. *Bull. geol. Soc. Am.* **80**, 135–42.

—— (1970). Offset of Late Mesozoic basement terrains by the San Andreas fault system. *Bull. geol. Soc. Am.* **81**, 3253–8.

Sutton, J. and Watson, J. (1969). Scourian-Laxfordian relationships in the Lewisian of northwest Scotland. *Geol. Assoc. Can. Spec. Paper* **5**, 119–28.

Suwa, K. (1961). Petrological and geological studies on the Ryoke metamorphic belt. *Nagoya Univ. Earth Sci. J.* **9**, 224–303.

Suzuki, M. and Kojima, G. (1970). On the association of potassium feldspar and corundum found in the Hida metamorphic belt. *J. Jap. Assoc. Mineral. Petrol. econ. Geol.* **63**, 266–74.

Suzuki, T. (1964). The Mikabu green rocks in Shikoku, I (in Japanese). *Kochi Univ. Res. Rep.* **13** (Natural Science Section), 94–102.

—— (1967). The Mikabu green rocks in Shikoku (in Japanese with English abstract). *J. geol. Soc. Japan* **73**, 207–16.

——, Sugisaki, R. and Tanaka, T. (1971). Geosynclinal igneous activity of the Mikabu green rocks of Ozu City, Ehime Prefecture (in Japanese with English abstract). *Mem. geol. Soc. Japan* **6**, 121–36.

Tagiri, M. (1971). Metamorphic rocks of the Hitachi district in the southern Abukuma Plateau. *J. Jap. Assoc. Mineral. Petrol. econ. Geol.* **65**, 77–103.

Takeuchi, H. and Uyeda, S. (1965). A possibility of present-day regional metamorphism. *Tectonophys* **2**, 59–68.

Taliaferro, N. L. (1943). Franciscan-Knoxville problem. *Bull. American Assoc. Petroleum Geol.* **27**, 109–219.

Talwani, M., Le Pichon, X. and Ewing, M. (1965). Crustal structure of the mid-ocean ridges 2. Computed model from gravity and seismic refraction data. *J. geophys. Res.* **70**, 341–52.

Taneda, S. (1965). 'Areal rock character' in Japan (in Japanese with English abstract). *Kyushu Univ. Fac. Sci. Rep., Geol.* **8**, 1–40.

Tayama, R. (1935). Topography, geology and coral reefs of the Yap Islands (in Japanese). *Tohoku Univ. Fac. Sci. Rep., Geol. Paleont. No. 19.*

Taylor, H. P., Jr. and Coleman, R. G. (1968). $^{18}O/^{16}O$ ratios of coexisting minerals in glaucophane-bearing metamorphic rocks. *Bull. geol. Soc. Am.* **79**, 1727–56.

Taylor, S. R. and White, A. J. R. (1965). Geochemistry of andesites and the growth of continents. *Nature, Lond.* **208**, 271–3.

Tazaki, K. (1964). Alkali amphibole-bearing metamorphic rocks in the Kamuikotan belt of the southwestern part of Asahikawa, central Hokkaido (in Japanese with English abstract). *Chikyu-kagaku No. 71*, 8–17.

Thomas, H. H. (1922). On certain xenolithic Tertiary minor intrusions in the island of Mull. *Q. J. geol. Soc. Lond.* **78**, 229–59.

Thompson, A. B. (1970a). A note on the kaolinite–pyrophyllite equilibrium. *Am. J. Sci.* **268**, 454–8.

—— (1970b). Laumontite equilibria and the zeolite facies. *Am. J. Sci.* **269**, 267–75.

—— (1971). Analcite-albite equilibria at low temperatures. *Am. J. Sci.* **271**, 79–92.

Thompson, J. B., Jr (1955). The thermodynamic basis for the mineral facies concept. *Am. J. Sci.* **253**, 65–103.

—— (1957). The graphical analysis of mineral assemblages in pelitic schists. *Am. Mineral.* **42**, 842–58.

—— (1970). Geochemical reaction and open systems. *Geochim. Cosmochim. Acta* **34**, 529–51.

—— and Norton, S. A. (1968). Paleozoic regional metamorphism in New England and adjacent areas. In *Studies of Appalachian Geology: Northern and Maritime,* ed. E-an Zen *et al.* 319–27. New York: Interscience.

——, Robinson, P., Clifford, T. N., and Trask, N. J. (1968). Nappes and gneiss domes in west-central New England. In *Studies of Appalachian Geology: Northern and Maritime,* ed. E-an Ze n *et al.* 203–18. New York: Interscience.

Tilley, C. E. (1924). Contact metamorphism in the Comrie area of the Perthshire Highlands. *Q. J. geol. Soc. Lond.* **80**, 22–70.

—— (1925). A preliminary survey of metamorphic zones in the southern Highlands of Scotland. *Q. J. geol. Soc. Lond.* **81**, 100–12.

—— (1947). The dunite-mylonites of St Paul's Rocks (Atlantic). *Am. J. Sci.* **245**, 483–91.

—— (1951). A note on the progressive metamorphism of siliceous limestones and dolomites. *Geol. Mag.* **88**, 175–8.

Tilling, R. I., Klepper, M. R. and Obradovich, J. D. (1968). K–Ar ages and time span of emplacement of the Boulder Batholith, Montana. *Am. J. Sci.* **266**, 671–89.

Tobschall, H. J. (1969). Eine Subfaziesfolge der Grünschieferfazies in den Mittleren Cévennen (Dép. Ardèche) mit Pyrophyllit aufweisenden Mineralparagenesen. *Contr. Mineral. Petrol.* **24**, 76–91.

Trommsdorff, V. (1966). Progressive Metamorphose Kieseliger Karbonatgesteine in den Zentralalpen Zwischen Bernina und Simplon. *Schweiz. Mineral. Petrogr. Mitt.* **46**, 431–60.

Tsuboi, C., Jitsukawa, A., Tajima, H., and Okada, A. (1953–6). Gravity survey along the lines of precise levels throughout Japan by means of a Worden gravimeter, Parts 1–8. *Earthquake Res. Inst. Tokyo Univ. Bull. Suppl.* **4.**

Tsuji, S. (1967). Petrology of the Higo metamorphic complex in the Kosa-Hamamati area, Kumamoto Prefecture, Kyushu. *Jap. J. Geol. Geogr.* **38**, 13–25.

Turner, F. J. (1948). Mineralogical and structural evolution of the metamorphic rocks. *Geol. Soc. Am., Mem. 30.*

— and Verhoogen, J. (1951, 1960). *Igneous and Metamorphic Petrology.* 1st and 2nd ed. New York: McGraw-Hill.

Tuttle, O. F. and Bowen, N. L. (1958). Origin of granite in the light of experimental studies in the system $NaAlSi_3O_8$–$KAlSi_3O_8$–SiO_2–H_2O. *Mem. Geol. Soc. Am. 74.*

— and Harker, R. I. (1957). Synthesis of spurrite and the reaction wollastonite + calcite $\rightleftharpoons$ spurrite + carbon dioxide. *Am. J. Sci.* **255**, 266–34.

Tyrrell, G. W. (1955). Distribution of igneous rocks in space and time. *Bull. Geol. Soc. Am.* **66**, 405–26.

Ueda, Y. and Onuki, H. (1968). K–Ar dating on the metamorphic rocks in Japan. I (in Japanese with English abstract). *J. Jap. Assoc. Mineral. Petrol. econ. Geol.* **60**, 159–66.

— , Yamaoka, K., Onuki, H. and Tagiri, M. (1969). K–Ar dating on the metamorphic rocks in Japan. II (in Japanese with English abstract). *J. Jap. Assoc. Mineral. Petrol. econ. Geol.* **61**, 92–9.

Ueta, S. (1961). Two contrasted regional metamorphic terrains in the Yatusiro district, Kumamoto Prefecture (in Japanese with English abstract). *J. geol. Soc. Japan* **67**, 526–39.

Uno, T. (1961). Metamorphic rocks of the Tukuba district, Ibaraki Prefecture (in Japanese with English abstract). *J. geol. Soc. Japan* **67**, 228–36.

Ustiyev, Y. K. (1965). Problem of volcanism and plutonism. *Internatn. Geol. Rev.* **7**, 1994–2016.

Utada, M. (1965). Zonal distribution of authigenic zeolites in the Tertiary pyroclastic rocks in Mogami district, Yamagata Prefecture. *Tokyo Univ. Coll. Gen. Educ. Sci. Paper* **15**, 173–216.

Uyeda, S. and Rikitake, T. (1970). Electrical conductivity anomaly and terrestrial heat flow. *J. Geomag. Geoelect.* **22**, 75–90.

— and Vacquier, V. (1968). Geothermal and geomagnetic date in and around the island arc of Japan. In *The Crust and Upper Mantle of the Pacific Area*, ed. L. Knopoff *et al., Geophys. Monogr.* **12**, 349–66. Washington, D.C.: American Geophys. Union.

Vallance, T. G. (1953). Studies in the metamorphic and plutonic geology of the Wantabadgery-Adelong-Tumbarumba district, N.S.W. I, II. *Proc. Linn. Soc. N. S. W.* **78**, 90–121, 181–96.

— (1960). Notes on metamorphic and plutonic rocks and their biotite from the Wantabadgery–Adelong–Tumbarumba district, N.S.W. *Proc. Linn. Soc. N.S.W.* **85**, 94–104.

— (1967). Mafic rock alteration and isochemical development of some cordierite-anthophyllite rocks. *J. Petrol.* **8**, 84–96.

van Bemmelen, R. W. (1949). *The Geology of Indonesia.* The Hague: Government Printing Office.

Vance, J. A. (1968). Metamorphic aragonite in the prehnite–pumpellyite facies, northwest Washington. *Am. J. Sci.* **266**, 299–315.

van der Linden, W. J. M. (1967). Structural relationships in the Tasman Sea and Southwest Pacific Ocean. *N. Z. J. Geol. Geophys.* **10**, 1280–301.

—— (1969). Extinct mid-ocean ridges in the Tasman Sea and in the Western Pacific. *Earth Planet. Sci. Letters* **6**, 483–90.

van der Plas, L. (1959). Petrology of the northern Adula region, Switzerland. *Leid. Geol. Mededel.* **24**, 411–602.

van Hise, C. R. (1904). *A Treatise on Metamorphism.* U.S. Geol. Survey Monogr. **47**.

Velde, B. (1965). Phengite micas: synthesis, stability and natural occurrence. *Am. J. Sci.* **263**, 886–913.

—— (1966). Upper stability of muscovite. *Am. Mineral.* **51**, 924–9.

Vidale, R. (1969). Metasomatism in a chemical gradient and the formation of calc-silicate bands. *Am. J. Sci.* **267**, 857–74.

Vine, F. J. (1966). Spreading of the ocean floor: new evidence. *Science, N.Y.* **154**, 1405–15.

Vinogradov, A. P. and Tugarinov, A. I. (1962). Problems of geochronology of the pre-Cambrian in eastern Asia. *Geochim. Cosmochim. Acta* **26**, 1283–300.

Vogt, Th. (1927). Sulitelmafeltets geologi og petrografi, *Norges Geol. Unders. No. 121.*

von Platen, H. (1965). Experimental anatexis and genesis of migmatites. In *Controls of Metamorphism,* ed. W. S. Pitcher and G. W. Flinn, 203–218. Edinburgh and London: Oliver & Boyd.

Wager, L. R. and Deer, W. A. (1939). The petrology of the Skaergaard intrusion, Kangerdlugssuaq, East Greenland. *Meddr. Grönland,* **105**, No. 4.

Walker, G. P. L. (1960). Zeolite zones and dike distribution in relation to the structure of the basalts of Eastern Iceland. *J. Geol.* **68**, 515–28.

Walter, L. S. (1963). Experimental studies on Bowen's decarbonation series – I and II. *Am. J. Sci.* **261**, 488–500, 773–9.

—— (1965). Experimental studies on Bowen's decarbonation series – III. *Am. J. Sci.* **263**, 64–77.

Ward, R. F. (1959). Petrology and metamorphism of the Wilmington complex, Delaware, Pennsylvania, and Maryland. *Bull. geol. Soc. Am.* **70**, 1425–58.

Warner, J. R., Doyle, R. G. and Hussey, A. M. (1967). Generalized map of regional metamorphic zones. In *Preliminary Geologic Map of Maine,* ed. R. G. Doyle. Maine Geol. Survey.

Washington, H. S. (1901). A chemical study of the glaucophane schists. *Am. J. Sci. 4th ser.* **11**, 35–59.

Waters, A. C. and Campbell, C. D. (1935). Mylonites from the San Andreas fault zone. *Am. J. Sci.* [5] **29**, 473–503.

Weeks, W. F. (1956). A thermochemical study of equilibrium relations during metamorphism of siliceous carbonate rocks. *J. Geol.* **64**, 245–70.

Weill, D. F. (1966). Stability relations in the Al_2O_3–SiO_2 system calculated from solubilities in the Al_2O_3–SiO_2–Na_3AlF_6 system. *Geochim. Cosmochim. Acta* **30**, 223–37.

—— and Fyfe, W. S. (1964). A discussion of the Korzhinskii and Thompson treatment of thermodynamic equilibrium in open systems. *Geochim. Cosmochim. Acta* **28**, 565–76.

Weill, D. F. and Fyfe, W. S. (1967). On equilibrium thermodynamics of open systems and the phase rule (a reply to D. S. Korzhinskii). *Geochim. Cosmochim. Acta* **31**, 1167–76.

Wenk, E. (1962). Plagioklas als Indexmineral in den Zentralalpen. *Schweiz. Mineral. petrogr. Mitt.* **42**, 139–52.

—— and Keller, F. (1969). Isograde in Amphibolitserien der Zentralalpen. *Schweiz. Mineral. Petrogr. Mitt.* **49**, 157–98.

White, A. J. R. (1964). Clinopyroxenes from eclogites and basic granulites. *Am. Mineral.* **49**, 883–8.

—— (1966). Genesis of migmatites from the Palmer region of South Australia, *Chem. Geol.* **1**, 165–200.

——, Compston, W. and Kleeman, A. W. (1967). The Palmer granite – a study of a granite within a regional metamorphic environment. *J. Petrol.* **8**, 29–50.

—— (1971). Comment on granulites. *Neues Jahrbuch Mineral. Monatshefte* (*1971*). 116–18.

White, R. W. (1966). Ultramafic inclusions in basaltic rocks from Hawaii. *Contr. Mineral. Petrol.* **12**, 245–314.

Williams, A. F. (1932). *The Genesis of the Diamond*, Vols. 1, 2. London: Ernest Benn.

Wilson, J. T. (1965). A new class of faults and their bearing on continental drift. *Nature, Lond.* **207**, 343–7.

—— (1966). Did the Atlantic close and then re-open? *Nature, Lond.* **211**, 676–81.

—— (1968). Static or mobile earth: the current scientific revolution. In *Gondwanaland Revisited: New Evidence for Continental Drift. Proc. Am. Phil. Soc.* **112**, No. 5, 309–20.

Windley, B. F. (1970). Anorthosites in the early crust of the earth and on the moon. *Nature, Lond.* **226**, 333–5.

Winkler, H. G. F. (1965, 1967). *Petrogenesis of metamorphic rocks*. 1st and 2nd eds. New York: Springer-Verlag.

Wiseman, J. D. H. (1934). The central and south-west Highland epidiorites: a study in progressive metamorphism. *Q. J. geol. Soc. Lond.* **90**, 354–417.

Wones, D. R. and Eugster, H. P. (1965). Stability of biotite: experiment, theory, and application. *Am. Mineral.* **50**, 1228–72.

Woodland, B. G. (1963). A petrographic study of thermally metamorphosed pelitic rocks in the Burke area, northeastern Vermont. *Am. J. Sci.* **261**, 354–75.

Wyllie, P. J. and Tuttle, O. F. (1961). Hydrothermal melting of shales. *Geol. Mag.* **98**, 56–66.

Wynne-Edwards, H. R. (1969). Tectonic overprinting in the Grenville province, southwestern Quebec. *Geol. Assoc. Canada Spec. Paper* **5**, 163–82.

—— and Hasan, Z. (1970). Intersecting orogenic belts across the North Atlantic. *Am. J. Sci.* **268**, 289–308.

Yagi, K., Kawano, Y. and Aoki, K. (1963). Types of Quaternary volcanic activity in Northeastern Japan. *Bull. volcanol,* **26**, 223–35.

Yamada, N. (1966). Nature of the Late Mesozoic igneous activities in and around southwest Japan (in Japanese with English abstract). *Chikyu-kagaku,* Nos. 85–6, 53–8.

——, Kawada, K. and Morohashi, T. (1971). The Nohi rhyolite as pyroclastic flow deposit (in Japanese with English abstract). *Chikyu-kagaku,* **25**, 52–88.

Yamada, N. and Nakai, Y. (1969). Geologic relations between the Nohi rhyolites and the so-called Ryoke granites in central Japan (in Japanese with English abstract). *Mem. geol. Soc. Japan* **4**, 51–60.

Yamaguchi, M. and Yanagi, T. (1970). Geochronology of some metamorphic rocks in Japan. *Eclogae geol. Helvetiae* **63**, 371–88.

Yamamoto, H. (1962). Plutonic and metamorphic rocks along the Usuki-Yatsushiro tectonic line in the western part of central Kyushu. *Fukuoka Gakugei Univ. Bull.* **12**, 93–172.

—— (1964). Metamorphic rocks of the Kiyama district, east of Kumamoto City, Japan (in Japanese). *Kyushu Univ. Fac. Sci. Rep. Geol.* **7**, 33–8.

Yen, T. P. (1959). Soda–amphibole–quartz schist from Taiwan. *Proc. geol. Soc. China* **2**, 153–6.

—— (1962). The grade of metamorphism of the Tananao schist. *Proc. geol. Soc. China* **5**, 101–8.

—— (1963). The metamorphic belts within the Tananao schist terrain of Taiwan. *Proc. geol. Soc. China* **6**, 72–4.

—— and Rosenblum, S. (1964). Potassium-argon ages of micas from the Tananao schist terrane of Taiwan – a preliminary report. *Proc. geol. Soc. China* **7**, 80–1.

Yoder, H. S., Jr (1952). The $MgO-Al_2O_3-SiO_2-H_2O$ system and the related metamorphic facies. *Am. J. Sci. (Bowen Vol.)*, 569–627.

—— (1959). Experimental studies on micas: a synthesis. *Proc. 6th natn. Conf. on Clays and Clay Minerals*, 42–60. Oxford: Pergamon Press.

—— and Eugster, H. P. (1954). Phlogopite synthesis and stability range. *Geochim. Cosmochim. Acta* **6**. 157–85.

—— and —— (1955). Synthetic and natural muscovite. *Geochim. Cosmochim. Acta* **8**, 225–80.

—— and Tilley, C. E. (1962). Origin of basalt magmas: an experimental study of natural and synthetic rock systems. *J. Petrol.* **3**, 342–532.

Yui, S. (1962). Notes on rock-forming minerals (24). Stilpnomelane from the Motoyasu mine, Sikoku. *J. geol. Soc. Japan* **68**, 597–600.

Zemann, J. (1962). Zur Kristallchemie der Granate. *Beitr. Mineral. Petrogr.* **8**, 180–8.

Zen, E-an (1960). Metamorphism of Lower Paleozoic rocks in the vicinity of the Taconic Range in west-central Vermont. *Am. Mineral.* **45**, 129–75.

—— (1961a). Mineralogy and petrology of the system $Al_2O_3-SiO_2-H_2O$ in some pyrophyllite deposits of North Carolina. *Am. Mineral.* **46**, 52–66.

—— (1961b). The zeolite facies: an interpretation. *Am. J. Sci.* **259**, 401–9.

—— (1966). Construction of pressure–temperature diagrams for multicomponent systems after the method of Schreinemakers – a geometric approach. *U.S. Geol. Survey Bull.* **1225**.

—— (1968). Introduction. In *Studies of Appalachian Geology: Northern and Maritime*, ed. E-an Zen *et al.*, 1–5. New York: Interscience.

—— and Albee, A. L. (1964). Coexisting muscovite and paragonite in pelitic schists. *Am. Mineral.* **49**, 904–25.

Zwart, H. J. (1962). On the determination of polymorphic mineral associations, and its application to the Bosost area (central Pyrenees). *Geol. Rundschau* **52**, 38–65.

—— (1963). Some examples of the relations between deformation and metamorphism from the central Pyrenees. *Geol. Mijnbouw* **42**, 143–54.

Zwart, H. J. (1967*a*). The duality of orogenic belts. *Geol Mijnbouw* **46**, 283–309.

—— (1967*b*). Orogenesis and metamorphic facies series in Europe. *Meddr. Dansk. Geol. Forening* **17**, 504–16.

—— (1969). Metamorphic facies series in the European orogenic belts and their bearing on the causes of orogeny. *Geol. Assoc. Can. Spec. Paper* **5**, 7–16.

—— , Corvalan, J., James, H. L., Miyashiro, A., Saggerson, E. P., Sobolev, V. S., Subramaniam, A. P. and Vallance, T. G. (1967). A scheme of metamorphic facies for the cartographic representation of regional metamorphic belts. *IUGS Geological Newsletter* (*1967*), No. 2, 57–72.

Index